Lecture Notes in Compu[illegible] 18

Commenced Publication in 1973
Founding and Former Series Editors:
Gerhard Goos, Juris Hartmanis, and Jan

Editorial Board

David Hutchison, UK
Takeo Kanade, USA
Josef Kittler, UK
Jon M. Kleinberg, USA
Alfred Kobsa, USA
Friedemann Mattern, Switzerland
John C. Mitchell, USA
Moni Naor, Israel
Oscar Nierstrasz, Switzerland
C. Pandu Rangan, India
Bernhard Steffen, Germany
Madhu Sudan, USA
Demetri Terzopoulos, USA
Doug Tygar, USA
Gerhard Weikum, Germany

FoLLI Publications on Logic, Language and Information

Subline of Lectures Notes in Computer Science

Subline Editors-in-Chief

Valentin Goranko, *Technical University, Lynbgy, Denmark*
Erich Grädel, *RWTH Aachen University, Germany*
Michael Moortgat, *Utrecht University, The Netherlands*

Subline Area Editors

Nick Bezhanishvili, *Imperial College London, UK*
Anuj Dawar, *University of Cambridge, UK*
Philippe de Groote, *Inria-Lorraine, Nancy, France*
Gerhard Jäger, *University of Tübingen, Germany*
Fenrong Liu, *Tsinghua University, Beijing, China*
Eric Pacuit, *Tilburg University, The Netherlands*
Ruy de Queiroz, *Universidade Federal de Pernambuco, Brazil*
Ram Ramanujam, *Institute of Mathematical Sciences, Chennai, India*

Maria Aloni Vadim Kimmelman
Floris Roelofsen Galit W. Sassoon
Katrin Schulz Matthijs Westera (Eds.)

Logic, Language and Meaning

18th Amsterdam Colloquium
Amsterdam, The Netherlands, December 19-21, 2011
Revised Selected Papers

Volume Editors

Maria Aloni
Floris Roelofsen
Galit W. Sassoon
Katrin Schulz
Matthijs Westera
University of Amsterdam, ILLC
Science Park 904
1098 XH Amsterdam
The Netherlands
E-mail: {m.d.aloni, k.schulz, m.westera}@uva.nl
E-mail: {floris.roelofsen, galitadar}@gmail.com

Vadim Kimmelman
University of Amsterdam, ACLC
Spuistraat 210
1012 VT Amsterdam
The Netherlands
E-mail: v.kimmelman@uva.nl

ISSN 0302-9743 e-ISSN 1611-3349
ISBN 978-3-642-31481-0 e-ISBN 978-3-642-31482-7
DOI 10.1007/978-3-642-31482-7
Springer Heidelberg Dordrecht London New York

Library of Congress Control Number: 2012940665

CR Subject Classification (1998): F.4.1, F.4.2, I.2.7, F.4, J.5, K.4.2, I.2.3

LNCS Sublibrary: SL 1 – Theoretical Computer Science and General Issues

© Springer-Verlag Berlin Heidelberg 2012

This work is subject to copyright. All rights are reserved, whether the whole or part of the material is concerned, specifically the rights of translation, reprinting, re-use of illustrations, recitation, broadcasting, reproduction on microfilms or in any other way, and storage in data banks. Duplication of this publication or parts thereof is permitted only under the provisions of the German Copyright Law of September 9, 1965, in its current version, and permission for use must always be obtained from Springer. Violations are liable to prosecution under the German Copyright Law.
The use of general descriptive names, registered names, trademarks, etc. in this publication does not imply, even in the absence of a specific statement, that such names are exempt from the relevant protective laws and regulations and therefore free for general use.

Typesetting: Camera-ready by author, data conversion by Scientific Publishing Services, Chennai, India

Printed on acid-free paper

Springer is part of Springer Science+Business Media (www.springer.com)

Preface

The 2011 edition of the Amsterdam Colloquium was the 18th in a series which started in 1976. The Amsterdam Colloquia aim at bringing together linguists, philosophers, logicians and computer scientists who share an interest in the formal study of the semantics and pragmatics of natural and formal languages. Originally an initiative of the Department of Philosophy, the colloquium is now organized by the Institute for Logic, Language and Computation (ILLC) of the University of Amsterdam.

These proceedings contain revised versions of a selection of the papers presented at the 18th Amsterdam Colloquium. The first section contains the invited contributions. The second, third and fourth sections contain submitted contributions to the three thematic workshops that were hosted by the colloquium: the Workshop on Inquisitiveness; the Workshop on Formal Semantics and Pragmatics of Sign Languages, and the Workshop on Formal Semantic Evidence. The final section consists of the submitted contributions to the general program.

For the organization of the 18th Amsterdam Colloquium, financial support was received from: the Institute for Logic, Language and Computation (ILLC); the NWO-funded projects: "The Inquisitive Turn. A New Perspective on Semantics, Logic, and Pragmatics" (coordinator: Jeroen Groenendijk); "Indefinites and Beyond: Evolutionary Pragmatics and Typological Semantics" (coordinator: Maria Aloni); "Vagueness — And How to Be Precise Enough" (coordinators: Robert van Rooij and Frank Veltman); the E.W. Beth Foundation and the Municipality of Amsterdam. This support is gratefully acknowledged.

The editors would like to thank the members of the Program Committee and the anonymous reviewers for their help in the preparation of this volume. Many thanks also to Peter van Ormondt, Eric Flaten and Haitao Cai for help with the organization of the conference.

March 2012

Maria Aloni
Vadim Kimmelman
Floris Roelofsen
Galit W. Sassoon
Katrin Schulz
Matthijs Westera

Organization

The 18th Amsterdam Colloquium was organized by the Institute for Logic, Language and Computation (ILLC) of the University of Amsterdam.

Organizing Committee

Maria Aloni (Chair)	ILLC, University of Amsterdam
Peter van Ormondt	ILLC, University of Amsterdam
Floris Roelofsen	ILLC, University of Amsterdam
Galit W. Sassoon	ILLC, University of Amsterdam
Matthijs Westera	ILLC, University of Amsterdam

Program Committee

General Program

Frank Veltman (Chair)	ILLC, University of Amsterdam
Paul Dekker	ILLC, University of Amsterdam
Hedde Zeijlstra	ACLC, University of Amsterdam

Workshop on Inquisitiveness

Jeroen Groenendijk	ILLC, University of Amsterdam
Floris Roelofsen	ILLC, University of Amsterdam
Matthijs Westera	ILLC, University of Amsterdam

Workshop on Formal Semantics and Pragmatics of Sign Languages

Vadim Kimmelman	ACLC, University of Amsterdam
Roland Pfau	ACLC, University of Amsterdam
Anne Baker	ACLC, University of Amsterdam

Workshop on Formal Semantic Evidence

Galit W. Sassoon	ILLC, University of Amsterdam
Katrin Schulz	ILLC, University of Amsterdam

Invited Speakers

Donka Farkas	University of California, Santa Cruz
Irene Heim	MIT
Chung-chieh Shan	Cornell University
Seth Yalcin	University of California, Berkeley
Manfred Krifka	Humboldt-Universität zu Berlin

Philippe Schlenker	Institut Jean-Nicod, Paris and New York University
Richard Breheny	University College London
Bart Geurts	Radboud University Nijmegen

Referees

Adrian Brasoveanu
Alessandro Zucchi
Alexis Dimitriadis
Angelika Kratzer
Anna Szabolcsi
Ariel Cohen
Bart Geurts
Chris Kennedy
Chris Potts
Christian Ebert
Chung-chieh Shan
Cleo Condoravdi
Cornelia Ebert
David Beaver
Donka Farkas
Dorit Abusch
Ede Zimmermann
Emmanuel Chemla
Elena Tribushinina
Fred Landman
Fritz Hamm
Gennaro Chierchia
Gerhard Jäger
Henriette de Swart
Henk Zeevat
Inge Zwitserlood
Irene Heim
Ivano Caponigro
Jakub Dotlacil
Jon Gajewski
Josep Quer
Kathryn Davidson
Kjell-Johan Saebo
Klaus von Heusinger
Louise McNally
Luis Alonso-Ovalle
Markus Steinbach
Martin Hackl
Martin Stokhof
Michael Franke
Nick Asher
Paul Portner
Pauline Jacobson
Philippe Schlenker
Raffaella Bernardi
Raquel Fernández Rovira
Regine Eckardt
Reinhard Blutner
Richard Breheny
Rick Nouwen
Robert van Rooij
Roger Schwarzschild
Roland Pfau
Ronnie Wilbur
Sabine Iatridou
Seth Cable
Seth Yalcin
Stefan Kaufmann
Stephanie Solt
Susan Rothstein
Tim Fernando
Uli Sauerland
Yael Sharvit
Yaron McNabb
Yoad Winter
Volker Gast

Table of Contents

Invited Speakers

Workshop on Inquisitiveness

Workshop on Sign Language

Workshop on Formal Semantic Evidence

General Program

Formal Indices and Iconicity in ASL*

Philippe Schlenker[1] and Jonathan Lamberton[2]

[1] Institut Jean-Nicod, CNRS; New York University
[2] CUNY - Queens College

Abstract. Iconic constraints play an important role in the semantics of sign language in general, and of sign language pronominals in particular (e.g. Cuxac 1999, Taub 2001, Liddell 2003, Lillo-Martin and Meier 2011). But the field is sharply divided among two camps: (a) specialists of formal linguistics (e.g. Neidle et al. 2000, Lillo-Martin 1991, Sandler and Lillo-Martin 2006) primarily attempt to integrate sign language pronominals to universal models of anaphora, giving iconic phenomena a peripheral position; (b) specialists who emphasize the centrality of iconicity (e.g. Taub 2001, Liddell 2003) do so within informal frameworks that are not considered as sufficiently explicit by the formalist side. We attempt to reconcile insights from the two camps within a *formal semantics with iconicity* (Schlenker 2011a). We analyze three kinds of iconic effects: (i) *structural iconicity*, in which relations of embedding among loci are directly reflected in their denotations; (ii) *locus-external iconicity*, in which the high position of a locus in signing space has a direct semantic reflex; and (iii) *locus-internal iconicity*, where different parts of a structured locus are targeted by different directional verbs, as was argued by Liddell. We suggest that these phenomena can be understood if the interpretive procedure can, at the level of variables, impose an 'iconic mapping' between loci and the objects they denote.

Keywords: anaphora, sign language, logic, iconicity.

1 Introduction: Sign Language Indexes as Formal Indices

In American Sign Language (ASL), the relation between a pronoun and its antecedent is often mediated by *loci*, which are positions in signing space that are usually

* Many thanks Jason Lamberton and Wes Whalen for help with the ASL data, to Oliver Pouliot for help with some transcriptions, as well as to Igor Casas for discussion of related LSF data (not reported here). Sign language consultants are not responsible for the claims made here, nor for any errors. The present work was supported by an NSF grant (BCS 0902671) and by a Euryi grant from the European Science Foundation ('Presupposition: A Formal Pragmatic Approach'). Neither foundation is responsible for the claims made here. The research reported in this piece also contributes to the COST Action IS1006.

Contribution of each author: Philippe Schlenker initiated the project and is responsible for all claims and errors. Jonathan Lamberton is a co-author for research in progress reported in Section 3 (iconic effects with directional verbs), and he was an ASL informant for all parts of the project.

Overlap with other works: Section 1 shares material with Schlenker 2011b. Section 2 is almost identical with Schlenker, to appear (with approval from *Snippets).* The analyses of Sections 3 and 4 are modified versions of ones sketched in Schlenker 2011a.

M. Aloni et al.(Eds.): Amsterdam Colloquium 2011, LNCS 7218, pp. 1–11, 2012.
© Springer-Verlag Berlin Heidelberg 2012

associated with NPs (e.g. Sandler and Lillo-Martin 2006). A pronoun that depends on an NP will thus point towards the locus introduced by that NP, as in (1):[1]

(1) a. IX-1 KNOW $_a$BUSH IX-1 KNOW $_b$OBAMA. IX-b SMART BUT IX-a NOT SMART.
'I know Bush and I know Obama. He [= Obama] is smart but he [= Bush] is not smart.'
b. IX-1 KNOW PAST SENATOR PERSON IX-a IX-1 KNOW NOW SENATOR PERSON IX-b. IX-b SMART BUT IX-a NOT SMART.
'I know a former senator and I know a current senator. He [= the current senator] is smart but he [= the former senator] is not smart.' (Inf 1, 4, 179)

Since there appears to be an arbitrary number of possible loci, it was suggested that the latter do not spell out morpho-syntactic features, but rather are the overt realization of indices (Lillo-Martin and Klima 1990, Sandler and Lillo-Martin 2006).

While pointing can have a variety of uses in sign language (Sandler and Lillo-Martin 2006, Schlenker 2011a), we will restrict attention to pronominal uses. Importantly, there are some striking similarities between sign language pronouns and their spoken counterparts – which makes it desirable to offer a unified theory: – Sign language pronouns obey at least some of the syntactic constraints on binding studied in syntax. For instance, versions of the following rules have been described for ASL (Lilla-Martin 1991, Sandler and Lillo-Martin 2006, Koulidobrova 2011): Condition A; Condition B; Strong Crossover. –In simple cases, the same ambiguity between strict and bound variable readings is found in both modalities (see Lillo-Martin and Sandler 2006):

(2) IX-1 POSS-1 MOTHER LIKE. IX-a SAME-1,a.
Ambiguous: I like my mother. He does too [= like my / like his mother] (Inf 1, 1, 108)

Still, an important fact has gotten in the way of an integration of sign language pronouns to the theories that were designed for spoken languages: sign language pronouns display numerous *iconic effects* that are absent from their spoken language counterparts. This has led some to take such a traditional grammatical approach to be hopeless (e.g. Liddell 2003). We will argue for a somewhat different position: iconic effects are real and central, but they can be integrated to a unified grammatical theory once we rethink some aspects of semantics and allow geometric considerations to play a central role in the meaning of variables. We analyze three kinds of iconic effects: (i) *structural iconicity*, in which relations of embedding among loci are directly reflected in their denotations; (ii) *locus-external iconicity*, in which the high position of a locus in signing space has a direct semantic reflex; and (iii) *locus-internal iconicity*, where different parts of a structured locus are targeted by different directional verbs, as was argued by Liddell 2003. We suggest that these phenomena can be understood if the interpretive procedure can, at the level of variables, impose an 'iconic mapping' between loci and the objects they denote.

---

[1] ASL sentences are glossed in capital letters. Subscripts correspond to the establishment of positions ('loci') in signing space. Pronouns are usually realized through pointing towards a locus, and they are also glossed as *IX-a*, *IX-b*, etc. Parentheses starting with *Inf* refer to videos.

2 Complement Set Readings and Structural Iconicity

2.1 The Phenomenon

We start with a type of iconicity which hasn't been much discussed in the sign language literature, but pertains to a traditional question in semantics: the availability of 'complement set' readings in examples such as (3)a. We argue that this option is highly restricted in English, but fully available in ASL *when* the signer makes use of 'structural iconicity' to ensure that a locus can denote the complement set.

(3) a. *Complement Set Anaphora:* (i) ?Few / (ii) #Most students came to class. They stayed home instead.
b. *Maximal Set Anaphora:* (i) Few / (ii) Most students came to class, (i) but / (ii) and they asked good questions.
c. *Restrictor Set Anaphora:* (i) Few / (ii) Most students came to class. They (i) aren't / (ii) are a serious group.

Recent dynamic approaches to anaphora are designed to account for cases of 'maximal set' anaphora as in (3)b, where a pronoun refers to the maximal group of individuals that satisfy both the restrictor and the nuclear scope of a generalized quantifier. Restrictor set anaphora, as in (3)c, is available as well: the plural pronoun just denotes the set of individuals that satisfy the restrictor. The question is whether complement set anaphora, as in (3)a, is genuinely available, and if so by what means. This option is notoriously restricted: it is often impossible with non-negative quantifiers, as in (3)a(ii); and some cases involving negative quantifiers can be re-analyzed in terms of a 'restrictor set' reading with a collective interpretation that tolerates exceptions. Following Nouwen 2003, we posit that when complement set anaphora is available it involves *inferred* discourse referents: no *grammatical* mechanism makes available a discourse referent denoting the complement set.

ASL signers can realize anaphora by using default loci (in front of the signers), or non-default ones. In the first case, English-style judgments can be replicated with maximal set anaphora (= (4)a-b) and restrictor set anaphora (=(4)a'-b').

(4)
a. 6.7 POSS-1 STUDENT FEW a-CAME CLASS.
'Few of my students came to class.'
IX-arc-a a-ASK-1 GOOD QUESTION
'They asked good questions.'
a'. 6 POSS-1 STUDENT FEW a-CAME.
'Few of my students came.'

IX-arc-a NOT SERIOUS CLASS.
'They are not a serious class.'
(Inf 1, 8, 198; 8, 199; 8, 204; 8, 222)

b. 6 POSS-1 STUDENT MOST a-CAME CLASS.
'Most of my students came to class.'
IX-arc-a a-ASK-1 GOOD QUESTION
'They asked good questions.'
b'. 6.7 POSS-1 STUDENT IX-arc-a MOST a-CAME CLASS.
'Most of my students came to class.'
IX-arc-a SERIOUS CLASS.
'They are a serious class.'
(Inf 1, 8, 200; 8, 201; 8 205; 8, 223)

(5)
POSS-1 STUDENT FEW a-CAME CLASS.
a. 3.6 [3.6] IX-arc-a a-STAY HOME
POSS-1 STUDENT MOST a-CAME CLASS.
b. 2.8 [2.7] IX-arc-a a-STAY HOME
Intended: 'Few/Most of my students came to class. They [the students that didn't come] stayed home.'
(Inf 1, 8, 225; 8, 226; 8, 285; 8, 300; 8, 305; 8, 348)

The high ratings in (4) were obtained from our main ASL consultant (1 = worst, 7 = best; average score over 3 iterations on separate days).The crucial data are in (5): they involve complement set anaphora, and were tested with our main consultant (3 iterations) and 2 further consultants (1 iteration each), with clearly degraded ratings (1st score: equal weight for each trial; 2nd score: equal weight for each consultant).

However, another anaphoric strategy can be used; it consists in establishing a large plural locus *A* for the restrictor set [= the set of all students], and a sublocus *a* for the maximal set [= the set of students who came]. Remarkably, this strategy automatically makes available a locus *A-a* for the complement set. As a result, all three readings become equally available, though with different indexings (importantly, all involve normal plural pronouns, and not the word *OTHER*). In (6), we provide our main consultant's judgments (3 iterations) based on this second anaphoric strategy ('embedded loci'). For perspicuity, we notate the large area *A* as *ab* to indicate that it comprises subloci *a* and *b* – although it is just signed as a large circular area:

(6) POSS-1 STUDENT IX-arc-ab MOST IX-arc-a a-CAME CLASS.
'Most of my students came to class.' (Inf 1, 8, 196; 8, 197; 8, 206; 8, 224)
a. 7 IX-arc-b b-STAY HOME 'They stayed home.'
b. 7 IX-arc-a a-ASK-1 GOOD QUESTION 'They asked good questions.'
c. 7 IX-arc-ab SERIOUS CLASS. 'They are a serious class.'

Data pertaining to complement set anaphora were also assessed in the same video as (5) (same 3 consultants); ratings confirm that with embedded loci complement set anaphora is acceptable ((7)b is similar to (6)a but was part of a different video):

(7)
a. 6.7 [6.5] POSS-1 STUDENT IX-arc-ab FEW IX-arc-a a-CAME. IX-arc-b b-STAY HOME
b. 6.3 [5.8] POSS-1 STUDENT IX-arc-ab MOST IX-arc-a a-CAME. IX-arc-b b-STAY HOME
'(a) Few/ (b) Most of my students came to class. They [= the students who didn't come] stayed home.'
(In 1, 8, 225; 8, 226; 8, 285; 8, 300 ; 8, 305; 8, 348)

2.2 An Analysis with Structural Iconicity

To account for these data, we hypothesize that assignment functions assign values to loci (Schlenker 2011a), and we further assume that: (a) geometric properties of plural loci (*qua* areas of space) guarantee that if a locus *A* and a sublocus *a* have been introduced, a complement locus *(A-a)* becomes *ipso facto* available; (b) relations of inclusion and subtraction among loci are preserved by the interpretation function *via* constraints on assignment functions - an instance of 'structural iconicity'. Specifically:

(8) Let LOC be the set of plural loci that appear in signing space, and let s an admissible assignment function that assigns values to loci. We view plural loci as sets of geometric points, and loci denotations as sets of individuals; we make the following assumptions:
a. *LOC:* for all a, b $\in$ LOC, (i) a $\subseteq$ b or b $\subseteq$ a or a $\cap$ b = Ø; (ii) if a $\subset$ b, (b-a) $\in$ LOC
b. *s:* for all a, b $\in$ LOC, if a $\subset$ b, (i) s(a) $\subset$ s(b); (ii) s(b-a) = s(b)-s(a)

In (4)-(7), we take the grammar to make available (i) a discourse referent for the maximal set and the restrictor set, but (ii) none for the complement set. When a default locus is used, ASL roughly behaves like English, and complement set anaphora is highly restricted (because of (ii)). In case embedded loci are used, ASL allows for complement set anaphora. Here is why: if *a* is a proper sublocus of a large locus *ab*, we can infer by (8)a(ii) that *(ab-a)* (i.e. *b*) is a locus as well; by (8)b(i), that s(a) $\subset$ s(ab); and by (8)b(ii), that s(b) = s(ab)-s(a). Complement set anaphora becomes available because ASL relies on an iconic property which is inapplicable in English.

3 High Loci and Locus-External Iconicity

3.1 The Phenomenon

In ASL, loci that are high in signing space can be used to refer to entities that are respectable, powerful, or tall – among others. This possibility is instantiated in (9), where a high locus *IX-a⁺* can be used to refer to the speaker's father (we do not have enough iterations of the judgment task to provide interesting numerical values; in cases of optionality, there is often a slight preference for the high locus):

(9) MY FATHER IX-a$^+$/ IX-a (SELF-a$^+$)[3] BUSINESSMAN. IX-a$^+$/ IX-a RICH.
'My father is a businessman. He is rich.' (Inf 1, 8, 17-18)

If the speaker intends to speak about his younger brother, the use of a high locus is dispreferred (hence *??* in (10)), unless he somehow wants to convey that his younger brother is more successful than him – in which case it is acceptable:

(10) MY BROTHER ??IX-a$^+$/ IX-a (??SELF-a$^+$)[2] BUSINESSMAN. ??IX-a$^+$/ IX-a RICH
Intended: 'My (younger) brother is a businessman. He is rich.' (Inf 1, 8, 19-20)

As can be seen in (11), reference to the government can also give rise to high loci:

(11) EVERY ONE-arc / IX-arc GIVE-GIVE-a$^+$/GIVE-GIVE-a MONEY GOVERNMENT IX-a$^+$/ IX-a. IX-arc-a$^+$/ IX-arc-a SHOULD RICH.
'Everybody gives money to the government. They should be rich!' (Inf 1, 8, 12-13)

Finally, a similar phenomenon holds when one refers to tall people:

(12) a. [EACH EACH GIANT GIANT GIANT]$_a$ THINK NONE ONE LIKE IX a$^+$/ IX-a.
'Every giant thinks that nobody likes him.' (Inf 1, 8, 119e,f-120)
b. [EACH EACH VERY TALL TALL TALL MAN]$_a$ THINK NONE ONE LIKE IX-a$^+$/ IX-a.
'Every very tall man thinks that nobody likes him.' (Inf 1, 8, 119c,d-120)

The word *GIANT* is produced high in signing space, hence the effect we see in (12)a could conceivably be morphological in nature. But this explanation won't carry over to (12)b: the word *MAN* is body-anchored, and hence does not introduce a locus of its own (unlike, say, the word for *ONE*, which is signed in a particular locus); furthermore, neither the sign for *VERY* nor the word for *TALL* is signed high. Nonetheless, use of a high locus is entirely natural, and appears to be licensed by semantic rather than by morphological considerations.

The examples in (12) make a further important point: use of high loci to refer to respectable, powerful or tall entities is by no means limited to deictic elements, since in these examples the pronouns are *bound* by universal quantifiers.

---

[2] *SELF-a⁺* was present in the 'high' version but not in the 'low' version.

3.2 An Analysis with Iconic Presuppositions

Following Lillo-Martin and Klima 1990, we take sign language indexes to be, in the cases under study, the overt realization of formal indices. Technically, we take assignment functions to assign values to loci as well as to standard (unpronounced) indices. For simplicity, we state our analysis in a framework in which gender features introduce presuppositions on the value of variables, as shown in (13)a:

(13) Let c be a context of speech and s be an assignment function (c_a = the author of c).
a. If *f* is a feminine feature and *i* is in index, $[[\text{pro-f}_i]]^{c,s} = \#$ iff $s(i) = \#$ or s(i) is not female. If $[[\text{pro-f}_i]]^{c,s} \neq \#$, $[[\text{pro-f}_i]]^{c,s} = s(i)$.
b. If *i* is a locus that appears high in signing space, $[[\text{IX-i}]]^{c,s} = \#$ iff $s(i) = \#$ or **s(i) is not powerful, respectable or tall relative to c_a.** If $[[\text{IX-i}]]^{c,s} \neq \#$, $[[\text{IX-i}]]^{c,s} = s(i)$.

(13)a states that a pronoun with feminine features *pro-f_i* yields a presupposition failure, i.e. denotes #, in case the assignment function s fails to assign to the formal index *i* a female individual. If no failure arises, *pro-f_i* denotes s(i). An analogous rule is stated in (13)b: if *i* is a locus that appears high in signing space, a pronoun *IX-i* realized by a pointing sign towards locus *i* give rise to a presupposition failure unless s(i) is powerful, respectable or tall relative to the speaker. Needless to say, future research will have to refine the properties that appear in the part in bold in (13)b.

As stated, the rule in (13)b makes it an arbitrary fact that high loci, rather than low ones, say, carry the meanings at hand. But intuitively this isn't an accident: in each case, a (real or metaphorical) projection seems to be established between the position of locus *i* relative to the signer, and the position of the denotation of *i* relative to the signer on some salient scale (of height, power or respectability). This iconic component is isolated in the following restatement of the rule:

(14) If *i* is a locus that appears high in the signing space,
$[[\text{IX-i}]]^{c,s} = \#$ iff $s(i) = \#$ or **<1, i> is not iconically projectable to <c_a, s(i)>** along the 'power', 'respectability' or 'height' dimension. If $[[\text{IX-i}]]^{c,s} \neq \#$, $[[\text{IX-i}]]^{c,s} = s(i)$.

The new rule is identical to the old one, except that it makes an explicit provision for an iconic component, which establishes a connection between the geometric properties of the loci and some (real or metaphorical) properties of the objects they denote; as before, a detailed analysis of the expression in bold is left for future work.

3.3 Further Application: Deictic Pronouns

When individuals are present in the discourse situation, one normally refers to them by pointing towards them. In other words, in such cases the locus must roughly correspond to the position of the person it denotes. The formalism we have just introduced can capture this fact:

(15) If *i* is a locus, and if s(i) is present in the discourse situation, $[[\text{IX-i}]]^{c,s} = \#$ iff $s(i) = \#$ or s(i) is present in the extra-linguistic situation and <1, i> is not iconically projectable to <c_a, s(i)> along the 'aligment' dimension. If $[[\text{IX-i}]]^{c,s} \neq \#$, $[[\text{IX-i}]]^{c,s} = s(i)$.

In words: if s(i) is present in the discourse situation, then <1, i>, i.e. the pair of positions corresponding to the position of the signer and the locus *i*, must be aligned with <c_a, s(i)>, i.e. the pair of the signer and the denotation of *i*. 1 and c_a are roughly identical, hence when the signer points towards *i* he must also point towards s(i).

This rule has some bite: there are cases in which the locus assigned to an individual 'moves' in signing space. Specifically, when an individual has been associated with a spatial location, one may (but need not) refer to him by pointing towards the locus *associated with that location*. This phenomenon, which we may call 'locative agreement', is illustrated in (16) (from Schlenker 2011a).

(16) JOHN $_a$[WORK IX-a FRENCH CITY] SAME $_c$[WORK IX-c AMERICA CITY]. IX-a IX-1 HELP IX-a, IX-c IX-1 NOT HELP IX-c.
'John does business in a French city and he does business in an American city. There [= in the French city] I help him. There [= in the American city] I don't help him.'
(Inf 1, 4, 66b; Inf 1, 4 67 & 10.05.06, 10.05.11, Inf 1 7, 214, 7,234)

The lexical entry in (15) predicts that for deictic pronouns, this possibility should be precluded. See Schlenker 2011a for preliminary evidence that part of this prediction might be borne out in ASL.

4 Directional Verbs and Locus-Internal Iconicity

4.1 The Debate about Directional Verbs

Directional verbs are realized as signs that target the locus of at least one of their arguments. Despite numerous disagreements, Liddell 2003 and Lillo-Martin and Meier 2011 both note that there are clear iconic effects with directional verbs, which *target different parts of a locus depending on their meaning* – e.g. Liddell 2003 writes that 'ASK-QUESTION$^{\rightarrow y}$ is directed toward the chin/neck area', while 'COMMUNICATE-TELEPATHICALLY-1$^{[RECIP]\rightarrow y}$ (...) is directed toward the forehead'. How should this observation be incorporated into a formal theory?

Lillo-Martin 1991 and Lillo-Martin and Meier 2011 posit that directional verbs have overt or null arguments whose features *and referential index* they copy. This gives rise to two cases, depending on whether the arguments are overt or null.

(17) (i) Overt arguments
a-MARY a-INFORM-b b-SUE a-IX PASS TEST (Lillo-Martin and Meier 2011)
'Mary$_1$ informs Sue$_2$ that she$_1$ passed the test.'
(ii) Null arguments
A. Did John send Mary the letter?
B. YES, a-SEND-b (Lillo-Martin and Meier 2001)
analyzed as: Ø-a a-SEND-b Ø-b
where *Ø-a* and *Ø-b* are null pronouns licensed by agreement.

One particularly striking argument for this analysis goes back to Lillo-Martin 1991, who showed that Strong Crossover effects (i) hold in ASL, (ii) are obviated by resumptive pronouns, and (iii) are *equally* obviated by directional verbs with no overt arguments (but displaying agreement). The latter point strongly suggests that directional verbs can always license null pronouns, which in these cases behave like resumptive pronouns. This appears to argue for a grammatical theory, one in which directional verbs are an instance of anaphoric constructions. Liddell 2003, by contrast, forcefully argues that iconic effects found with directional verbs remain unexplained unless one gives a central place to their iconic dimension:

"Each individual verb has specific gestural characteristics associated with it. (...) For those that do point, if they are directed at a person, they are directed at specific parts of the person (e.g. forehead, nose, chin, sternum). These are not general characteristics of gestural 'accompaniments' to signing. These are specific, semantically relevant, properties of individual verbs."

We argue that the two sides of the debate can be reconciled, but only if a semantics is introduced in which (i) loci are structured areas of space, as Liddell argued, and (ii) semantic rules can make reference to iconic requirement.

4.2 Iconic Effects with Directional Verbs

On an empirical level, we show (a) that Liddell's claims also hold of donkey and bound pronouns, and (ii) that the particular part of a locus which is targeted by agreement *depends on the position (upright or hanging) of the person referred to.* Thus in (18)a, 8 sentences were rated in which (i) *ASK-QUESTIONS vs.* (ii) *COMMUNICATE-TELEPATHICALLY* were rated depending on the height of the loci they targeted. Preferred height was 'medium low' for (i) and 'high' for (ii) - and all examples involved 'donkey' pronouns, or rather agreement markers (7-point scale). In (18)b, a single directional verb was investigated, *ASK-QUESTIONS.* When its (existentially bound) direct object denoted a person *hanging* from a branch (roughly, with ⊤, where horizontal = index of the non-dominant hand, and vertical = index+middle finger of the dominant hand), the position of the vertical that was preferably targeted was 'low'. When the object denoted a person *standing* on a branch (with ⊥), the position of the vertical that was preferably targeted 'high'. Importantly, in (18)b we only report ratings that concern *homogeneous* height assignments for the pronoun *IX-a* and for the target of *1-ASK-a.* Since our focus is on the latter, we leave further subtleties concerning the full pronoun for future research.[3]

(18) a. Height differences

YESTERDAY [LINGUIST PERSON]$_a$ MEET [PHILOSOPHY PERSON]$_b$. THE-TWO-a,b
1. ASK-QUESTIONS-a,b
2. COMMUNICATE-BY-TELEPATHY-a,b (Inf 1, 8, 320-1)

Condition:	1. ASK-a,b	2. COMMUNICATE-a,b
a. high	5	7
b. medium high	6	4
c. medium low	7	2
d. low	3	1

'Yesterday a linguist met a philosopher. They 1. asked each other questions / 2. communicated by telepathy.'

b. The role of classifier position

Context: Several of my friends were hanging from a branch.
TREE BRANCH APPROXIMATE CHEST-HIGH.
SEVERAL POSS-1 FRIEND
1. HANG HANG HANG
2. STAND STAND STAND STAND (Inf 1, 9, 8-10; 9, 9-11)
ONE$_a$ (IX-arc-a) a-TELL-1 IX-a WANT IX-1 1-ASK-QUESTIONS-a

Condition:	1. Hanging	2. Standing
a. High IX-a and a	3, 4	7, 7
b. Medium IX-a and a	5, 6	7, 7
c. Low IX-a and a	7, 7	2, 2

'There was a chest-high tree branch. Several of my friends were 1. hanging / 2. standing. One of them told me that he wanted me to ask him questions.'

[3] Some examples in (18)a2 might involve Role Shift, which should be better controlled for.

The data in (18)a are just a replication of Liddell's observations. The data in (18)b show that it would be misguided to state a lexical entry in which the target of the movement is at different relative heights for different verbs: when the position of the classifier is reversed, from 'standing' to 'hanging', the preferred target of the movement is changed as well – preferably high in the 'standing' position and low in the 'hanging' position. This is entirely in line with Liddell's observation: one should view the locus which is targeted as a simplified picture of the person denoted; and since *ASK* targets the head of this person, it is only natural that it should be high in the 'standing' position and low in the 'hanging' condition. This appears to be a genuinely iconic effect, which ought to be captured by an appropriate formal semantic analysis.

4.3 An Analysis with Structured Loci

Concentrating on directional verbs (rather than the pronouns they co-occur with), we propose an analysis which is based on two main ideas: (i) loci are structured, as Liddell claimed; (ii) the lexical semantics of directional verbs has an iconic component, which requires that the target of the movement should correspond to a particular part of the object denoted by the locus. For simplicity, we implement this requirement in terms of a presuppositional semantics.

We start by revising our earlier assumptions and propose that assignment functions assign values to areas of space rather than to points:

(19) Assignment functions assign values to **areas** of space ('area loci') rather than to **points** of space ('point loci'). Lower-case letters (e.g. *i*) designate point-loci; capital letters (e.g. *I*) designate area-loci;assignment functions assign values to variables and capital letters.

Our analysis posits that directional verbs have a presuppositional component. To illustrate it on a more familiar case, we use a toy example: we assume that person marking on verbs has a presuppositional semantics. On this view, *am working* should give rise to a presupposition failure unless its argument denotes the speaker:

(20) For any objects x and y of type e, for any context c and assignment function s, $[\![\text{am\_working}]\!]^{c,s}(x) = \#$ iff $x = \#$ or $x \neq c_a$. If $\neq \#$, $[\![\text{am\_working}]\!]^{c,s}(x) = 1$ iff x is working.

We posit a similar presuppositional semantics for directional verbs, but with an explicit iconic requirement (which for simplicity we take to hold of both arguments):

(21) For any objects x and y of type e, for any context c and assignment function s, $[\![\text{i-ASK-QUESTIONS-j}]\!]^{c,s}(y)(x) = \#$ iff $x = \#$ or $y = \#$ or $s(J) \neq y$ or $s(I) \neq x$ or <I, i> is not iconically projectable to <body(s(I)), chin(s(I))> along the 'position' dimension or <J, j> is not iconically projectable to <body(s(J)), chin (s(J))> along the 'position' dimension. If $\neq \#$, $[\![\text{i-ASK-QUESTIONS-j}]\!]^{c,s}(y)(x) = 1$ iff x ask questions to y.

In words: when *i-ASK-QUESTIONS-j* is a movement from a point i of an area locus I to a point j of an area locus J; a failure is obtained unless i is with respect to I in the same position as the chin of s(I) with respect to the body of s(I); we abbreviate this as: <I, i> is not iconically projectable to <body(s(I)), chin(s(I))> along the 'position' dimension. The same condition holds of j, J. Needless to say, this characterization is just a placeholder for a properly geometric analysis; our goal at this point is just to indicate where in the interpretive procedure this condition plays a role.

For *COMMUNICATE-BY-TELEPATHY*, the rule is the same, except that (i) there is no temporal asymmetry between i, j (both are targeted at the start) and (ii) the part of the structured locus which is targeted is the forehead rather than the chin.

(22) For any objects x and y of type e, for any context c and assignment function s, $[\![\text{i-j-COMMUNICATE-BY-TELEPATHY}]\!]^{c,s}(y)(x) = \#$ iff $x = \#$ or $y = \#$ or $s(J) \neq y$ or $s(I) \neq x$ or $\langle I, i\rangle$ is not iconically projectable to $\langle body(s(I)), \boxed{forehead(s(I))}\rangle$ along the 'position' dimension or $\langle J, j\rangle$ is not iconically projectable to $\langle body(s(J)), \boxed{forehead(s(J))}\rangle$ along the 'position' dimension. If $\neq \#$, $[\![\text{i-j-COMMUNICATE-BY-TELEPATHY}]\!]^{c,s}(y)(x) = 1$ iff x communicates by telepathy with y.

To illustrate, consider the sentence *John asks me questions*, i.e. *JOHN i-ASK-QUESTIONS-k* $\emptyset_k$, with *k* targeting the signer. At the first step of the semantic derivation, we just 'feed' the verb its arguments. *JOHN* denotes the individual John, and $\emptyset_k$ denotes whatever the assignment function specifies that the area-locus K denotes – in this case, the signer, hence $s(K) = c_a$. Thus we have the step in (23):

(23) $[\![\text{JOHN i-ASK-QUESTION-k } \emptyset_k]\!]^{c,s} = [\![\text{i-ASK-QUESTION-k}]\!]^{c,s}([\![\emptyset_k]\!]^{c,s})([\![\text{JOHN}]\!]^{c,s})$
$= [\![\text{i-ASK-QUESTION-k}]\!]^{c,s}(s(K))(John) = [\![\text{i-ASK-QUESTION-k}]\!]^{c,s}(c_a)(John)$

We can now apply the lexical entry in (21), with $y = c_a$ and $x =$ John. It is immediate that $x \neq \#$ and $y \neq \#$, and by assumption s(K) is the signer c_a. Taking these facts into account, we derive the desired presupposition, i.e. that *ASK-QUESTIONS* targets what corresponds to the 'chin' position of the signer (as well as of I):

(24) $[\![\text{JOHN i-ASK-QUESTIONS-k } \emptyset_k]\!]^{c,s} = \#$ iff $[\![\text{i-ASK-QUESTION-k}]\!]^{c,s}(c_a)(John) = \#$, iff $\langle K, k\rangle$ is not iconically projectable to $\langle body(c_a), chin(c_a)\rangle$ along the 'body' dimension or $\langle I, i\rangle$ is not iconically projectable to $\langle body(j), chin(j)\rangle$ along the 'body' dimension.

When this presupposition is satisfied, we derive the desired truth conditions:

(25) If $[\![\text{JOHN i-ASK-QUESTIONS-k } \emptyset_k]\!]^{c,s} \neq \#$, $[\![\text{JOHN i-ASK-QUESTION-k } \emptyset_k]\!]^{c,s} = 1$ iff $[\![\text{i-ASK-QUESTION-k}]\!]^{c,s}(c_a)(John) = 1$, iff John asks questions to c_a.

In conclusion, we hope to have shown that a formal semantics with iconicity holds good promise for the analysis of sign language anaphora, in particular of complement set readings, high loci, and directional verbs. The key idea is that in the end loci are *both* variables (or discourse referents) and simplified pictures of what they denote.

References

Cuxac, C.: French Sign Language: Proposition of a Structural Explanation by Iconicity. In: Braffort, A., Gibet, S., Teil, D., Gherbi, R., Richardson, J. (eds.) GW 1999. LNCS (LNAI), vol. 1739, pp. 165–184. Springer, Heidelberg (2000)

Koulidobrova, E.: SELF: Intensifier and 'long distance' effects in American Sign Language (ASL). Manuscript, University of Connecticut (2011)

Liddell, S.K.: Grammar, Gesture, and Meaning in American Sign Language. Cambridge University Press, Cambridge (2003)

Lillo-Martin, D.: Universal grammar and American Sign Language: Setting the null argument parameters. Kluwer Academic Publishers, Dordrecht (1991)

Lillo-Martin, D., Klima, E.S.: Pointing out Differences: ASL Pronouns in Syntactic Theory. In: Fischer, S.D., Siple, P. (eds.) Theoretical Issues in Sign Language Research. Linguistics, vol. 1, pp. 191–210. University of Chicago Press, Chicago (1990)

Lillo-Martin, D., Meier, R.: On the linguistic status of 'agreement' in sign language. Theoretical Linguistics 37(3-4), 95–141 (2011)

Neidle, C., et al.: The Syntax of American Sign Language: Functional Categories and Hierarchical Structure. MIT Press (2000)

Nouwen, R.: Plural pronominal anaphora in context. Number 84 in Netherlands Graduate School of Linguistics Dissertations, LOT, Utrecht (2003)

Sandler, W., Lillo-Martin, D.: Sign Language and Linguistic Universals. Cambridge University Press (2006)

Schlenker, P.: Iconic Agreement. Theoretical Linguistics 37(3-4), 223–234 (2011a)

Schlenker, P.: Donkey Anaphora: the View from Sign Language (ASL and LSF). Linguistics & Philosophy 34(4), 341–395 (2011b)

Schlenker, P.: Complement Set Anaphora and Structural Iconicity in ASL. To appear in Snippets (to appear)

Taub, S.F.: Language from the body. Cambridge University Press (2001)

Context Probabilism

Seth Yalcin

University of California, Berkeley, USA
yalcin@berkeley.edu

Abstract. We investigate a basic probabilistic dynamic semantics for a fragment containing conditionals, probability operators, modals, and attitude verbs, with the aim of shedding light on the prospects for adding probabilistic structure to models of the conversational common ground.

Keywords: dynamic semantics, common ground, conditionals, epistemic modals, attitudes, probability operators, probability.

1 Introduction

Our story begins with a core notion of pragmatics:

Definition 1. It is COMMON GROUND that ϕ in a group if all members presuppose that ϕ, and it is common knowledge in the group that all members presuppose that ϕ.

Broadly following Stalnaker and Lewis building on Grice, we take it that the mutually understood proximal rational aim of speech acts is generally to change the common ground of the conversation—to update, if you like, the conversational scoreboard. Our more distal communicative objectives (of transferring knowledge, raising questions, convincing, misleading, etc.) are generally achieved by way of changing what is presupposed. Presupposition is a public attitude, in the sense that the presuppositional states of a set of interlocutors are correct just in case they match, having the same content and structure. Contexts in which agents are not presupposing the same things are defective, though the defect may never emerge explicitly. In theorizing we largely restrict attention to the nondefective case.

We do not assume that the common ground coincides with common belief. Generally, it does not. What one comes away from a conversation actually believing depends on subtleties about trust and authority, and about what the mutually understood import of the communicative exchange was. Such matters depends on the particular goals an interests of the interlocutors, and do not fall within the scope of linguistic pragmatics *per se*.

A formal pragmatics investigates (*inter alia*) possible models for the common ground, and investigates possible ways the common ground might be updated by agents who share knowledge of the semantics and pragmatic rules of a language. A natural course is to begin with a relatively unstructured model of the common ground, adding structure just as required by the phenomena. Thus we might

M. Aloni et al. (Eds.): Amsterdam Colloquium 2011, LNCS 7218, pp. 12–21, 2012.
© Springer-Verlag Berlin Heidelberg 2012

begin by taking the common ground to be an unstructured set of propositions, the presupposed propositions. Assertion (for instance) could then be modeled as a sort of proposal to add a proposition to the common ground. Still more austere we might, following Stalnaker, take the common ground to be a set of possible worlds, the possible worlds not yet eliminated by what is presupposed (what Stalnaker calls a *context set*). On this model propositions are identified with sets of possible worlds, and the propositions which are common ground are taken to be those true throughout the context set. The set of presupposed propositions is thus assumed to be closed under entailment. The effect of successful assertion, on Stalnaker's view, is to eliminate possibilities incompatible with the proposition asserted from the common ground.

Various dynamic semantic systems can be seen as adding structure to a broadly Stalnakerian conception of the common ground. The file change semantics of [Heim 1982] famously added assignment functions and discourse referents; [Veltman 1996] added expectation patterns; [Hulstijn 1997] added partitions; and so on. My interest in the paper is in the question of adding probabilistic structure. I call the thesis I want to investigate *context probabilism*:

Context Probabilism. The common ground of a conversation may have probabilistic structure.

I will suggest this thesis (broached initially in [Yalcin 2005]) has at least the following three motivations: first, adopting it paves the way for a plausible semantics and pragmatics of probability operators (*probably, likely, is as likely as*, etc.); second, it enables us, together with other assumptions, to capture a nontrivial connection between indicative conditionals and the corresponding conditional probabilities; third, it permits a model of linguistic communication congenial to a broadly Bayesian perspective. Further motivations tied to particular linguistic phenomena will accrue as we proceed.

There are a variety of conceivable ways of implementing context probabilism in a semantic-pragmatic theory. I elect in this paper to approach the issue from the perspective of a relatively 'textbook' dynamic semantic system for a propositional language, asking how that textbook system might be conservatively extended to incorporate context probabilism. We begin with a review of the textbook system.

2 A Textbook Propositional Dynamic Semantics

Assume a propositional language $\mathcal{L}$ containing negation, conjunction, epistemic *might* ($\Diamond$), an indicative conditional operator ($\rightarrow$), a belief operator (B), a knowledge operator (K) and a presupposition operator (∂).

Definition 2. A MODEL $\langle\mathcal{W},\mathcal{I}\rangle$ is a pair of a space of worlds $\mathcal{W}$ and a valuation function $\mathcal{I}$ mapping propositional letters of $\mathcal{L}$ to subsets s of $\mathcal{W}$ (INFORMATION STATES).

Definition 3. A MODEL WITH ATTITUDES is a triple of a model $\mathcal{M}$ together with a doxastic accessibility relation $\mathcal{B}^w$ and an epistemic accessibility relation $\mathcal{K}^w$ (construing these as functions from worlds to sets of worlds within $\mathcal{W}_{\mathcal{M}}$).

Definition 4. An UPDATE FUNCTION $[\cdot]$ is a function from wffs of $\mathcal{L}$ to functions from information states to information states (in some given model with attitudes) subject to the following constraints, where α is any propositional letter, ϕ and ψ are any wffs:

$$s[\alpha] = s \cap \mathcal{I}(\alpha)$$

$$s[\neg\phi] = s - s[\phi] \qquad \text{[Heim 1983]}$$

$$s[\phi \wedge \psi] = s[\phi][\psi] \qquad \text{[Heim 1982], [Stalnaker 1974]}$$

$$s[\Diamond\phi] = s \text{ iff } s[\phi] \neq \emptyset, \text{ else } \emptyset \qquad \text{[Veltman 1996]}$$

$$s[\phi \rightarrow \psi] = s \text{ iff } s[\phi] = s[\phi][\psi], \text{ else } \emptyset \qquad \text{[Gillies 2004]}$$

$$s[B\phi] = \{w \in s : \mathcal{B}^w[\phi] = \mathcal{B}^w\} \qquad \text{[Heim 1992]}$$

$$s[\partial\phi] = s \text{ iff } s[\phi] = s; \text{ else undefined} \qquad \text{[Beaver 2001]}$$

$$s[K\phi] = \{w \in s[\partial\phi] : \mathcal{K}^w[\phi] = \mathcal{K}^w\}$$

Definition 5. s ACCEPTS ϕ iff $s[\phi] = s$.

Definition 6. $\phi_1, ..., \phi_n \vDash \psi$ iff $\forall s$: if s accepts each of $\phi_1, ..., \phi_n$, s accepts ψ.

This supplies our starting point. I assume familiarity with ideas reflected in the above semantics. See the works cited above for introduction and some initial motivation (see also [Yalcin 2007]).

3 Discontents of the Textbook Semantics

We should like to extend this system so as to make room for probability operators. As it is, it is difficult to see how this might be achieved if, as seems intuitive, we have it that $\triangle\phi \vDash \Diamond\phi$ (abbreviating *probably* with $\triangle$). For if $\Diamond\phi$ performs a 'test' on a state of information, so too should $\triangle\phi$; but what test would be suitable? It seems clear we need information states to incorporate further structure, structure which the test corresponding to $\triangle\phi$ can be sensitive to. We should also like to extend this system so as to supply some intelligible connection between the acceptability of an indicative conditional and its corresponding conditional probability, given empirical evidence that the two are highly correlated (see [Douven and Verbrugge 2010] and references therein).

Aside from these limitations of coverage, the textbook semantics also suffers from the following two quite nontrivial predictive failures, failures our probabilistic upgrade will eventually overcome.

Negation Problem. According to the semantics, (1) $\vDash$ (2):

1. It is not the case that John believes it might be raining. ($\neg B\Diamond\phi$)
2. John believes that it is not raining. ($B\neg\phi$)

This is not plausible in general. One can fail to believe it might be that ϕ without believing $\neg\phi$. The failure might, for instance, be the result of not being appropriately sensitive to the question whether ϕ. (This worry is discussed in detail in [Yalcin 2011]; see also [Willer 2010].)

Defective Logic of Knowledge and Belief. Suppose it is raining. From this, it of course follows that it is compatible with what John knows that it is raining. Hence it follows, according to the semantics above, that John knows it might be raining. This result is absurd.

We get a yet greater absurdity when we add the plausible assumption that knowledge entails belief (understood as $K\phi \vDash B\phi$). When we add this assumption, we get the result that belief in falsehoods is not possible. As an example, suppose (for *reductio*) that it is raining, but that John believes it is not raining. By the reasoning above, John knows it might be raining. By the assumption that knowledge entails belief, John believes it might be raining. But this precludes, given the semantics, the possibility that John believes it is not raining.[1]

4 Sharp Context Probabilism

To make room for probability operators and conditional probabilities, let us enrich the structure of the common ground. An obvious first idea is to embrace:

Sharp Context Probabilism. A conversational common ground can be modeled as a probability space.

We let probabilities spaces of a certain sort displace our earlier, non-probabilistic information states:

Definition 7. A SHARP INFORMATION STATE i is a pair of a set s, $s \subseteq W$ (call it the DOMAIN of i) and a probability function Pr on the elements of some Boolean algebra A of subsets of W such that: (i.) $Pr(s) = 1$; and (ii.) $Pr(p \cup q) = Pr(p) + Pr(q)$, for all disjoint p, q in A.

In essence, we are simply equipping our earlier information states with probability measures. Reinterpret $\mathcal{B}^w$ and $\mathcal{K}^w$ so that they correspond to sharp information states (given a world). Let our update function now be defined on sharp information states as follows:[2]

Definition 8. A PROBABILISTIC UPDATE FUNCTION $[\cdot]$ is a function from wffs of $\mathcal{L}$ to functions from information states to information states subject to the following constraints, where α is any propositional letter, ϕ and ψ are any wffs:

$$i[\alpha] = \langle s_i \cap \mathcal{I}(\alpha), Pr_i(x|s_i \cap \mathcal{I}(\alpha))\rangle$$
$$i[\neg\phi] = \langle s_i - s_{i[\phi]}, Pr_i(x|s_i - s_{i[\phi]})\rangle$$
$$i[\phi \wedge \psi] = i[\phi][\psi]$$
$$i[\Diamond\phi] = i \text{ iff } s_{i[\phi]} \neq \emptyset, \text{ else } \langle\emptyset, Pr_i(x|\emptyset)\rangle$$

[1] I am indebted here to conversations with Stephen Yablo and Rohan Prince.

[2] Thanks here to Justin Bledin for essential suggestions.

$$i[\triangle\phi] = i \text{ iff } Pr_i(s_{i[\phi]}) > .5, \text{ else } \langle\emptyset, Pr_i(x|\emptyset)\rangle$$
$$i[\phi \to \psi] = i \text{ iff } i[\phi] = i[\phi][\psi], \text{ else } \langle\emptyset, Pr_i(x|\emptyset)\rangle$$
$$i[B\phi] = \langle\{w \in s_i : \mathcal{B}^w[\phi] = \mathcal{B}^w\}, Pr_i(x|\{w \in s_i : \mathcal{B}^w[\phi] = \mathcal{B}^w\})\rangle$$
$$i[\partial\phi] = i \text{ iff } i[\phi] = i; \text{ else undefined}$$
$$i[K\phi] = \langle\{w \in s_{i[\partial\phi]} : \mathcal{K}^w[\phi] = \mathcal{K}^w\}, Pr_i(x|\{w \in s_i : \mathcal{K}^w[\phi] = \mathcal{K}^w\})\rangle$$

Definition 9. A sharp information state i ACCEPTS ϕ iff $i[\phi] = i$.

Definition 10. $\phi_1, ..., \phi_n \vDash \psi$ iff $\forall i$: if i accepts each of $\phi_1, ..., \phi_n$, i accepts ψ.

(Definition 10 now supplants Definition 6.) In general, updates to the common ground involving ordinary factual information proceed by eliminating worlds from the domain of the context (as before) *and* by conditionalizing the probability measure. Probability operators now perform tests, akin to epistemic modals and conditionals. Whereas $\Diamond$ serves to perform a test on the domain of a sharp information state, $\triangle$ serves to perform a test on the measure. It is obvious how to add *is as likely as* ($\succeq$) to this system, representing it, too, as a test.

The present account addresses the two limitations of coverage noted for the textbook account. For the connection between indicatives and conditional probabilities, we can note that, for instance,

3. John believes that if the door is ajar, Bob is probably in his office.

is in the present framework a matter of whether John's credence in the proposition that Bob is in his office, conditional on the proposition that the door is ajar, is sufficiently high (above .5). In general, the system tells us that an indicative conditional is accepted relative to a sharp information state i just in case, roughly, the consequent is accepted relative to the state resulting from conditionalizing i on the antecedent.

As for the semantics of probability operators, the system delivers an array of desirable entailment predictions. I will just list a few: (i) $\Box\phi \vDash \triangle\phi \vDash \Diamond\phi$; (ii) $\phi \vDash \triangle\phi$; (iii) $\neg\Diamond\phi \vDash \neg\triangle\phi$; (iv) $\{\phi \to \psi, \triangle\phi\} \vDash \triangle\psi$; (v) $\{\phi \to \psi, \neg\triangle\psi\} \vDash \neg\triangle\phi$; (vi) $\triangle\phi \wedge \neg\phi \vDash \bot$. Adding $\succeq$, we could include (vii) $\phi \to \psi \vDash \psi \succeq \phi$; (viii) $\{\phi \succeq \psi, \triangle\psi\} \vDash \triangle\phi$.

It is natural to ask whether the resources of probability theory are strictly required to cover these and other inference patterns. For some further relevant discussion, see [Yalcin 2010], [Lassiter 2011]. Whether or not such resources are strictly required, the system is anyway perspicuous, and it offers a relatively conservative extension of the textbook picture. Where probabilities are not at issue, the additional probabilistic structure can safely be ignored, and one can work with the ordinary textbook account.

Indeed, the present system is *too* conservative an extension of the textbook picture, for it inherits the two predictive failures described in the last section 3 above: as the reader can verify, it also suffers from the negation problem, and it also yields a defective logic for knowledge and belief.

To these problems we can add a more basic difficulty: the representation of the common ground the present system delivers is too rich. The common ground is supposed to reflect the information that participants in conversation mutually presuppose. But the present system *requires* each interlocutor to assign all the propositions which reflect live possibilities in the conversation a probability, *regardless* of whether or not that probability has come up in conversation. This makes it impossible to 'read off', given an agent's state of presupposition, which probability values are those which the agent has explicitly coordinated on, vitiating to a significant degree the very idea of the common ground. It also means, given the semantics in place, that for every open proposition p, one is *required* to presuppose either that p is likely, or that p is not likely. But it is just false that one is required to presuppose one of these two things for every open proposition—especially when nothing has been said on the matter.

5 Blunt Context Probabilism

The discontents of sharp context probabilism can be addressed by retreating to a weaker, but still probabilistic, representation of the common ground, and by making a supervaluationist move within the semantics. We drop the thesis of sharp context probabilism in favor of:

Blunt Context Probabilism. A conversational common ground can be modeled as a set of probability spaces.

We introduce the notion of a blunt information state:

Definition 11. A BLUNT INFORMATION STATE is a set of sharp information states.

We assume that a state of presupposition, hence a common ground, is representable as a blunt information state; and we assume the same for states of belief and of knowledge. Correspondingly, we reinterpret $\mathcal{B}^w$ and $\mathcal{K}^w$ in our model so that they correspond to blunt information states (given a world).

The idea of representing a state of belief as a set of probability spaces (or measures) has been investigated extensively in the formal epistemology literature. The idea of representing a state of knowledge as a set of probability spaces has been less discussed, but it is perfectly intelligible; it is moreover recommended by the considerations reviewed in [Moss 2011].

Turning back to semantics, we need not start over, redefining every clause now for blunt information states. Rather, we can simply take our probabilistic update function as defined above, and extend it to blunt states by adding the following clause: for all blunt information states I:[3]

$$I[\phi] = \{i \in I : i[\phi] = i\}$$

To update a blunt information state I with a sentence ϕ, eliminate all those $i \in I$ which don't accept ϕ. We extend the notion of acceptance to blunt information

[3] Compare the structurally analogous approach of [Willer 2010].

states, and we redefine consequence in terms of preservation of acceptance with respect to blunt states:

Definition 12. Blunt state I ACCEPTS ϕ iff $I[\phi] = I$.

Definition 13. $\phi_1, ..., \phi_n \vDash \psi$ iff $\forall I$: if I accepts each of $\phi_1, ..., \phi_n$, I accepts ψ.

6 Problems Solved

Blunt context probabilism, together with the dynamic semantics given above, solves all our problems.

First, the negation problem is solved: (1) $\nvDash$ (2). Likewise, (4) $\nvDash$ (5):

4. It is not the case that John believes that it is probably raining. $(\neg B\triangle\phi)$
5. John believes that it is not likely to be raining. $(B\neg\triangle\phi)$

For John to fail to believe it is probably raining, his doxastic state need only leave open some sharp information state which associates the rain outcome with some probability less than or equal to .5. By contrast, for John to believe it is not likely to be raining, *all* of the sharp information states compatible with his doxastic state must be such as to associate the rain outcome with some probability less than or equal to .5.

Second, we are no longer driven to absurdities concerning the relation between belief and knowledge. Given the truth of some factual ψ, it does *not* follow that $K\Diamond\psi$; hence we cannot use it to infer, by the schema $K\phi \vDash B\phi$, that $B\Diamond\psi$; hence it doesn't follow, merely from the assumption of ψ, that $\neg B\neg\psi$. In short, false belief is once again compatible with $K\phi \vDash B\phi$. In general, we can allow for a picture which comports with the one we find in familiar possible worlds models of belief and knowledge. In that picture, one's belief worlds form a subset of one's knowledge worlds; one believes everything one knows, and (typicially) more besides. Likewise, we can say that one's blunt doxastic state forms a subset of one's blunt epistemic state.

Third, blunt context probabilism delivers a much more plausible representation of the common ground than its sharp cousin. One need not presuppose probabilities for propositions that one has no reason to believe one's interlocutor is presupposing. And it is not the case that for every open proposition p, one is required to either presuppose that p is likely, or that p is not likely. One might leave the whole interval of probability values for p open.

There is a further advantage of the system worth mentioning. The textbook semantics for $\Diamond$ and $\rightarrow$ faces a problem explaining how exactly epistemic modal and conditional utterances can be informative, despite the fact that they do not serve to add information to the common ground in virtue of their context change potentials. Our sharp probabilistic dynamic semantics has the same problem—indeed, if anything, it makes the problem worse, as it adds sentences of the form $\triangle\phi$ to the list of problematic sentences. By contrast, the semantics based around blunt information states has no such explanatory burden. Sentences of the form $\Diamond\phi$, $\phi \to \psi$, and $\triangle\phi$ all have the potential to change, without destroying, the common ground. Characteristically their role is to eliminate sharp information states from the common ground, just like ordinary factual claims. These sentences differ from ordinary factual claims insofar as the may eliminate information states as a function of global properties of their domains, or as a function of their probability measures *per se*.

7 Probability Conditions

Our system in effect replaces the traditional notion of a truth-condition (a condition on possible worlds) with the more general notion of a probability condition (a condition on probability spaces). Anything you can do with truth-conditions, you can do with probability conditions; but as the system illustrates, with probability conditions one can do more besides. In addition to eliminating possibilities in conversation, we can shift the admissible probabilities over the possibilities, perhaps without eliminating any possibilities.

The conception of information this picture recommends is not as radical as it may appear. On the contrary, it is not far from the conception already found in standard information theory. For in information theory, the amount of information a signal carries is not just a function of the possibilities it eliminates. It is also a function of how it shifts the probabilities over the open possibilities. Information is a fundamentally probabilistic notion.

8 Dynamicness

A semantics which associates sentences with probability conditions might in principle be given statically or dynamically. We have adopted a dynamic formulation, but it can be asked: is such a formulation strictly required to achieve exactly the update effects the system delivers?

We are in effect asking whether the update system supplied by our blunt dynamic semantics is "fundamentally dynamic." We should clarify this question. The general notion of an update system may be defined as follows:

Definition 14. An UPDATE SYSTEM is a triple $\langle L, C, [\cdot]\rangle$, where L is a set of sentences, C is a set of contexts, and $[\cdot]$ is a function from L to a set of unary operations on C.

We can then define the notion of a *static* update system by reference to a highly general notion of a *static semantics*, as follows:

Definition 15. A STATIC SEMANTICS is a triple $\langle L, W, [\![\cdot]\!]\rangle$, where L is a set of sentences, W is a set of points, and $[\![\cdot]\!]$ is an interpretation function, with $[\![\cdot]\!] : L \to \mathcal{P}(W)$.

Definition 16. An update system $\langle L, C, [\cdot]\rangle$ is STATIC if and only if there exists a static semantics $\langle L, W, [\![\cdot]\!]\rangle$ and a one-to-one function f from C to $\mathcal{P}(W)$ such that for all $c \in C$ and $s \in L$: $f(c) \cap [\![s]\!] = f(c[S])$.

It is fair to say this reflects a fairly standard sense of 'static'.

What makes for staticness so conceived? [Rothschild and Yalcin 2012] answer this question by proving the following theorem:

Theorem (static representability). An update system $\langle L, C, [\cdot]\rangle$ is static iff for all $s \in L$ and $c \in C$: (i) $c[s][s] = c[s]$ (*idempotence*); (ii) $c[s][s'] = c[s'][s]$ (*commutativity*).

Given this result, it is easy to see that the blunt dynamic system is static, because it is easy to see that it is idempotent and commutative. To make it plain, we can rewrite the extension of the semantics to blunt information states as:

6. $I[\phi] = I \cap \{i \in \mathbb{I} : i[\phi] = i\}$

(Where $\mathbb{I}$ is the space of all sharp information states for the given model.) So in a relatively deep sense, the blunt dynamic system is only superficially dynamic.

My central interest here has not been to adjudicate between dynamic versus static implementations of the above ideas. Perhaps ultimately a static formulation (see, for instance, [Yalcin 2007], [Yalcin 2012a]) is to be preferred. Or perhaps the (superficially) dynamic statement of the system is to be preferred, not because an overtly static formulation is impossible, but simply because the dynamic statement is relatively elegant and perspicuous, or has other explanatory virtues. The matter deserves further investigation.

I wish to close by noting one way of reintroducing bonafide dynamicness. Consider the idea of rewriting the extension of the semantics to blunt states as follows:[4]

7. $I[\phi] = \{i : \text{for some } i' \in I : i'[\phi] = i \text{ and } i \neq \langle \emptyset, Pr_i(x|\emptyset)\rangle\}$

Here we update a blunt state I, not by tossing out those $i \in I$ which fail to accept ϕ, but rather by updating each i with ϕ (and by setting aside any updates that result in the null context).

The proposal (7.) allows for commutativity failures, and (hence) gives rise to a non-static update system. For instance, it would not be true in general that $I[\neg\phi][\Diamond\phi] = I[\Diamond\phi][\neg\phi]$. The update corresponding to the former order will always yield the empty set; not so the latter order. Likewise, it is not generally the case that $I[\neg\phi][\triangle\phi] = I[\triangle\phi][\neg\phi]$.

It is not at all obvious that this failure of commutativity is an empirical virtue. On the contrary, (6.) makes it easier to understand these data:

8. It is probably raining. # It is not raining.
9. # If it is probably raining, then if it is not raining, we don't need an umbrella.

(See [Yalcin 2012b] for some additional discussion.) In support of (7.), however, consider a conversation accepting $\Diamond\phi$ and $\Diamond\neg\phi$. Doesn't this context make explicit provision for the possibility of eventually incorporating the information that ϕ? But such a context would be crashed by ϕ, according to (6.).

A conservative compromise would be to take (6.) as the primary rule, letting (7.) reflect the appropriate recourse where (6.) would yield the empty set.

[4] Compare again [Willer 2010].

References

Beaver 2001. Beaver, D.: Presupposition and Assertion in Dynamic Semantics. Studies in Logic, Language and Information. CSLI Publications (2001)

Douven and Verbrugge 2010. Douven, I., Verbrugge, S.: The Adams family. Cognition 117(3), 302–318 (2010)

Gillies 2004. Gillies, A.: Epistemic conditionals and conditional epistemics. Nous 38, 585–616 (2004)

Heim 1982. Heim, I.: The Semantics of Definite and Indefinite Noun Phrases. PhD thesis, University of Massachusetts (1982)

Heim 1983. Heim, I.: On the projection problem for presuppositions. In: Flickinger, D., et al. (eds.) Proceedings of the Second West Coast Conference on Formal Linguistics, pp. 114–125. Stanford University Press

Heim 1992. Heim, I.: Presupposition projection and the semantics of attitude verbs. Journal of Semantics 9(3), 183–221

Hulstijn 1997. Hulstijn, J.: Structured information states. In: Benz, A., Jager, G. (eds.) Proceedings of MunDial 1997. University of Munich (1997)

Lassiter 2011. Lassiter, D.: Measurement and Modality: The Scalar Basis of Modal Semantics. PhD thesis, PhD dissertation, New York University (2011)

Lewis 1976. Lewis, D.K.: Probabilities of conditionals and conditional probabilities. Philosophical Review 85, 297–315 (1976)

Moss 2011. Moss, S.: Epistemology formalized. University of Michigan (unpublished manuscript)

Rothschild and Yalcin 2012. Rothschild, D., Yalcin, S.: Dynamics. Oxford University and the University of California, Berkeley (unpublished manuscript)

Stalnaker 1974. Stalnaker, R.: Pragmatic presuppositions. In: Munitz, M.K., Unger, P. (eds.) Semantics and Philosophy. NYU Press (1974)

Stalnaker 1978. Stalnaker, R.: Assertion. In: Cole, P. (ed.) Syntax and Semantics 9: Pragmatics. Academic Press (1978)

Stalnaker 1984. Stalnaker, R.: Inquiry. MIT Press, Cambridge (1984)

Stalnaker 2002. Stalnaker, R.: Common ground. Linguistics and Philosophy 25, 701–721 (2002)

Veltman 1996. Veltman, F.: Defaults in update semantics. Journal of Philosophical Logic 25(3), 221–261 (1996)

Willer 2010. Willer, M.: Modality in Flux. PhD thesis, University of Texas, Austin (2010)

Yalcin 2005. Yalcin, S.: Epistemic modals. In: Gajewski, J., Haquard, V., Nickel, B., Yalcin, S. (eds.) New Work on Modality. MIT Working Papers in Linguistics, vol. 51, pp. 231–272 (2005)

Yalcin 2007. Yalcin, S.: Epistemic modals. Mind 116(464), 983–1026 (2007)

Yalcin 2010. Yalcin, S.: Probability operators. Philosophy Compass 5(11), 916–937 (2010)

Yalcin 2011. Yalcin, S.: Nonfactualism about epistemic modality. In: Egan, A., Weatherson, B. (eds.) Epistemic Modality, pp. 295–332. Oxford University Press (2011)

Yalcin 2012a. Yalcin, S.: A counterexample to Modus Tollens. Journal of Philosophical Logic (forthcoming, 2012)

Yalcin 2012b. Yalcin, S.: Dynamic semantics. In: Russell, G., Fara, D. (eds.) Routledge Encyclopedia of the Philosophy of Language (forthcoming, 2012)

Free Choice in Deontic Inquisitive Semantics (DIS)*

Martin Aher

University of Osnabrueck, Institute of Cognitive Science

Abstract. We will propose a novel solution to the free choice puzzle. The approach is driven by empirical data from legal discourse and does not suffer from the same problems as implicature-based accounts. Following Anderson's violation-based deontic logic, we will demonstrate that a support-based radical inquisitive semantics will correctly model both the free choice effect and the standard disjunctive behaviour when disjunctive permission is embedded under negation. An inquisitive semantics also models the case when disjunctive permission is continued with "but I do not know which" which coerces an ignorance reading. We also demonstrate that a principled approach to negation provides a monotonic but restricted definition of entailment, which solves the problem of strengthening with a conjunct that is used as a counterargument against violation-based accounts.

1 Introduction

(1) A country may establish a research center or a laboratory.

When (1) is law, its salient reading gives permission to establish a research center and it gives permission to establish a laboratory. Although, it does not necessarily give permission to establish both. The problem lies in that a classical analysis of modality as quantification over worlds and disjunction as set union predicts that permission is given to do one or the other, which is a less salient reading. This so called free choice effect has become one of the better documented puzzles in semantics since it was investigated by Hans Kamp [13].

The puzzle is exacerbated by the observation made by Alonso-Ovalle [3], and Simons [19, p. 8] that embedding disjunctive permission under negation reverts disjunction to classical behaviour.

(2) A country may not establish a research center or a laboratory.

The salient reading of (2) says that permission is not granted to establish a research center nor is it granted to establish a laboratory. As this reading follows from standard accounts of modality and disjunction, a modification of disjunction to account for (1) will lead to problems with the salient reading of (2).

* Thank you to Maria Aloni, Regine Eckardt, Stefan Hinterwimmer, Jeroen Groenendijk, Floris Roelofsen, Mandy Simons, Carla Umbach and my anonymous referees.

M. Aloni et al. (Eds.): Amsterdam Colloquium 2011, LNCS 7218, pp. 22–31, 2012.
© Springer-Verlag Berlin Heidelberg 2012

A third puzzle arises from the fact that (1) can be coerced into a standard boolean reading of disjunction by adding "...but I do not know which."

(3) A country may establish a research center or a laboratory but I do not know which."

To utter (3) one must know that permission is granted to establish either a research center or laboratory, but it is not specified which of the two is permitted. This has become known as the ignorance reading.

This paper will propose a model that captures the truth conditions of all three of (1), (2) and (3), or in other words models the free choice effect, classical behaviour of disjunctive permission when embedded under negation and the ignorance reading. This sets it apart from prior accounts as is demonstrated in the next section.

2 Previous Accounts

Zimmermann [20] reignited interest in free choice by positing a pragmatic mechanism that reinterprets disjunction as a conjunctive list of epistemic possibilities: $\Diamond A \wedge \Diamond B$. Unfortunately a reinterpretation of disjunction as conjunction fails to provide the salient reading of (2) as the negation of $\Diamond A \wedge \Diamond B$ comes out as $\overline{\Diamond A} \vee \overline{\Diamond B}$.

Many following accounts accepted that free choice is essentially a pragmatic effect and suggested implicature-based accounts. This approach is supported by (3), as there appears to be a way to cancel the free choice effect in disjunctive permission utterances. These accounts include Schulz [18], Eckardt [7], Fox [8] and the game theoretic implicature account by Franke [9]. As approaches to free choice have been extensively discussed in the literature, for example by Schulz [17] or more recently by Barker [5], we will concentrate on examining [7] to expatiate on general issues with implicature-based solutions.

Eckardt derives the free choice effect utilizing an implicature through the maxims of manner and quality. A simplified version of her account says that if an informed speaker uses disjunction $\Diamond(\varphi \vee \psi)$ then either disjunct would have been more economical. From this we infer that the governing permissions are best described by disjunction because either disjunct alone would be false. This provides the free choice effect: there must be some worlds where $\Diamond\varphi \wedge \overline{\Diamond\psi}$ and others where $\overline{\Diamond\varphi} \wedge \Diamond\psi$.

The weakness of this account lies in its conclusion as the intuition behind the deontic free choice effect in (1) is that the speaker believes that A and B are permitted. The reason why $\Diamond\varphi \wedge \overline{\Diamond\psi}$ and $\overline{\Diamond\varphi} \wedge \Diamond\psi$ worlds are the case is that there exist some worlds in the speaker's information state in which either φ or ψ is not permitted. But this is contrary to the intuition outlined above.

Simons [19, p. 14] argues generally against implicature based accounts on the grounds that there does not seem to be a distinction between what is said and what is implicated in examples such as (1). Compare this to a classic example of generalized implicature from Grice [10, p. 32].

(4) X is meeting a woman this evening.

Grice states that such a statement generally implicates that the woman being met is not X's wife, mother, sister, etc. Thus, there exists a clear distinction between that which is said (X will meet a woman) and that which is implicated (X will meet a potential romantic acquaintance). The lack of such distinctions in free choice sentences poses a challenge to any implicature based account.

Barker [5, p. 16] casts doubt on the existence of another marker of implicatures, namely cancellability. Observe the following example.

(5) You may eat an apple or a pear, although in fact you may not eat an apple.

When an implicature in cancelled, the utterance only has the meaning of what is said. If (4) were cancelled by "... but it's only her mother." then the utterance would lose the implicature that the woman is a romantic acquaintance. Yet, instead of reverting the phrase to that which is said, the added phrase in (5) appears to make the statement contradictory or offers a correction of the preceding information.

There appear to be other possible routes for cancellation, which is to utter either of the following continuations.

(6) You may eat an apple or a pear, although in fact you may not eat both.

The consequence of uttering (6) does not cancel the free choice effect. Permission is given to eat an apple and permission is given to eat a pear. Yet, the continuation provides the additional information that eating both an apple and a pear is prohibited. The additional information does not conflict with free choice readings.

But contrary to these facts, the ignorance reading in (3) does affect the free choice effect. Adding "...but I do not know which." intuitively suggests that the speaker does not know the governing permissions and thus such utterances do not give permission for both disjuncts. We will show that the ignorance reading can be accounted for as a scope effect, similar to one that is in effect in the following example.

(7) There isn't an apple or a pear on the table, but I do not know which is missing.

Assuming that it was expected that there would be an apple and a pear on the table, the utterance of (7) says that one of them is missing, but it is not necessary that both of them are missing as would be the case if the continuation "...but I do not know which." were omitted.

Barker [5] proposes a semantic approach similar to the one pursued here, by following Kanger [14] in positing a normative ideality δ such that if φ is obligatory, then if φ then δ. This view is a contrapositive view of Anderson's reduction [4] and, thus, similar to the proposal to follow, but in terms of details, a prior analysis of World Trade Organization (WTO) examples in [1] suggests that legal reasoning does not concern idealities but rather violations. While this might be contingent on the deontic context, in terms of legal language, the violation-based solution remains preferable.

Also, while Barker's account of the free choice effect is entailment based, his semantics fail to predict the salient reading of negated disjunctive permission sentences such as (2) and he is forced to tell a pragmatic story to account for it. This observation also holds for the semantic account of Aloni [2]. The next section will show that the salient reading of disjunctive permission under negation can also be incorporated into the semantic account.

3 The Proposal

We observed that many previous accounts of free choice fail to capture the effect of negating disjunctive permission utterances. This led us to base the model on an independently motivated prior version of inquisitive semantics that focuses on the effects of negation - Radical Inquisitive Semantics. An earlier version of the language used here was developed and explored by Groenendijk and Roelofsen [12] and Sano [16]. Our proposal adds clauses for deontic permission and discusses entailment in the radical environment. Due to space constraints we must assume familiarity with standard inquisitive semantics and the above proposals.

We shall only consider a propositional language of a finite set of propositional variables and the operators: $\varphi, \wedge, \vee, \rightarrow$. We also need to define worlds as binary valuations for atomic sentences and states as sets of worlds. σ and τ are variables that range over states, w is the variable that ranges over worlds and W is the set of all (classical) valuation functions.. Propositions expressed by sentences are defined through a support and reject relation. When a state supports φ then we write $\sigma \models^{+} \varphi$ and when a state rejects φ then we write $\sigma \models^{-} \varphi$.

Definition 1. *Radical inquisitive semantics (DIS).*

1. $\sigma \models^{+} p$ iff $\forall w \in \sigma : w(p) = 1$
   $\sigma \models^{-} p$ iff $\forall w \in \sigma : w(p) = 0$
2. $\sigma \models^{+} \overline{\varphi}$ iff $\sigma \models^{-} \varphi$
   $\sigma \models^{-} \overline{\varphi}$ iff $\sigma \models^{+} \varphi$
3. $\sigma \models^{+} \varphi \vee \psi$ iff $\sigma \models^{+} \varphi$ or $\sigma \models^{+} \psi$
   $\sigma \models^{-} \varphi \vee \psi$ iff $\sigma \models^{-} \varphi$ and $\sigma \models^{-} \psi$
4. $\sigma \models^{+} \varphi \wedge \psi$ iff $\sigma \models^{+} \varphi$ and $\sigma \models^{+} \psi$
   $\sigma \models^{-} \varphi \wedge \psi$ iff $\sigma \models^{-} \varphi$ or $\sigma \models^{-} \psi$
5. $\sigma \models^{+} \varphi \rightarrow \psi$ iff $\forall \tau \subseteq \sigma.(\ \tau \models^{+} \varphi$ implies $\tau \models^{+} \psi)$
   $\sigma \models^{-} \varphi \rightarrow \psi$ iff $\exists \tau.(\tau \models^{+} \varphi$ and $\forall \tau' \supseteq \tau.(\tau' \models^{+} \varphi$ implies $\sigma \cap \tau' \models^{-} \psi))$

Sano has also shown that in the propositional setting there are always maximal states under the $\subseteq$-relation that support or reject a sentence.

Proposition 1. *(Persistence).*
If $\sigma \models^{+} \varphi$ and $\tau \subseteq \sigma$ then $\tau \models^{+} \varphi$ and if $\sigma \models^{-} \varphi$ and $\tau \subseteq \sigma$ then $\tau \models^{-} \varphi$

We refer to the maximal states that support a sentence as the possibilities for that sentence, and denote it by $[\varphi]^+$. We refer to the maximal states that reject a sentence as the counter-possibilities for that sentence, and denote it by $[\varphi]^-$. Given persistence, this means that the meaning of a sentence φ is fully characterized by $\langle[\varphi]^+, [\varphi]^-\rangle$.

Definition 2. *Informativeness and inquisitiveness.*

1. *φ is informative iff $\bigcup[\varphi]^+ \neq W$.*
2. *φ is inquisitive iff $\bigcup[\varphi]^+ \notin [\varphi]^+$.*

It follows from these definitions that φ is informative if it eliminates worlds and φ is inquisitive if there are at least two possibilities for φ. The above definitions are standard in inquisitive semantics and have been used to characterize assertions, questions and hybrids in the way stated below (see for example [12]).

Definition 3. *Assertions, questions and hybrids.*

1. *φ is an assertion iff φ is not inquisitive.*
2. *φ is a question iff φ is not informative.*
3. *φ is a hybrid iff φ is inquisitive and informative.*

Unlike in standard inquisitive semantics, where negations are always assertions, in radical inquisitive semantics, negations can be inquisitive. Even when negation is applied to an assertion φ, the resulting sentence $\overline{\varphi}$ can be an inquisitive sentence. We can define a new characterization for sentences that are assertions both on the positive and negative side.

Definition 4. *Radical assertions.*
φ is a radical assertion iff both φ and $\overline{\varphi}$ are not inquisitive.

The following figures illustrate the clauses in the definitions. We draw possibilities on our figures such that any world outside of the connected line is eliminated and for counter-possibilities, any world outside of the dashed line is eliminated. For example, the possibility for the atom p is the set of worlds where p is the case and its counter-possibility is the set of worlds where p is not the case. The clause for negation flips between possibilities and counterpossibilities, so that the negation of p is the set of worlds that reject p. Atoms are radical assertions.

Disjunction is the source of inquisitiveness in standard inquisitive semantics. Figure 2 shows the possibilities and counter-possibilities for disjunction. As there can be more than one possibility for disjunction it can be inquisitive on the positive side but its negation is an assertion. Disjunction also eliminates the worlds where neither disjunct is the case, which makes it a hybrid.

In standard inquisitive semantics, conjunction is a radical assertion but in DIS the clause for the negation of conjunction allows inquisitiveness (see figure 3). This accounts for the fact that after a conjunction is rejected, one can ask: "Why?" and the other interlocutor can specify which conjunct was unacceptable. So, while a conjunction is an assertion, the negation of conjunction is a hybrid. The current clauses make $p \wedge q$ symmetric with $\overline{p} \vee \overline{q}$.[1]

[1] In standard inquisitive semantics, the treatment of negation as a complement creates a situation where $p \wedge q \not\models \overline{p} \vee \overline{q}$.

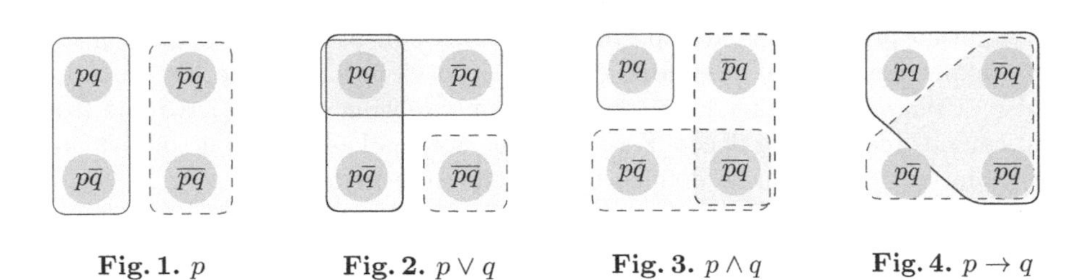

Fig. 1. p **Fig. 2.** $p \vee q$ **Fig. 3.** $p \wedge q$ **Fig. 4.** $p \rightarrow q$

The clauses for conditionals are inspired by Ramsey Test literature and the original motivation is discussed at length elsewhere [12, pp. 18-23, 28-30]. The crucial idea is that the negation of conditionals should not make the antecedent the case. This means that unlike in classical models, a state that supports a conditional and a state that rejects a conditional do not have an empty intersection. This is demonstrated on figure 4. The negation of conditionals allows for inquisitiveness and thus conditionals are assertions.

The negation of conditionals also requires us to adapt the definition of entailment. Lewis and Langford [15] provided the intuition that if φ entails ψ then it should be impossible that $\varphi \wedge \overline{\psi}$. Yet, this does not hold under our clause for conditionals. According to standard inquisitive entailment, $\overline{p} \vee q \models p \rightarrow q$ but their intersection is $\overline{p}$ and thus not impossible.

Fortunately there is a principled way to characterize entailment in DIS. Standarly if $\varphi \models \psi$ then $\overline{\psi} \models \overline{\varphi}$. We want the implication to also hold in DIS as when you reject the weaker sentence, you should be able to reject the stronger one as well. As DIS looks both at the positive and negative side, then our definition of entailment must do the same and we must add $\neg\psi \models \neg\varphi$ to our definition of entailment.

Definition 5. *Radical inquisitive entailment.*

1. $\varphi \models \psi$ iff $\forall\sigma$ if $\sigma \models^{+} \varphi$ then $\sigma \models^{+} \psi$ and $\sigma \models^{-} \psi$ then $\sigma \models^{-} \varphi$.

This definition also allows us to characterize equivalence as mutual entailment. Radical inquisitive entailment restricts the number of available inferences as compared to standard inquisitive semantics and we will see that such a restriction invalidates some of the inferences that cause trouble for deontic semantics. But before we can discuss this, we need to introduce deontics into the model.

A violation-based deontic logic gravitates around the question whether an act violates a specific law. A permission sentence in a law text provides information on what is not a violation. Following Anderson [4] and the way in which WTO judges reason, we take permission statements to provide information about what is not a violation. This can be captured via introducing the atom v that provides the information that a specific violation has occurred.

Generally, v shall designate a specific law or regulation that is being violated. To account for different types of violations that can occur within a single legal framework, one can designate v_1, v_2, etc. for each specific violation. For example, v_1 may be taken as the proposition "Violation of law number 1 has occurred."

As violation propositions are specific, violations can be reasoned about in the same manner as any other information. So the violation of one law does not lead to violations of other laws, nor does not violating one law save one from indictments due to other deeds. For simplicity, we shall assume that there is only one violation.[2] This will be defined in the semantics as follows.

Definition 6. *Permission.*

1. $\sigma \models^{+} \Diamond\varphi$ iff $\forall\tau \subseteq \sigma.(\ \tau \models^{+} \varphi$ implies $\tau \models^{-} v)$
   $\sigma \models^{-} \Diamond\varphi$ iff $\forall\tau \subseteq \sigma.(\tau \models^{+} \varphi$ implies $\tau \models^{+} v)$

As can be seen in figure 5, $\Diamond p$ coincides with the conditional: $p \rightarrow \overline{v}$. This similarity is not a general feature, though, as the negation of conditionals can be inquisitive but the negation of permission sentences cannot. So, permission is a radical assertion like atoms. This is intuitively correct as permission statements are generally made with authority, and thus the salient reading should be of an assertion.

Also, a permission sentence does not predetermine whether p is in fact the case. A state that supports $\Diamond p$ includes the world $< \overline{pv} >$ in which p is not the case. This accounts for the intuition that permission sentences do not require one to in fact perform the act that is permitted. Furthermore, the world $< \overline{p}v >$ is not eliminated. This world allows for a more fine-grained analysis of interaction between different permissions and prohibitions, as a permission for one thing, in this case p does not guarantee that a violation may not occur when another thing, for example q, is the case.

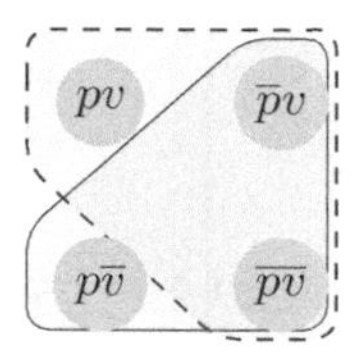

Fig. 5. $\Diamond p$

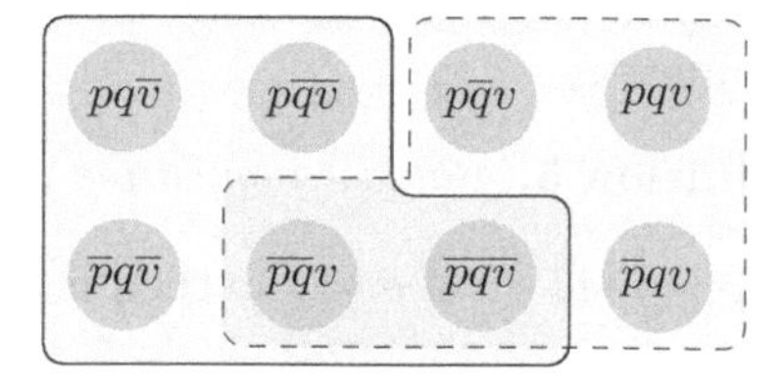

Fig. 6. $\Diamond(p \vee q)$

4 Puzzles Solved

For this account to provide a solution to the free choice puzzle, we must assume that permission takes scope over disjunction. And, indeed, this is supported by general observations regarding disjunction and scope. Following Eckardt [7, pp. 9-10] we assume that in case of ambiguities, one chooses the strongest of the alternatives. As $\Diamond(p \vee q) \models \Diamond p \vee \Diamond q$, permission scoping over disjunction provides the stronger reading. This is illustrated in figures 6 and 8.[3]

[2] This formulation should be tested on the Chisholm's paradox and the gentle murder paradox. Yet, as these fall out of the scope of describing the natural language semantics of permission, it will not be discussed in this article.

[3] This scope movement also accounts for examples of wide disjunction [20, p. 278].

The semantics predicts that $\diamondsuit(p \vee q)$ eliminates three worlds: $< p\overline{q}v >$, $< \overline{p}qv >$ and $< pqv >$. The result is an assertion that includes the remaining worlds as shown in figure 6. A comparison with figure 7 shows that $[\diamondsuit(p \vee q)]^+$ is the same as $[\diamondsuit p \wedge \diamondsuit q]^+$, yet their negations differ and thus it follows that $\diamondsuit p \wedge \diamondsuit q \models \diamondsuit(p \vee q)$ but $\diamondsuit(p \vee q) \not\models \diamondsuit p \wedge \diamondsuit q$.

This appears to be in line with our intuitions regarding permission being granted for both disjuncts. Whether one enacts p or q, $\diamondsuit(p \vee q)$ guarantees that a violation does not occur. Note that while disjunctive free choice provides the information that doing either disjunct is permitted, inferring $\diamondsuit p$ from $\diamondsuit(p \vee q)$ is blocked by radical inquisitive entailment. This is due to the fact that disjunctive permission does not guarantee that doing both p and q simultaneously is not prohibited, which makes permission for either disjunct a contingent fact. By allowing $\diamondsuit(p \vee q) \models \diamondsuit p$ one gives each disjunct independence from the other, so that one can derive $\diamondsuit(p \wedge q)$ and $\diamondsuit p \wedge \diamondsuit q$ which we would not want.

The second puzzle concerns the fact that disjunctive permission embedded under negation as in (2) has the salient reading that neither disjunct is permitted. This result also straightforwardly follows from our clause for the negation of permission. As permission is a radical assertion, then $\overline{\diamondsuit(p \vee q)}$ is an assertion that eliminates all worlds in which doing either p or q would not result in a violation. This can can be seen by looking at the counter-possibility in figure 6.

The third puzzle concerned the fact that appending a disjunctive permission sentence with "... but I do not know which." such as is done in (3) gives it an ignorance reading. We do not take this effect to be cancellation of an implicature but rather one of blocking the modality from taking strongest scope - scoping over the disjunction. The result is a translation of (3) as a wide scope reading in which disjunction takes scope over permission such that "may" distributes into the disjuncts: $\diamondsuit p \vee \diamondsuit q$.

As disjunction is a hybrid, when disjunction scopes over permission it raises an issue for the speaker to solve, modelled as two possibilities as shown in figure 8 below.[4] It is no longer guaranteed that doing p or q will not incur a violation.

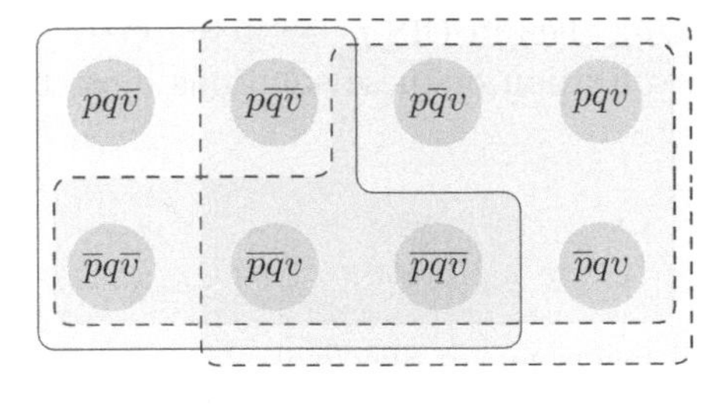

Fig. 7. $\diamondsuit p \wedge \diamondsuit q$

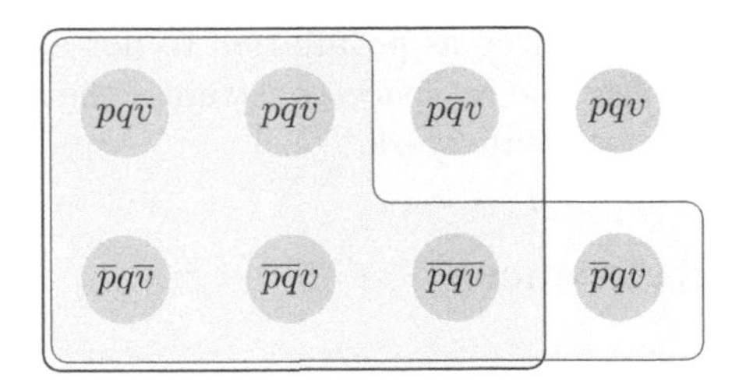

Fig. 8. $\diamondsuit p \vee \diamondsuit q$

[4] The counter-possibilities have been omitted for clarity.

5 Counterargument Countered

One might expect this account to suffer from the problem of strengthening with a conjunct. It should not follow from permission to eat an apple, that eating an apple and killing a postman does not incur a violation. As one can see in the following figures, this inference is blocked by radical inquisitive entailment.

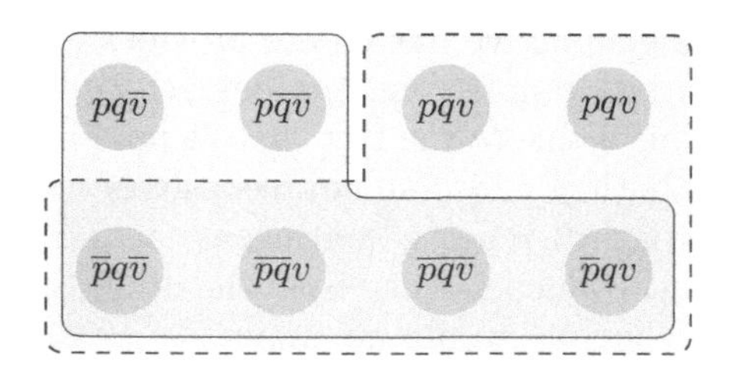

Fig. 9. $\Diamond p$

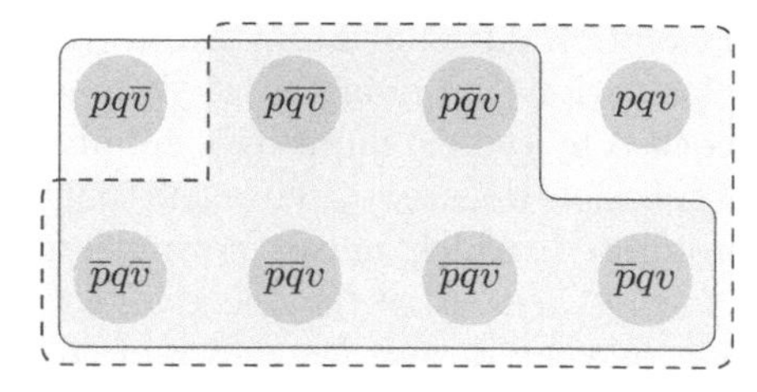

Fig. 10. $\Diamond(p \wedge q)$

As the comparison of figures 9 and 10 demonstrates, the counter-possibility for $\Diamond(p \wedge q)$ is not contained in the counter-possibility for $\Diamond p$, which means that according to radical inquisitive entailment $\Diamond(p \wedge q)$ is not entailed by $\Diamond p$.

6 Future Work

There are a number of puzzles regarding inferences associated with deontic modality, among them Ross's paradox and the interaction of modality with conditionals (for discussion, see Cariani [6]). The treatment of modality and the introduction of radical inquisitive entailment in DIS provides a promising avenue for dealing with these issues.

For Ross's paradox, the solution will follow from the fact that $\Diamond p \not\models \Diamond(p \vee q)$, which can be observed by studying figures 9 and 6. For conditionals, the puzzle lies in the fact that in Kratzer semantics $p \rightarrow \Box p \models \Box p$, yet intuitively everything that is the case does not have to be the case. Assuming that obligation can be characterized as prohibition to not do p ($\overline{\Diamond\overline{p}}$), then in DIS $p \rightarrow \Box p \not\models \Box p$.

Yet, due to space constraints the full explication of these solutions must be left for future work.

References

1. Aher, M.: Inquisitive semantics and legal language. In: Slavkovik, M. (ed.) Proceedings of the 15th Student Session of the European Summer School for Logic, Language and Information, pp. 124–131 (2010)
2. Aloni, M.: Free Choice, Modals, and Imperatives. Natural Language Semantics 15, 65–94 (2007)
3. Alonso-Ovalle, L.: Equal right for every disjunct! Quantification over alternatives or pointwise context change? Presentation at Sinn und Bedeutung 9 (2004)

4. Anderson, A.R.: Some Nasty Problems in the Formal Logic of Ethics. Nous 1, 345–360 (1967)
5. Barker, C.: Free choice permission as resource-sensitive reasoning. Semantics and Pragmatics 3, 1–38 (2010)
6. Cariani, F.: 'Ought' and resolution semantics. Forthcoming in Nous, http://bit.ly/wNBTd1
7. Eckardt, R.: Licencing 'or'. In: Presupposition and Implicature in Compositional Semantics, pp. 34–70. Palgrave MacMillan (2007)
8. Fox, D.: Free choice disjunction and the theory of scalar implicature. In: Presupposition and implicature in compositional semantics, pp. 71–120. Palgrave MacMillan (2007)
9. Franke, M.: Free Choice from Iterated Best Response. In: Aloni, M., Bastiaanse, H., de Jager, T., Schulz, K. (eds.) Logic, Language and Meaning. LNCS, vol. 6042, pp. 295–304. Springer, Heidelberg (2010)
10. Grice, H.P.: Logic and conversation. Studies in the Way of Words, ch. 2. Harvard University Press (1989)
11. Groenendijk, J.A.G., Roelofsen, F.: Inquisitive Semantics and Pragmatics. In: Standford Workshop on Language, Communication and Rational Agency (2009)
12. Groenendijk, J.A.G., Roelofsen, F.: Radical Inquisitive Semantics. Preliminary version, presented at the Colloquium of the Institute for Cognitive Science, University of Osnabrueck (2010)
13. Kamp, H.: Free choice permission. Aristotelian Society Proceedings N.S 74, 57–74 (1973)
14. Kanger, S.: New Foundations for ethical theory. In: Hilpinen, R. (ed.) Deontic Logic: Introductory and Systematic Readings, ch. 12, Reidel Publishing Company (1971)
15. Lewis, C.I., Langford, C.I.: Symbolic Logic. Century Company. Reprinted by Dover Publications, New York (1959)
16. Sano, K.: A Note on Support and Rejection for Radical Inquisitive Semantics (2010) (unpublished)
17. Schulz, K.: You may read it now or later: A Case Study on the Paradox of Free Choice Permission. Master thesis, University of Amsterdam (2003)
18. Schulz, K.: A pragmatic solution for the paradox of free choice permission. Synthese: Knowledge, Rationality and Action 147, 343–377 (2005); Natural Language Semantics 13, 271–316 (2005)
19. Simons, M.: Semantics and Pragmatics in the Interpretation of or. In: Proceedings of SALT XV, pp. 205–222 (2005)
20. Zimmermann, T.E.: Free Choice Disjunction and Epistemic Possibility. Natural Language Semantics 8, 255–290 (2000)

Negative Inquisitiveness and Alternatives-Based Negation

Robin Cooper[1] and Jonathan Ginzburg[2]

[1] Department of Philosophy, Linguistics and Theory of Science, University of Gothenburg, Box 200, 405 30 Göteborg, Sweden
[2] Univ. Paris Diderot, Sorbonne Paris Cité CLILLAC-ARP (EA 3967), 75004 Paris, France

Abstract. We propose some fundamental requirements for the treatment of negative particles, positive/negative polar questions, and negative propositions, as they occur in dialogue with questions. We offer a view of negation that combines aspects of alternative semantics, intuitionist negation, and situation semantics. We formalize the account in TTR (a version of type theory with records) [6,8]. Central to our claim is that negative and positive propositions should be distinguished and that in order to do this they should be defined in terms of types rather than possible worlds. This is in contrast to [10] where negative propositions are identified in terms of the syntactic or morphological properties of the sentences which introduce them.

Keywords: interrogatives, negation, dialogue, type theory.

1 Introduction

In the classical formal semantics treatments for questions the denotation of a positive polar interrogative (PPInt) p? is identical to that of the corresponding negative polar (NPInt) $\neg p$? [15,14, for example]. This is because the two interrogatives have identical exhaustive answerhood conditions. Indeed Groenendijk and Stokhof (1997), p. 1089 argue that this identification is fundamental.

However, other evidence calls the identification of PPInt and NPInt denotations into question. (1a,b) based on examples due to [17] seems to describe distinct cognitive states. Hoepelmann, in arguing for this distinction, suggests that (1a) is appropriate for a person recently introduced to the odd/even distinction, whereas (1b) is appropriate in a context where, say, the opaque remarks of a mathematician sow doubt on the previously well-established belief that *two is even.* (1c,d) seem to describe distinct investigations, the first by someone potentially even handed, whereas the second by someone tending towards DSK's innocence.

(1) a. The child wonders whether 2 is even.
b. The child wonders whether 2 isn't even.
c. Epstein is investigating whether DSK should be exonerated.
d. Epstein is investigating whether DSK shouldn't be exonerated.

M. Aloni et al. (Eds.): Amsterdam Colloquium 2011, LNCS 7218, pp. 32–41, 2012.
© Springer-Verlag Berlin Heidelberg 2012

That root PPInts and NPInts are appropriate in distinct contexts is well recognized in the literature since Hoepelmann and [19]. However, it is not merely the background that differs, it is also the *responses* triggered that are markedly and systematically different. A corpus study of the British National Corpus, whose results are displayed in Table 1, reveals that the two types of interrogatives exhibit almost a mirror image distribution:[1] it suggests that PPInts $p?$ are significantly biassed to eliciting p, whereas NPInts $\neg p?$ are almost identically biassed to eliciting $\neg p$:

Table 1. Distribution of responses to Positive ('Did..?')/Negative ('Didn't..?') polar interrogatives in the British National Corpus

Question type	Positive answer	Negative answer	No answer	Total
`Positive polar`	53%	31%	16%	n = 106
`Negative polar`	23%	54%	22%	n = 86

[13], who developed a view of questions as propositional abstracts, showed how such an account, combined with a theory of negative situation types developed in [5], can distinguish between PPInts and NPInts denotations and presuppositions while capturing the identity of resolving answerhood conditions. Their account relied on a complex *ad hoc* notion of simultaneous abstraction. In this paper we consider a number of phenomena relating negation and dialogue, on the basis of which we develop an account of propositional negation in the framework of Type Theory with Records (TTR) [7,8]. This account extends the earlier results in a type theoretic framework, based on standard notions of negation and abstraction. An important part of the analysis is that we distinguish semantically between positive and negative propositions. This is possible because our type theory is intensional and we have a more fine-grained notion of proposition developed from the conception of propositions as types than, for example, the notion of proposition in inquisitive semantics which is developed from the conception of propositions as sets of possible worlds. Part of our argument for making the distinction between positive and negative propositions is based on data which Farkas and Roelofsen [10] analyze in terms of inquisitive semantics where they rely on syntactic features of utterances in order to distinguish those propositions which are to count as negative.

[1] Our pilot corpus study searched the BNC using SCoRE [21]. For the NPInts the sample reported below consists of all the NPInts of the form 'Didn't ...?' that we found. For the PPInts we found 1500 hits of the form 'Did ...?'. From these we selected a random sample of 106. The 'no answer' category includes cases where either no response concerning the question was forthcoming or where it was difficult to understand how the information provided resolved the question.

2 Negation in Simple Dialogue

(2) a. [[child B approaches socket with nail]
A:(1) No. (2) Do you want to be electrocuted?
(2') Don't you want to be electrocuted?
B: (3) No.
A: (4) No.]

b. [A: (1) Did Merkel threaten Papandreou?
B:(2) No.
A: (3) That can't be true.
C(4): No.]

c. [A: Marie est une bonne étudiante? B: Oui / #Si.
A: Marie n'est pas une bonne étudiante? B: #Oui / Si.]

From (2a,b,c) one can extract some fundamental requirements for a theory of negation in dialogue. In (2a(1)) B's initial action provides the background for A's initial utterance of 'No', in which A ultimately expresses a wish for the negative situation type ¬StickIn(B,nail,socket). More generally, we argue that this type of use ('Neg(ative)Vol(itional) 'No') involves the specification of a negative situation/event type, thereby providing motivation for (3a). Additional motivation for this is provided by complements of naked infinitive clauses discussed below and the large body of work on the processing of negation, reviewed recently in [18]. Kaup offers experimental evidence that comprehending a negative sentence (e.g. *Sam is not wearing a hat*) involves simulating a scene consistent with the negated sentence. She suggests that indeed initially subjects simulate an "unnegated" scene (e.g. involving Sam wearing a hat). [25] offer additional evidence supporting the simulationist perspective. However, they argue against the "two step" view of negation (viz. unnegated and then negated), in favour of a view driven by dialogical coherence, based on QUD.

In the aftermath of (2a(1)), (2a(2)) would be a reasonable question to ask, whereas (2a(2')) would be grounds for summoning the social services. This, together with our earlier discussion on PPInts and NPInts motivates (3b). Assuming (2a(2)) were uttered, B's response asserts the negation of the proposition $p_{\text{Want(B,electr(B))}}$. A can now *agree* with B by uttering 'No'. That is, 'propositional' No always resolves to a negative proposition. This partly motivates (3c). Additional motivation for this is the existence in many languages, such as French and Georgian, of dialogue particles which presuppose respectively a positive (negative) polar question as the maximal element in QUD (MaxQUD), as in (2c).

In (2b(2)) B retorts with $\neg p_1$ (p_1 = **Threaten(Merkel,Papandreou)**), whereas in (2b(3)) A disagrees with B and affirms $\neg\neg p_1$. Clearly, we need (2b(3)) to imply p_1, but this should not be *identified* with p_1—C's utterance (2b(4)) can be understood as agreement with A, not with B, hence motivating (3d).

(3) *Informal intuitive desiderata for a theory of positive and negative situation types and propositions*
a. **Negative situation types evoke precluding positive types**: If a situation s is of a negative type $\neg T$, then s is of some positive type T' which precludes T, that is no situation can be both of type T and T'

b. **Positive/negative polar question distinction**: If p is a proposition and $p?$ is the question whether p then $p?$ should query whether p is true; The question derived from the corresponding negative proposition $\neg p$, $\neg p?$, should query whether $\neg p$ is true; these questions are distinct though have equivalent resolving answerhood conditions.
c. **Negative propositions**: negative propositions are recognizably distinct from positive propositions.
d. **Equivalence but non-identity of p and $\neg\neg p$**: The propositions p and $\neg\neg p$ should be distinct but nevertheless truthconditionally equivalent.

In the following we will attempt to characterize a system which meets these informal criteria and makes the notions involved more precise.

3 Negation and Types

Our discussion builds on what Luo [20] calls "modern" type theory and what we call *rich* type theory, since it presents a much larger selection of types than the simple type theory used, for example, by Montague. Central to type theory is the notion of a judgement that objects are of certain types. The judgement that a is of type T is written in symbols as $a : T$. We will also express this by saying that a is a *witness* for T. Ranta [22] suggested that non-mathematical declarative sentences in natural language correspond to types of Davidsonian events. This idea has been taken up and elaborated in [6] and elsewhere where the term *situation* . The discussion here builds on the type theoretical dictum: "propositions as types". The idea is that we can consider propositions to be types of situations (possibly among other things). If a type has at least one witness it corresponds to a true proposition. A type with no witnesses corresponds to a false proposition.

In [9] we considered various options for treating negation in TTR considering negation as complement in possible worlds, intuitionistic negation, classical negation as a variant of intuitionistic negation, infonic negation, and negation in simulation semantics.

In our version of *intuitionistic negation* the negation of type T is viewed as the type of functions $(T \rightarrow \bot)$ where $\bot$ is a necessarily empty type. In terms of TTR we say that $\{a \mid a : \bot\} = \emptyset$ no matter what is assigned to the basic types, thus giving $\bot$ a modal character: it is not only empty but *necessarily* empty. If T is a type then $\neg T$ is the function type $(T \rightarrow \bot)$. This works as follows: if T is a type corresponding to a proposition it is "true" just in case there is something of type T (i.e. a witness or proof) and "false" just in case there is nothing of type T. Now suppose there is a function of type $\neg T$. If there is something a of type T then a function f of type $\neg T$ would have to be such that $f(a) : \bot$. But $\bot$, as we know, is empty. Therefore there cannot be any function of type $\neg T$. The only way there can be a function of type $\neg T$ is if T itself is empty. Then there can be a function which returns an object of type $\bot$ for any object of type T, since, T being empty, it will never be required to return anything.

This gives us a notion of negative type, that is a function type whose range type is $\bot$, which can be made distinct from positive types (which could be anything other than a negative type, though in practice we use record types as the basis for our propositions). In this way we fulfil (3) by making negative types distinct from non-negative types. However, the proposals made in [9] did not yet give us a *type* of negative propositions. The problem is that for any type T there are infinitely many corresponding negative types $(T \to \bot)$, $((T \to \bot) \to \bot)$ and so on. All of these are types and therefore, if we allow a type *Type* of types[2] they will all be of type *Type*. Things become a little more complicated when we want to talk of some particular collection of types closed under negation as we do below. If $\mathcal{T}$ is a type of types then we shall use $cl_\neg(\mathcal{T})$ to represent the type of types whose witnesses are the closure of the set of witnesses of $\mathcal{T}$ under $\neg$. We shall also use $map_\neg(\mathcal{T})$ to represent the type $\mathfrak{T}$ such that $\neg T : \mathfrak{T}$ iff $T : \mathcal{T}$. This gets us a type whose witnesses involve one iteration of negation over the types belonging to $\mathcal{T}$, leaving out the types we started with, that is, a type of negative types.

Given this, and following [12], we introduce situation semantics style Austinian propositions into TTR [6,8]. These are objects of type (4a). (4a) is a *record type*, that is a set of pairs consisting of a label (represented to the left of the colon) and a type (represented to the right of the colon).[3] An object is of a record type if it is a record containing fields with the same labels as in the record type with objects in those fields which are of the types specified in the record type. An example of an Austinian proposition of this type would be (4b). Here *RecType* is the type of record types as defined in [6,8] and 'run(sam)' is the type of situation in which 'sam' runs. Denoting (4a) by *AusProp*, the type of Austinian propositions, we can say that *NegAusProp*, the type of negative Austinian propositions, is (4c).

(4) a. $\begin{bmatrix} \text{sit} & : \mathit{Rec} \\ \text{sit-type} & : \mathit{cl}_\neg(\mathit{RecType}) \end{bmatrix}$ b. $\begin{bmatrix} \text{sit} & = s \\ \text{sit-type} & = \begin{bmatrix} \text{c}_{\text{run}} : \text{run(sam)} \end{bmatrix} \end{bmatrix}$

c. $\begin{bmatrix} \text{sit} & : \mathit{Rec} \\ \text{sit-type} & : \mathit{cl}_\neg(\mathit{map}_\neg(\mathit{RecType})) \end{bmatrix}$

Truth for these Austinian propositions involves a notion of Austinian witness which in turn involves a notion of incompatible types. Two types T_1 and T_2 are *incompatible* just in case for any a not both $a : T_1$ and $a : T_2$, no matter what assignments are made to basic types. Incompatibility thus means that there is necessarily no overlap in the set of witnesses for the two types. In order to be fully viable *incompatibility* needs to be further restricted using a notion of *alternativehood* [4]. In some cases what the alternatives amount to is fairly straightforward and even lexicalized—classifying the table as *not black* requires evidence that it is green or brown or blue, say. But in general, figuring out the alternatives, as Cohen illustrates, is of course itself context dependent, relating to QUD (Questions Under Discussion).

---

[2] We can do this if we are careful to avoid paradoxes, for example by stratifying the types as we do in [6,8].

[3] The types may be dependent. See [6,8] for details.

Using the notion of "model" defined in [8], that is, an assignment of objects to basic types and to basic situation types constructed from a predicate and appropriate arguments, we can characterize the set of witnesses for a type T with respect to "model" M, $[\check{}T]^M$, to be $\{a \mid a :_M T\}$ where the notation $a :_M T$ means that a is a witness for type T according to assignment M. We can then say that two types T_1 and T_2 are *incompatible* if and only if for all M, $[\check{}T_1]^M \cap [\check{}T_2]^M = \emptyset$.

We define a notion of *Austinian witness* for record types closed under negation:

(5) a. If T is a record type, then s is an Austinian witness for T iff $s : T$
b. If T is a record type, then s is an Austinian witness for $\neg T$ iff $s : T'$ for some T' incompatible with T
c. If T is a type $\neg\neg T'$ then s is an Austinian witness for T iff s is an Austinian witness for T'

The intuitions behind clauses (5b) and (5c) are based on the intuitive account of intuitionistic negation. (5b) is based on the fact that a way to show that s being of type T would lead to a contradiction is to show that s belongs to a type that is incompatible with T. (5c) is based on the fact that if you want to show that a function of type $(T \rightarrow \bot)$ would lead to a contradiction requires finding a witness for T.

We say that an Austinian proposition p is *true* iff p.sit is an Austinian witness for p.sit-type. Notice that if p is true in this sense then p.sit-type will be non-empty, that is, "true" in the standard type-theoretical sense for propositions as types. If p is an Austinian proposition as in (6a), then the negation of p, $\neg p$, is (6b):

(6) a. $\begin{bmatrix} \text{sit} & = s \\ \text{sit-type} & = T \end{bmatrix}$ b. $\begin{bmatrix} \text{sit} & = s \\ \text{sit-type} & = \neg T \end{bmatrix}$

We obtain the desideratum (3) in virtue of the requirement involving an incompatible type in (5b). We obtain the desideratum (3) because negative propositions are distinct from positive propositions. We obtain (3) because double negations of propositions will be distinct from the original proposition but they will now (contrary to intuitionistic propositions) be truth-conditionally equivalent (that is, an Austinian proposition will be true just in case its double negation is true in virtue of (5c)).

4 Negation and Inquisitiveness

In unpublished work Farkas and Roelofsen [10] and Brasoveanu, Farkas and Roelofsen [2] deal with a range of examples related to (2c). In order to treat these examples it is important to distinguish between negative and positive assertions and questions. One initial problem that arises is that questions correspond to sets of sets of possible worlds corresponding to the positive possibility and the negative possibility and thus the positive and negative questions are not distinguished. In order to solve this they make use of highlighting as introduced in [24]. According to this view compositional semantics introduces something more like what we might represent as a record structure, that is, an interpretation which is divided into two components: a highlighted proposal and a set of possibilities. However, there is still a remaining problem of determining which of the highlighted possibilities are negative. As they point out sets of possible worlds do not distinguish between negative and positive propositions. For example, they discuss *John failed the exam* and *John did not pass the exam* as corresponding to the same set of possible worlds. For us these would be two distinct propositions. In order to make the distinction between positive and negative propositions, they use the syntax of the sentences which introduce them. This we see as problematic. As we point out in [9], languages use various ways of expressing negation. In addition to standard negative particles, languages have a variety of ways of expressing negation and we run the risk of listing an arbitrary set of morphemes or constructions if we cannot characterize semantically the fact that they engender negative propositions.

5 Alternatives

It is widely recognized that positive Naked Infinitive (NI) sentences describe an agent's perception of a situation/event, one which satisfies the descriptive conditions provided by the NI clause, as in (7a,b). More tricky is the need to capture the 'constructive' nature of negation in negative NI sentences such as (7c,d). These reports mean that s actually possesses information which rules out the descriptive condition (e.g. for (7c) Mary avoiding contact with Bill), rather than simply lacking concrete evidence for this (e.g. Ralph shutting his eyes.). As [5] points out, Davidsonian accounts (e.g. [16]), are limited to the far weaker (7f):

(7) a. Ralph saw Mary serve Bill. b. Saw(R,s) $\wedge$ s : Serve(m,b).
c. Ralph saw Mary not serve Bill. d. Ralph saw Mary not pay her bill.
e. Saw(R,s) $\wedge$ s : $\neg$ Serve(m,b). f. Saw(R,s) $\wedge$ $s \not{:} Serve(m, b)$

[5] provides axioms on negative SOAs (infons) in situation semantics that attempt to capture this, as in (8a,b). (8a) states that if a situation s supports the dual of σ, then s also supports positive information that precludes σ being the case. (8b) tells us that if a situation s supports the dual of σ, then s also supports information that defeasibly entails that σ is the case.

(8) a. $\forall s, \sigma[s : \overline{\sigma}$ implies $\exists(Pos)\psi[s : \psi$ and $\psi \Rightarrow \overline{\sigma}]]$
b. $\forall s, \sigma[s : \overline{\sigma}$ implies $\exists(Pos)\psi[s : \psi$ and $\psi > \sigma]]$

(5) accounts for (8a). In order to cover (8b) we could refine (5) as in (9).

(9) If T is a record type, then s is an Austinian witness for $\neg T$ iff $s : T'$ for some T' incompatible with T and there is some T'' such that $s : T''$ and if any situation is of type T'' this creates the expectation that it is also of type T.

We do not at this point have a precise proposal for treating the notion "creates the expectation". However it is done, it would mean that T' (the type that is incompatible with T) is regarded as an alternative for T given T''. One way of handling these defeasible inferences is in terms of Aristotelian enthymemes as discussed in [3]. We regard these as being resources available to agents either in particular limited types of contexts or as part of their general knowledge. For example, Fillmore's [11] examples (10), uttered out of context, depend on such general knowledge.

(10) a. Her father doesn't have any teeth.
 b. # Her husband doesn't have any walnut shells.
 c. Your drawing of the teacher has no nose/#noses.
 d. The statue's left foot has no #toe/toes.

We generally assume that people have teeth but not walnut shells and that humans have one nose but many toes. Such resources may also be local to a restricted domain or even a single dialogue or even part of a dialogue. So, for example, a previous turn in a dialogue is sufficient to create an association between husbands and walnut shells, thus making (10b) acceptable.

(11) A: My husband keeps walnut shells in the bedroom.
 B: Millie's lucky in that respect. Her husband doesn't have any walnut shells.

This particular resource is quite likely not going to be used beyond this particular dialogue.

6 Polar Interrogatives

We are left with the desideratum (3). We follow [13] in analyzing polar questions as 0-ary propositional abstracts. We rely on a standard type theoretic notion of abstraction, couched in terms of functional types. For instance, (2a(2)) and (2a(2')) would be assigned the 0-ary abstracts in (12a) and (12b) respectively. These are *distinct* functions from records of type [] (in other words from all records) into the corresponding Austinian propositions, which do not depend on the particular record chosen as an argument (that is, r does not occur in the notation for the resulting type). This use of vacuous abstraction corresponds to the proposal in [13] to treat polar questions as vacuous abstracts and *wh*-questions as non-vacuous abstracts. This accords with the need to distinguish the distribution of their expected responses and the information states of questioners asking or agents investigating the corresponding issues:

(12) a. $\lambda r{:}\mathit{Rec}\ (\begin{bmatrix}\text{sit} = \text{s} \\ \text{sit-type} = \begin{bmatrix}\text{c : want(B(electrocute(B)))}\end{bmatrix}\end{bmatrix})$

 b. $\lambda r{:}\mathit{Rec}\ (\begin{bmatrix}\text{sit} = \text{s} \\ \text{sit-type} = \begin{bmatrix}\text{c : }\neg\text{want(B(electrocute(B)))}\end{bmatrix}\end{bmatrix})$

Given the witnessing conditions introduced in (9), wondering about the question $\lambda r{:}Rec\ (\begin{bmatrix}\text{sit} = \text{s} \\ \text{sit-type} = \neg\text{T}\end{bmatrix})$ involves wondering about whether s has the characteristics that typically involve T being the case, but which—nonetheless, in this case—fail to bring about T. The *simple answerhood* relation of [13] recast in TTR will ensure that the exhaustive answer to p? are $\{p, \neg p\}$, whereas to $\neg p$? they are $\{\neg p, \neg\neg p\}$, so the exhaustive answers are equivalent, as needed. (Recall that p and $\neg\neg p$ are required to be truth-conditionally equivalent by (5c)). [4]

7 Conclusion

We have proposed that positive and negative questions are semantically distinct and, in order to achieve this, that there is a corresponding semantic distinction between positive and negative propositions. On our analysis, for a negative proposition $\neg p$ to be true, an alternative must be true, a proposition which precludes the truth of p. Distinguishing positive and negative propositions means that we can have a semantic account of the use of polarity particles in dialogue without relying on syntactic properties to characterize negative propositions. While positive and negative questions are distinct, we are still able to characterize an equivalence between their exhaustive answers.

Acknowledgements. This research was supported in part by VR project 2009-1569, Semantic analysis of interaction and coordination in dialogue (SAICD) and by a visiting professorship to Cooper at the LabEx Empirical Foundations of Linguistics, Paris Sorbonne Cité. Some portions of this paper were presented at SemDial 2011 in Los Angeles. We thank the audience there as well as the reviewers for Amsterdam Colloquium for their comments.

References

1. Artstein, R., Core, M., DeVault, D., Georgila, K., Kaiser, E., Stent, A. (eds.): SemDial 2011 (Los Angelogue): Proceedings of the 15th Workshop on the Semantics and Pragmatics of Dialogue (2011)
2. Brasoveanu, A., Farkas, D., Roelofsen, F.: Polarity particles and the anatomy of n-words (ms). In: Sinn und Bedeutung (2011)
3. Breitholtz, E., Cooper, R.: Enthymemes as rhetorical resources. In: Artstein, et al. (eds.) [1]
4. Cohen, A.: How are alternatives computed? Journal of Semantics 16(1), 43 (1999)

[4] In an extended version of this paper, we address the issue of how to accommodate an additional understanding/reading NPInts manifest, one that has been known since [19] as the *outside negation* reading, in which there is actually a positive bias to the question. We find the arguments of [23] that such a reading has a metalinguistic nature convincing, though we do not adopt his proposal that such utterances are complex assertion/query speech acts.

5. Cooper, R.: Austinian propositions, Davidsonian events and perception complements. In: Ginzburg, J., Khasidashvili, Z., Levy, J.J., Vogel, C., Vallduvi, E. (eds.) The Tbilisi Symposium on Logic, Language, and Computation: Selected Papers, pp. 19–34. CSLI Publications (1998)
6. Cooper, R.: Austinian truth, attitudes and type theory. Research on Language and Computation 3, 333–362 (2005)
7. Cooper, R.: Records and record types in semantic theory. Journal of Logic and Computation 15(2), 99–112 (2005)
8. Cooper, R.: Type theory and semantics in flux. In: Kempson, R., Asher, N., Fernando, T. (eds.) Handbook of the Philosophy of Science. Philosophy of Linguistics, vol. 14, Elsevier, Amsterdam (2012)
9. Cooper, R., Ginzburg, J.: Negation in dialogue. In: Artstein, et al. (eds.) [1]
10. Farkas, D., Roelofsen, F.: Polarity particles in an inquisitive discourse model (ms), Manuscript, University of California at Santa Cruz and ILLC, University of Amsterdam
11. Fillmore, C.J.: Frames and the semantics of understanding. Quaderni di Semantica 6(2), 222–254 (1985)
12. Ginzburg, J.: The Interactive Stance: Meaning for Conversation. Oxford University Press, Oxford (2012)
13. Ginzburg, J., Sag, I.A.: Interrogative Investigations: the form, meaning and use of English Interrogatives. CSLI Lecture Notes, vol. 123. CSLI Publications, Stanford, California (2000)
14. Groenendijk, J., Stokhof, M.: Questions. In: van Benthem, J., ter Meulen, A. (eds.) Handbook of Logic and Linguistics. North Holland, Amsterdam (1997)
15. Hamblin, C.L.: Questions in montague english. In: Partee, B. (ed.) Montague Grammar. Academic Press, New York (1973)
16. Higginbotham, J.: The logic of perceptual reports: An extensional alternative to situation semantics. Journal of Philosophy 80(2), 100–127 (1983)
17. Hoepelmann, J.: On questions. In: Kiefer, F. (ed.) Questions and Answers. Reidel (1983)
18. Kaup, B.: What psycholinguistic negation research tells us about the nature of the working memory representations utilized in language comprehension. Trends in Linguistics Studies and Monographs 173, 313–350 (2006)
19. Ladd, R.: A first look at the semantics and pragmatics of negative questions and tag questions. Papers from the 17th Regional Meeting of the Chicago Linguistics Society, pp. 164–171 (1981)
20. Luo, Z.: Contextual Analysis of Word Meanings in Type-Theoretical Semantics. In: Pogodalla, S., Prost, J.-P. (eds.) LACL 2011. LNCS, vol. 6736, pp. 159–174. Springer, Heidelberg (2011)
21. Purver, M.: SCoRE: A Tool for Searching the BNC. Tech. Rep. TR-01-07, King's College, London (2001)
22. Ranta, A.: Type-Theoretical Grammar. Clarendon Press, Oxford (1994)
23. Reese, B.: The meaning and use of negative polar interrogatives. Empirical Issues in Syntax and Semantics 6, 331–354 (2006)
24. Roelofsen, F., van Gool, S.: Disjunctive Questions, Intonation, and Highlighting. In: Aloni, M., Bastiaanse, H., de Jager, T., Schulz, K. (eds.) Logic, Language and Meaning. LNCS, vol. 6042, pp. 384–394. Springer, Heidelberg (2010)
25. Tian, Y., Breheny, R., Ferguson, H.: Why we simulate negated information: A dynamic pragmatic account. The Quarterly Journal of Experimental Psychology 63(12), 2305–2312 (2010)

Where Question, Conditionals and Topics Converge

Edgar Onea and Markus Steinbach*

University of Göttingen,
Courant Research Centre "Text Structures"
Nikolausberger Weg 23, 37073 Göttingen, Germany
edgar.onea@zentr.uni-goettingen.de,
Markus.Steinbach@phil.uni-goettingen.de

Abstract. One puzzling fact about German is that yes-no questions that surface as verb-first structures can be interpreted as conditionals in a topic position. We provide an analysis using the basic idea of inquisitive semantics that questions and assertions can be treated on a par as denoting sets of possibilities. The key assumption is that in topic position, questions can be interpreted as conditionals if and only if they contain exactly one highlighted alternative possibility. The analysis correctly predicts the distribution of wh-questions and the distribution of so called irrelevance-conditionals containing *auch* ('too') as well.

Keywords: Topic, Conditionals, Inquisitive Semantics, Questions.

1 Introduction

One difference between assertions and yes-no questions in German is that the former require a verb-second (V2) (1-a) and the latter a verb-first (V1) (1-b) syntactic structure. Wh-questions, like assertions, exhibit a V2 structure but contain a sentence initial wh-word, (1-c).

(1) a. Er kommt nach Hause.
he comes to home
'He comes home.'
b. Kommt er nach Hause?
comes he to home
'Does he come home?'
c. Wer kommt nach Hause?
who comes to home
'Who comes home?'

* This research has been supported by the German Initiative for Excellence funded by the DFG (German Science Foundation), which we gratefully acknowledge. We thank the audience of the Amsterdam Colloquium and especially Manfred Krifka, Floris Roelofsen, Chungmin Lee and the anonymous reviewers for helpful comments. All shortcommings are, of course, our own.

M. Aloni et al. (Eds.): Amsterdam Colloquium 2011, LNCS 7218, pp. 42–51, 2012.
© Springer-Verlag Berlin Heidelberg 2012

However, a V1-clause can also be interpreted as a conditional, whenever it appears in sentence initial position, embedded into a V2 clause, as in (2-a). Crucially, the same applies neither to plain V2 constructions nor to wh-questions, as shown in (2-b) and (2-c) respectively.

(2) a. Kommt er, gehe ich.
comes he go I
'If he comes, I go.'
b. *Er kommt, gehe ich.
comes he go I
intended: 'If he comes, I go.'
c. *Wer kommt, gehe ich.
comes he go I
intended: 'If he comes, I go.'

The main question this paper addresses, is accordingly: What is it about yes-no questions that makes a conditional interpretation possible in a topical position, as in (2-a). The answer will be that yes-no questions can be analyzed as sets of alternatives containing exactly one highlighted alternative. The rest of the analysis comes for free, if one makes the right assumptions about the function of two different sentence initial topic positions in German, one for *frame setting* topics and one for *aboutness* topics.

In the following section, we develop an analysis that correctly predicts the distributional facts in (1) vs. (2). For this we will use the idea that yes-no questions can come with so-called highlighted alternatives [Roelofsen and van Gool, 2010], and some general assumptions about the discourse function of two different topic positions in German. In particular, we argue that yes-no questions can be aboutness but not frame-setting topics. In section 3, we extend the set of data under consideration and show how the addition of the discourse particle *auch* ('too') makes wh-questions acceptable in a higher topical position, which triggers a frame setting topic interpretation. In addition, *auch* also successfully combines with yes-no questions, still preserving a conditional interpretation.

2 Questions as Conditionals

In this section we present the basic analysis that predicts that in German yes-no questions can be interpreted as conditionals in a topical position. For this, we first briefly introduce the idea of inquisitive semantics including some relevant details. We then discuss the differences between the two topic positions and finally we turn to the distribution of topical conditionals in German.

2.1 Inquisitiveness and Highlighted Alternatives

The first main ingredient of our analysis is the recently developed theory of inquisitive semantics [Groenendijk and Roelofsen, 2009, Ciardelli, 2009]. One of the interesting features of inquisitive semantics is the unified treatment of questions and propositions. Traditionally, questions have been assumed to denote sets of sets of worlds, while propositions denote sets of worlds. By contrast, in inquisitive semantics the notion of a proposition is shifted to sets of sets of worlds called possibilities. In the standard version of inquisitive semantics, possibilities are not properly included in each other. A proposition is *inquisitive* if it contains more than one possibility, and *informative*, if it rules out at least some possibilities. *Hybrids*, like (3-b), are both inquisitive and informative:

(3) a. Peter smokes. atomic assertion: [+informative] [-inquisitive]
$\{\lambda w.\text{Peter smokes in } w\}$
b. Peter or Mary smokes. disjunction: [+informative] [+inquisitive]
$\{\lambda w.\text{Peter smokes in } w, \lambda w.\text{ Mary smokes in } w\}$
c. Does Peter smoke? yes-no question: [-informative] [+inquisitive]
$\{\lambda w.\text{Peter smokes in } w, \lambda w.\text{ Peter does not smoke in } w\}$

We assume with [Roelofsen and van Gool, 2010] and [Brasoveanu et al., 2011] that the alternative possibilities may or may not be highlighted. An alternative is highlighted if it is explicitly mentioned. We signal 'highlighting' by underlining the corresponding alternative. In simple assertions, we get only one alternative which is mentioned and, hence, highlighted. In yes-no questions, the alternative explicitly mentioned is, again, highlighted. We assume that we get this result in the process of composition, although we don't discuss the compositional details.

(4) a. Does Peter smoke?
$\{\underline{\lambda w.\text{Peter smokes in } w}, \lambda w.\text{ Peter does not smoke in } w\}$
b. Doesn't Peter smoke?
$\{\lambda w.\text{Peter smokes in } w, \underline{\lambda w.\text{ Peter does not smoke in } w}\}$

We define the operator *Highlight* as taking a set as an argument and returning the unique highlighted element in a set. *Highlight* is undefined if either no such element exists or more than one element is highlighted:

(5) $Highlight([\![\text{Does Peter smoke?}]\!]) = \lambda w.\text{Peter smokes in } w.$

Under the assumptions above, a V1 sentence in German denotes a set containing exactly two possibilities, as already suggested in [Lohnstein, 2000] and [Truckenbrodt, 2006], but we now take it that exactly one of them is highlighted.

2.2 The Topic Issue

We assume with [Frey, 2004b], and [Ebert et al., 2008] that in German there are at least two types of left dislocated topic positions. The so called hanging topic position is available for frame setting topics, as in (6), whereas the position known as German left dislocation or fronting[1], shown in (7), is reserved for aboutness topics.

[1] We ignore structural differences between fronting and left dislocation as irrelevant for the current purposes of this paper.

(6) Der/den Minister, den liebt nur seine Frau.
the.NOM/the.ACC Minister, the.ACC loves only his wife
'The minister, only his wife loves him.'

(7) Den Minister (den) liebt nur seine Frau.
the.ACC Minister (the.ACC) loves only his wife
'The minister, only his wife loves him.'

It is known at least since [Schlenker, 2004] that in German if-clauses can appear in both topic positions. [Ebert et al., 2008] argue that in the hanging topic position the interpretation of conditionals is the one known under the label of *biscuit conditionals* and exemplified in (8). Such conditionals are special because the truth of the consequence does not depend on the truth of the antecedent. Conditionals in the left dislocation position behave as expected: The truth of the consequence typically depends on the truth of the antecedent, as in (9).[2]

(8) Wenn du Hunger hast, es gibt Kekse im Kühlschrank.
If you hunger have there exist biscuit in-the fridge
'If you are hungry, there are biscuits in the fridge.'

(9) Wenn du welche gekauft hast, gibt es Kekse im Kühlschrank.
If you some bought have there exist biscuit in-the fridge
'If you bought some, there are biscuits in the fridge.'

[Ebert et al., 2008], building on [Schlenker, 2004], analyze conditionals as definite descriptions over possible worlds. Schlenker claims that p, in *if p then q* interpreted relative to a world w_0, denotes the single world w_1, most similar to w_0 such that $w_1 \in p$. The whole sentence is then analyzed as saying that $w_1 \in q$. It is easy to see, that this captures the truth conditions of conditionals, since this predicts that if a world w_2 exists in which p holds true but q does not, this world will be less similar to w_0, and, hence, the intuition is captured that if the antecedent holds true but the consequence does not, independent reasons will apply that do not hold in the world of evaluation. An example clarifies this: (10-a) is true if in the most similar world to the world of evaluation w_0 in which the hearer puts the glass on the table, the hearer is happy. This does not predict that (10-b) also comes out as true, as the world in which the hearer puts the glass on the table such that it breaks into pieces might be less similar to the world of evaluation. Hence, (10-a) is silent about the truth of the consequence in it. Note that ι in Schlenker's system is interpreted as a choice function involving a parameter in subscript, and not as a classical Russelian ι operator.

[2] We use the position of the verb as a test. We assume that in German the verb always appears in the second position in assertions. If the conditional counts in determining the second position of the verb, as in (9) we get left dislocation or fronting, whereas if the conditional does not count we get hanging topic, as in (8). Practically, in a hanging topic construction, the verb appears in third position. Intonational and further clues also help distinguishing, see [Ebert et al., 2008] and [Frey, 2004a] for details.

(10) a. If you put the glass on the table, I will be happy.
$\rightarrow$ $happy(Speaker)(\iota w_{w_0}.put(Speaker, Glas, Table)(w))$

b. If you put the glass on the table such that it breaks into pieces, I will not be happy.

[Ebert et al., 2008] argue that conditionals interpreted as aboutness topics are ultimately interpreted as the world argument of the consequence. In particular, the topics are established via an independent speech act noted as the REF operator, formally defined in [Endriss, 2009]. This speech act singles out one possible world, and the assertion takes up on this world as in (11-a). In the hanging topic position, i.e. frame setting topics such as (8), the world argument of the consequence is not (necessarily) the world depicted by the antecedent, but rather the world of evaluation, roughly as in (11-b).

(11) a. $REF_X(\iota w_{w_0}.Buy(Speaker, biscuit)(w)) \wedge ASSERT(\exists biscuit(X))$

b. $REF_X(\iota w_{w_0}.Hungry(Speaker)(w)) \wedge ASSERT(\exists biscuit(w_0))$

2.3 Our Analysis

We follow the main line of attack pursued in [Ebert et al., 2008], but we see no evidence for the treatment of conditionals as definite descriptions over worlds. For one thing conceptual problems are already noticed in [Lewis, 1973]: The selection of *the* most similar world is problematic. In addition, we do not see how such an analysis could be extended to V1-conditionals. Instead, we assume the more traditional analysis of conditionals as restricting the quantification of overt or covert modals. In particular, this means that a conditional interpretation is only possible if at LF at least a covert modal is available in the consequent.

A proposition interpreted in an aboutness topic position will simply end up in its standard function (though mediated by the topic operator). The difference is just that we get some contrastive marking in the sense of [Büring, 2003]. By contrast, a frame setting topic ends up as a referential act independent of the logical structure of the assertion, much like in [Ebert et al., 2008], cf. (11-b).

We also follow [Ebert et al., 2008] in assuming that both aboutness topics and frame setting topics must be, in a sense, referential. We note, however, that propositional arguments generally can appear in these positions. A few examples are given in (12). The referentiality of topics cannot imply that propositions must denote one single world in this case, otherwise one would need to postulate that every embedded clause should refer to one single world, which seems unwanted.

(12) a. Weil Peter klug ist, (deshalb) geht er nach Hause.
Because Peter clever is for-that goes he to home
'Peter goes home, because he is clever.'

b. Dass Peter klug ist, wissen wir.
That Peter clever is know we
'We know that Peter is clever'

c. Dass Peter klug ist, das wissen wir.
That Peter clever is that know we
'We know that Peter is clever'

More important seems, however, that inquisitive expressions such as disjunctions or non specific indefinites cannot appear in the topic position, cf. [Lee, 2006] for discussion of English data:

(13) a. #Peter oder Paul, die kommen spät.
Peter or Paul they come late
intended: 'Peter or Paul come late'
b. #Peter oder Paul, der kommt spät.
Peter or Paul he comes late
intended: 'Peter or Paul come late'
c. #Irgendein Mann, der kommt spät.
Some man he comes late
intended: 'Some man comes late'

All these data can be accounted for by one simple assumption: The topic operator in German presupposes that its argument contains one unique highlighted alternative, which is then established as the topic.

So we basically incorporate the *Highlight* operator into the REF operator of [Ebert et al., 2008]: The REF operator works as before, but it will only consider the unique highlighted alternative of its argument.

(14) Inquisitive topic operator: $REF_X(P) \rightsquigarrow REF_X(HIGHLIGHT(P))$

The highlighting requirement is very similar to the referentiality requirement but is more general. Assume that referential expressions denote one explicitly mentioned alternative, hence they can be topics. Also assume that non-specific indefinites have no highlighted alternatives. Therefore, they cannot be topics. Likewise, disjunction may give more than one highlighted alternative, hence, it cannot be topical. In contrast, yes-no questions can be topical, for they have exactly one highlighted alternative. In the same vein, wh-clauses denote multiple alternative possibilities, as in (15). This does not contain a highlighted alternative, hence, wh-clauses cannot be interpreted in a topic position. This prediction is correct, as illustrated in (2-c).

(15) Wer kommt nach Hause?
who comes to home
'Who comes home?'
a. $\{\lambda w.$ Peter comes in $w, \lambda w.$ John comes in $w, \lambda w.$ Max comes in $, \lambda w.$ Dan comes in $w...\}$

Let us now, turn to the interpretation: The semantics of the aboutness topic operator is vacuous as far as the assertion is concerned, but it does actually get rid of any non-highlighted alternatives. In other words, only the one highlighted alternative gets established as a topic. This leads to the interpretation of the V1 conditional as a proposition (i.e. the highlighted alternative and no longer a question) that somehow has to be combined with another full proposition.

Assuming that the second proposition has some overt or covert modal, the V1-highlighted proposition will end up in the restrictor of that modal and hence yield a plain conditional interpretation. Additional discourse functional effects amounting to contrast may apply, see [Büring, 1997].

The semantics of the frame setting topic amounts to asserting the matrix clause, drawing the attention of the hearer to the topical proposition beforehand, which is interpreted as a biscuit conditional. Indeed, this interpretation is available. In fact, it is the only possible interpretation. This yields the contrast between (16-a) and (16-b), since world knowledge blocks a biscuit conditional interpretation for (16-a) but not for (16-b).

(16) a. *Kommst du, ich gehe.
come you I go
intended: 'If you come I go.'
b. Hast du immer noch Hunger, es gibt was im Kühlschrank.
'If you are still hungry, there is something in the fridge.'

One seemingly unfortunate prediction is that V2-sentences (plain assertions) can appear in aboutness topic-position and receive a conditional interpretation whenever the matrix clause contains some (covert) modal. Similarly, it is predicted that V2-sentences can appear in the hanging topic position, yielding a biscuit conditional interpretation. This is because simple V2 assertions may denote one single alternative possibility. Of course the contrary is the case. We assume that the reason for the unavailability of these readings is that V2-clauses in German come with very direct assertoric power which blocks their interpretation as referential acts, as induced by the topic operator [Lohnstein, 2000, Truckenbrodt, 2006]. The details we leave to further research at this point.

3 The Presence of *auch*

In German, it is possible to add the particle *auch* ('too') to a conditional, hence yielding an *irrelevance* conditional, as shown in (17):

(17) Wenn Peter auch schläft, wir tanzen weiter.
If Peter too sleeps we dance on
'Even if Peter sleeps, we keep on dancing.' .

Once we add *auch* ('also') a V1 conditional can appear in a hanging topic position. This we can see, again since it is not 'counted' when checking the V2 constraint, see (18-a). Interestingly, leaving the V1-conditional with *auch* in the dislocation position we have analyzed as aboutness topic, yields a marked or even unacceptable structure as in (18-b).

(18) a. Endet es auch vor Gericht, wir zahlen die Miete
Ends it also in-front court we pay the rent
(trotzdem) nicht.
(nevertheless) not
'Even if we end up in front of the court, we will not pay the rent.'

b. ??Endet es auch vor Gericht, zahlen wir die Miete
Ends it also in-front court pay we the rent
(trotzdem) nicht.
(nevertheless) not
'Even if we end up in front of the court, we will not pay the rent.'

Moreover, adding *auch* actually makes even wh-structures acceptable, but, again, only in the hanging topic position,compare (19) vs. (20).

(19) a. *Wer kommt, ich gehe
who comes I go
intended: 'Whoever comes, I go.'
b. *Wer kommt, gehe ich
who comes go I
intended: 'Whoever comes, I go.'

(20) a. Wer auch kommt, ich gehe
who too comes I go
intended: 'Whoever comes, I go.'
b. *Wer auch kommt, gehe ich
who too comes go I
intended: 'Whoever comes, I go.'

The arising puzzle is: Why does the presence of *auch* change the acceptability of V1 and wh-clauses in the different topic positions?

We assume that the particle *auch* in German acts as a non-inquisitive closure operator in such constructions, i.e. it acts as the union of the alternative possibilities denoted by the inquisitive propositions it occurs in. Using a version of inquisitive semantics, in which the non-inquisitive closure of a proposition, if generated compositionally, is added to its representation, we end up with structures such as (21) vs. (22). We call such non-inquisitive propositions attentive propositions, and we assume with [Ciardelli et al., 2009] that their pragmatic function is to draw the attention of the hearer to the possibilities, which now are sub-possibilities of the 'big' possibility, added by the non-inquisitive closure. In addition, we assume that the 'big' possibility is highlighted.

(21) $[\![$ Wer auch kommt $]\!] = \{\lambda w.$ Peter comes in w, $\lambda w.$ John comes in w, $\lambda w.$ Max comes in, $\underline{\lambda w. \exists x. x \text{ comes in } w}\}$

(22) $[\![$ Kommst du auch $]\!] = \{\lambda w.$ hearer comes in w, $\lambda w.$ hearer doesn't come in w, $\underline{\mathbf{W}}\}$

The assumptions above that frame setting topics have the discourse function to draw the attention of the hearer to a certain possibility, and that topics require exactly one highlighted alternative we end up with the prediction, that both wh-questions and yes-no questions can appear in the frame setting topic function in a hanging topic position, whenever *auch* is added.

Finally, we have to answer one additional question: Why can questions with *auch* not appear in an aboutness topic position, i.e. left dislocated. Why is it that e.g. (20-b) is bad in German.

As for now, we don't have a fully satisfactory answer to this question but we conjecture that the answer involves the pragmatics of aboutness-topics: In principle, an irrelevance conditional interpretation is possible in the aboutness topic position, however, typically, the additional marking has to be pragmatically justified. This happens, whenever an alternative aboutness topic lends itself. However, in these cases, the aboutness topic is always the entire set of worlds (excluding those in which noone comes for wh-questions), and it is very hard to imagine any alternative possibility to such an unlimited possibility. This correctly predicts the oddity of such examples. Unfortunately, it predicts also that any tautology in the aboutness topic position is just as bad, although it seems that (20-b) is much worse than (23). Using Roothian focus semantics, one could probably find conceivable ways to deal with the problem.

(23) ?Wenn Wasser Wasser ist, gehe ich nach Hause.
'If water is water, I go home.'

A final remark is that our theory correctly predicts that disjunctive polar questions such as (24) are not acceptable in topic position, for they have two highlighted alternatives, see [Pruitt and Roelofsen, 2011] for further discussion.

(24) *Kommt er oder kommt er nicht, gehe ich.
intended: 'Whether he comes or not, I go.'

4 Conclusion

In this paper we have provided an analysis of V1-conditionals, i.e. yes-no questions with conditional interpretation in a topic position in German. Our analysis is based on the idea that topics can pick out the highlighted alternative of a polar question. We have shown that the analysis sketched above not only correctly predicts the possibility to use V1-questions as conditionals in German but also correctly predicts the distributional facts regarding the presence and absence of the particle *auch* and the interaction with the hanging topic position and the so called German left dislocation or fronting, which typically host frame setting topics and aboutness topics respectively.

Our analysis converges with insights from the literature in inquisitive semantics and it provides evidence that inquisitive semantics gives useful tools for the analysis of topic constructions by allowing for a unified treatment of inquisitive items at the DP-level (definites vs. indefinites) and the clausal level without using referential analyses of conditionals. Also the notions of highlighted alternatives have proven useful, since they allow for a fairly simple and parsimonious generalization.

The analysis leaves some questions open, for now. For instance, we still have to examine the binding observations enumerated in [Ebert et al., 2008], which distinguish between hanging topics and left-dislocation, the presence or absence of the resumptive pronoun *dann* and its role in the interpretation, and finally, some of our more controversial data must be backed up with experimental research.

References

Brasoveanu et al., 2011. Brasoveanu, A., Farkas, D., Roelofsen, F.: Polarity particles and the anatomy of n-words. In: Proceedings of SUB 12, Utrecht (2011)

Büring, 1997. Büring, D.: The meaning of topic and focus: the 59th Street Bridge accent. Routledge (1997)

Büring, 2003. Büring, D.: On B-trees, beans, and B-accents. Linguistics & Philosophy 26(5), 511–545 (2003)

Ciardelli, 2009. Ciardelli, I.: Inquisitive semantics and intermediate logics. Master's thesis. University of Amsterdam (2009)

Ciardelli et al., 2009. Ciardelli, I., Groenendijk, J., Roelofsen, F.: Attention! *Might* in inquisitive semantics. In: Cormany, E., Ito, S., Lutz, D. (eds.) Proceedings of SALT 2009, pp. 91–108. OSU (2009)

Ebert et al., 2008. Ebert, C., Endriss, C., Hinterwimmer, S.: Topics as speech acts: An analysis of conditionals. In: Abner, N., Bishop, J. (eds.) Proceedings of the 27th West Coast Conference on Formal Linguistics (WCCFL 27), Sommerville, MA, Cascadilla Proceedings Project, pp. 132–140 (2008)

Endriss, 2009. Endriss, C.: Quantificational topics - A scopal treatment of exceptional wide scope phenomena. Studies in Linguistics and Philosophy. Springer (2009)

Frey, 2004a. Frey, W.: A medial topic position for German. In: Linguistische Berichte, pp. 153–190 (2004a)

Frey, 2004b. Frey, W.: Notes on the syntax and the pragmatics of German left dislocation. In: Lohnstein, H., Trissler, S. (eds.) The syntax and semantics of the left periphery, pp. 203–233. Mouton de Gruyter (2004b)

Groenendijk and Roelofsen, 2009. Groenendijk, J., Roelofsen, F.: Inquisitive semantics and pragmatics. Manuscript University of Amsterdam (2009)

Lee, 2006. Lee, C.: Contrastive topic/focus and polarity in discourse. In: von Heusinger, K., Turner, K. (eds.) Where semantics meets pragmatics. CRiSPI, vol. 16, pp. 381–429. Elsevier (2006)

Lewis, 1973. Lewis, D.: Counterfactuals. Harward University Press (1973)

Lohnstein, 2000. Lohnstein, H.: Satzmodus - kompositionell. Akademie Verlag, Berlin (2000)

Pruitt and Roelofsen, 2011. Pruitt, K., Roelofsen, F.: Prosody, syntax and semantics of disjunctive polar questions. Manuscript University of Massachusetts Amherst and University of Amsterdam (2011)

Roelofsen and van Gool, 2010. Roelofsen, F., van Gool, S.: Disjunctive Questions, Intonation, and Highlighting. In: Aloni, M., Bastiaanse, H., de Jager, T., Schulz, K. (eds.) Logic, Language and Meaning. LNCS, vol. 6042, pp. 384–394. Springer, Heidelberg (2010)

Schlenker, 2004. Schlenker, P.: Conditionals as definite descriptions (a referential analysis). Research on Language and Computation 2(3), 417–462 (2004)

Truckenbrodt, 2006. Truckenbrodt, H.: On the semantic motivation of syntactic verb movement to C in German. Theoretical Linguistics 32(3), 257–306 (2006)

Inquisitive Knowledge Attribution and the Gettier Problem*

Wataru Uegaki

Linguistics, Massachusetts Institute of Technology
wuegaki@mit.edu

Abstract. A disjunctive belief cannot be described as knowledge if the subject does not justifiably believe a true disjunct, even if the whole disjunctive belief is true and justified (Gettier 1963). This phenomenon is problematic if the verb *know* semantically operates on a (classical) proposition, as standardly assumed. In this paper, I offer a solution to this problem using Inquisitive Semantics, arguing that *know* operates on the set of alternative possibilities expressed by its complement. It will also be shown that the proposed semantics for *know* provides a novel account of its compatibility with both declarative and interrogative complements.

Keywords: disjunction, *know-that*, *know-wh*, attitude verb, question-embedding, Gettier problem, Inquisitive Semantics, Alternative Semantics.

1 Introduction

The attitude verb *know* can embed either a declarative or an interrogative complement, in contrast to other attitude verbs, such as *believe* or *ask/wonder*, which take only one of the two complement types, as shown in (1).

(1) a. John **knows** {that Sue came / who came} to the party.
b. John **believes** {that Sue came / *who came} to the party.
c. John **asked** me/**wonders** {*that Sue came / who came} to the party.

One of the basic issues in the semantics of question-embedding concerns this selection property of *know* and other verbs that behave similarly (e.g., *forget*, *tell*). Namely, how we can semantically account for the compatibility of *know* (and other verbs) with both a declarative and an interrogative complement.

The standard answer to this question states that the basic denotation of *know* selects for a proposition, which is the meaning of declarative clauses, and assumes some form of reduction from the meaning of embedded interrogatives to propositions (e.g., Karttunen 1977, Groenendijk and Stokhof 1984). However,

* I thank Maria Aloni, Danny Fox, Ben George, Irene Heim, Floris Roelofsen, Yasutada Sudo and an anonymous reviewer, as well as the audience at 18th Amsterdam Colloquium, for helpful discussion and criticism. Of course, they need not agree with the claims made in this paper, and all errors are my own.

M. Aloni et al. (Eds.): Amsterdam Colloquium 2011, LNCS 7218, pp. 52–61, 2012.
© Springer-Verlag Berlin Heidelberg 2012

such an account wrongly predicts that a *believe*-type verb should be able to embed an interrogative complement unless further stipulations.[1]

In this paper, I propose an alternative approach to the issue that avoids this problem, focusing on a puzzling interpretation of a disjunction in a declarative complement of *know*, known as the GETTIER PROBLEM. Specifically, I will propose that *know* always operates on a *set of alternative possibilities*, which is typically the type of an interrogative meaning, even when *know* takes a declarative complement. I will argue that the solution to the Gettier problem crucially requires the proposed view of the meaning of *know*, and implement the analysis using the treatment of disjunction in Alternative Semantics and Inquisitive Semantics (Groenendijk 2009, Groenendijk and Roelofsen 2009).

2 The Gettier Problem (Gettier 1963)

Knowledge has been traditionally analyzed as a JUSTIFIED TRUE BELIEF (JTB) in epistemology. This traditional view is also underlying in the lexical entry for *know* in the standard semantic theory, as in (2), which basically treats the meaning of *know* as that of *(justifiably) believe* + factivity.[2,3]

(2) $[\![\text{know}]\!] = \lambda p \in D_{\langle s,t\rangle} \lambda x \lambda w\text{: } [p(w)=1].\mathsf{JDOX}_{x,w} \subseteq p$
where $\mathsf{JDOX}_{x,w} = \{w' \mid w' \text{ is compatible with } x\text{'s justified belief in } w\}$

In his famous 1963 paper, Gettier presents counterexamples to this JTB analysis of knowledge. In the situation described in (3), the knowledge attribution in (4) is intuitively false even though the proposition 'Jones owns a Ford or a BMW' is a true and justified belief of Smith.

(3) Situation: Smith justifiably believes that Jones owns a Ford. (He saw Jones with the key of a Ford, driving a Ford etc.) He justifiably deduces from this belief that Jones owns a Ford or a BMW although he is unopinionated about whether Jones owns a BMW. However, it turns out that Jones in fact does not own a Ford, but he owns a BMW.

(4) Smith knows that [Jones owns a Ford or he owns a BMW].

A Gettier example need not involve a disjunction. The following is a case where the belief in question involves existential quantification.

[1] An exception is Ginzburg (1995), who has a reduction in terms of coercion, but avoids this problem by positing an ontological distinction between the objects *believe* and *know* select for. Unfortunately, limited space prevents me from going into an extensive comparison between Ginzburg's and the current proposal, but it is important to note that the current proposal accounts for the difference in the selection restrictions in an ontology that is more conservative than Ginzburg's.

[2] In this paper, a presupposition is captured in terms of partial functions. A clause in square brackets after a colon in a lambda term indicates a restriction on the domain of the function that the lambda term expresses.

[3] The reference to a *justified* belief (JDOX) instead of a mere belief (DOX) might not be strictly standard in semantics. What I mean by 'standard' in the text here is the analysis of *know* as a straightforward extension of *believe* with additional conditions on its arguments and accessibility relation.

(5) Situation: "James, who is relaxing on a bench in a park, observes a dog that, about 8 yards away from him, is chewing on a bone. So he believes [that there is a dog in the park]. [However,] what he takes to be a dog is actually a robot dog so perfect that, by vision alone, it could not be distinguished from an actual dog. [...] But, just a few feet away from the robot dog, there is a real dog. Sitting behind a bush, he is concealed from James's view." (Steup 2009, pp.7-8)

(6) James knows that there is a dog in the park.

Again, the knowledge attribution in (6) is intuitively false although the proposition 'there is a dog in the park' is a true and justified belief of James.

This famous problem in epistemology is also a problem for the standard semantic analysis of *know* in (2), which incorrectly predicts sentences (4) and (6) to be true in the given contexts.[4] In the next section, I will offer a solution to this puzzle using the treatment of disjunction in Inquisitive Semantics.

3 Analysis in Inquisitive Semantics

In this section, after briefly setting up the theoretical framework of Inquisitive Semantics in Section 3.1, I propose a solution to the Gettier problem by arguing that *know* operates on the *set* of possibilities denoted by its complement, unlike in the standard analysis where it operates on a classical proposition. In the last subsection, I show that the proposed meaning for *know* can be used with *wh*-complements with necessary modifications to the analysis.

3.1 Disjunction in Alternative/Inquisitive Semantics

In Inquisitive Semantics (Groenendijk 2009, Groenendijk and Roelofsen 2009), the semantic value of a sentence is conceived of as a set containing one or more ways of updating the common ground. Each update possibility (referred to simply as a POSSIBILITY) is modeled as a set of indices (of type $\langle s, t \rangle$) (Groenendijk and Roelofsen 2009). A sentence denotes a set (of type $\langle st, t \rangle$) of such alternative possibilities qua index-sets.

In the context of this paper, particularly important is the treatment of disjunction. Along with the proposals in Alternative Semantics (Kratzer and Shimoyama 2002, Alonso-Ovalle 2006, a.o.), Inquisitive Semantics treats a sentential disjunction as set union (Groenendijk and Roelofsen 2009).

---

[4] Despite the surface similarity, the behavior of disjunction under *know* is crucially different from the free choice effect of disjunction under imperative or a possibility modal (e.g., Kamp 1973) in licensing the inference pattern in (i), the failure of which is the hallmark of free choice, e.g., (ii).

(i) John knows that Sue came, but he does not know that Mary came. $\models$ John knows that Sue or Mary came. (Disjunction under *know*)

(ii) You may take a pear, but you may not take an apple. $\not\models$ You may take a pear or an apple. (Free choice permission)

(7) $[\![\alpha \text{ or } \beta]\!] = [\![\alpha]\!] \cup [\![\beta]\!]$

That is, the meaning of a disjunction is the union of the sets of possibilities conveyed by each disjunct. In this way, Inquisitive Semantics captures the *hybrid* nature of disjunction: a disjunction is INQUISITIVE in proposing alternative possibilities just like a question, while it is also INFORMATIVE in eliminating a possibility (namely, one where neither disjunct is true) from the common ground just like a classical assertion.[5]

3.2 First Attempt: Point-Wise Function Application

Having set up the theoretical background, let us return to the Gettier problem. One immediate question that arises is how an attitude verb semantically composes with its complement, given that the complement denotes a *set* of alternative possibilities. The standard approach to subsentential composition in this kind of setup is to use the rule of POINT-WISE FUNCTION APPLICATION (PFA) in (8) following the literature of Alternative Semantics after Hamblin (1973).

(8) If $[\![\alpha]\!] \subseteq D_{\langle\sigma,\tau\rangle}$ and $[\![\beta]\!] \subseteq D_{\sigma}$, then
$[\![\alpha\ \beta]\!] := \{d \in D_{\tau} \mid \exists a \in [\![\alpha]\!] \exists b \in [\![\beta]\!] [d = a(b)]\}$ (**Point-wise FA; PFA**)

A natural denotation for *know* one would expect in a compositional system like Alternative Semantics that utilizes PFA is simply the singleton set of its standard denotation, as given in (9).

(9) $[\![\text{know}]\!] = \{\lambda p \in D_{\langle s,t\rangle} \lambda x \lambda w{:}[w \in p].\mathsf{JDOX}_{x,w} \subset p\}$

If we apply the denotation of *know* in (9) to the set of possibilities in (10) via PFA, we get the set in (11) as the meaning of the Gettier example in (4).

(10) $[\![\text{Jones owns a Ford or he owns a BMW}]\!] = \{F\} \cup \{B\} = \{F, B\}$

(11) $[\![\text{Smith knows that Jones owns a Ford or he owns a BMW}]\!]$
$= \{\lambda w{:}\ [w \in F].\mathsf{JDOX}_{\mathbf{s},w} \subseteq F,\ \lambda w{:}\ [w \in B].\mathsf{JDOX}_{\mathbf{s},w} \subseteq B\}$

In this setup, the set of possibilities for sentence φ, i.e., $[\![\varphi]\!]$, is defined so that φ is true in world w iff there is some possibility for φ that contains w, as in (12).

(12) φ is true in w iff there is some $p \in [\![\varphi]\!]$ such that $w \in p$

Given this, we predict (11) to be false in the Gettier scenario in (3). This is so because the first possibility in (11) is undefined for the situation, as the presupposition triggered by the factivity of *know*, i.e., F, is not satisfied. The latter possibility does not include a world where the Gettier scenario holds, as Smith does not believe B in the scenario. Thus, what we end up with is the correct prediction that the Gettier example in (4) is false.

However, this analysis is clearly too strong as it predicts that a disjunction under *know* is always equivalent to a matrix disjunction of *know* i.e., disjunction

[5] A sentence φ is INQUISITIVE iff $[\![\varphi]\!]$ contains at least two possibilities (none of which contains the other). A sentence φ is INFORMATIVE iff $\bigcup [\![\varphi]\!] \neq \mathcal{W}$ (i.e., for some index, φ is false.)

taking wider scope than *know*. This means that we cannot capture 'purely disjunctive' knowledge, which does not require knowledge of any specific disjunct, as described in the following sentence.

(13) Smith knows that Jones owns a Ford or he owns a BMW, but he doesn't know which.

In order to capture the meaning of (13) in the current system, we seem to need an operation to 'collapse' the multiple possibilities denoted by the complement into a single possibility, along the lines of the NON-INQUISITIVE CLOSURE in (14).

(14) $[\![!\,\alpha]\!] := \{\bigcup [\![\alpha]\!]\}$ **(Non-inquisitive closure of $[\![\alpha]\!]$)**

By applying this operation to the complement, we can capture purely disjunctive knowledge in (13), as shown below.

(15) $[\![$Smith knows **!**(that Jones owns a Ford or he owns a BMW)$]\!]$
$= \{\lambda w{:}[w \in F \cup B].\mathsf{JDOX}_{\mathbf{s},w} \subseteq F \cup B\}$

The problem is how to constrain this closure operation. Allowing the operation freely would get us back to the original Gettier problem, as (15) is true also in the Gettier scenario. How can we allow attribution of purely disjunctive knowledge in a sentence like (13) while blocking it in the Gettier case? The next section presents an answer to this question by analyzing knowledge as requiring a *strongest* true belief.

3.3 Solution: Knowledge Requires a Strongest Justified Belief

What distinguishes a Gettier case from purely disjunctive knowledge as in (13) is that the subject believes a false disjunct in the former while she believes no specific disjunct in the latter. The descriptive generalization, therefore, is that purely disjunctive knowledge attribution, or 'collapsing' of alternative possibilities, is valid only if the subject believes no specific disjunct.

We can account for the generalization if purely disjunctive knowledge *entails* that the subject has no justified belief about a specific disjunct. That is, the generalization is captured if we analyze the meaning of *x knows p or q* as the disjunction of the following two cases described in classical logic.

(16) a. $(w \in p \wedge \mathsf{JDOX}_{x,w} \subseteq p) \vee (w \in q \wedge \mathsf{JDOX}_{x,w} \subseteq q)$
b. $w \in p \cup q \wedge \mathsf{JDOX}_{x,w} \subseteq p \cup q \wedge \mathsf{JDOX}_{x,w} \not\subseteq p \wedge \mathsf{JDOX}_{x,w} \not\subseteq q$

The case in (16a) is equivalent to a matrix disjunction, in which the possibilities in the complement are not collapsed by the closure. On the other hand, the case in (16b) captures purely disjunctive knowledge, in which the possibilities are collapsed by the closure. The crucial point here is the third and forth conjuncts in (16b), which require the subject *not* to believe either disjunct specifically. This condition falsifies *Smith knows that Jones own a Ford or he owns a BMW* in the Gettier scenario, as the subject believes the false disjunct, i.e., that Jones owns a Ford in the scenario. As we saw above, the sentence is false in the Gettier scenario in the case of (16a), too. Thus, we succeed in predicting the sentence to be false in the Gettier scenario. Generalizing the disjunction case, a Gettier case involving existential quantification as in (5-6) above can be accounted for, as will be shown shortly in the formal analysis.

Thus, the claim is that every standard Gettier case involves multiple possibilities induced by a disjunction/existential quantification over individuals, times or places. Knowledge attribution is invalid in those cases when the subject believes no true possibility while believing a false one. In Section 4, I discuss other putative Gettier cases that the analysis sketched here is not applicable to, and claim that they should be treated separately.

Now, let us move on to the compositional implementation of this analysis. To derive the two readings in (16) in a unified way, I encode in the meaning of *know* that knowledge requires a *strongest* justified belief among the relevant possibilities. Specifically, I propose the following denotation for *know*. Crucially, this denotation directly operates on the set of alternative possibilities in its complement, via ordinary Function Application instead of Point-wise FA.

(17) $[\![\text{know}]\!] = \lambda Q \in D_{\langle st,t\rangle}\{\lambda x \lambda w\colon [\exists p' \in Q[w \in p']].$
$\exists p \in uniQ[w \in p \wedge \mathsf{JDOX}_{x,w} \subseteq p \wedge strong(p, \mathsf{JDOX}_{x,w}, uniQ)]\}$

- $uniQ := \{\bigcup s \mid s \in \mathcal{P}(Q)\}$ (the closure of Q under union)
- $strong(p, q, S)$ iff $\neg\exists p' \in S[p' \subset p \wedge q \subseteq p']$
  (p is the smallest among the sets in S that are as big as q)

Below, I illustrate how this entry for *know* works in the specific example *Smith knows that Jones owns a Ford or he owns a BMW*. First, the closure-under-union (*uni*) of the possibility-set denoted by the complement is (18).

(18) $uni[\![\text{J. owns a Ford or he owns a BMW}]\!] = uni\{F, B\} = \{F, B, F \cup B\}$

Applying (17) to this set, and supplying the subject via PFA, we get the meaning for the sentence in (19) below. The meaning in (19) contains a single possibility which presupposes that either F or B is true, and entails that there is a true possibility among (18) such that Smith justifiably believes it and that he believes no stronger possibility in (18).

(19) $[\![\text{Smith knows that Jones owns a Ford or he owns a BMW}]\!]$
$= \{\lambda w\colon [\exists p' \in \{F, B\}[w \in p']].\ \exists p \in \{F, B, F \cup B\}$
$[w \in p \wedge \mathsf{JDOX}_{\mathbf{s},w} \subseteq p \wedge strong(p, \mathsf{JDOX}_{\mathbf{s},w}, \{F, B, F \cup B\})]\}$

In the following, I illustrate the evaluation of (19) under four possible belief states of the subject one by one, in all of which the actual world is contained in B but not in F. The four states are ones where Smith justifiably believes (i) F but not B, (ii) $F \cup B$ but neither F nor B, (iii) B but not F, and (iv) both F and B. The evaluation for each belief state is summarized in the following list, where the (a)-clauses describe which possibility in (18) is Smith's strongest justified belief in each state, and (b)-clauses describe whether this possibility contains the actual world. Only if there is a possibility that meets these two conditions, is (19) evaluated true.

(i) $\mathsf{JDOX}_{\mathbf{s},w} \subseteq F$ and $\mathsf{JDOX}_{\mathbf{s},w} \not\subseteq B$ **(A Gettier state)**

a. F is Smith's strongest justified belief in $\{F, B, F \cup B\}$
b. $w \notin F$ (since $w \in \neg F \cap B$) $\Rightarrow$ **(19) false.**

(ii) $\mathsf{JDOX}_{\mathbf{s},w} \subseteq F \cup B \wedge \mathsf{JDOX}_{\mathbf{s},w} \not\subseteq F \wedge \mathsf{JDOX}_{\mathbf{s},w} \not\subseteq B$ **(Purely disj. belief)**
a. $F \cup B$ is Smith's strongest justified belief in $\{F, B, F \cup B\}$
b. $w \in F \cup B$ (since $w \in \neg F \cap B$) $\Rightarrow$ **(19) true.**

(iii) $\mathsf{JDOX}_{\mathbf{s},w} \subseteq B$ and $\mathsf{JDOX}_{\mathbf{s},w} \not\subseteq F$
a. B is Smith's strongest justified belief in $\{F, B, F \cup B\}$
b. $w \in B$ (since $w \in \neg F \cap B$) $\Rightarrow$ **(19) true.**

(iv) $\mathsf{JDOX}_{\mathbf{s},w} \subseteq F \cap B$
a. B and F are Smith's strongest justified beliefs in $\{F, B, F \cup B\}$
b. $w \notin F$, but $w \in B$ (since $w \in \neg F \cap B$) $\Rightarrow$ **(19) true.**

As illustrated above, we predict sentence (4) to be true in states (ii-iv) but false in state (i), namely the Gettier scenario where the subject justifiably believes a false disjunct but does not believe a true one. This is exactly the judgment pattern that we observe for a sentence with a disjunction under *know*.[6]

In the remainder of this section, I illustrate the interpretation of an example involving existential quantification, as in (6), *James knows that there is a dog in the park.* Although I have to gloss over the compositional details for the sake of space, the denotation of the complement in (6) will look like the following, generalizing the semantics for disjunction in Inquisitive Semantics.

(20) $[\![\text{there is a dog in the park}]\!] = \{\{w | \mathbf{dog}(x)(w) \wedge \mathbf{inPark}(x)(w)\} | x \in D_e\}$

The meaning for the sentence in (6) will thus be the following, which contains the single possibility entailing that there is a true possibility in $uni(20)$ such that James believes it and that there is no stronger possibility in $uni(20)$.

(21) $[\![\text{James knows that there is a dog in the park}]\!] = \{\lambda w \colon [\exists p' \in (20)[w \in p']]. \exists p \in uni(20)[w \in p \wedge \mathsf{JDOX}_{\mathbf{j},w} \subseteq p \wedge strong(p, \mathsf{JDOX}_{\mathbf{j},w}, uni(20))]\}$

As the reader can check, (21) predicts that sentence (6) is false in the Gettier scenario described in (5), while it is true in other states parallel to those in (ii-iv) in the disjunction case above.

3.4 Extension to Interrogative Complements

So far, I have been discussing only the embedding of a declarative complement by *know*. What about the embedding of an interrogative complement, as in (22)?

(22) Smith knows {whether Jones or Lee/who} came to the party.

[6] Perhaps, the condition on the relevant possibility as the subject's strongest belief in the union-closure (i.e., *strong* in (17)) has to be weakened with respect to a particular justification. That is, the condition specifies a possibility to be such that there is no stronger possibility that the subject believes *with the same justification.* This is needed to account for the fact that x *knows* p *or* q is intuitively true if x has an independent justification for believing $p \cup q$ even if he falsely believes p.

If we assumed that the denotation of a *wh*-complement *whether Jones or Lee came* is $\{J,L\}$ (i.e., the Hamblin denotation), and directly applied the meaning of *know* in (17), we would face a problem: (22) is predicted to be true when Smith knows that either Jones or Lee came, but does not know which.

The problem is that purely disjunctive knowledge, which we allowed in the case of declarative embedding, should somehow be blocked in the case of interrogative embedding (See below for the ambiguity between the alternative-question reading and the polar-question reading). Remember that in the previous section, purely disjunctive knowledge is made possible by the union operation *uni* encoded in the meaning of *know*. To extend the analysis to interrogative embedding, however, I revise the analysis so that the *uni* operation is contributed *not* by the meaning of *know*, but by *the declarative clause*.[7] Also, I propose that *wh*-complements contribute the *partition operation* instead of *uni*. More specifically, I propose a revised meaning for *know* in (23) and the entries for the declarative and interrogative complementizers, as in (24).

(23) $[\![\text{know}]\!] = \lambda Q \in D_{\langle st,t \rangle} \{\lambda x \lambda w\colon [\exists p' \in Q[w \in p']].$
$\exists p \in Q[w \in p \wedge \mathsf{JDOX}_{x,w} \subseteq p \wedge \mathit{strong}(p, \mathsf{JDOX}_{x,w}, Q)]\}$

(24) a. $[\![\text{that}]\!] = \lambda Q.\mathit{uni}(Q)$ b. $[\![\text{wh(ether)}]\!] = \lambda Q.\mathit{part}(Q)$, where
$\mathit{part}(Q) := \{\{w' \sim_Q w\} \mid w \in \mathcal{W}\}$ and $w' \sim_Q w$ iff $\forall p \in Q[w \subset p \leftrightarrow w' \in p]$

(25) $[\![\text{whether Jones or Lee came to the party}]\!]$
$= \mathit{part}(\{J, L\}) = \{J \cap L, \neg J \cap L, J \cap \neg L, \neg J \cap \neg L\}$

Thus, (22) is true iff Smith has a justified true belief about the true cell in the partition in (25), just as in Groenendijk and Stokhof's (1984) semantics for *wh*-embedding. Note that the condition on the strongest belief is here vacuous since cells in a partition do not contain each other. Purely disjunctive knowledge is not allowed since, if Smith believes that either J or L is true, but neither believes J nor L specifically, there is no possibility in (25) that Smith believes.

As for the ambiguity between the 'alternative-question' reading and the 'polar-question' reading of (22), I follow Roelofsen and van Gool (2010) in attributing the difference to the focus structure. The basic claim is that a focus has an effect of collapsing possibilities. Thus, if the sentence has a 'block' focus on the whole disjunction, the two possibilities are collapsed into one, applying the partition operation to which derives the polar-question reading. On the other hand, if there are multiple foci on the two disjuncts, we retain multiple possibilities, and thus we get the alternative-question reading I described above.

4 Other Putative Gettier Cases

In the literature, authors discuss other cases of putative Gettier problems which cannot be accounted for by the current analysis. One paradigm case is the fake barn case by Goldman (1976):

[7] Ciardelli et al. (2010) speculate that natural language declaratives in general involve the inquisitive closure in Unrestricted Inquisitive Semantics, defined as $!\varphi := \varphi \vee \neg\neg\varphi$, which is similar to my *uni*. However, note that a Gettier case involving more than two disjuncts requires us to adopt *uni* instead of Ciardelli et al.'s !.

(26) Situation: While Henry was driving through a certain country, he saw a building, and identified it as a barn. His sight is excellent and he is perfectly justified in identifying it as a barn. However, the country he was driving through was a strange country that has many fake barns. But, by accident, the building he saw was one of the few *real* barns. In fact, if he had seen a different one, he might have wrongly identified the fake barn as a real barn.

(27) Henry knows that what he saw was a barn.

The judgment that (27) is false in the situation in (26) cannot be accounted for by the current analysis since there *is* a true possibility in the denotation of the complement which is a strongest justified belief of Henry, namely that the particular building he saw was a barn.

Such cases are different from the standard cases discussed in the previous section in that they involve a skeptical information that is independent of the truth of the complement of *know* or the subject's belief state. In fact, as Kratzer (2002) notes, a case like (26-27) is judged less clearly as false if the skepticism in the context is weaker (e.g., only few of the barns are fake in (26)). In contrast, the judgment of the falsity in the standard Gettier cases is not affected by contextual factors which are external to the subject's belief state. Thus, I claim that these other Gettier cases should be treated with an argument separate from the one accounting for the standard cases discussed in the previous section. More specifically, I argue that the skeptical information in the context of these cases raise the standard of belief *justification*, along the lines of Epistemic Contextualism (cf. e.g., DeRose 1992).

5 Conclusions and Remarks on Other Attitude Verbs

In this paper, I argued that the Gettier problem can be given a solution by analyzing *know* as operating on a *set of alternative possibilities* denoted by its declarative complement, using proposals in Alternative Semantics and Inquisitive Semantics. Also, we saw that this meaning for *know* can be easily extended to the case of interrogative complements. The difference between a declarative and an interrogative is that the former involves a closure under union, while the latter involves a partition. In sum, the Gettier problem provides support for the view that *know* always operates on a set of possibilities even when it combines with a declarative complement. Putting it differently, the problem motivates the view that 'knowing p' is to be able to resolve an *issue* raised by p. The case where the complement of *know* is non-inquisitive is then a special case in which the issue is trivial.

Lastly, I make a brief remark about how the selection restrictions of other attitude verbs can be accounted for. Under the current approach, the selection restrictions of attitude verbs such as *believe* and *ask/wonder* are accounted for in terms of the properties of the set of possibilities they select for. As for *believe*, I propose the following entry, which entails that the subject believes the union of the possibility set denoted by the complement.

(28) $[\![\text{believe}]\!] = \lambda Q \in D_{\langle st,t\rangle}\{\lambda x \lambda w.\mathsf{DOX}_{x,w} \subseteq \bigcup Q\}$

The entry in (28) is incompatible with a non-informative complement because it would derive a trivial entailment. This correctly accounts for the fact that *believe* does not take an interrogative complement since the union of a partition is equal to the entire set of worlds. As for verbs such as *ask* or *wonder*, I claim that they require a partition of the whole worlds as their complement, which makes them incompatible with the closure property of declaratives. I have to leave a further explanation of these restrictions for future research.

References

Alonso-Ovalle, L.: Disjunction in Alternative Semantics. Ph.D. thesis, University of Massachusetts at Amherst (2006)

Ciardelli, I., Groenendijk, J., Roelofsen, F.: Information, issues, and attention. Ms., ILLC, University of Amsterdam (2010)

DeRose, K.: Contextualism and knowledge attributions. Philosophy and Phenomenological Research 52(4), 913–929 (1992)

Gettier, E.: Is justified true belief knowledge? Analysis 23(6), 121–123 (1963)

Ginzburg, J.: Resolving questions. Linguistics and Philosophy 18(5), 459–527 (Part I) and 567–609 (Part II) (1995)

Goldman, A.: Discrimination and perceptual knowledge. Journal of Philosophy 73(20), 771–791 (1976)

Groenendijk, J.: Inquisitive Semantics: Two Possibilities for Disjunction. In: Bosch, P., Gabelaia, D., Lang, J. (eds.) TbiLLC 2007. LNCS, vol. 5422, pp. 80–94. Springer, Heidelberg (2009)

Groenendijk, J., Roelofsen, F.: Inquisitive semantics and pragmatics. In: Larrazabal, J.M., Zubeldia, L. (eds.) Meaning, Content, and Argument: Proceedings of the ILCLI International Workshop on Semantics, Pragmatics, and Rhetoric (2009)

Groenendijk, J., Stokhof, M.: Studies on the Semantics of Questions and the Pragmatics of Answers. Ph.D. thesis. University of Amsterdam (1984)

Hamblin, C.L.: Questions in Montague English. Foundations of Language 10(1), 41–53 (1973)

Kamp, H.: Free choice permission. Proceedings of the Aristotelian Society 74, 57–74 (1973)

Karttunen, L.: Syntax and semantics of questions. Linguistics and Philosophy 1(1), 3–44 (1977)

Kratzer, A.: Facts: Particulars or Information units? Linguistics and Philosophy 25(5-6), 655–670 (2002)

Kratzer, A., Shimoyama, J.: Indeterminate pronouns: The view from Japanese. In: Otsu, Y. (ed.) The Proceedings of the Third Tokyo Conference on Psycholinguistics, pp. 1–25. Hituji Shobo, Tokyo (2002)

Roelofsen, F., van Gool, S.: Disjunctive Questions, Intonation, and Highlighting. In: Aloni, M., Bastiaanse, H., de Jager, T., Schulz, K. (eds.) Logic, Language and Meaning. LNCS, vol. 6042, pp. 384–394. Springer, Heidelberg (2010)

Steup, M.: The analysis of knowledge. In: Zalta, E.N. (ed.) Stanford Encyclopedia of Philosophy. The Metaphysics Research Lab and CSLI (2009)

When Wide Scope Is Not Enough: Scope and Topicality of Discourse Referents

Gemma Barberà

Universitat Pompeu Fabra
Roc Boronat, 138, Barcelona, Spain
gemma.barbera@upf.edu
http://parles.upf.edu/llocs/gbarbera/

Abstract. This paper analyses the semantic attributes discourse referents in Catalan Sign Language may have in order to have a corresponding location established in sign space. It is argued that a combination of scope and topicality is required when analysing the correlation between the introduction of entities into the discourse and assigning a spatial location.

Keywords: Catalan Sign Language (LSC), location, modal subordination, scope, sign space, specificity.

1 Introduction

As natural languages in a visual-spatial modality, sign languages use the space in front of the signer's torso to articulate signs. Sign space is morphosyntactically relevant, since signs are spatially modulated for grammatical purposes to express number, person, and arguments of the verb. It is also relevant at the discourse level because it is commonly assumed that entities introduced into the discourse model are identified with certain spatial locations established on the horizontal plane [12], which is the plane that extends parallel to the floor [2]. However, not all the entities introduced have a corresponding spatial location and the semantic attributes that discourse referents should have in order to be spatially localised have not been thoroughly analysed. Moreover, the frontal plane, which extends parallel to the signer's body, has not been analysed when considering the spatial establishment of entities.

Catalan Sign Language (LSC) makes systematic use of signs directed to the horizontal plane, as commonly assumed for other sign languages (SLs) but also to the frontal plane. This paper focuses on the grammatical distinction denoted by the establishment of discourse referents within the two parts of the frontal plane, namely upper and lower. It is argued that this distinction is relevant for LSC grammar and it is explained in terms of scope behavior as well as topicality. My main claims are interrelated: (i) The expression of narrow scope quantifiers leads to a lack of spatial location establishment; however when focusing on specificity contexts, (ii) narrow scope related to specificity is overtly encoded in LSC grammar and, more particularly, it is expressed with marked spatial locations

M. Aloni et al. (Eds.): Amsterdam Colloquium 2011, LNCS 7218, pp. 62–71, 2012.
© Springer-Verlag Berlin Heidelberg 2012

established on the upper part of the frontal plane; yet (iii) narrow scope variables can also establish a lower spatial location as long as they denote a prominent discourse referent. Although claim (ii) and (iii) may seem to be contradictory, I show that they are in fact complementary once discourse structure is included in the analysis.

The rest of the paper is structured as follows. §2 analyses the relation between dependent variables and sign space in LSC. §3 focuses on specificity marking. §4 presents contexts of modal subordination where variables attached to narrow scope quantifiers discursively behave as wide scope ones. §5 presents the interaction between scope and prominence, and §6 concludes.

2 Dependent Variables

Discourse referents (DRs) are semantic objects which denote the object of thought or the thing the discourse is about. Once established in the discourse they can be referred back to by a pronoun or retrieved by a definite description [11]. In dynamic semantics, variables are the construct which correspond to DRs. Inspired by [6], dependent variables are introduced into the model the values assigned to which co-vary with those assigned to another variable. Here I consider contexts where a universal quantifier or an operator binds the variable. Classical Discourse Representation Theory [10] considers that donkey sentences include universal quantification which takes scope over the entire sentence, and unselectively binds all the free variables in it. In LSC only DRs attached to wide scope quantifiers are spatially localised. The expression of narrow scope quantifiers leads to the lack of establishment of a spatial location, as contexts of dependent variables such as donkey sentences, genericity and quantified noun phrases (NPs) show.

In LSC donkey sentences, [1] nominals do not occur with a determiner index sign directed to space to establish a location, but rather are uttered as bare nouns and hardly ever localised. As shown by [15], verb agreement is realised in a neutral location (1). [2]

(1) IF TOWN FARMER HORSE THERE-IS, SURE 1-TAKE-CARE-3_c.
'If a farmer owns a horse, he certainly takes care of it'.

Correspondingly, in the DRT semantic representation of (1) the variable is represented under an embedded context.

---

[1] Cf. [18] for an analysis of donkey sentences in American SL and French SL, where it is argued in line with [13] that SL variables are overtly expressed.

[2] Glossing conventions: Manual signs are represented by the capitalised word corresponding to the translation of the sign; IX3 (pointing sign directed to the lateral parts of space); #-VERB-# (verb agreeing with subject and object: the numbers refer to the grammatical person); subindices mark direction towards space: l (low), u (up), ip (ipsilateral) cl (contralateral), ce (centre); +++ (reduplication of signs).

(2)

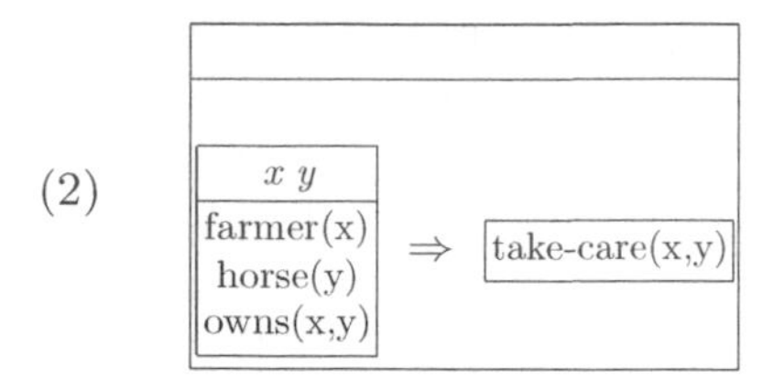

Another context of a dependent variable is that of genericity. Generic statements express general claims about kinds, rather than claims about particular entities. In LSC bare nouns assume a generic interpretation when they are not localised [14], as shown in (3). Any attempt to localise the DR in space is understood as referential, i.e. as denoting a specific man (5). Generic statements are represented according to the idea that a generic operator binds particular variables in its scope. As shown in the corresponding DRS of (3), variables appear in the complex construction represented by a subordinate DRS bound by the generic operator (4).

(3) MAN PLAY LIKE
'Men like to play'

(4)

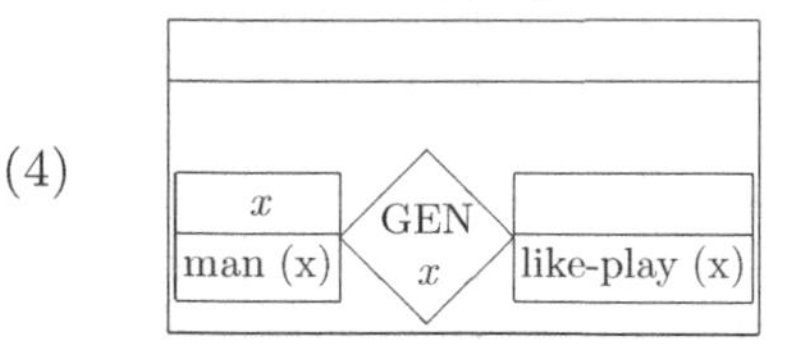

(5) MAN $IX3_{ip}$ PLAY LIKE
'A/the man likes to play'

(6)

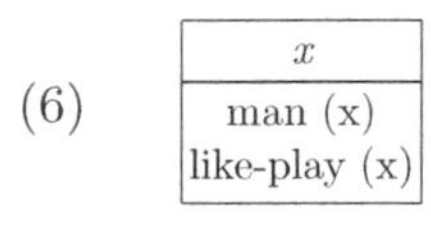

The third argument comes from quantified NPs. In American SL quantified expressions do not establish a spatial location, and sign space is only used to quantify over the domain [12]. In LSC the verbal morphology influences the quantificational interpretation of the bare noun STUDENT [14], as shown in (7). In the corresponding DRS the variable is embedded under the scope of the quantifier (8).

(7) STUDENT EACH-ONE+++ TEACHER ASK+++
'Each pupil asked his/her teacher.'

(8)

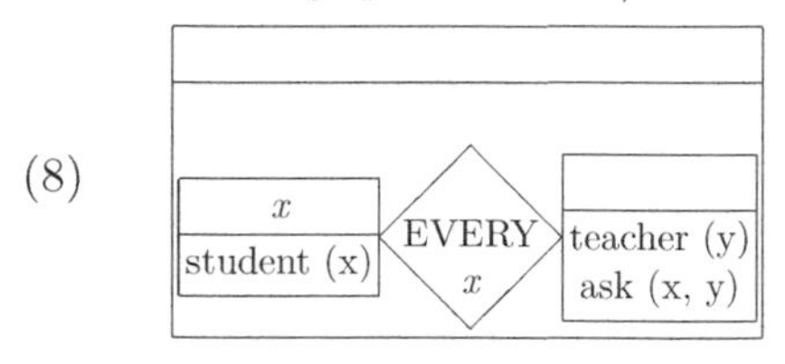

As seen in these examples, when the variable is bound there is a lack of spatial location establishment in LSC. Hence, spatial locations can be defined as the overt manifestation of the DR attached to a quantifier that has wide scope only. Nevertheless, narrow scope does not entail a lack of spatial location establishment but rather a marked location is established on the upper part of the frontal plane for non-specific DRs.

3 Specificity Marking

Whether definiteness is grammatically encoded in SLs is still a matter of debate among SL linguists. While some works argue that an index sign directed to space is the formal marking of definiteness [1], other works have questioned the definiteness marking of index signs [5]. In contrast, in LSC there is no formal marking to distinguish (in)definiteness. As shown in (9), an NP co-occurring with an index sign directed to space is ambiguous between having a definite or an indefinite interpretation.

(9) TODAY IX1 INTERVIEW $IX3_{ip}$ WOMAN.

a. Today I have an interview with a woman.

b. Today I have an interview with the woman.

In LSC the localisation of indefinite NPs spatially differs according to the specificity interpretation. Signs can be localised on the lower and the upper part of the frontal plane and this distinction corresponds to the overt marking of specificity (10) and non-specificity (11), respectively.

(10) IX1 INTERVIEW $IX3_{l}$ WOMAN
I have an interview with a $woman_{spec}$

(11) IX1 INTERVIEW $IX3_{u}$ WOMAN
I have an interview with a $woman_{nonspec}$

The properties specificity encompasses, namely scope and partitivity, can be distinguished in the two localisation processes towards the frontal plane. Scopal specificity is defined in terms of the interpretation of the indefinite NP outside the scope of an operator. According to this view specificity is equated with wide scope [6,8,9]. Hence indefinite NPs which are outside the scope of an operator are considered to have wide scope, and indefinite NPs under the scope of an operator are treated as narrow scope ones.

In LSC indefinite NPs are not ambiguous between having a specific or a non-specific reading. Specific NPs are established on the lower frontal plane (12), whereas non-specific NPs are established on the upper part (13).[3]

(12) I want to buy **a cat**.
It is very obedient.

(13) I want to buy **a cat**.
It must be obedient.

[3] For the interest of space, these examples are provided with the English counterpart of the LSC sentence. The NP localised in space shown in the still is marked with boldface.

The implementation of specificity marking is formally represented with a variable appearing in the main DRS. This variable has wide scope over the other possible embedded variables in the subordinated DRS (14). Non-specificity is implemented with a subordinate variable embedded under the modality operator (15).

The other specificity property considered in the present analysis is partitivity. Partitive indefinite NPs receive a semantic partitive interpretation when the denotation of the NP is included within a given set and they have a restricted set as a possible value. The quantification ranges over some specific, non-empty, contextually fixed set. [4] views specificity as partitivity, since in Turkish NPs ambiguity is resolved through case marking. NPs with overt case morphology are partitive and they introduce into the domain of discourse entities from a previously given set. Partitive NPs denote a specific DR. In contrast, NPs without case morphology are non-partitive which denote a non-specific DR.

In LSC there is a difference between NPs which have a restriction of the quantified NP and those which do not have such a restriction. This is marked in LSC with a difference on the two opposed directions on the frontal plane [15]. Under the restriction of the quantified NP, LSC locations are established on the lower frontal plane (16). When there is no such restriction, the upper frontal plane is used (17), as shown in the LSC counterparts below.

(16) **Some of the friends** were hidden there for two years.

(17) **Someone** denounced they were there.

The quantifier in (16) is an element of the group denoted by the NP. This is shown in the corresponding DRS by the relation x ∈ X, where X corresponds to a non-atomic variable that is projected to the main universe. x is an atomic variable and it is a subset of X. Although x is not projected into the main DRS, it belongs to the set (18). In contrast, the sentence in (17) denotes a non-specific DR which does not belong to a contextually determined set. In the corresponding

DRS, this is represented with an embedded variable which does not belong to any set from the main DRS (19).

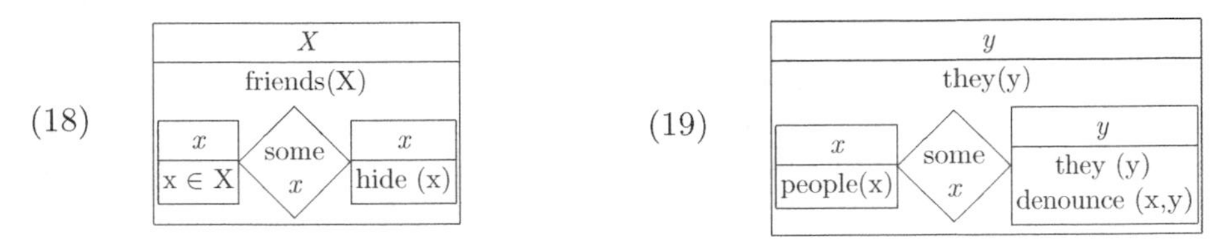

Previously, in §2 it has been shown that when the variable is bound there is a lack of spatial location establishment. However, this section has shown that when considering specificity marking, binding of an operator can also establish a marked location on the upper frontal plane. Upper locations denote that there is no restriction of the quantified NP and they occur with scopally non-specific DRs.

4 Modal Subordination

This section is devoted to the analysis of narrow scope variables which behave as wide scope ones. Here I focus on modal subordination contexts which consist of noteworthy DRs which are introduced into the model and the existence of which is not presupposed. Modal subordination are anaphoric contexts which are under the scope of a modal operator or a propositional attitude predicate, but display anaphoric relations that appear at first glance to violate generalisations about scope operators and anaphoric potential [17].

In LSC modal subordination contexts, the variable is attached to a narrow scope quantifier but behaves as a wide scope variable. This behaviour is overtly expressed in LSC with localisation of signs. In (20) the DR 'person' refers to a non-specific and non identifiable entity. As indicated in the subindices, it is localised towards the upper part of the frontal plane.

(20) IX1 THINK IX3 BOOK 1-OFFER-3_{cl-u} ADEQUATE PERSON-3_{u}... MUST PERSON-3_{c-u} LIKE HOBBY$_{cl-u}$ IS/SAME TRADITIONAL PAST SAME/ALWAYS.
IX3_{ip-l} IX1 1-OFFER-3_{ip-l} PERSON-3_{ip-l} IX3_{ip-l}
'I think that I would offer this book to a person$_{nonspec}$...
It must be someone who likes traditional things.
Definitely, I would offer it to him/her.'

While introducing the antecedent ('someone who likes traditional things') the signer directs a darting eyegaze to an upper direction that goes from the ipsilateral, center and to the contralateral part (21). This eyegaze moves around along the upper frontal plane, without being directly fixed to an area. It functions as an overt operator denoting a *de dicto* mode. Once the intensional context is established by this *de dicto* mode, every subsequent sentence is anchored to this mode even across sentence boundaries and all the variables in the semantic representation are bound by the operator.

(21)

However, as shown in the third utterance in (20) the signer directs pronominal signs and an agreement verb towards a lower spatial location. Hence in subsequent sentences resumptive pronouns referring back to an antecedent which is bound by an operator may also establish a lower spatial location. As shown in (22), both antecedent and consequent are bound by the necessity operator. As long as the variable is under the scope of the corresponding operator, resumptive pronouns are thus felicitous.

(22)

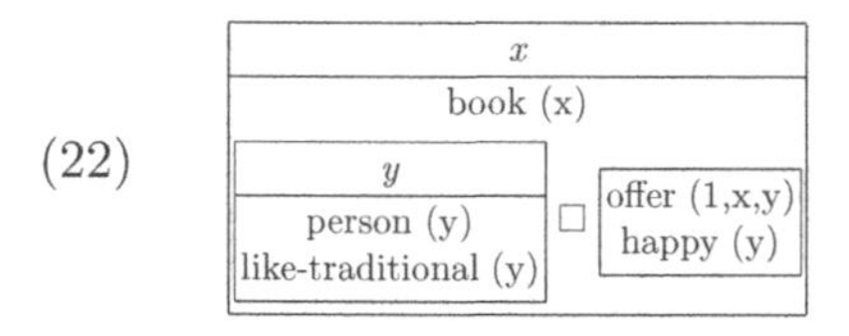

In LSC modal subordination contexts, a lower spatial location is established for a narrow scope variable once the intensional context has been set. Although modal subordination contexts seem to contradict the hypothesis presented in §3 where narrow scope variables have been analysed as being overtly expressed on the upper frontal plane, this apparent puzzle is resolved once we introduce the discursive notion of 'prominence'. Hence when studying the establishment of DRs in sign space, the analysis in terms of scope is not enough, but discourse structure and topicality of entities must also be incorporated.

5 Prominent Narrow Scope Variables

Prominence is defined as the degree of relative salience of a unit of information, at a specific point in time, in comparison to the other units of information [3]. Centering Theory represents probably the most influential account of entity-based prominence in discourse [7]. This processing model relates the local utterance-by-utterance context and discourse anaphoric reference. It constitutes a basis to theorise about local coherence, prominence and choice of referring expressions. Centering has a set of basic notions, which are defined and adapted to the present account in what follows. A discourse model contains:

(i) a set of forward looking DRs (henceforth, DRf(U_k)), which appear in the DRS of K and that can be referred to in subsequent utterances;

(ii) a backward looking DR (DRb(U_k)), which is a unique entity defined for each utterance U_k (except for the intial segment) that refers back to a forward looking DR of the preceding utterance U_{k-1}, and that intuitively represents the DR which is the center of attention at utterance U_k;

(iii) a preferred DR (DRp(U_k)), which is the one that is on the top of the hierarchy of the set of DRs in the main DRS.

Although in most cases topics tend to represent old information, this is neither a sufficient nor a necessary condition for topicality. Topics are better analysed in terms of their effect on the ongoing discourse and considering the effects of previous discourse on the given utterance, rather than as old information [16,19]. A DR is linked to the discourse topic of the fragment of discourse (i.e. it is the most prominent entity of that specific fragment) if it verifies the following formula:

(23) $\mathrm{DRb}(K_n)=\mathrm{DRb}(K_{n-1}) \wedge \mathrm{DRb}(K_n)=\mathrm{DRp}(K_n)$

Topicality is thus here verified when the variable corresponds to the intersection between the DR relations in previous utterances, as specified in the first argument of the formula, and the DR relations in subsequent utterances, as specified in the second argument. More concretely, in (23) the corresponding variable is the one that intersects between the backwards DR of the current utterance being the same as the backwards DR of the preceding utterance, and the backwards DR of the current utterance being the same as the preferred DR of the current utterance.

The variable which verifies the formula in (23) will be connected to the discourse topic and will be thus the most prominent DR at a specific point in discourse. The set of forward looking variables DRf(U_k) are not only restricted to the ones appearing on the main DRS of U_k, but also to subordinated variables as long as they are embedded under the corresponding operator. In LSC, the DRb(U_k) among the DRf(U_k) will be correlated with a lower spatial location as long as it verifies (23) and independently of the scope of the quantifier attached to the variable. This explains why modal subordination contexts in LSC although referring to non-specific DRs establish a lower spatial location. An example of a narrow scope variable which is prominent at a specific fragment of discourse is shown below.

(24) I would offer the book to someone who likes traditional things. He would be very happy, and he would enjoy it a lot.

In (24) the variable that verifies (23) is z, as shown in (25). In the corresponding DRS a subindex p is assigned to the most prominent variable in the specific fragment of discourse.

(25) $[\mathrm{DRb}(K_n)=\mathrm{DRb}(K_{n-1}) \wedge \mathrm{DRb}(K_n)=\mathrm{DRp}(K_n)] \equiv z$

(26)

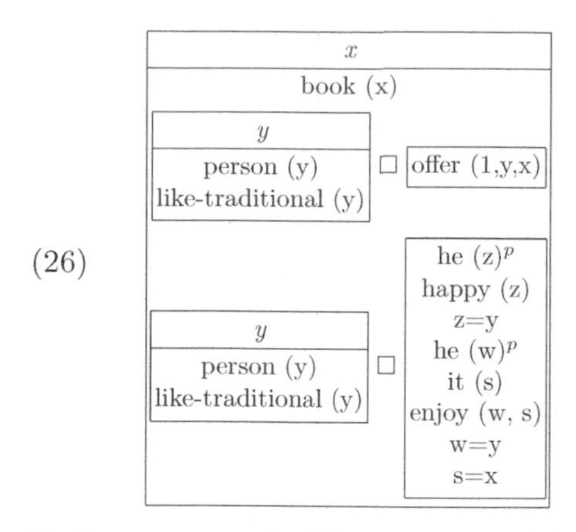

Both narrow and wide scope quantifiers attached to variables can be linked to the discourse topic and hence represent the most prominent DR. The assignment of the subindex into the corresponding variable allows to have the semantic representation for a fragment of discourse linked to its prominent structure. When the subindex is assigned, a lower spatial location in LSC sign space is established.

6 Conclusions

The proposal offered has determined the properties that DRs may have, which lead to the establishment of a location in sign space. It has offered a novel dynamic semantics account, which previous non-dynamic analyses have not been able to propose. Also a representational semantic level which integrates a theory of discourse structure with special focus on prominence has been offered. It has shown that the binding of an operator leads to a lack of spatial location establishment. Nevertheless, narrow scope does not entail the lack of spatial location establishment, but rather a marked location is established on the upper part of the frontal plane for non-specific DRs. LSC has an overt marking of specificity on the two parts of the frontal plane. But scope is not enough when studying the semantic attributes DRs need in order to have a corresponding spatial location in space, since the scope of the quantifier attached to the variable must be combined with the prominence of the variable at a point in the discourse. In cases of narrow scope marking the variable can establish a lower spatial location as long as it is connected to the prominent DR. In the future, cases of intermediate scope will be incorporated in the specificity analysis, as well as a refinement of the hierarchy motivations of prominence.

Acknowledgments. I am grateful to Josep Quer, Berit Gehrke and the audience at FEAST-Venice and at 18th Amsterdam Colloquium for interesting comments. Also my deaf colleagues Santiago Frigola and Delfina Aliaga deserve special credit for stimulating discussions. This research was partly made possible thanks to the Spanish Ministry of Science and Innovation (FFI2009-10492),

Generalitat de Catalunya (URLING-2009SGR00763) and SignGram Cost Action IS1006. Of course, the remaining errors are all mine.

References

1. Bahan, B., Kegl, J., MacLaughlin, D., Neidle, C.: Convergent evidence for the structure of determiner phrases in American Sign Language. In: Gabriele, L., Hardison, D., Westmoreland, R. (eds.) Proceedings of the Sixth Annual Meeting of the Formal Linguistics Society of Mid-America, FLSMVI. Syntax & Semantics/Pragmatics, vol. II. Indiana University Linguistics Club (1995)
2. Brentari, D.: A prosodic model of sign language phonology. The MIT Press, Cambridge (1998)
3. Chiarcos, C., Claus, B., Grabski, M.: Salience in linguistisc and beyond. In: Chiarcos, C., Claus, B., Grabski, M. (eds.) Salience. Multidisciplinary perspectives on its function in discourse. Mouton de Gruyter, Berlin (2010)
4. Enç, M.: The semantics of specificity. Linguistic Inquiry 22(1), 1–25 (1991)
5. Engberg-Pedersen, E.: Space in Danish Sign Language: The semantics and morphosyntax of the use of space in a visual language. Signum Press, Hamburg (1993)
6. Farkas, D.F.: Dependent indefinites. In: Corblin, F., Godard, D., Marandin, J.M. (eds.) Empirical Issues in Formal Syntax and Semantics, pp. 243–267. Peter Lang Publishers (1997)
7. Grosz, B., Joshi, A., Weinstein, S.: Centering: A framework for modelling the local coherence of discourse. Computational Linguistics 2(21), 203–225 (1995)
8. von Heusinger, K.: Specificity and definiteness in sentence and discourse structure. Journal of Semantics 19, 245–274 (2002)
9. Ionin, T.: This is definitely specific: Specificity and definiteness in article systems. Natural Language Semantics 14, 175–234 (2006)
10. Kamp, H., Reyle, U.: From discourse to logic. Introduction to modeltheoretic semantics of natural language, formal logic and discourse representation theory. Kluwer Academic Press, Dordrecht (1993)
11. Karttunen, L.: Discourse referents. In: McCawley, J. (ed.) Syntax and Semantics: Notes from the Linguistic Underground, pp. 363–386. Academic Press, New York (1976)
12. Klima, E., Bellugi, U.: The signs of language. Harvard University Press, Cambridge (1979)
13. Lillo-Martin, D., Klima, E.: Pointing out differences: ASL pronouns in syntactic theory. In: Fischer, S., Siple, P. (eds.) Theoretical Issues in Sign Language Research. Linguistics, vol. 1, pp. 191–210. University Chicago Press, Chicago (1990)
14. Quer, J.: Quantifying across language modalities: generalized tripartite structures in signed languages. Presentation at the I Workshop on Sign Languages, EHU Vitoria-Gasteiz (2005)
15. Quer, J.: Signed agreement: Putting some more arguments together. Presentation at TISLR10. Purdue University, Indiana (2010)
16. Reinhart, T.: Pragmatics and linguistics. an analysis of sentence topics. Philosophica 27, 53–94 (1981)
17. Roberts, C.: Modal subordination and pronominal anaphora in discourse. Linguistics and Philosophy 12, 683–721 (1989)
18. Schlenker, P.: Donkey anaphora: the view from sign language (ASL and LSF). Linguistics & Philosophy (to appear)
19. Vallduví, E.: The Informational component. Garland Press, New York (1992)

When Disjunction Looks Like Conjunction: Pragmatic Consequences in ASL

Kathryn Davidson

University of California, San Diego
9500 Gilman Drive. #0108
La Jolla, CA 92093-0108
kdavidson@ling.ucsd.edu

Abstract. In American Sign Language (ASL), conjunction and disjunction are often conveyed by the same coordinators (transcribed as "COORD"). So the sequence of signs WANT TEA COORD COFFEE can be interpreted as "I want tea or coffee" or "I want tea and coffee" depending on contextual or world knowledge or other linguistic information such as prosodic marking and the addition of disambiguating lexical material. In this paper I show that these general use coordinators in ASL can be a test case for understanding the role of the lexicalization of scalar items in the semantic/pragmatic phenomenon known as scalar implicature by collecting quantitative data from 10 adult native signers of ASL and 12 adult speakers of English using a felicity judgment paradigm. Results show a significant difference in interpretation of the general use coordination scale from other lexically-based scales in ASL and the lexically-based coordination scale in English, suggesting that lexical contrast between scalemates increases scalar implicature calculation.

Keywords: sign languages, experimental pragmatics, scalar implicature, conjunction, disjunction.

1 Introduction

In languages like English, the disjunctive coordinator "or" can occasionally be interpreted as conjunction ("and"), as shown by the paraphrase in parentheses in (1).

(1) You can have coffee or tea.
(You can have coffee and you can have tea.)

This paper focuses on a different type of relationship between these operators that is found in American Sign Language (ASL), where two very common strategies of coordination work as *general use coordinators* which can be interpreted as either disjunction or as conjunction in the very same sentence (unlike (1), where the paraphrase is structurally different) and in a wide variety of contexts.

Although the focus of this paper is general use coordination, it is important to note that there are many forms of coordination in ASL, of which only a subset are general use coordination. For example, there is a sign *AND* which is frequently used when

M. Aloni et al.(Eds.): Amsterdam Colloquium 2011, LNCS 7218, pp. 72–81, 2012.
© Springer-Verlag Berlin Heidelberg 2012

translating titles into ASL directly from English, but is also grammatical when used as a connective in its own right in ASL. There are also two ways to uniquely signal disjunction in ASL: (a) through fingerspelling using the manual alphabet letters "O" and "R" consecutively, which is clearly related to the English "or", and (b) through a mostly antiquated sign that is homophonous with the Wh-question sign *WHICH*. In addition to these specific conjunctive and disjunctive coordinators, ASL very frequently employs a general use coordination strategy, which is described in more detail in section 2. In section 3, I present new data from a felicity judgment experiment showing how this way of lexicalizing coordination is accompanied by a decreased calculation of scalar implicature readings of disjunction. Section 4 discusses consequences for pragmatic theory and concludes. All grammaticality judgments in this paper were provided by a deaf signer whose parents and grandparents were all deaf, and all signed in ASL. Each judgment was confirmed by at least one other fluent signer of ASL, although these signers varied much more in their language background.

2 Two Types of General Use Coordinators in ASL

2.1 COORD-Shift

Of general use coordination in ASL, there are two types, which I label COORD-shift and COORD-L, respectively. The first, COORD-shift, involves moving the body (a combination of torso, head, and/or eyes) slightly for each coordinated element and signing each of the coordinated items in separate places in the signing space (Fig. 1a). For ease of reading I have transcribed it in the way that manual signs are transcribed using the convention of capitalized English words representing a sign in ASL which can be roughly translated with that English word. The phonetic properties of COORD-shift, however, involve only a change in position from one side of the body to the other; the placement of `COORD-shift' marks the timing of this change.

As shown in (2)-(3), COORD-shift is a *general use coordinator*: it can convey either disjunction (2) or conjunction (3), depending on the context.

(2) MARY HAVE COFFEE COORD-shift TEA, DON'T-KNOW WHICH.
`Mary had coffee or tea, I'm not sure which.'

(3) MARY HAVE COFFEE COORD-shift TEA, SHE[1] THIRSTY.
`Mary had coffee and tea, she was thirsty.'

In (2), the clause containing the coordinator is followed by an elided clause, specifically, a clause whose complement is an embedded sluiced constituent interrogative introduced by a D-linked wh-word. This clause is compatible only with a disjunctive interpretation of the preceding coordination phrase. On the other hand,

[1] I have also attempted to increase readability by glossing indexical points in ASL with English pronouns, although neither gender nor nominative case are encoded in the sign in ASL.

in (3) the clause containing the coordinator is followed by a phrase biasing the interpretation of the coordinator towards a conjunctive interpretation.

As disjunction, COORD-shift can appear in alternative questions (4a) as well as both exclusive (4b) and inclusive (4c) disjunctive statements, although signers report that ideally its use in alternative questions would also co-occur with further clarifying linguistic information, such as a sentence-final wh-word WHICH (4a').

(4) a. ?HER PARENTS WILL BUY HER CAR COORD-shift SHE WILL TRAVEL?
`Will her parents buy her a car, or will she [use the money to] travel?'

a'. HER PARENTS WILL BUY HER CAR COORD-shift SHE WILL TRAVEL, WHICH?
`Will her parents buy her a car, or will she [use the money to] travel?'

b. HER PARENTS WILL BUY HER CAR COORD-shift SHE WILL TRAVEL, (DON'T-KNOW WHICH)
`Her parents will buy her a car or she will travel, I'm not sure which.'

c. HER PARENTS WILL BUY HER CAR COORD-shift SHE WILL TRAVEL, (MAYBE BOTH)
`Her parents will buy her a car or she will travel, maybe both.'

Syntactically, COORD-shift can coordinate clauses (4), verbs (5) and noun phrases (6), in any of the three sentence types shown in (4).

(5) MARY SWIM COORD-shift RUN, (DON'T-KNOW WHICH).
`Mary swims or runs, I'm not sure which.'

(6) MARY HAVE COFFEE COORD-shift TEA, (DON'T-KNOW WHICH).
`Mary had coffee or tea, I'm not sure which.'

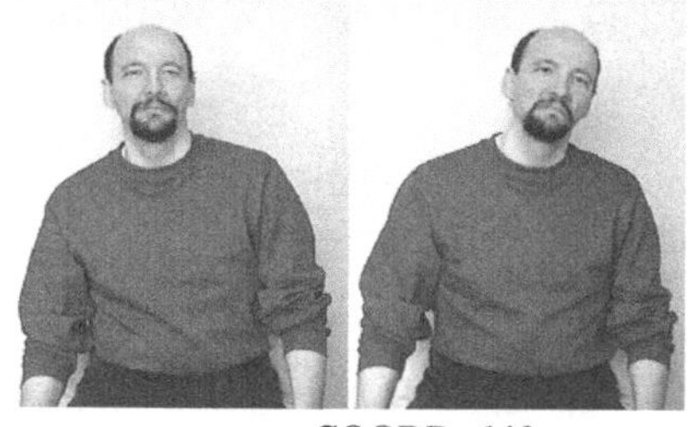

a. COORD-shift

b. COORD-L

Fig. 1. COORD-L and COORD-shift, general use coordinators in ASL, from [1]

2.2 COORD-L

The second form of general use coordination in ASL, COORD-L, consists of the signer's dominant hand pointing with a G "extended index finger" (or sometimes a B "mitten-shaped") handshape to successive fingers on the non-dominant hand, beginning with the thumb. Possible number of fingers on the non-dominant hand range from two (the thumb and forefinger) to all five fingers, which is also the range of coordinated items that this strategy allows. An illustration of COORD-L can be seen in Fig. 1b. Another name for this strategy of coordination is List Buoy [2].

Like COORD-shift, COORD-L can convey either conjunction or disjunction, and when used as disjunction can occur in alternative questions and inclusive and exclusive statements. Syntactically, COORD-L can connect clauses, verbs, and nouns. For some but not all signers (notated as %), the use of COORD-L is reported to feel too heavy prosodically to connect two small light nouns as shown in (8), contrasted with prosodically heavier noun phrases or the full clauses in (7).

(7) HER PARENTS WILL BUY HER CAR COORD-L SHE WILL TRAVEL, (DON'T-KNOW WHICH).
`Her parents will buy her a car or she will travel, I'm not sure which.

(8) %MARY HAVE COFFEE COORD-L TEA, (DON'T-KNOW WHICH).
`Mary had coffee or tea, I'm not sure which.'

Overall, COORD-L seems to share the syntactic and semantic properties of English coordinators, but may sometimes prefer a more restricted set of prosodic environments. Because of the potential interference of this restriction, COORD-shift is used as the general use coordinator in the experimental design in section 3, as the experiment involves coordination of prosodically light nouns.

2.3 General Use Coordination in Other Languages

Both COORD-shift and COORD-L pattern semantically and syntactically much like English disjunction in coordinating a wide range of constituent types and being used for various semantic purposes, and depending on the context they can also both be interpreted as conjunction. Although this is rare from the point of view of English, there are reports of other languages in which conjunction and disjunction are similarly disambiguated by context. One is example is Maricopa, a Yuman language, which juxtaposes items for coordination ([3] as reported in [4]) similarly to COORD-shift in ASL. In (9) the verb is marked with the plain future tense, and it is believed with a higher certainly than the sentence in (10), in which the verb has the additional marking of a modal/evidential element. Consequently, (9) is interpreted conjunctively while (10) is interpreted disjunctively, even though in both cases the constituents are simply juxtaposed.

(9) John-s Bill-s vʔaawuum.
John-NOM Bill-NOM 3.come.PL.FUT
"John and Bill will come."

(10) John-s Bill-s vʔaawuumsaa.
John-NOM Bill-NOM 3.come.PL.FUT.INFER
"John or Bill will come."

Japanese is another language in which coordinates can simply be juxtaposed, and where meaning also depends on the surrounding context or by adding *to* ('and') or *ka* ('or') to disambiguate [5]. Although [4] and [6] provide an extensive typological investigation of conjunction, very little attention is given to disjunction. The pattern in ASL presented here suggests that further investigation of disjunction in ASL and other languages like Maricopa and Japanese may lead to uncovering more generalizations concerning the relationship between conjunction and disjunction, and especially the ways that these logical relationships can be conveyed by juxtaposing items in a list without the use of overt lexical items like English 'and' or 'or'.

3 Quantitative Measures of Scalar Implicatures Based on General Use Coordination

This section presents data from an experiment involving native signers of ASL and native speakers of English using a felicity judgment task. Included is a comparison of the disjunctive/conjunctive ("coordination") scale in ASL with the lexically contrastive coordination scale in English, as well as a comparison of the coordination scale in ASL with a lexically contrastive scale within ASL ("quantifiers"). There are many examples in the literature testing scalar implicature calculation in languages other than English ([7][8][9], among others) and all find similar behavioral results among their participants with implicatures based on prototypical, lexically-contrastive scales like English quantifiers. So, we can expect that where ASL makes a similar lexical distinction to English, as in the case of its quantifier scale (<ALL, SOME>), there should be a similar rate of scalar implicatures.

However, it seems that no previous work has directly compared one language to another that makes a different lexical distinction of the same scalar items, and this is the first experiment testing scalar implicatures in a language that completely lacks a lexical distinction between potential scalar items. If calculation by deaf native signers on the coordination scale in ASL has similar rates to the quantifier scale in ASL, and the coordination scale in English, then we can conclude that having contrasting lexical items is not an important feature of a semantic scale. On the other hand, if there is less scalar implicature calculation on the coordination scale in ASL than other scales in ASL, or than the coordination scale in English, this would suggest that lexical contrast is an important and as-yet-undocumented aspect of scalar implicature calculation. The experiment below tests these predictions, and finds that in fact there is less scalar implicature calculation based on coordination in ASL.

3.1 Methods

Participants were 22 adults from the greater San Diego area. Ten were adults who self-identify as deaf and have been learning and using American Sign Language from birth because they had at least one deaf parent. All were unable to hear normal speech, and all used ASL in their home, at work, or both. These participants were recruited directly through email requests from a laboratory database of interested participants or indirectly through recommendations by friends. All received reimbursement in cash or gift cards. These ten participants will be referred to as "native signers of ASL". The twelve remaining participants were typically hearing undergraduate students at the University of California who are native speakers of English and have had no exposure to ASL. These participants received course credit for participating in the experiment. These participants will be referred to as "native speakers of English".

3.2 Procedures and Stimuli

Each testing session lasted 30-35 minutes. Both the instructions and the task itself were presented on the laptop in video form, by a native signer of ASL (ASL version) or a native speaker of English (English version). Participants were instructed that for each trial of the experiment, a picture will appear on the screen, and that after looking at the picture, they should press the Space Bar key and a video description will begin to play next to the picture. Participants were told to press the smile face if they were "satisfied that the description matches the picture." If they were "not satisfied, and think that the description does not match the picture", they were instructed to press the frown face. It was impossible in both the ASL version and in the English version to replay a video.

Participants saw three practice trials to acquaint them with the task. These practice trials were followed by further instructions, and a confirmation that the task was understood. Finally, 48 experimental trials were presented. Of these, 24 were fillers used as experimental conditions for other studies, and 24 were experimental conditions in the current study. The current experiments' 24 trials consisted of 12 trials of each of two sentence types: (a) Quantifiers, which are a prototypical scale in ASL and in English and (b) Coordination, which has a lexical contrast (e.g. “and” vs “or”) in English but not in ASL. Responses were recorded using *Psyscope* experimental software.

The quantifier scale was used as the baseline case for scalar implicature calculation in this experiment for ASL, both compared to the coordination scale in ASL and compared to the quantifier scale in English. ASL has multiple signs that can be translated into English as "some" or "all"; in this experiment, the version of the quantifiers SOME and ALL that are shown in Fig. 2 were used. These quantifiers can serve as a prototypical scale in ASL because they contrast lexically in the same way in ASL as they do in English.

SOME ALL

Fig. 2. SOME and ALL used in the Quantifier sentence type

In Quantifier trials, each picture consisted of a set of three objects of which either some or all of the objects fulfilled a characterization about that object (e.g. red cans, lit candles, full glasses, etc.). Under the Match condition (a total of 4 trials), the characterization applied to all of the objects (e.g. three cans, all red), and the description was accurate (e.g. CANS, ALL RED "All of the cans are red."). Under the Mismatch condition (4 total trials), the characterization applied to only two of the objects (e.g. three cans, only two are red), and the description was not accurate (e.g. CANS, ALL RED "All of the cans are red."). Finally, under the Test condition (4 total trials), the characterization applied to all of the objects (e.g. three cans, all red), and the description was not maximally informative (e.g. CANS, SOME RED "Some of the cans are red."). Thus, SOME was only evaluated by participants in the Test condition, and was not directly compared to a maximally informative use of the term. Trials for all sentence types were counterbalanced so that each sentence frame (e.g. red cans) appeared in only one trial type (Match, Test, Mismatch) for each participant, and each third of participants saw the sentence frame in a different trial type.

COORD-shift was used to study the affects of general use coordination. When interpreted as disjunction, there is an additional brow-raising nonmanual marking on the disjuncts (see Fig. 3) and I labeled this use of the coordinator COORD-shift(or). The other use, with a different intonational ("nonmanual") marking[2] conveying conjunctive meaning, was labeled COORD-shift(and). In each trial the picture consisted of two different objects (e.g. a mug and a bowl), and then either one or two of the same type of object (e.g. spoons) in relation to the first objects.

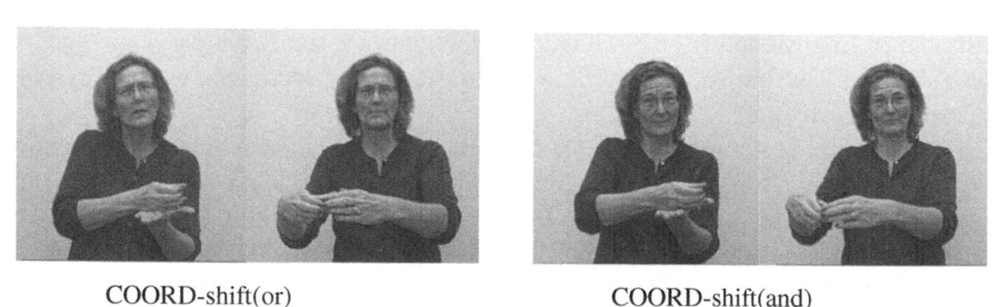

COORD-shift(or) COORD-shift(and)

Fig. 3. COORD-shift(or) and COORD-shift(and) used in the Coordination sentence type

[2] Although an extensive phonetic investigation is beyond the scope of this paper, one reviewer noticed an eye squint, which may have a relationship with the expression of modal meanings.

In the Coordination Match condition, each of the two different objects were related to one of the similar objects, and the description was accurate (e.g. HAVE SPOON IN CUP COORD-shift(and) SPOON IN BOWL. "A spoon is in the mug and a spoon is in the bowl."). Under the Mismatch condition, only one of the two different objects were related to one of the similar objects, but the description said that they both were equally related (e.g. HAVE SPOON IN CUP COORD-shift(and) SPOON IN BOWL. "A spoon is in the mug and a spoon is in the bowl."). Finally, under the Test condition, each of the two different objects were related to one of the similar objects, but the description was not maximally informative due to the disjunctive nonmanual marking on the general use coordinator (e.g. HAVE SPOON IN CUP COORD-shift(or) SPOON IN BOWL "A spoon is in the mug or a spoon is in the bowl").

3.3 Results

First, regarding scales based on prototypical lexical contrast, native speakers of English showed no significant difference in their acceptance of the Test conditions for the Quantifier scale and the Coordination scale (in fact, the mean accuracy rate (here, *rejection* rate because the description was *underinformative*) for each scale was each 0.77, with a standard deviation of 0.36 for coordination and 0.38 for quantifiers). Both were also accepted (i.e. indicated with a smile face) significantly less often than the control Match sentence in each sentence type ($t(11)=7.10$, $p<0.0001$ for quantifiers; $t(11)=6.76$, $p<0.0001$ for coordination), indicating that scalar implicature calculation was occurring for both scales in English, at the same rate, which is what we would expect from previous research on scalar implicatures. As for ASL, native signers accepted the Test condition of the prototypical quantifier scale in ASL significantly less often than the Match condition of the quantifier scale ($t(9)=15.38$, $p<0.0001$), indicating that scalar implicatures were being calculated for the Test condition, as expected because these were based on a prototypical scale. Moreover, they behaved just like their English counterparts: there was no significant difference on scalar implicature calculation in the quantifier scale between native signers of ASL and native speakers of English ($t(20)=1.03$, $p>0.1$). This is a first indication that on a prototypical scale, signers exhibit prototypical rates of scalar implicature calculation.

This contrasts with coordination in ASL, which exhibited a different pattern: native signers show significantly less calculation of scalar implicatures based on the coordination scale in ASL compared to native speakers on the coordination scale in English ($t(20)=25$, $p<0.01$)(Fig. 3). Since there was no difference between the acceptance of the Quantifier scales in both languages, we cannot attribute this to a general behavioral difference in the populations of native ASL signers and native English speakers. Furthermore, within ASL there was significantly less scalar implicature calculation on the Coordination scale than on the Quantifier scale ($t(9)=7.57$, $p<0.0001$). These results are striking, considering that they are both responses to the same trial type (the underinformative Test conditions) by the same set of native signers; nonetheless, there were more rejections (i.e. pragmatic strengthening of underinformative utterances) in the Quantifier sentence type than the Coordination sentence type. Together with the difference seen between ASL and English on the coordination scale in the two languages, this supports the view that the instantiation of scale members as separate lexical items is important for a high rate of scalar implicature calculation. Disambiguation via nonmanual marking is not enough to trigger a normal rate of scalar implicature calculation.

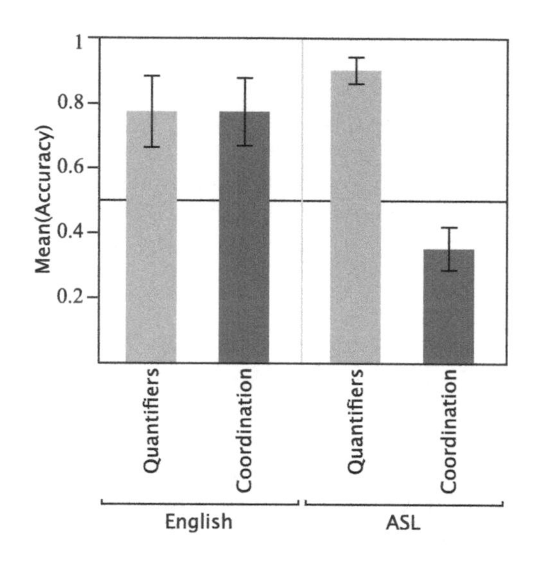

Fig. 4. Calculation of implicatures (i.e. rejection rates for underinformative descriptions) based on quantifier and coordination scales in ASL and English

Recall that the general use coordinator COORD-shift is ambiguous between disjunction or conjunction. One explanation of the data presented so far could be that signers chose to interpret the coordinator in the most charitable way: they were accepting the Match sentences, but also the Test sentences, which had disjunctive nonmanual marking but which under a coordination interpretation would be felicitous. For this point we can turn to the Mismatch data, where under a charitable interpretation the participants would be expected to accept the descriptions. Instead, participants overwhelmingly reject the descriptions in the coordination Mismatch case (M=0.83, SD=0.17), which is not significantly different from the percent of accepted trials in the Match case (M=0.80, SD=0.20)($t(9)=0.29$, $p>0.1$), but is significantly different from rejections in the Test case (M=0.35, SD=0.21)($t(9)=5.02$, $t<0.001$). We can conclude that while participants are not likely to reject Test trial descriptions, this is specific to underinformative descriptions not triggering a scalar implicature, and not due to an overall charitable answering strategy for coordination in ASL.

4 Conclusions

General use coordination in ASL is similar to strategies reported in Maricopa and in Japanese in allowing both disjunctive and conjunctive interpretations depending on the surrounding context. However, ASL is unique among these language in having multiple forms of general use coordination, COORD-shift and COORD-L, in being a signed language, and in differentiating these terms via nonmanual marking.

General use coordination allowed us to test the role of lexical contrast in the calculation of scalar implicatures. Coordination has been studied in many previous experimental studies of scalar implicature due to its status a prototypical scale in English, although it is not so prototypical in ASL. We compared scalar implicature calculation based on coordination in ASL and in English, and found less implicature calculation on the ASL scale. Furthermore, within ASL the coordination scale triggered less implicatures than the lexically-contrastive quantifier scale. Together, results suggest that the semantics relationship of the scalemates is not enough to order items on a scale. In addition to being "in salient opposition: of the same form class, in the same dialect or register, and lexicalized to the same degree" [10], scalemates seem to also be required to be separate lexical items that are linked to each other, via a learned scalar relationship. Further testing on similar structures in Japanese, or even Maricopa, as well as more complex structures within these languages and in ASL can help us further understand the role of the lexicon in scalar implicature calculation. Finally, it was crucial to the current work that the conjunctive and disjunctive readings were able to be disambiguated via nonmanual marking. Future work should explore exactly how the nonmanual marking interacts to create these readings.

Acknowledgments. Much gratitude to Marla Hatrak, Ivano Caponigro, Rachel Mayberry, Carol Padden, Brandon Scates, Peggy Lott, Cami Miner, Corinne Brion, members of the UCSD Sign Language Reading Group, Multimodal Language Development lab, Semantics Babble, and the 2011 Amsterdam Colloquium audience, especially the workshop on formal semantics and pragmatics of sign languages.

References

1. Lifeprint.com, http://www.lifeprint.com/dictionary.htm
2. Liddell, S.: Grammar, Gesture, and Meaning in American Sign Language. University Press, Cambridge (2003)
3. Gil, D.: Aristotle Goes to Arizona and Finds a Language Without "And". In: Zaefferer, D. (ed.) Semantics Universals and Universal Semantics, pp. 96–130. Walter de Gruyter (1991)
4. Haspelmath, M.: Coordinating Constructions. J. Benjamins, Philadelphia (2004)
5. Ohori, T.: Coordination in Mentalese. In: Haspelmath, M. (ed.) Coordinating Constructions. J. Benjamins, Philadelphia (2004)
6. Haspelmath, M.: Coordination. In: Shopen, T. (ed.) Language Typology and Syntactic Description. Complex Constructions, vol. II. Cambridge University Press (2007)
7. Noveck, I.: When Children are More Logical Than Adults: Experimental Investigations of Scalar Implicature. Cognition 78, 165–188 (2001)
8. Papafragou, A., Musolino, J.: Scalar Implicatures: experiments at the semantics-pragmatics interface. Cognition 86, 253–282 (2003)
9. Slabakova, R.: Scalar Implicatures in Second Language Acquisition. Lingua 120, 2444–2462 (2010)
10. Levinson, S.: Presumptive Meanings: the Theory of Generalized Conversational Implicature. MIT Press, Cambridge (2000)

Quantificational Strategies across Language Modalities

Josep Quer

ICREA-Universitat Pompeu Fabra, Barcelona
josep.quer@upf.edu

Abstract. The study of quantification traditionally focused on structures where quantificational meanings are encoded in determiners. Only as a later development attention was paid to quantificational strategies that rely on adverbs, or affixes. In this paper I discuss three varieties of quantificational strategies attested in two sign languages (ASL and LSC) and argue that even the apparent instances of determiner quantification in those languages make use of the more "constructional" way of encoding quantificational meanings that partially reflect the mapping onto tripartite structures overtly. Further, lexical quantification is addressed in the domain of distributivity.

Keywords: Quantification, sign languages, D-quantification, A-quantification.

1 Introduction

The study of quantification in natural language traditionally focused on structures where quantificational meanings are encoded in determiner-like elements in the nominal domain. This bias results from circumscribing the empirical range of inquiry to Indo-European languages, and mostly to English. This tendency was only countered relatively recently by cross-linguistic reasearch by formal linguists on less studied or undescribed languages, the most prominent examples of it being Bach et al.'s (1995) and Matthewson's (2008) volumes. The works presented there make a clear case for reconsidering the inherited research agenda on the basis of theoretical analysis of a broader and typologically and genetically more diverse set of languages, as descriptive grammars usually lack the level of detail and theoretical insight required for solid crosslinguistic semantic inquiry.

Sign languages (SLs) have only become the object of systematic linguistic analysis in the past few decades, with very irregular coverage of description and a rather limited language sample. American Sign Language (ASL) has been the most thoroughly studied one. However, semantics is the least well-known area of SL grammars, let alone from a formal semantics perspective.[1] This applies to the grammar of quantification, as well, with the exception of Partee (1995) and Petronio (1995) on ASL. The present paper aligns with those two works in two senses:

---

[1] For a sample of the few exceptions in the study of different topics in the semantics of sign langages, see the works by J. Quer, Ph. Schlenker, R. Wilbur, A. Zucchi, for instance, as well as those referenced in the text on quantification in SLs.

M. Aloni et al.(Eds.): Amsterdam Colloquium 2011, LNCS 7218, pp. 82–91, 2012.
© Springer-Verlag Berlin Heidelberg 2012

- It explores the ways in which language modality (visual-gestural vs. aural-oral) might have an impact on the expression of quantificational meanings in natural languages. It resorts to the main divide proposed in Bach et al. (1995), namely quantification expressed in the nominal domain by the determiner system vs. quantification conveyed through other means like adverbials or affixes (D- vs. A-quantification, see below for details). Following Heim (1982) and Partee (1992, 1995), tripartite structures are taken as the unifying generalization across those two types of quantifying strategies in natural language.
- It brings SLs to the forefront of the discussion by providing new and unpublished data from Catalan SL (LSC) (Quer 2005) and by comparing it with parallel phenomena described and analyzed in ASL.

This study confirms that the two SLs under discussion display very similar ways of encoding quantification, both of the D- and A-types. It is suggested that the more constructional way of expressing quantified meanings in SLs could be related to the discourse-oriented character of their surface structures, which makes mapping into tripartite structures partially transparent.

Section 2 presents the basic ingredients of the D- vs. A-quantification divide and sets up the stage for the sections to follow. Section 3 adresses A-quantification structures in LSC that support the hypothesis that tripartite quantificational structures are a common means to encode quantificational meanings overtly. Section 4 tackles D-quantification in LSC and argues that the relevant part of the restriction gets overt marking. Section 5 discusses affixal quantification in the verb and points towards an analysis as pluriactional marking. The paper concludes with general considerations drawn from the empirical and theoretical discussion.

2 D-Quantification vs. A-Quantification

The classical analysis of quantification relied on the properties that quantified nominal expressions of the form '*every/most/some* N' have in languages like English. The most prominent representative of this view is probably Generalized Quantifier Theory in Barwise & Cooper (1981). However, a whole line of inquiry into quantification was initiated by Lewis and developed by Kamp (1981) and Heim (1982) on the basis of quantificational readings triggered by adverbs like *always* or *usually*, as in (1a), which can be paraphrased as (1b). What (1b) does is make the quantificational structure explicit. The quantificational relation is represented in a more abstract or general format in a tripartite structure as in (2).

(1) a. A quadratic equation usually has two different solutions.
    b. Usually, x is a quadratic equation, x has two different solutions.

(2) Operator [Restrictor] [Nuclear scope]

In a Kamp/Heim type of analysis, the indefinite NP *a quadratic equation* introduces an open variable without quantificational force. The open variable is unselectively bound by the quantificational adverb (Q-adverb) *usually*. This operator binds the unbound instances in the restrictor and in the nuclear scope. The virtual equivalence of (1a) to a sentence like *Most quadratic equations have two different solutions*, featuring a D-quantifier instead of a Q-adverb, has led to developing a very fruitful avenue of research on quantification from both a theoretical and a descriptive point of view, as natural languages turn out to vary significantly in the ways they realize quantification, well beyond enconding it solely in the NP/DP domain.

Partee et al. (1987) coined the terms *D-quantification* to refer to the "classical" nominal quantifed nominal phrases that resort to the determiner system, next to *A-quantification*. The latter is used to denote a cluster of other quantificational coders, namely Adverbs, Auxiliaries, Affixes, and Argument-structure adjusters that "can be thought of as alternative ways of introducing quantification in a more 'constructional' way" (Partee 1995: 544).

A further development in this approach has been the attempt to understand the interaction of information structure partitioning and quantificational structure, as in Partee (1992, 1995) and Bach et al. (1987). The main thrust behind this connection is to understand how topic/focus determines the projection of material onto a tripartite structure. Partee (1995: 545-546) expresses this generalized view of tripartite structure as in Figure 1, featuring "a number of hypothesized syntactic, semantic, and pragmatic structures that can be argued to be correlated with each other and with the basic tripartite scheme." The underlying motivation of this analytical tool lies in the fact that a broad range of quantificational structures show focus-sensitivity. It should be kept in mind, though, that syntactic constraints override the generality of tripartite structures, thus leading to more complex mappings from syntax to semantics.

In what follows it will be argued that relevant evidence from SLs (LSC and ASL) provides further support for the proposed view. With this limited sample, it is shown that languages in the visual-gestural modality remain within the limits attested crosslinguistically in this specific domain of the syntax/semantics-pragmatics mapping.

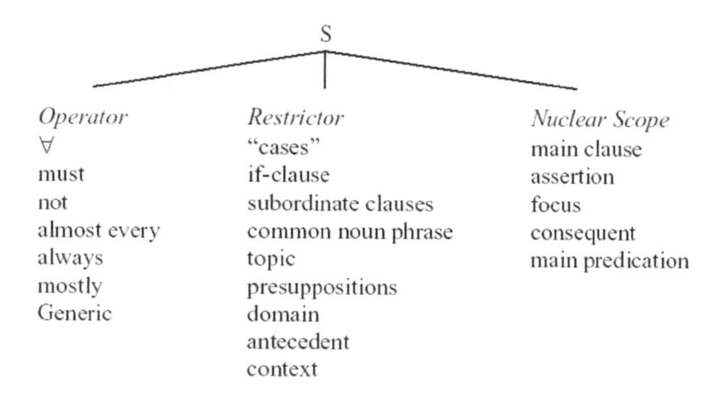

Fig. 1. Tripartite structures generalized (Partee 1995)

3 A-Quantification Structures in SLs

3.1 With Overt Operators

Conditional and generic statements with indefinite descriptions are typical instances where quantificational variablity is attested in the presence of overt Q-adverbs. LSC is no exception in this respect. Note that the language marks familiar or specific DPs by means of an accompanying index (cf. also Barberà 2011): nominal descriptions bound unselectively by a generic or habitual operator appear as bare nouns in the conditional antecedent, as in (3).[2] The arguments in the consequent of (3) are not realized by pronouns and the potentially inflecting verb TAKE-CARE displays a neutral form without overt marking for object agreement.

_____________________________br[3]
(3) [IF PEASANT HORSE THERE-BE] SURE TAKE-CARE
'If a farmer has a horse, he certainly takes care of it.'

It seems quite uncontroversial to posit that antecedent and consequent instantiate here the restrictor and the nuclear scope of a tripartite structure, respectively (restrictor indicated with square brackets). Following much work in the dynamic approaches to such cases, we take SURE in (3) to lexicalize a modal epistemic necessity operator that raises at the level of LF and unselectively binds the argument variables within its scope. The same kind of interpretation is triggered by Q-adverbs, as in (4):

_____________________________br
(4) [IF PEASANT HORSE THERE-BE] ALWAYS TAKE-CARE
'If a farmer has a horse, he always takes care of it.'

Generic statements (understood as characterizing sentences in the sense of Krifka et al. 1995, thus also encompassing habitual predications) have been argued to be overtly marked in ASL by the manual sign TEND (Wilbur 1998, Wilbur & Patschke 1999), as exemplified in (5). Parallel structures in LSC in (6) feature the sign ÉS that characterizes this type of statements. It can be argued that such specialized signs lexicalize the generic operator in these languages.

____br
(5) LION SELF:CL1 #PREDATORY TEND [ASL]
'The lion is a predatory cat.'

_________br
(6) a. IX JAPAN EARTHQUAKE ÉS++ [LSC]
'Japan is in a seismic area.'

---

[2] For a recent detailed treatment of donkey-sentences of the 'bishop' variety in ASL and LSF (French SL), see Schlenker (2011).

[3] Standard conventions for glossing SL data are followed here: manual signs are represented with a gloss in capital letters roghly corresponding to the sign; a tier above the manual signs indicates the scope in the coarticulation of non-manual signals like 'br' (brow raise), 'bf' (brow forrowing) or 'hs' (headshake). +++ indicates reduplication. # signals fingerspelling.

____br
b. LLEÓ DEPREDAR+++ ÉS
'The lion is a predator.'

3.2 Without Overt Operators

As predicted in the framework of generalized tripartite structures, no overt operator is required for them to obtain if the relevant contextual and morphosyntactic conditions are met. LSC example (7) is an instance of a generic or habitual predication. The nominal description in the antecedent and the lack of any indication of episodicity in the main predication allow for it. Note that the antecedent has no overt manual introducer and is only flagged by brow raise (see Wilbur & Patsche 1999 and Wilbur 2011 for extensive arguments in favor of analyzing brow raise as the marker of restrictions of non-Wh operators in ASL).

_____________________br
(7) [FRIEND PERSON COME] IX1 3-INVITE-1
'If/When a friend comes, I treat him/her.'

Non-conditional generic predications essentially rely on the same ingredients: non-episodicity, lack of index marking of the generic subject and presence of brow raise on it, as in (8):

______br _______hs
(8) [WOMAN] PLAY LIKE NOT
'Women don't like to play.'

In both types of examples it is legitimate to argue that the interpretation is the result of the licensing of a covert generic operator GEN that unselectively binds the open argument variables (cf. Krifka et al. 1995).

The structures reviewed in this section constitute an example of transparent mapping of quantificational sentences onto tripartite structures: the conditional antecedent and the brow-raise marked subjects correspond to the material in the restrictive clause of the covert operator (indicated with square brackets), while the rest is trivially projected into the nuclear scope. What is often trivially dubbed as topic-marking with brow raise is claimed to actually flag the restriction of the tripartite structure.

4 D-Quantification and Partial Overt Realization of Tripartite Structures

The SLs under study display D-quantifiers which can form regular quantificational phrases. A strong tendency to split the noun and the quantifier has been identified in both ASL and LSC, where the nominal restrictor occurs either in argument position or in a left-peripheral position marked with brow raise, as in ASL (9a) (examples adapted from Boster 1996, Partee 1995, Petronio 1995). Note that when the quantifier restriction appears in situ, the quantifier cannot be left-detached (9b).

_____br
(9) a. BOOK I WANT ALL/SOME/THREE
_______________br
b. *ALL/SOME/THREE I WANT BOOK

A related, though different, kind of split is observed with restricted Wh-phrases, where syntactic rightward movement of the Wh-sign in LSC strands the restriction in argument position or else the restriction appears in the left periphery marked with brow raise again, as in (10a) and (10b) respectively. Similar structures have been attested in ASL (Boster 1996) and Indian SL (Aboh et al. 2005).

__br ___________________bf
(10) a. IX2 CAR BUY WHICH
'Which car did you buy?'
_________br _________________bf
b. BOOK IX3 READ HOW-MANY
'How many of the books did s/he read?'

However, the type of quantifiers exemplified in (9) and (10) is weak and thus symmetrical in their two arguments, so at face value they do not require a tripartite structure for the computation of their truth conditions under the cardinal interpretation. The Q-N split observed here, though, is arguably the result of their strong or proportional interpretation, which is often taken to arise from the presuppositional set encoded in the nominal restrictor. Under this reading, they are trivially mapped onto tripartite structures. Information structure considerations mark the presuppositional material in the restrictor with brow raise in (9a) and (10a). In addition, in LSC, an SOV language, the default right-edge position of the Wh-element is what determines the surface split of operator and restrictor, thus yielding only a partial resemblance to proper tripartite structured quantification. It remains to be explored whether a covert Focus operator or a Q-typing particle is involved in the quantificational structure, and if so, how.

There is one especially interesting case involving negation. LSC has no proper negative determiners (see Matthewson 2008, and especially Zerbian & Krifka 2008 on Bantu), and the structures we get can be more adequately characterized as cases of A-quantificantion.[4] Despite the appearance of a negative determiner in a fragment answer like (11B), it can be shown that the several existing negative markers are sentential operators.

__________bf
(11) A: THIS MORNING STUDENT COME WHO
'Which students came this morning?'
B: NO-RES2
'None.'

[4] One note of caution should be included here, as a sign NINGÚ 'no(ne)' is attested as a negative adnominal determiner in utterances that are perceived as influenced by Spanish or Catalan. Further research is needed to ascertain whether this is a contact borrowing or it has become integral part of the quantificational system of LSC.

It can be shown that negative operators behave as adverbial unselective binders that sit in the Specifier of NegP (Quer 2003, Quer & Boldú 2007). As negative operators, they bind the unbound variables of the predicate, be it the event argument (12), the subject (13) or both subject and object arguments in one of the readings of (14). In these examples there is no overt material to be mapped onto the restrictor. Nevertheless, covert contextual information might provide it.

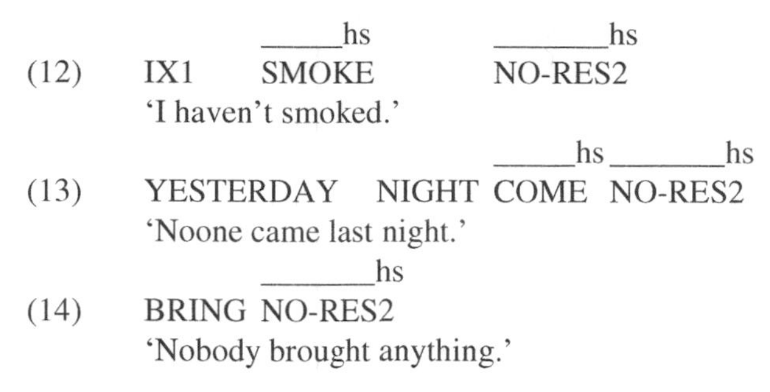

(12) IX1 SMOKE (hs) NO-RES2 (hs)
'I haven't smoked.'

(13) YESTERDAY NIGHT COME (hs) NO-RES2 (hs)
'Noone came last night.'

(14) BRING NO-RES2 (hs)
'Nobody brought anything.'

If an argument of the predicate does not contribute an open variable to the structure, it cannot be unselectively bound by negation, as observed in (15), where only the object slot can be bound by the negative operator (the subject slot is saturated independently):

(15) PEOPLE+++ SOME (br) BRING (hs) NO-RES2 (hs)
'Some people didn't bring anything.'

It is worthwile noting cases like (16), where ALL appears in the subject description of a negative statement and scopes over negation. The interpretation that obtains confirms the recurrent observation that 'all', as opposed to 'every', is not a proper operator, but rather a predicate indicating exhaustification of the intended set.

(16) FRIEND ALL (br) COME (hs) NO-RES2 (hs)
'No friends came along.'

Still, we do find bona fide strong/asymmetric D-quantifiers that only yield proportional readings like MAJORITY, the equivalent of English *most* in LSC, in (17).

(17) STUDENT (br) MAJORITY EXAM PASS
'Most students passed the exam.'

(18) MOSTx [student(x)] [pass(x)]

This case makes it clear that, when compared to A-quantification, proper cases of D-quantification in SLs impose a less straighforward mapping from overt quantificational statements onto tripartite structures. For basic cases like (17), though, one simply needs to assume that the quantifier will raise at LF to take scope over the restrictor (overtly encoded through brow raise above the nominal constituent in the left periphery) and the nuclear scope (the remaining material).

5 Lexical Quantification

Under the label A-quantification, a rather heterogenous set of resources to encode quantificational meanings are included in Bach et al. (1995). Verbal affixes of quantificational nature are among them, next to Q-adverbs. Partee (1995) herself actually suggests that A-quantification might not form a natural class and it might need to be further split into true unselective quantifier structures, on the one hand, and lexical quantification applied directly to a verb or other predicate type, on the other. She argues that morphology as operator on the verb can be quite different from unselective binding, because the operator is directed to a specific argument or arguments of the verbs.

In their seminal work on ASL, Klima & Bellugi (1979) analyze and classify a whole set of verbal inflections related to aspect, number and distributivity. Among the last group we find the following:

- [dual]: action with respect to an argument of cardinality two;
- [multiple]: action to many, viewed as a single episode;
- [exhaustive]: distributed action to each individual in a set.

Mostly coinciding with the characterization drawn for ASL by Petronio (1995), the following LSC paradigm illustrates the behaviour of quantificational inflection on the lexical verb and the corresponding interpretations. All of them quantify over the internal argument of the verb. Such inflections are clearly related to what is known as pluriactionals in many spoken languages, but a detailed analysis along these lines is still pending. For a more recent analysis of reduplication in ASL, see Wilbur (2005).

(19) PERSON++ STUDENT IX^TWO IX1 1ASK3[dual]
'I asked the two students.'

(20) PERSON++ STUDENT IX^THREE IX1 1ASK3[mult]
'I asked the three students.'

(21) PERSON++ STUDENT IX^THREE IX1 1ASK3[exh]
'I asked each one of the three students.'

The exhaustive inflection is also known as *distributive* marking. A crucial fact in this connection is that in LSC the same morphological mechanism is at play in the expression of distributivity on lexical and functional nominal items: the short reduplication along an arc attested in (21), for instance, is the same morpheme that marks distributive-key and the distributive-share in the language (Gil 1995). Typically the reduplicated form of the numeral ONE appears with both functions, but it can also appear on other nominal elements like the possessive (22):

          ____________________________________br
(22) STUDENT ONE+++ TEACHER POSS+++ ASK+++
'Each student asked his/her teacher.'

Such cases suggest that what might be seen as verbal inflection encoding quantificational interpretations has a more widespread use in the grammar of the language, as a general way of expressing distributivity and pluriactionality.

6 Conclusion

The overview of the data on quantified structures in LSC and ASL tentatively allows to confirm that languages in the visual-gestural modality resort to essentially the same kinds of mechanisms attested for the expression of quantification in spoken languages, namely A- and D-quantification, and A-quantificational structures seem to be widespread. At the same time, this study validates tripartite structures as an important heuristic and analytical tool that helps capture the correlations and correspondences in the different strategies languages employ to encode quantificational meanings.

Acknowledgments. The original work reported in this paper was presented at conferences in 2004/2005. The reviewers of this version and the audience at the 18th Amsterdam Colloquium are gratefully thanked for their feedback and comments. Delfina Aliaga, Santiago Frigola, Josep Boronat and Joan Frigola have contributed as Deaf LSC experts with their intuitions in invaluable ways. This research has been partly made possible by the grants awarded to the author by the Spanish Ministry of Science and Innovation (FFI2009-10492) and to UR-LING by the Govern de la Generalitat de Catalunya (2009SGR00763), as well as by COST Action IS 1006 SignGram.

References

1. Aboh, E., Pfau, R., Zeshan, U.: When a wh-word is not a wh-word: The case of Indian Sign Language. In: Bhattacharya, T. (ed.) Yearbook of South Asian Languages and Linguistics 2005, pp. 11–43. Mouton de Gruyter, Berlin (2005)
2. Bach, E., et al. (eds.): Quantification in natural languages. Kluwer, Dordrecht (1995)
3. Barberà, G.: When Wide Scope Is not Enough: Scope and Topicality of Discourse Referents. In: Aloni, A., et al. (eds.) Amsterdam Colloquium 2011. LNCS, vol. 7218, pp. 62–71. Springer, Heidelberg (2012)
4. Barwise, J., Cooper, R.: Generalized Quantifiers and Natural Language. Linguistics and Philosophy 4, 159–219 (1981)
5. Boster, C.T.: On the Quantfier-Noun Phrase Split in American Sign Language and the Structure of Quantified Noun Phrases. In: Emondson, W.H., Wilbur, R.B. (eds.) International Review of Sign Linguistics, vol. 1, pp. 159–208. Lawrence Erlbaum, Mahwah (1996)
6. Gil, D.: Universal Quantifiers and Distributivity. In: Bach, E., et al. (eds.) Quantification in Natural Languages, pp. 321–362. Kluwer, Dordrecht (1995)

7. Heim, I.: The Semantics of Definite and Indefinite Noun Phrases. Doctoral dissertation, University of Massachusetts/Amherst (1982)
8. Kamp, H.: A Theory of Truth and Semantic Representation. In: Groenendijk, J., et al. (eds.) Truth, Representation and Information, pp. 1–41. Foris, Dordrecht (1981)
9. Klima, E.S., Bellugi, U.: The signs of language. Harvard UP, Harvard (1979)
10. Krifka, M., et al.: Genericity: An Introduction. In: Carlson, G.N., Pelletier, F.J. (eds.) The Generic Book, pp. 1–124. The University of Chicago Press, Chicago (1995)
11. Matthewson, L.: Quantification: A Crosslinguistic Perspective. Emerald, Bingley (2008)
12. Partee, B.H.: Topic, Focus, and Quantification. In: SALT I Proceedings, pp. 159–189. DMLL Publications, Cornell University (1992)
13. Partee, B.H.: Quantificational structures and compositionality. In: Bach, E., et al. (eds.) Quantification in Natural Languages, pp. 487–540. Kluwer, Dordrecht (1995)
14. Petronio, K.: Bare noun phrases, verbs and quantification in ASL. In: Bach, E., et al. (eds.) Quantification in Natural Languages, pp. 603–618. Kluwer, Dordrecht (1995)
15. Quer, J.: Operadores negativos en Lengua de Signos Catalana. Ms. ICREA & Universitat de Barcelona (2003)
16. Quer, J.: Quantifying across Language Modalities: Generalized Tripartite Structures in Signed Languages. Invited lecture at the 1st Workshop on Sign Language. University of the Basque Country/EHU, Gasteiz-Vitoria (2005)
17. Quer, J., Boldú, R.M.: Lexical and morphological resources in the expression of sentential negation in Catalan Sign Language (LSC). In: Actes del 7è Congrés de Lingüística General, Universitat de Barcelona. CD-ROM, Barcelona (2006)
18. Schlenker, P.: Quantifiers and Variables: Insights from Sign Language (ASL and LSF). In: Partee, B.H., Glanzberg, M., Skilters, J. (eds.) Formal Semantics and Pragmatics: Discourse, Context, and Model. The Baltic International Yearbook of Cognition, Logic and Communication, vol. 6 (2011)
19. Wilbur, R.B.: Generic and habitual structures in ASL: The role of brow raise. Talk at Theoretical Issues in Sign Language Research, Gallaudet University, Washington, D.C. (1998)
20. Wilbur, R.B., Patschke, C.: Syntactic Correlates of Brow Raise in ASL. Sign Language and Linguistics 2(1), 3–41 (1999)
21. Wilbur, R.B.: A reanalysis of reduplication in American Sign Language. In: Hurch, B. (ed.) Studies in Reduplication, pp. 593–620. De Gruyter, Berlin (2005)
22. Wilbur, R.B.: Nonmanuals, semantic operators, domain marking, and the solution to two outstanding puzzles in ASL. Sign Language & Linguistics 14, 148–178 (2011)
23. Zerbian, S., Krifka, M.: Quantification across Bantu languages. In: Matthewson, L. (ed.) Quantification: A Crosslinguistic Perspective, pp. 383–414. Emerald, Bingley (2008)

Degree Modification and Intensification in American Sign Language Adjectives

Ronnie B. Wilbur[1], Evie Malaia[1,2], and Robin A. Shay[1]

[1] Purdue University
{wilbur,shayra}@purdue.edu
[2] University of Texas, Arlington
malaia@uta.edu

Abstract. Scalar adjectives lacking closed upper boundaries (like *far*) can be coerced to have a closed upper boundary reading when combined with degree modification with *too*, e.g. *too far to walk*. Parallel to the mapping of event structure to scalar structure in adjectives [4], we observe that scalar adjectives are *end-marked* in ASL. These adjectives receive marking similar to telic verbs, indicating that, like the visibility of event structure in verbs, scalar structure, or at least the upper boundary, is also visible in ASL. The Event Visibility Hypothesis (EVH) was formulated based on the observation that telic verb signs are distinguished from atelics by end-marking reflecting final states of telic events. Here, it is extended to a general Visibility Hypothesis for sign languages.

Keywords: gradable adjectives, American Sign Language, degree modification, intensification.

1 Introduction

To date, adjectives in ASL have not received detailed investigation comparable to that of verbs and nouns. Two studies both focused on syntax [1-2]. MacLaughlin [1] explored the distinction between *attributive* and *predicative* adjectives and related word order. Bernath [2] investigated their syntax and suggested that different word orders result from movement of the noun. He further raised the question of whether adjectives like SICK[1] should be considered as adjectives at all, given that they can be aspectually modified, and suggested that they should instead be treated as verbs, e.g. BE-SICK. This project takes a different perspective by focusing on the semantics of the adjectives. In particular, it focuses on gradable adjectives and interaction with degree modification, reporting new observations on how such modification is marked.

Section 2 introduces gradable adjectives and degree modifications. Section 3 presents examples of ASL gradable adjectives and how they are marked under degree modification. Section 4 considers interaction of gradable adjectives with the semantics of *too* in the form *too Adjective to Verb* (e.g., *too hot to eat*, *too far to walk*). Section 5 ties the pieces together. We relate the marking of adjectives in *too A to V* to

[1] The glosses for signs are traditionally written in capital letters.

M. Aloni et al.(Eds.): Amsterdam Colloquium 2011, LNCS 7218, pp. 92–101, 2012.
© Springer-Verlag Berlin Heidelberg 2012

marking of end-state boundaries in ASL signs denoting telic events, previously discussed under the rubric of the Event Visibility Hypothesis (EVH) [3]. This similarity of marking is not coincidence but related to the existence of scalar boundaries/limits in both cases. We extend the EVH to a more general Visibility Hypothesis (VH).

2 Gradable Adjectives, Scales, and Degree Modification

Following Kennedy and McNally [4], we take a relational approach and assume that gradable adjectives denote a relation (G) between individuals (x) and degrees (d) on an appropriate scalar dimension for that adjective. For example, the adjective *expensive* could be represented as a relation between objects and degrees of cost so that the cost of x equals d.

$$[[\text{expensive}]] = \lambda d \lambda x.\textbf{expensive}\ (x) = d \quad (1)$$

This representation does not take into account the idea that an object could have a cost d that would not be considered expensive but rather normal, fair, or cheap. To decide that something should be called expensive, there needs to be a way of determining when a cost is big enough to be considered expensive.

The notion of a scale for a dimension such as cost requires that the degrees of cost be ordered in such a way that it is always possible to tell whether one particular degree is above or below another. The variation along this scale is what allows us to talk about an adjective being gradable. Let us assume that there is a **standard of comparison** value (s) on the scale above which the cost of something is expensive. That is, the cost of x must be greater than (>) the standard d_S on the scale of degrees of cost.

$$[[\text{expensive}]] = \lambda d \lambda x.\textbf{expensive}\ (x) > d_S \quad (2)$$

What is expensive for a cup of coffee is different from what is expensive for a new car, that is, the standard of comparison may vary by context. Thus, the application of a gradable adjective to an object (deciding to call something expensive) always requires a comparison, which is sometimes contextually dependent (*relative* adjectives) and sometimes fixed (*absolute* adjectives) [5], even if it is not overtly mentioned.

Now assume that as the distance between the cost of x and the standard of comparison d_s becomes greater, we want to talk about larger degrees of expensive. In English, this can be done with degree intensifiers such as *very* and *too*. Kennedy and McNally [6] provide a semantic analysis of *very* as in (3).

$$[[\text{very}]] = \{\langle G, \langle d_{S(G)}, x\rangle\rangle \mid \exists d[G(x) \geq d_{S(G)} + d \wedge \text{LARGE}(d)]\} \quad (3)$$

Very applies to a gradable adjective and has the effect of increasing (*boosting*) its value by a contextually-determined large amount; in (3) *very* is a function G that applies to the value of the adjective to ensure that it exceeds the normal comparison d_s by a contextually LARGE degree.

3 ASL Gradable Adjectives and Their Marking Under Degree Modification

To begin, the signing of an ASL lexical adjective is similar to that of any other lexical sign, in that the sign components (handshapes, place of articulation, movement, etc.) are lexically specified (as outlined in [7]). There are prosodic contextual effects, so that actual production depends on position in its phrase (Phrase Final Lengthening), relative degree of stress or emphasis, and current signing rate [8].[2]

While there is a sign for *very* (Fig.1), it is considered 'English register' rather than ASL, and has extremely limited use. Except for discussion of it, it does not occur in our ASL corpus.[3] Intensification is seen in alternate ways.

Fig. 1. VERY *very*; rejected as ASL degree intensifier

3.1 Plain Adjectives

The typical production of an adjective sign is the baseline against which intensified productions must be compared. Baseline production is the one in which the standard for application of the adjective has been met, as in (2) above. Typically, the signs are accompanied by mouthing of the English word or by mouth positions that do not change during the movement of the sign, as seen in FAR (Fig. 2).

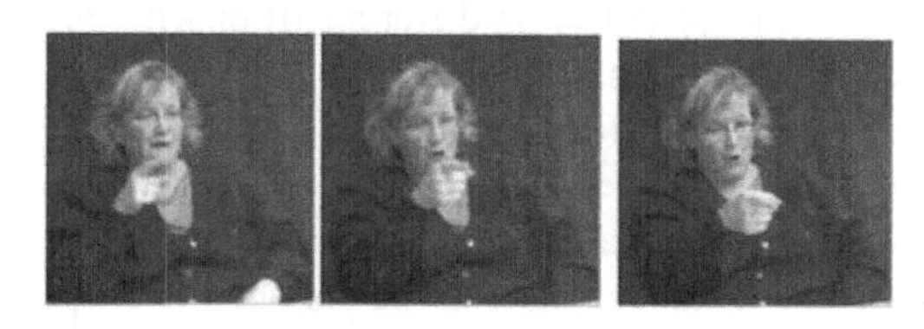

Fig. 2. The sign FAR

Among the adjectives we investigated specifically for this paper are: BIG, CLOSE, FAR, HARD, HEAVY, LATE, LONG, NICE, OBVIOUS, SMALL, SOON, SORRY. Data includes elicited and natural productions, the latter coming from our lab archives, online videologs, published and youtube videos.

---

[2] We emphasize *lexical* here because there are also classifier-based constructions that are contextually-dependent. The form of the ASL translation of *thick* depends on whether one means *thick liquid*, *thick horizontal object* (book lying down), or *thick vertical object* (book standing up), and so on. That is, there is no lexical sign for *thick*.

[3] Our ASL data has been collected over 30+ years and includes more than 50 signers.

3.2 Phonological Marking of Intensification

An intensified ASL adjective must meet two criteria. First, it must contain the semantic degree *boosting effect* given in (3). Second, a sign language specific criterion: the intensified adjective must be visually distinct from the baseline form. This leads to a somewhat unusual topic in a semantics paper, namely phonological marking that represents degree morphology.[4] Our prior work in semantics-phonology interface of sign languages and the visibility of event structure in formation of predicate signs was formalized as the Event Visibility Hypothesis (EVH) [3] [9]. EVH, in its original formulation, was concerned only with the boundaries expressed in verbs, stipulating that the end-points (boundaries) of events were marked at points in space by rapid deceleration in hand movement. Experimental investigations of production and perception confirmed that in ASL, the boundaries are marked by kinematic properties of hand movement [14, 15]. The expressive means for boundaries in adjectival scales have not, to date, been considered in terms of their phonologico-semantic properties.

In our data, we observed modifications to adjective signs under intensification:

- Overall increase in tension of the hands and face;
- Movement modifications;
  - Add or enlarge movement trajectory;
  - [delayed release] of the start of the movement;
- Non-manual modifications (face, head, body);
  - Frown on face;
  - Head tilt away from neutral.

Many intensified adjectives with [delayed release] have a prefixal hold prior to the onset of movement and, if there is mouth position change, it occurs with the onset of hand movement.[5] To illustrate, Figure 3 presents a sequence of stills from the signing of FAR-intensified. Production is distinct from plain FAR in Fig. 2. Mouth and hand position are held at the beginning (pictures 1-2), hand and mouth movement begin, and the sign ends with the end of hand movement and no further change in the mouth position (it stays open). The head is tilted for the entire sign.

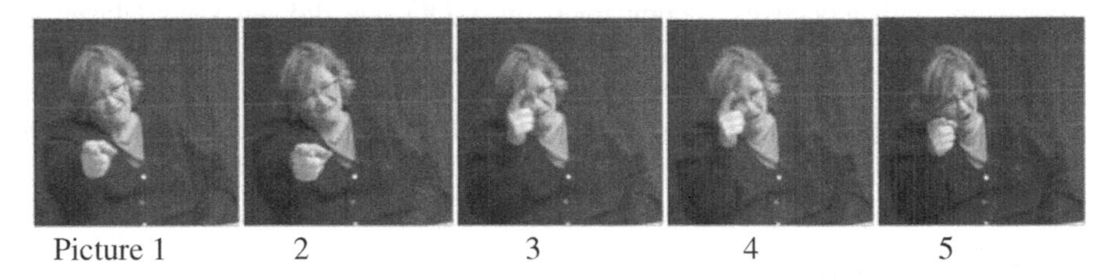

Picture 1 2 3 4 5

Fig. 3. FAR-intense with [delayed release], mouth opens with start of hand movement (3), mouth open at end (5); note head tilt, frown eyes and forehead

[4] A kindred phenomenon from spoken languages is vowel lengthening ("faaaaar away") – we thank an anonymous reviewer for the example.

[5] Some forms such as HEAVY prefix additional movement rather than the hold seen in e.g. FAR, CLOSE, LITTLE. Both types constitute delay of onset. Adjectives without path movement, e.g. HARD or SMALL, are modified by the intensification adverb Y –OO (section 3.3. below).

While [delayed release] and mouth change timing are reasonably regular, we observed the same head tilt behavior noted by Liddell [19], namely that variation in intensity depends on how much contrast the signer wants to provide.

3.3 Adverbial Intensifiers

Intensifier semantics can also be conveyed by combining adjective signs with adverbial intensifiers and/or non-manuals. Observations show that some adjective signs are produced with the simultaneous mouthing of *too*. Other adjectives are preceded by the loan sign #TOO (derived from fingerspelling the English word *too*). We are unable to predict which adjectives take which form, nor can we rule out both markings occurring with the same adjective. These two forms occur in both elicited and corpus examples, but only in intensification contexts, not in the *too A to V* discussed below.

Our investigation led to the realization that there is another sign in ASL that could be considered to have the meaning of *so, very* (Fig. 4). This sign has no known conventional gloss, and since we have not fully tested its semantics, we have dubbed it Y-OO, reflecting its use of Y handshape and circling movement.[6] Y-OO can be made with one or both hands. One observation is that Y-OO cannot modify all gradable adjectives. For example, it cannot occur with the sign SORRY, which readily takes the intensification modifications described in 3.2. But it does occur with HARD, FAST, HEAVY, BRIGHT/CLEAR/OBVIOUS, AWESOME, among others, covering a range of phonological forms with and without path movement.

Fig. 4. The adverb sign Y-OO *so, very*

Furthermore, with adjectives accompanied by Y-OO, we did not see [delayed release], which might mean the two are in complementary distribution; we do not have enough data to fully substantiate this possibility. Another adjective structure we investigated was *Adj like a N*, for example, *hard like a brick*. This translated as Y-OO HARD SAME-AS BRICK *so/very hard like (a) brick.*

4 'Too Adjective to Verb'

To understand the interesting behavior of ASL adjectives in the context *too A to V*, we need to consider the meaning of *too*, and the role of the infinitive *to V*. *Too* is often described as a form of degree morphology, along with English *-er*, *more*, *so*, and

[6] This sign should not be confused with the sign glossed SILLY, which has diagonal linear movement or bending at the wrist.

others. In one sense, like other intensifiers, it implies a greater degree. But *too* differs in containing the notion of excess, that is, beyond a limit, especially in the context *too A to V*. Meier [10] analyzes this construction by treating *too* as a comparative quantifier relating two values, the extent of the gradable adjective and an incomplete conditional provided by the complement *to V* (treated as *extent* predicates [11-12]). Consider (4).

$$\text{The food is too good to throw (it) away.} \tag{4}$$

Meier argues that (4) has the form in (5), where two values are compared, the actual goodness of the food, and the maximal value of food-goodness to be allowed to throw it away (from the sentential complement represented as modalized proposition).

$$x \text{ is too adjective MODAL } p. \tag{5}$$

The paraphrase she suggests for *too good to throw away* is "the value *v* such that the food is *v*-good is *greater* than the maximum of all values *v** such that if the food is *v*-good, we *are allowed* to throw it away". Note that this paraphrase has (a) a statement that the food is *v*-good, (b) a statement that this value is greater than the maximum of the conditional, (c) the conditional in the form *if food is v-good, we are allowed to throw it away*, and (d) the modal *be allowed to*. Similarly, for negative adjectives, e.g. *too young to date*, the paraphrase would be the same but the comparison in (b) would require that the value be smaller than the minimum of the conditional. The paraphrases in (c) and (d) provide the contribution of the complement 'to V'.

Let us turn now to the ASL structure of interest. What happens to an adjective like FAR when it is put into 'too far to walk' and why does it happen? Given that *too* is a degree intensifier, we should not be surprised to find that TOO-FAR exhibits the [delayed release] of the hand movement identified in section 3 (Fig. 5).[7]

Fig. 5. 'Too far to walk' : TOO-FAR with [delayed release] (pictures 1-2), mouth opens at *end* of hand movement (picture 5)

However, there are at least three differences in the *too A to V* productions compared to the A-intensified productions seen in Fig. 3. Note first that the head tilt seen in Fig. 3 is not used.[8] Second, there is a striking difference in the mouth behavior, with the mouth opening at the start of the hand movement in Fig. 3 but not until the *end* of the

---

[7] [5] refers to *too* as a sufficiency morpheme, while [10] separates it from sufficiency *enough* and refers to it as a morpheme of excess.

[8] Optionality follows observations in [19]. When it does occur, it can be easily represented by merging the EVH with the phonological representation of articulators on independent tiers in Brentari [7]; here, we do not go further into the topic for considerations of space.

hands movement in Fig. 5. Third, what cannot be seen in a sequence of still pictures is the sharp ending of the adjective sign movement in the *too A to V* contexts. The mouth position change at the end of the movement and the sharp ending of the movement itself caught our attention, because we have seen these two linguistic features together before, namely at the end of movements in verb signs that denote a telic event [9][13]. The mouth position change is referred to as a *transition non-manual* (T-NM). The sharp ending of the movement is due to a rapid deceleration from the peak velocity to the end stop [14-15]. This end-marking is considered to reflect the end state of a telic event, and led to the formulation of the Event Visibility Hypothesis: In the predicate system, the semantics of event structure is visible in the phonological form of the predicate sign. Its presence in the *too A to V* context suggests that it may have a broader function.

Kennedy and McNally [4] discuss the parallelism between adjectives derived from verbs (e.g., *closed, written*) and aspectual properties with respect to their common scales, noting there is a general correlation between event structure and scale structure. Their suggestion is that deverbal adjectives inherit scale structure either from the event denoted by the source verb or from the entity to which the adjective applies. In particular, deverbal gradable adjectives derived from state and activity denoting atelic verbs tend to be associated with scales that are open on the upper end (have no maximum value) because atelic events have no natural endpoint. In contrast, deverbal adjectives derived from achievement and accomplishment (those with incremental themes) denoting telic verbs are associated with closed upper scales (do have a maximum value) because telic events have endpoints reflecting the end state of the event.[9] Thus, their work on derived deverbal adjectives uses event structure analysis of boundaries in spoken languages, and establishes linguistic significance of the adjectival scale, that is, of the structure of ordered set of degrees in adjectival comparison.

What we have found is that our gradable adjectives display the same end marking that we observe on telic predicates in ASL. That is, they behave as though they had end states, or for adjectives, closed upper boundaries. Yet only one of our current adjectives, HARD, passes the tests for closed upper boundary,[10] and indeed it is totally closed [6] [16]. The other adjectives, e.g., FAR, which themselves have open upper scales, display the behavior of a closed upper scale in the *too A to V* context even though alone they do not have maximum values.

5 Putting the Pieces Together

How do we compose *too A to V*? Viewing *too* as a quantifier relating two extent values, the adjective and the complement verb, Meier [10] requires there to be a maximum value for the adjective scale which she suggests is provided through the hidden incomplete conditional analysis. Thus, scalar adjectives lacking closed upper

---

[9] The ASL adjectives that we investigated for this report are not deverbal, but it would be interesting to see how ASL adjectival predicates behave in this regard.

[10] Adverbial modifiers like *half* and *mostly* or *most of the way* are only acceptable with adjectives that have a closed scale (boundary), and unacceptable with those that do not.

boundaries (like *far*) must be coerced to have a closed upper boundary reading with a type of measure phrase providing the limit in the construction *too A to V*.

Beginning with sentence (6), we use rule (2) and treat the distance of the restaurant as being greater than a distance that justifies the use of *too*.

The restaurant is too far to walk (to). (6)

$$\lambda d \lambda x \mathrm{far}(x) > d_{\mathrm{too}} \quad (7)$$

This needs to be combined with the value that is *walkable*, which we get from the modalized conditional suggested by [10], 'possible walkable distance'.

If x is d-distance, we can walk to it. (8)

The distance to the restaurant exceeds this maximum value:

$\lambda d \lambda x$ too-far $(x) >$ MAX d such that if x is d-distance, we can walk to it. (9)

That is, the distance is greater than the maximum value of the scale of *walkable*, and this maximum is where the closed upper boundary comes from in *too far to walk*. Thus, following [10], *the restaurant is at a distance that is greater than the maximum distance that we can walk (to)* or *at a distance that is so far that we cannot walk to it*.

The ASL structure has the following pieces. The adjective FAR takes the prefix [delayed release] for intensification; this affects both the hand movement and the non manual (mouth) change. [7] notes that [delayed release] in the *delayed completive aspect* attaches to the first timing slot of a telic verb; here we suggest that it attaches to the first timing slot of the adjective.[11] The adjective also takes *end marking,* a sharp movement to a stop, and the T-NM [closed -> open]. The end marking attaches to the second timing slot of telic verbs, and here to the second slot of the adjective, and is aligned to the right edge. Thus, the mouth does not open until the end of the movement. The sharp movement is the result of earlier peak velocity and greater deceleration than in plain signs. This end marking indicates the upper boundary/limit of the extent to which the complement verb (e.g. *walk*) is possible. Thus, the extent of the walkable distance stops at a boundary that the restaurant is located beyond. As [10] notes, the modal 'possible' is contributed by the hidden incomplete conditional; this can be epistemic or deontic but is usually covert. When signers were offered *too A to V* with explicit modals (TOO-FAR WALK CAN'T, TOO-HARD EAT CAN'T), they accepted them but did not produce them on their own. One paraphrase structure was suggested: CAN'T LIFT WHY, BOX HEAVY-intens *I can't lift the box because it's too heavy*, but this is a different structure - the wh-cleft. In all cases in the data so far, there is no intensity placed on the complement verb itself nor is there any indication of negation.

The end result is that the adjective has the same motion characteristics as a delayed completive telic verb (but without the tongue wagging that [7] notes). This same form is seen with other open scale adjectives in our set, e.g. HEAVY, but not with HARD,

[11] The typical sign is monosyllabic, consisting of two timing slots associated to the movement.

which is a closed scale adjective and which already ends with contact of the two hands. Instead, HARD is modified by Y-OO, and there is no mouth change or noticeably different movement pattern compared to its plain form. A more systematic investigation is needed but it may be that final contact as in HARD or final deceleration as added to other adjectives both mark a closed boundary.

However, the parallel between end-marked verb signs and end-marked *too-A-to-V* adjectives suggests that the original formulation of the Event Visibility Hypothesis, based on verbs only, is too narrow. We propose an extended version, the Visibility Hypothesis, which reflects the idea that sign languages have grammaticalized resources from physics and geometry for perceptual and production purposes to convey meanings that humans wish to express. Extension of the EVH to adjectival scales yields specific predictions for how sign languages express scalar boundaries, and the (modality specific) impact on the grammatical system of sign languages. The Visibility Hypothesis formulation for the manual component of the sign follows:

Sign languages express the boundaries of semantic scales by means of phonological mapping. (10)

However, this is not the 'iconicity' that many people think is there. The semantics-phonology interface goes well beyond the typical notion of 'iconicity' ('guessability') [9]. End-marking is an example of grammaticalization of physics (deceleration) for linguistic purposes. Thus, whereas [4], among others, noted the parallel between event predicates and scalar adjectives in spoken languages, in ASL and probably other sign languages, it is, in fact, *visible.*[12]

Acknowledgements. The work was partially supported by NIH grants DC00524 and DC011081 to R. B. Wilbur.

References

1. MacLaughlin, D.: The Structure of Determiner Phrases: Evidence from American Sign Language. Dissertation, Boston University (1997)
2. Bernath, J.: Adjectives in ASL. In: 10th Theoretical Issues in Sign Language Research, pp. 162–163. Purdue University, West Lafayette (2010)
3. Wilbur, R.B.: Complex Predicates Involving Events, Time and Aspect: Is This Why Sign Languages Look So Similar? In: Quer, J. (ed.) Signs of the Time, pp. 217–250. Signum, Hamburg (2008)
4. Kennedy, C., McNally, L.: Scale Structure, Degree Modification, and the Semantics of Gradable Predicates. Lang. 81, 345–381 (2005)
5. Kennedy, C.: Vagueness and Grammar: Semantics of Relative and Absolute Gradable Adjectives. Ling & Philo 30, 1–45 (2007)

[12] Visible to a signer – as multiple perceptual studies suggest, signers' visual perception differs from that of non-signers in multiple dimensions, including ability to attend to the visual periphery and high spatial frequencies [17, 18].

6. Kennedy, C., McNally, L.: From Event Structure to Scale Structure: Degree Modification in Deverbal Adjectives. In: Mathews, T., Strolovitch, D. (eds.) SALT IX, pp. 163–180. CLC Publications, Ithaca (1999)
7. Brentari, D.: Prosodic Model of Sign Language Phonology. MIT Press, Cambridge (1998)
8. Wilbur, R.B.: Effects of Varying Rate of Signing on ASL Manual Signs and Nonmanual Markers. Lang. and Sp. 52(2/3), 245–285 (2009)
9. Wilbur, R.B.: The Semantics-Phonology Interface. In: Brentari, D. (ed.) Cambridge Language Surveys: Sign Languages, pp. 357–382. Cambridge University Press, Cambridge (2010)
10. Meier, C.: The Meaning of Too, Enough, and So... That. Nat. Lang. Seman. 11, 69–107 (2003)
11. von Stechow, A.: Comparing Semantic Theories of Comparison. J. Seman. 3, 1–77 (1984a)
12. von Stechow, A.: My Reaction to Cresswell's, Hellan's, Hoeksema's and Seuren's Comments. J. Seman. 3, 183–199 (1984b)
13. Schalber, K.: Phonological Visibility of Event Structure in Austrian Sign Language: A Comparison of ASL and OGS. MA Thesis, Purdue University (2004)
14. Malaia, E., Wilbur, R.B.: Kinematic Signatures of Telic and Atelic Events in ASL Predicates. Lang. and Sp. 55(3), November 21 (2011), doi:10.1177/0023830911422201
15. Malaia, E., Ranaweera, R., Wilbur, R.B., Talavage, T.M.: Neural Representation of Event Structure in American Sign Language: fMRI Comparison of Cortical Activations in Deaf Signers and Hearing Non-signers. Neuroimage 59, 4094–4101 (2012)
16. Demonte, V.: Adjectives. In: Maienborn, C., von Heusinger, K., Portner, P. (eds.) Semantics: An International Handbook of Natural Language Meaning. Mouton de Gruyter, Berlin (2011)
17. Bosworth, R., Dobkins, K.: Visual Field Asymmetries for Motion Processing in Deaf and Hearing Signers. Brain and Cognition 49, 170–181 (2002)
18. Bosworth, R., Bartlett, M., Dobkins, K.: Image Statistics of American Sign Language: Comparison with Faces and Natural Scenes. J. Optical Society America 23(9), 2085–2096 (2006)
19. Liddell, S.K.: Nonmanual signals and relative clauses in American Sign Language. In: Siple, P. (ed.) Understanding Language through Sign Language Research, pp. 59–90. Academic Press, New York (1978)

Experimenting with the King of France

Márta Abrusán[1] and Kriszta Szendrői[2]

[1] Lichtenberg Kolleg, Göttingen, Germany
[2] University College London, UK
abrusan@alum.mit.edu, k.szendroi@ucl.ac.uk

Abstract. Definite descriptions with reference failure have been argued to give rise to different truth-value intuitions depending on the local linguistic context in which they appear. We conducted an experiment to investigate these alleged differences. We have found that pragmatic strategies dependent on verification (Lasersohn 1993,von Fintel 2004) and topicalization (Strawson 1964, Reinhart 1981), suggested in the context of trivalent theories, both play a role in people's subjective judgments. We suggest that a way to reconcile this finding is to assume that verification of a sentence –where possible – proceeds through a *pivot* constituent, and that this concept is relevant for the proper description of how speakers understand semantic meaning.At the same time, it seems that trivalent theories cannot easily account for the full pattern of the results found. We speculate that our findings are best explained by combining these pragmatic strategies with an approach that assumes that definite descriptions have a bivalent semantics, as well as a pragmatic presupposition attached to them.

Keywords: definite descriptions, presupposition, topics, verifiability, experimental pragmatics.

1 Introduction

According to Russell (1905, 1957), definite descriptions assert the existence of a unique individual that satisfies the description. When such an individual does not exist, as in (1), the sentence makes a false assertion. Famously, Strawson (1950, 1964) argued that Russell's theory cannot predict why speakers (like him) feel "squeamish" about assigning the truth-value 'false' to sentences such as (1). His proposal (which can be traced back to Frege 1892) was that definite descriptions instead of asserting, presuppose the existence of a unique individual that satisfies the description: when this presupposition is not met, the question of truth or falsity does not even arise and so the sentence does not have a truth-value.

(1) The king of France is bald.

Strawson (1964) has noted however that truth-value intuitions change when the same definite description is put in different contexts. He conceded that according to his intuition examples such as (2) do appear to be straightforwardly false. This in turn requires some explanation from the view according to which (2) should not have a truth-value.

A. Aloni et al.(Eds.): Amsterdam Colloquium 2011, LNCS 7218, pp. 102–111, 2012.
© Springer-Verlag Berlin Heidelberg 2012

(2) The exhibition was visited yesterday by the king of France.

Since Strawson's observation, various factors have been identified in the literature that might influence our truth-value intuitions about sentences with reference failure noun phrases. Strawson (1964) himself (cf. also Reinhart 1981 and many others) have identified topic-comment structure as a factor. Others (cf. Fodor 1979, Lasersohn 1993, von Fintel 2004) stressed the importance of background knowledge based on which the sentence could be verified (or not), independently of the existence of the problematic referent. Sometimes, conflicting judgments have been asserted, which makes it hard to judge the relative import of the theories.

We have conducted a behavioral experiment that was designed to capture people's intuitions about sentences like (1)-(2). We havefound that the above-mentioned theories all identify relevant pragmatic factors that influence truth-value intuitions, even though none of them can fully predict the behavioral pattern we found: The sentence's topic-comment structure (as suggested by Strawson 1964, Reinhart 1981, etc) and the sentence's verifiability (as suggested by Lasersohn 1993, von Fintel 2004) are *both* important factors influencing truth-value judgments. At the end of this paper, we offer a brief discussion of how to reconcile these seemingly divergent factors.

2 Possible Factors behind Wavering Truth-Value Judgments

Strawson (1969) has proposed that one factor behind difference in (1) and (2) is the topic-comment structure of the sentence.Topics are understood to be the constituents that the sentence is pragmatically about. Strawson proposed that when the definite description is not in a topic position, it is "absorbed" into the meaning of the predicate, and since it is not a referring expression anymore its presupposition is turned into an existential statement. This predicts the difference between (1) and (2): The noun phrase the king of France is in topic position in (1) hence in a context where it is known that France has no king, it leads to a presupposition failure, associated with "squeamishness". In (2) the same noun phrase is not in a topic position and so the sentence is simply false. The topic approach has been adopted and developed further by many researchers, most importantly Reinhart (1981, 1995).(See also Hajicova 1984, Gundel 1977, Lambrecht 1994, Erteschik-Shir 1997, Atlas 2004, Geurts 2007, Shoubye 2009, among others).

However, the importance of topichoodhas been questioned (cf. von Fintel2004). One reason to doubtthat the existential presupposition is absorbed into the predicate (or is just non-existent) when the definite description is not in topic is that the existential presupposition of definites seems to project out of embedded contexts, such as the antecedents of conditionals, whether or not the definite is in topic position. Thus (3) still seems to imply the existence of a French king, which suggests that the definite is still presuppositional, despite not being in topic.

(3) If the exhibition was visited by the king of France, the organizers must be happy.

This latter observation highlights a very important point, identified most clearly by von Fintel. Namely, that our intuitions about accepting or rejecting a sentence as true or false, and the sentence's actual semantic truth-value (and hence its presuppositionality) are two separate things. Speakers might feel that a sentence is false or true even wheno semantically it has no truth-value, as long as they can find some reason based on which they can reject (or accept) the sentence. The feeling of "squeamishness" arises only when all pragmatic repair strategies for dealing with a truth-valueless sentence fail.

The first proposal in this spirit was due to Lasersohn (1993). His main focus were examples such as (4) which are said to be judged false. Lasersohn's observation was that in a situation where the chair in front of the speaker is empty, or when it is occupied by somebody other than the king of France, speakers have enough grounds to reject the sentence as false: They can look at the chair and see that the king of France (whether or not he exists) is not in it.

(4) The king of France is sitting in this chair.

In the case of (1), in the absence of background knowledge about the hairstyle of French royals, speakers do not have enough grounds to reject or accept the sentence, and are left with the feeling of squeamishness. Lasersohn's theory rests on the following tenets: **(a)** Assume two kinds of truth-values: (i) semantically assigned values 1,0, and a third value #, which corresponds to 'neither 0 or 1' (ii) pragmatically assigned values TRUE, FALSE which represent the status of a sentence with respect to a given body of information, and correspond to acceptance and rejection.**(b)**Once we are faced with presupposition failure (neither 0 nor 1), there are fall-back strategies to fill in the gap and arrive at TRUE and FALSE.**(c)**Lasersohn'sfall-back strategy: <u>Step 1:</u> revise the given body of information to remove the knowledge that there is no king of France. <u>Step 2:</u> See if the given body of information can be consistently extended to include the target proposition.

A consequence of Lasersohn's analysis is that only those propositions are predicted to have the truth-values TRUE or FALSE that are in direct conflict (or in accordance) with what can be concluded from the given body of information. This conclusion has been argued to be too weak by von Fintel (2004), based on examples such as (5), which he argues is felt to be FALSE, even in the absence of any information about who is on a state visit in Australia this week.

(5) The king of France is on a state visit to Australia this week.

He proposes to add another fall-back strategy, besides (c) above:[1,2] (d)Rejection/acceptance might (also) be based on the possibility of examining the intrinsic properties of a contextually salient independent entity (that everyone agrees exists).

---

[1] That both (c) and (d) are needed is suggested by examples that fall under (c) but not (d): *The king of France can jump 100 feet into the air unaided.*(Example from intro by Bezuidenhout and Reimer to von Fintel 2004). However, if independent footholds for rejection/verification can also be general laws, then Lasersohn's account can be subsumed under von Fintel's, and thus (c) is a subcase of (d). It seems to us however that this would make it very hard to track what predictions von Fintel's proposal actually makes.

This suggestion explains (5), even if the given body of information does not contain anything about who was visiting Australia. In principle, we could examine the properties of Australia and see whether the king of France is in it or not. *Australia* thus serves as a salient foothold for verification,based on which the truth of the sentence in (5) can be evaluated.

3 The Experiment

We tested 33 native speakers of English (mostly British English, all of them familiar with basic elements of British culture), aged 20-55, most of whom participated for a small fee. We investigated how participants judge different types of sentences with reference failure noun phrases. The participants first read instructions given to them on the computer screen, reproduced in (6).

(6) *In this experiment, statements will appear on your screen. If you think a statement is true,you should click on the 'TRUE' button. If you think a statement is false, you should click on the 'FALSE' button. Sometimes, it may happen that you cannot decide. In those cases, you should click on the 'CAN'T SAY' button. Please do not dwell on your decision for too long. There is no right or wrong answer!*

After a short practice session, participants were left alone with a program which presented the test items one-by-one on the screen. Each item contained one sentence, as shown in (7). Participants could use the mouse to click on the buttons. After they chose an answer, the next item appeared automatically.

(7) *Example of an experimental trial:*

The king of France is bald.
FALSE CAN'T SAY TRUE

There were eleven test conditions, with eight test items in each condition. The test items were obtained by placing 8 definite descriptions that lack referents (listed in (8)) in eleven types of sentential contexts, (the test conditions), illustrated in (9). More on the test conditions below.

(8) the king of France; the emperor of Canada; the Pope's wife; Princess Diana's daughter; the beaches of Birmingham; the Belgian rainforest; the coral reefs of Brighton; the volcanoes of Kent.

(9) Examples of test conditions, illustrated here with *the king of France*
0 The king of France is bald.
1 France has a king and he is bald.
2 The king of France is on a state visit to Australia this week.
3 The king of France is married to Carla Bruni.
4 The king of France, he was invited to have dinner with Sarkozy.
5 Sarkozy, he was invited to have dinner with the king of France
6 The king of France isn't bald.

7 The king of France is not on a state visit to Australia this week.
8 The king of France is not married to Carla Bruni.
9 The king of France, he wasn't invited to have dinner with Sarkozy.
10 Sarkozy, he wasn't invited to have dinner with the king of France.

The 88 test items were supplemented by almost twice as many filler items containing true and false controls. Altogether, there were 253 items presented in three blocks. The items were pseudo-randomised: there were no items from the same condition, or with the same NP, closer than 4 trials. The statements were presented in three separate blocks.

All three theories agree that sentences like *The king of France is bald*, i.e. Condition 0, should lead to truth-value gaps and predict FALSE judgments for the non presuppositional assertions in Condition 1. The theory of von Fintel predicts a higher number of FALSE judgments in Condition 2, 3, 4, 5 than in Condition 0 because the former contain an independent NP alongside the referentially challenged NP. It also does not predict any difference among Conditions 2, 3, 4, 5. Lasersohn predicts a higher percentage of FALSE answers for Condition 3: The items in this condition were individually paired with 8 true control items, e.g. the control item for (12-3) and (12-8) was that Sarkozy is married to Carla Bruni.It was predicted that participants who judge control items TRUE would have the necessary knowledge to reject the corresponding items in Condition 3. Strawson and Reinhart predicted that items in Condition 4 lead to truth-value gaps because in these the referentially challenged NP is tropicalized, while their variants in Condition 5 in which the same NP is not in topic are judged FALSE.

Following pilot studies, we have discovered that a large proportion of our participants judged items in our base condition, Condition 0, as FALSE. For this reason, we included a further five conditions, Conditions 6-10, which corresponded to Conditions 0, 2-5, respectively, but which involved sentential negation. Our assumption was that a FALSE judgment that reflects a rejection of a particular statement based on it being semantically false should turn into a TRUE judgment once the statement in question is negated. If a negated sentence such as *The king of France is not bald* is not judged to be TRUE, then likely the corresponding positive sentence is judged to be FALSE for some other reason than having the semantic value 0. Further, if *The king of France is not bald* is not judged to be TRUE, the non-Russellians are still in business and we can test the validity of the theories presented above. In particular,we make the following predictions. We would get a higher number of TRUE judgments in Condition 7, the negated version of Condition 2, compared to Condition 6, the negated version of Condition 0, to support von Fintel's theory. Lasersohn predicted a higher number of TRUE judgments in Condition 8, the negated version of Condition 3, than in Condition 6. While Reinhart/Strawson predicted higher number of TRUE judgments in Condition 10 than in Condition 9, which are the negated equivalents of Conditions 6 and 5 respectively.

4 Results

4.1 Data

All but one of our participants successfully finished the task. This participant was excluded.We compared the proportion (%) at which subjects replied FALSE to the test items in Conditions 0-5 with ANOVA, following checks for normal distribution and homogeneity of variance. We found only weak or nearly significant differences between any of the conditions 0-5 (p>0.05, Bonferroni post-hoc test), i.e. the positive conditions. Our subjects said FALSE to most of these most of the time. This was counter to our predictions, according to which we expected a low number of FALSE judgments in Condition 0, and a high number of FALSE judgments in Condition 1. Neither of the conditions 0-5 differed significantly from our FALSE-controls either.

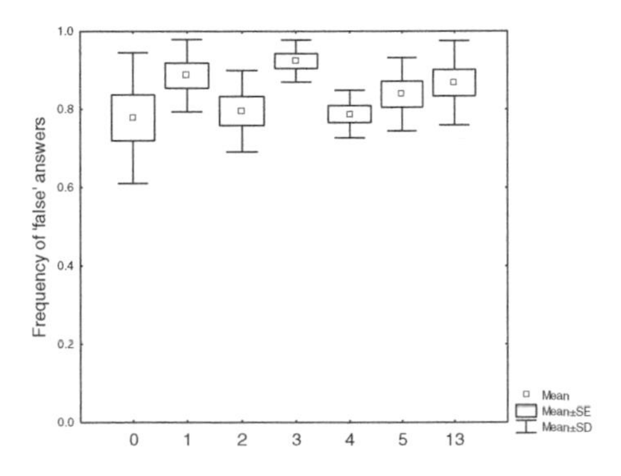

Fig. 1. Frequencies of FALSE answers for conditions 0-5, and the control condition 13 for which we expected FALSE answers.

Since there was no significant difference in the percentage of FALSE judgments between any of the conditions 0-5, we could not verify either a von Fintel effect (0 vs. 2), nor a Lasersohn effect (0 vs. 3) nor a topic-effect (4 vs. 5) by looking at positive sentences.

The negated versions of Conditions 0-5 (except Condition 1), namely Conditions 6-10 turned out to be more informative. We compared the proportion (%) of TRUE responses in these conditions with ANOVA, (Bonferroni post-hoc test). As Figure 2 illustrates, we found that speakers responded TRUE at a significantly higher proportion to the items in Conditions 7, 8, 9, 10 than to the items in Condition 6 (p<0.003 in all cases), our baseline. The significant difference between Conditions 6 and 7 (and also 6 vs. 8, 9, 10) indicates that there is a von Fintel-effect. The significant difference between Conditions 6 and 8 supports Lasersohn's theory. Condition 8 also differed significantly from Condition 7 (p=0.0012). In addition, we also found a significant difference (p= 0.037) between Conditions 9 vs. 10: speakers responded with TRUE at a significantly higher proportion to Condition 10 than to Condition 9, which is in accordance with Strawson/ Reinhart's predictions, and indicates that topichood also plays a part in subjective evaluations of the truth of sentences with reference failure definite descriptions. The difference between conditions 7 vs. 9 and 8 vs. 10 was not significant (p>0.05).

When comparing Conditions 6-10 with true controls, we found that all the conditions 6-10 (except condition 8) differed significantly from true controls as well, even if weakly (p<0.045). It is hard however to interpret this result, as Conditions 6-10 contained negation, but our control sentences did not.

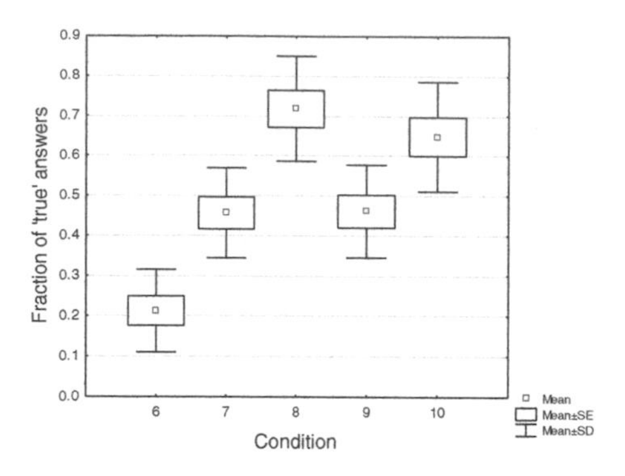

Fig. 2. Frequencies of TRUE answers for conditions 6-10. Box represents standard errors of the sample, whiskers standard deviation.

4.2 Interpreting the Results

Positive vs. Negative Conditions. We found that the responses to Condition 0 did not differ significantly from Condition 1, or from our control condition involving non-problematic false statements. At a first glance, this result might seem to support the Russellian view. However, in the light of the answers we got for Conditions 6-10, the negative conditions, the Russellian position is hard to maintain as this view has no room to predict the systematic variation in conditions 6 to 10.

The truth-value gap account of Strawson (also assumed by Lasersohn 1993 and von Fintel 2001) seems to have more room for maneuvers that might explain the overall pattern of data we found. On this account, all the sentences in conditions 0, 2-10 are truth-valueless.Nevertheless it is possible to explain the overapplication of FALSE judgments that are not semantically false by assuming that our subjects interpreted the response option FALSE'I do not think it is true' or 'I am rejecting this sentence' rather than 'I think this sentence is false'. Given that none of the theories presented (Strawson, Lasersohn, von Fintel) expected such a high number of FALSE judgments as was found for the moment we have to contend ourselves with this potential explanation. We will return to this issue in the final section. But first, let us discuss our test conditions.

The von Fintelvs the Lasersohn-Effect. We found a von Fintel effect: Condition 7 (and also Conditions 8, 9 and 10) received a significantly higher proportion of TRUE responses than Condition 6. At the same time, we found that there was a significant difference between the proportion of TRUE responses to Condition 7, the 'pure' von Fintel condition, and Condition 8, the one that aimed to test Lasersohn's theory. We take this result to indicate that there is indeed a separate Lasersohn and von Fintel-effect, and both effects measurably influenced truth-value judgments. For the von Fintel effect it is enough that an NP is present on the basis of which it would be possible to verify the sentence (given some appropriate knowledge dataset, say Wikipedia). For the Lasersohn effect, the properties of the NP provide a basis for verification given the speaker's actual dataset. This explains the finding that the Lasersohn effect is stronger: Our sentences that satisfied Lasersohn's criteria also satisfied von Fintel's criteria, but not the other way around.

Topic Effect. We also found that there was a significantly higher number of TRUE responses in Condition 10, where an existentially sound NPwas topicalised, than in Condition 9, where the NP with referential failure wastopicalised.Our results are thus consistent with Strawson's and Reinhart's prediction.

5 Further Discussion

5.1 Topics and Verifiability

We think that overall our findings support the position that all the three proposals discussed above were right to some degree, in that they all identified relevant factors for truth value intuitions. One question that arises is whether the above are independent factors, or whether there is some connection among them.

We believe that verification and topichood are not unrelated notions. However, there is also a fair amount of confusion around the notion of topic: sometimes it is understood as a discourse based notion (cf.Kuno 1972, Gundel 1974,Givon 1983, von Fintel 1996, Roberts *toappear,* among others) but sometimes it is (also) couched in verificationalist terms (cf. Strawson 1964, Reinhart 1981, Lambrecht 1994,Erteschik-Shir 2007, among others). However, even the latter employ tests for topichood that are discourse based. We suggest to distinguish two related but different concepts from each other: a discourse based notion of *topic*, which is what the sentence is pragmatically about, and what we call *pivot,* which is the constituent in the sentence based on which the sentence is verified, in other words, the contextually salient entity that provides the foothold for verification in von Fintel's sense. Thus pivothood is a semantic notion that is concerned with the process of understanding a sentence. This involves– at least in some cases – the possibility of verifying the sentence. The pivot is what the sentence is semantically about, if such an entity can be found, in this verificational sense of aboutness. It is not an obligatory element of understanding a sentence: In some cases it might not be possible to identify a unique pivot or indeed any pivot. Further, pivots do not have to be constituents of the sentence. This might be the case in examples with focus marking or clefting, where the set of alternatives provided by focus (or the background question) can serve as the pivot.

5.2 A Problem for the Truth-Value Gap Approach

Recall that the truth-value gap account of Strawson (also assumed by Lasersohn 1993 and von Fintel 2001) does not provide an immediate explanation for the findings in the positive condition, especially Condition 0. We offered a speculation to explain the overapplication of FALSE judgments that are not semantically false by assuming that our subjects interpreted the response option FALSE'I do not think it is true' or 'I am rejecting this sentence' rather than 'I think this sentence is false'.For conditions 2-5, we may also invoke Lasersohn's/von Fintel'sproposal according to which speakers opt for FALSE in order to avoid the conversational impasse created by a presupposition failure, but only if they have independent reasons to reject the sentence. The same reasoning also predicts a difference between conditions 6 vs. 7-10, which was also confirmed by our results. The problem is when we compare our answers in Condition 0 and Condition 6: To the extent that the above approaches would go along with our explanation for the high number of FALSE judgments in Condition 0 as instances of overapplication of FALSE to truth-valueless sentences, the same reasoning should apply to Condition 6. However, a post hoc comparison of the number of FALSE results for Conditions 0 and 6 revealed a significant difference (t-test p=0.0002), our subjects were much more likely to reply FALSE to the examples in Condition 0 than to the examples in Condition 6. We do not see how to reconcile this finding with the truth-value gap approach.

The theory of pragmatic presuppositions, according to which definite descriptions both assert and presuppose the existence of a unique individual that satisfies the description, such as Stalnaker's (1974) theory can explain the full pattern of results (cf. also Karttunen and Peters 1979; Abbott 2000; Simons 2001; Schlenker 2008, 2009; (among others)). As for the positive sentences (Condition 0), speakers say these are false, because they are indeed semantically false, and the fact that these sentences at the same time exhibit a presupposition failure is not in itself a reason to change this judgment, at least as long as false sentences do not have to be added to the context set. As for the sentences in Condition 6 (the negative baseline sentences), although these are semantically true, speakers are reluctant to mark these sentences as such because they also manifest a presupposition failure, and therefore one should avoid adding them to the context set. Suppose TRUE implies a commitment to add the sentence to the context set: this suffices to prevent people from saying that such sentences are TRUE. But when there is a good pragmatic strategy based on which the presupposition failure can be ignored, as in Conditions 7-10, the sentences with negation (that are semantically true) will be felt as pragmatically TRUE as well. This predicts the difference we found between Condition 6 vs. conditions 7, 8, 9 10.

Here is how this reasoning might explain our findings: (i) Sentences with reference failure NP's are semantically false (or true, if negated), but infelicitous because of presupposition failure. (ii) The presupposition failure can be ignored (and so the predicted infelicity can disappear) if a potential pivot that is independent from the reference failure NP is present in the sentence (≈von Fintel). (iii) The above effect is even stronger, if the independent, referentially sound NP is topicalised (≈Strawson/Reinhart). This latter fact is because topics are default pivots, and hearers tend to verify sentences based onthe sentence's topic. (iv) The effect is also stronger if the speaker has direct knowledge about the properties of the potential pivot, which contradicts the proposition irrespective of whether the entity denoted by the reference failure NP exists or not (≈Lasersohn).

6 Conclusion

To sum up, we have found that experimenting with King of France sentences may impact on theoretical proposals: our experimental results found evidence for Strawson/Reinhart's, von Fintel's and Lasersohn's theories. We also found that speakers, perhaps unexpectedly from the point of view of these three proposals, reject sentences such as *The King of France is bald* without too much "sqeamishness". We speculated that a Stalnakerian view, marrying a semantic and pragmatic view of presuppositions,might be the way to account for this.

Acknowledgements. Thanks to GyörgyAbrusán for his invaluable help with statistics. We are grateful to Nathan Klinedinst and Daniel Rothschild for insightful comments on an earlier draft, as well as Klaus Abels, Bob Borsley, JuditGervain, Vanessa Harrar, Ad Neeleman and the audiences at the UCL LingLunch and the Departmental Colloquium at Essex University for their helpful questions and suggestions. The first author acknowledges financial support by the Mellon Foundation and the Lichtenberg Kolleg, Georg-August-UniversitätGöttingen. All remaining errors are our own.

References

1. Erteschik-Shir, N.: Information structure. Oxford University (2007)
2. von Fintel, K.: Would you believe it? The King of France is back! (Presuppositions and truth-value intuitions.). In: Bezuidenhout, A., Reimer, M. (eds.) Descriptions and Beyond: an Interdisciplinary Collection of Essays on Definite and Indefinite Descriptions and Other Related Phenomena, pp. 315–341. Oxford University Press (2004)
3. Givon, T.: Topic continuity in discourse: An introduction. In: Givon, T. (ed.) Topic Continuity in Discourse: A Quantitative Cross-Language Study. Amsterdam/Philadelphia, John Benjamins (1983)
4. Karttunen, L., Peters, S.: Conventional Implicatures in Montague Grammar. In: Oh, C.-K., Dineen, D. (eds.) Syntax and Semantics 11: Presupposition, pp. 1–56. Academic Press, New York (1979)
5. Kuno, S.: Functional sentence perspective: A case study from Japanese and English. Linguistic Inquiry 3, 269–320 (1972)
6. Lasersohn, P.: Existence presuppositions and background knowledge. Journal of Semantics 10(2), 113–122 (1993)
7. Lambrecht, K.: Information structure and sentence form: A theory of topic, focus and the mental representations of discourse referents. Cambridge University Press (1994)
8. Reinhart, T.: Pragmatics and linguistics: An analysis of sentence topics. Philosophica 27, 53–94 (1981)
9. Roberts, C.: Topics. In: Maienborn, C., von Heusinger, K., Portner, P. (eds.) Semantics: An International Handbook of Natural Language Meaning. Mouton de Gruyter (to appear)
10. Russell, B.: On Denoting. Mind 14, 479–493 (1905)
11. Russell, B.: Mr. Strawson on Referring. Mind 66, 385–389 (1957)
12. Schlenker, P.: Be articulate: A pragmatic theory of presupposition. Theoretical Linguistics 34, 157–212 (2008)
13. Schoubye, A.: Descriptions, Truth-value Intuitions, and Questions. Linguistics and Philosophy 32(6), 583–617 (2009)
14. Simons, M.: On the conversational basis of some presuppositions. In: Hastings, R., Jackson, B., Zvolensky, Z. (eds.) Proceedings of Semantics and Linguistics Theory 11, pp. 431–448. CLC Publications, Ithaca (2001)
15. Stalnaker, R.C.: Pragmatic Presuppositions. In: Munitz, M., Unger, P. (eds.) Semantics and Philosophy: Essays. New York University Press (1974)
16. Stalnaker, R.C.: Assertion. In: Cole, P. (ed.) Syntax and Semantics 9, pp. 315–332. New York Academic Press, New York (1978)
17. Strawson, P.F.: On Referring. Mind 59, 320–344 (1950)
18. Strawson, P.F.: Identifying reference and truth-values. Theoria 30(2), 96–118 (1964)

Adjectives as Saturators vs. Modifiers: Statistical Evidence

Gemma Boleda[1], Stefan Evert[2], Berit Gehrke[1], and Louise McNally[1]

[1] Universitat Pompeu Fabra, Roc Boronat 138,
08018 Barcelona, Spain
{gemma.boleda,berit.gehrke,louise.mcnally}@upf.edu
http://www.upf.edu/

[2] Technische Universität Darmstadt, Hochschulstrasse 1,
64289 Darmstadt, Germany
evert@linglit.tu-darmstadt.de
http://www.tu-darmstadt.de/

Abstract. This paper reports on a large-scale, statistical analysis of corpus data to support the null hypothesis that ethnic adjectives (EAs, e.g. *French*) are ordinary adjectives, rather than argument-saturating "nouns in disguise" (in, e.g., *French agreement*). In particular, EAs are argued to simply modify the noun they combine with; their special properties in inducing argument-like behavior arises from the interaction between the semantics of event nominals and that of the adjective.

Keywords: adjective, argument structure, kind modifier, modification, nominalization, lexical semantics, logistic regression model, statistical analysis.

1 Introduction

Though surprising given standard assumptions about the formal semantics of adjectives, denominal relational adjectives, such as *molecular* and the so-called ethnic adjective (EA) subclass of these, e.g. *French*, have been repeatedly claimed in the syntactic literature to be able to saturate the presumably e-type arguments of the nouns they combine with [16,6,10,1]. On such a view, the compositional contribution of e.g. *French* in (1-a) would be identical to that of the PP in (1-b), saturating the agent argument of *agreement*, though specific semantic analyses for how this is done are not offered in this literature.

(1) a. French agreement to participate in the negotiations
    b. agreement by France to participate in the negotiations

[3], in contrast, focusing specifically on EAs, defend the null hypothesis that EAs are ordinary adjectives that simply modify the noun they combine with, such that the argument-like interpretation arises from the interaction between the semantics of the nominalization and that of the adjective. The goal of this paper is to show that large-scale, statistical corpus data analysis supports the latter analysis.

M. Aloni et al. (Eds.): Amsterdam Colloquium 2011, LNCS 7218, pp. 112–121, 2012.
© Springer-Verlag Berlin Heidelberg 2012

2 The Two Competing Analyses

We begin by sketching the main features of the argument-saturating analysis and the modifier analyses. For reasons of space, we only discuss [1] to exemplify the former and [3] for the latter, but our comments should generalize to other analyses that preserve the key features of each type of analysis.[1]

2.1 The Argument-Saturating Analysis

[1], working in the framework of Distributed Morphology, treat thematically-used EAs such as those in (1-a)[2] as covert nominals whose nominal source is visible at the level of interpretation. They bear the agent thematic role assigned to them by the (deverbal) noun they modify but lack case. Since every noun needs case, these nouns are deficient and become adjectives in the course of the syntactic derivation. The analysis of a (Greek) example of theirs is shown in (2).

(2) a. germaniki epithesi
German attack

b. $[_{\mathrm{DP}}\ [\ldots]\ [_{\mathrm{FP/AGRP}}[_{\mathrm{a(sp)P}}[_{\mathrm{a(sp)'}}[_{\mathrm{a(sp)^0}}\ \mathrm{german}_1\ [_{\mathrm{a(sp)^0}}\text{-ik}\ ]]]]$
$[_{\mathrm{F'}}\ \mathrm{F}\ [_{nP}[_{\mathrm{DP}}t_1[_{\mathrm{n'}}\ \mathrm{n}\ [_{vP}\ \mathrm{v}\ \sqrt{\mathrm{EPITH}}]]]]]]$

German- starts out as a DP in the specifier of the noun phrase *epithesi* 'attack', represented in (2-b) via the root $\sqrt{\mathrm{EPITH}}$.[3] In this position, *german-* is necessarily assigned the agent theta role by the underlying verb, on analogy to genitive DPs, which are also generated in this position. Since *german-* is not valued for case, it is forced to move up and to adjoin as a head to a(sp), the head of an adjectival projection that generally occupies the specifier position of a functional category between D and N, where it is spelled out as an adjective. On this account, both thematically-used EAs and genitive DPs (e.g. *the Germans' (attack)*) or PPs (e.g. *(the attack) of the Germans*) are base-generated in the same position, hence their relation to the nominal they combine with is the same, namely they saturate the agent argument of the nominal.

One argument put forward for nominal approaches to EAs is the fact that they do not behave like typical adjectives in some respects. For example, EAs cannot be used predicatively ((3-a)), are not gradable ((3-b)) and cannot be coordinated with 'normal' adjectives ((3-c)) (examples from [1]).

(3) a. *The intervention in Cyprus was American.
b. *the very / more American invasion
c. *the immediate / quick / possible and American intervention

[1] This section draws heavily on [3]; see that work for further details and discussion.

[2] EAs have been argued to have a second, classificatory use (e.g. *French* classifies wine in *French wine*); see references cited for discussion. [1] treat such adjectives as merely homophonous to thematically-used EAs, whereas e.g. [10] and [3] provide (contrasting) uniform accounts.

[3] Presumably, the nominalizing suffix *-esi* sits in *n*; these details are left out in [1].

However, a serious problem for the nominal account is that the EA does not behave like a typical noun, either. Already [18] noted that EAs are 'anaphoric islands': the alleged nominal underlying the EA does not license anaphora ((4-a)). This fact is also acknowledged by [1], and they provide further examples that show the failure of EAs to bind reflexives ((4-b)), antecede personal pronouns, or control a relative pronoun (see [1] for examples).

(4) a. *The American$_i$ proposal to the UN reveals its$_i$/her$_i$ rigid position. America$_i$'s proposal to the UN reveals its$_i$/her$_i$ rigid position.
b. The Albanian destruction (*of itself) grieved the expatriot community.

[1] argue that the status of EAs as anaphoric islands follows from the fact that the underlying noun is morphologically deficient, which results in it becoming an adjective in the course of the derivation. They stipulate that the resulting 'adjective' is deprived of typical nominal anaphoric properties, and that anaphoric rules are sensitive to surface structure configurations only, even though for argument-saturation purposes the nominal nature of EAs is still visible.

We consider this an inelegant solution at best. The facts in (3) do not force the abandonment of an adjectival analysis of EAs in favor of a nominal one, given that many indisputable adjectives (e.g. *alleged*, *main*, *other*) display similar properties (see also [11]); and since there is ample evidence that non-quantificational saturators of e-type arguments quite systematically license discourse anaphora to token individuals, [1]'s analysis effectively renders EAs an anomaly. There are therefore good reasons to explore alternative analyses.

2.2 The Modifier Analysis

Building on [17] and others, [3] posit that nouns denote descriptions of kinds of individuals. EAs modify these descriptions, introducing a contextually-valued relation (R in (5-b)) between the kind described by the noun and the country, ethnicity, etc.[4] associated with the EA ($\mathbf{R}$ is [8]'s realization relation). Number turns the resulting property of kinds into a property of token entities ((5-d,e)).

(5) a. *agreement*: $\lambda x_k[\mathbf{agreement}(x_k)]$
b. *French*: $\lambda P_k \lambda x_k[P_k(x_k) \wedge R(x_k, \mathbf{France})]$
c. $[_{NP}$ *French agreement*$]$: $\lambda x_k[\mathbf{agreement}(x_k) \wedge R(x_k, \mathbf{France})]$
d. Num^0: $\lambda P_k \lambda y_o \exists x_k[P_k(x_k) \wedge \mathbf{R}(y_o, x_k)]$
e. $[_{NumP}$ *French agreement*$]$: $\lambda y_o \exists x_k[\mathbf{agreement}(x_k) \wedge R(x_k, \mathbf{France}) \wedge \mathbf{R}(y_o, x_k)]$

The relation R is the source of the argument-saturating effect, as nothing in principle prevents it from corresponding to a thematic relation when the modified nominal describes a kind of eventuality. Since EAs do not denote entities, the

[4] For brevity, hereafter we refer to this entity simply as the *country*, since our study involved only names of countries.

analysis predicts their general failure to introduce discourse referents, and thus their failure to license discourse anaphora.

Essential to understanding the relative distribution of EAs vs. PPs is the observation that the value of R is restricted, as seen in the strong preference of a PP to describe a visit to Canada out of context, as in (6-a), and in the oddness of an EA in the same context ((6-b)).

(6) a. Yeltsin met the prospective Democratic presidential candidate Bill Clinton on June 18. His itinerary also included **an official visit to Canada**. *(BNC)*
b. ... His itinerary also included ??**an official Canadian visit**.

To account for this and similar restrictions on the classificatory use of EAs, [3] argue that in the default case, R is the **Origin** relation defined in (7).

(7) **Origin**(x, y) iff x comes into existence within the spatial domain of y.

Crucially, they posit that not only kinds of concrete objects (e.g. *French bread*) but also kinds of eventualities participate in this relation, by suggesting that the agent(-like) participant in a kind of eventuality can be considered its origin. **Origin** is a *default* value for R because other interpretations are possible when prior discourse makes it clear what that specific relation is between the country and the referent of the head noun (see [3] for examples). Conceiving of the agent relation as a subcase of a default **Origin** value for R allows [3] to provide a unified account of both the basic semantics for the thematic and classificatory uses of EAs and the restrictions on the specific interpretations under these uses.

In our empirical study, we looked at three contrasting predictions of the two analyses that could be translated into features that could be automatically extracted from a corpus containing only morphosyntactic annotation.

2.3 Predictions

Prediction 1. The argument-saturating analysis predicts the distribution of EAs vs. PPs to be roughly the same with event nominals, all other things being equal, given that both are treated as nominals that saturate an argument of the noun they modify. In contrast, the modifier analysis predicts event nominals to combine less often with EAs because when the event nominal has argument structure that must be saturated (i.e. is complex in the sense of [12]), the EA will not be able to do the job, and thus a PP will be required.

Prediction 2. Since, in nondefault cases, the modifier analysis relies on context to supply the identity of the relation between the referent of the head nominal and the country, this analysis predicts the distribution of EAs to be more restricted than that of PPs, as the latter make the relation explicit via the preposition, not relying on context in the same way (recall (6)). Specifically, EAs should occur only when the relevant relation (R in (5-b)) is default or entailed by prior context. In contrast, the argument-saturating analysis predicts, all things being equal, no sensitivity to context in the distribution of EAs vs. PPs.

Prediction 3. A third difference in prediction is specific to analyses such as [3]'s that treat the EA as a modifier of kind descriptions as opposed to token descriptions. A modifier of kind descriptions produces a description of subkinds of the modified kind description. For example, *French bread* and *Italian bread* describe subkinds of bread. We assume that there must be nontrivial criteria that motivate the use of a subkind: a certain number of recognizable instances, an "act of baptism", a recognizable property that characterizes the subkind in contrast to other subkinds, etc. We therefore expect the use of EAs to be concentrated in a comparatively smaller number of nouns than the PPs, i.e. those that meet these criteria, rather than being thinly and evenly distributed across all nouns, as would be expected if the EA could denote a modifier of any token individual description. Conversely, though we see no reason in principle for PPs whose DP complements denote token individuals to be prohibited from serving as a modifier of kind descriptions, we also see no reason for them not to be used as complements or modifiers to descriptions of token entities. The argument-saturating analysis, again all things being equal, does not predict this asymmetric distribution, as it does not provide any basis to distinguish EAs and PPs in terms of the sorts of descriptions they can combine with.

In a previous study on the British National Corpus that did not employ a statistical model [5], we found that this prediction was in fact borne out and that the effect was even more pronounced with EAs with a low frequency and with event nominals. From this we concluded that use of the EA positively correlates with concept stability, i.e. the degree to which the full noun phrase describes a well-established subkind of (abstract or concrete) entities. We posited 1) that stable subkinds describable with EAs are unlikely to be formed for events (e.g. we do not classify agreements according to who makes them) and 2) that fewer stable concepts are formed for those countries we talk about less. However, as will be discussed in Section 4, with the statistical model we obtain different results, so this is a parameter that needs further exploration.

These different predictions grounded our decisions about which features in the corpus to include in the statistical analysis. For Prediction 1, the feature was whether the noun was an event nominal or not. For Prediction 2, since our corpus lacks any semantic annotation, we approximated prior contextual entailment of the value for R with features that correlate with prior mention of the relation: the definiteness of the DP containing the target EA/PP, and prior mention of the EA, the head noun, and the corresponding country noun. We also could not test Prediction 3 directly because it is sensitive to the number of *types* of lemmata, and our model operates on the *token* level. To approximate type frequency, we chose the frequency of the head noun as a factor, on the hypothesis that well-established concept descriptions will tend to be formed with nouns of a higher frequency, and thus that EAs will occur more often with these nouns than with low frequency nouns, once possible collocational relations between the EA and head noun as well as effects due to variation in the overall token distribution of the EA vs. the corresponding country noun are controlled for. We will see, however, that this hypothesis turned out to be incorrect.

3 Method

We conducted a study on the British National Corpus (BNC)[5] in order to determine the factors influencing the choice between an EA and a PP, for a sample of 44 different countries whose adjective (e.g., *French*) and proper noun (*France*) forms occur between 1,000 and 30,000 times in the BNC.[6] We tested the predictions outlined in the previous section by defining *features* that could be automatically extracted from a corpus (by running computer programs on the information contained in the BNC) and tested as *factors* in a statistical model. For instance, to test for definiteness of the NP containing the EA or PP (for Prediction 2), we searched for the words *the*, *this*, *that*, *these*, and *those*, followed by at most 4 (for EAs) or 5 (for PPs) arbitrary words excluding verbs, nouns, prepositions, pronouns, determiners, subordinating conjunctions, and punctuation (but optionally allowing for a comma after each adjective) preceding the relevant EA or PP. This type of approach is *noisy*, that is, the information thus gathered is just an approximation of the real syntactico-semantic information that we want to model. However, it has the advantadge that it can be applied on a large scale and that it can be refined and extended with very little effort, so that very different types of information can be explicitly coded and tested.

Our model contains information for the 74,094 occurrences of the relevant adjectives and prepositional phrases (*target expressions* from now on) found in the BNC. These data were analyzed with a logistic regression model [13], which predicts the probability of an adjective realization based on the specified factors and their interactions. Logistic regression has recently become a popular approach for the analysis of similar binary choice problems in quantitative linguistic studies, such as the English dative alternation [7].[7] For model fitting and analysis, we use the R package *rms* [14].

Our best model used 9 factors, which are listed below and grouped according to the theoretical predictions they are connected to. The factor labels shown here will also be used in the presentation and discussion of the results in Section 4.

Prediction 1: **tcol**: the semantic sort of the head noun, according to the Top Concept Ontology resource (TCO, [2]). This resource restructures the noun hierarchy in WordNet 1.6[8] into a coarse-grained ontology. We only use the highest level of the TCO concept hierarchy, which divides the nominal domain into, roughly, object, event, and abstract nouns.

Prediction 2: **definite**: the definiteness of the NP containing the target expression, defined as explained above; **recent-mod**: distance, in number of words, to the last mention of the target expression in the same discourse,

---

[5] `http://www.natcorp.ox.ac.uk`

[6] The prepositions considered in this study (all those occurring at least 100 times with the tested countries) are the following, ordered by decreasing frequency: *in*, *of*, *to*, *with*, *from*, *against*, *for*, *between*, *on*, *by*, *into*, *over*, *like*, *about*.

[7] See [13] for a detailed introduction to logistic regression, or [4, Sec. 6.3.1] for practical examples of its application in linguistics.

[8] A lexical semantic resource for English, see `http://wordnet.princeton.edu/`.

rescaled as a "decaying activation" 10/(9 + *distance*), or 0 if no previous mention is found; **last-mod-equiv**: whether the previous mention is of the same form (adjective or noun) as the target expression; **recent-head**: same as *recent-mod*, but for the head noun and rescaled as 1/*distance*.[9]

Prediction 3: **collocAN**: the collocational strength between head noun and EA, measured by a conservative estimate of pointwise mutual information[10]; **nhead**: total frequency of the head noun (log-transformed); **ntotal**: total frequency of the target expression (both adjective and noun form, log-transformed); **log-odds-ea-country**: ratio between the frequencies of adjective and noun forms of the target expression in the corpus (log-transformed).

4 Results

From a logistic regression analysis, two basic insights can be gained: (i) which factors or combinations of factors ("interactions") play a significant role in the choice between EA and PP; and (ii) for each significant factor, to what extent and in which manner it increases or decreases the likelihood of an adjective realization (the "partial effect" of the factor).

Table 1. Logistic Regression Model: Results of an ANOVA test on the model (Wald Statistics, response: EA). Data distribution: EAs: 51,946 datapoints (70%), nouns (PPs): 22,148 datapoints. See Section 3 for the interpretation of each factor.

Prediction	Factor	Chi-Square	d.f.	P
1	tcol	2507.40	2	<.0001
2	definite	42.19	1	<.0001
	recent-mod	658.74	1	<.0001
	last-mod-equiv	46.39	1	<.0001
	recent-head	20.76	1	<.0001
3	collocAN	14.30	1	0.0002
	nhead	648.24	1	<.0001
	ntotal	8.37	1	0.0038
	log-odds-ea-country	1230.93	1	<.0001
	TOTAL	5598.08	10	<.0001

Following [13, Ch. 10], we test the significance of factors and interactions by analysis of variance based on the asymptotic standard errors of coefficient estimates (so-called Wald statistics). Table 1 shows highly significant effects for all 9 predictive factors included in the model, lending initial support to

[9] The different scaling formulae for *recent-mod* and *recent-head* were found by manual experimentation and resulted in a better fit of the logistic regression model.

[10] This measure compares the observed number of co-occurrences of two words against their expected number of co-occurrences assuming independence; see [15] and [9, p. 86ff.] for details.

the modifier analysis. There is no clear evidence for an interaction between the factors: most interaction terms are not significant or only weakly significant (not shown in the table). Considering the large sample size, we feel that inclusion of such interactions in the model is not justified at this point.

Figure 1 displays a graphical representation of the partial effect of each factor on the likelihood of an adjective realization. In the log odds scale, a value of 0 corresponds to equal likelihood of adjective and noun; positive values indicate that an EA is more probable than a PP. The baseline adjective likelihood of 70% corresponds to a log odds slightly below 1. For example, the middle left panel shows that speakers are more likely to use an adjective with low-frequency head nouns. The same holds if there was a prior mention of the target expression in the same discourse (top center panel). Prior mention of the head noun has an opposite effect (top left panel), but the shaded confidence band around the line indicates considerable uncertainty.

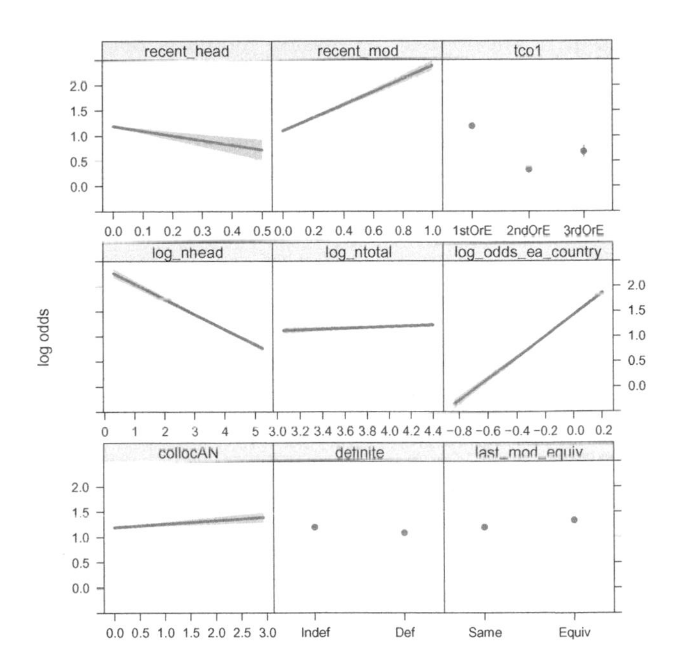

Fig. 1. Logistic Regression Model: graphical representation of the effect of each factor

Prediction 1 is borne out, as event-denoting nouns strongly disprefer EAs: In the top right panel of Figure 1, adjective likelihood is considerably lower for 2nd Order Entities (the TCO equivalent of event-related nouns) than for 1st Order Entities (object-denoting nouns in the TCO). This is also the most significant effect in Table 1. The results also support Prediction 2, as (a) NPs with definite determiners (bottom center panel in Figure 1) have a slight preference for PPs, (b) a recent mention of the target expression (modifier) favors the use of an EA (top center panel), and (c) a prior mention of the head noun is associated with a PP rather than EA realization (top left panel).[11]

However, the results do not support Prediction 3, as (a) the collocational strength between the EA and head noun has only a very small effect (bottom left panel), (b) frequent head nouns typically prefer the PP realization rather than the EA (contrary to what we expect based on a modifier analysis; middle left panel), and (c) the overall frequency of the target expression has virtually no effect (whereas we expected more familiar countries to combine more readily with EAs than infrequent countries; middle center panel). However, manual inspection of the corpus data revealed that many of the apparent counterexamples to Prediction 3 involve descriptions of unique individuals (e.g., *Gulf of Mexico*), which have been predicted to resist EAs [3]. Further examination of these data may thus lead to better results. Finally, lexical effects also seem to play an important role, as the overall corpus ratio of adjective/noun expressions has a very large effect on the EA likelihood (middle right panel).

The goodness-of-fit of the logistic regression model is not quite satisfactory yet, with a Nagelkerke R^2 of 11.5% and c index of 67.5% indicating low discriminative power [13, p. 247].[12] As explained in the previous section, all the data used in this study were automatically obtained, a method that, for semantic analysis in particular, adds noise to the data set insofar as the factors are sometimes imperfectly correlated with meaning. Moreover, we are testing the effect of only 9 factors for more than 74,000 tokens. As the choice between EA and PP is a multi-faceted and intricate problem, we expect that adding more factors will improve the predictive power of the model. In particular, we plan to add factors such as posterior mention of the EA/country noun (as discourse topichood of the country may be a factor) and the presence of other modifiers in the target NP (since syntax allows only one EA per noun phrase).

5 Conclusions

We presented statistical support for an account of ethnic adjectives as modifiers, as opposed to saturators. We tested three predictions of the modifier analysis on automatically extracted data for over 70,000 phrases and showed support for two of them; however, the statistical model is still not predictive enough. We have discussed ways in which the model can be enhanced in the future. Given that automatically extracted information is noisy, another enhancement we are planning is to build a supplementary statistical model on a much smaller, manually annotated set of data points. The annotation will include information that cannot be automatically extracted and that has been found to be relevant in the analysis of the results, e.g., whether the NP denotes a unique entity.

[11] It has to be noted that (a) and (c) are relatively small effects despite their high significance, so they do not provide strong evidence for either analysis.

[12] The c index shows how well the model can discriminate between an example of an EA realization and an example of a PP realization, with a value of 0.5 corresponding to random guesses. It has been shown that c is identical to another popular evaluation measure, the area under ROC curve.

Acknowledgments. This work was supported by grants from the Spanish Ministry of Science and Innovation (FFI2010-15006, TIN2009-14715-C04-04 (KNOW-II), JCI-2010-08581), as well as by the European Union via the EU PASCAL2 Network of Excellence (FP7-ICT-216886) and by the Fundació ICREA.

References

1. Alexiadou, A., Stavrou, M.: Ethnic adjectivs as pseudo-adjectives: A case study in syntax-morphology interaction and the structure of DP. Studia Linguistica 65, 1–30 (2011)
2. Álvez, J., Atserias, J., Carrera, J., Climent, S., Laparra, E., Oliver, A., Rigau, G.: Complete and consistent annotation of WordNet using the Top Concept Ontology. In: Proceedings of 6th Language Resources and Evaluation Conference (LREC 2008), Marrakech, Morocco (2008)
3. Arsenijević, B., Boleda, G., Gehrke, B., McNally, L.: Ethnic adjectives are proper adjectives. In: Proceedings of Chicago Linguistic Society 46 (to appear)
4. Baayen, R.H.: Analyzing Linguistic Data. A Practical Introduction to Statistics Using R. Cambridge University Press, Cambridge (2008)
5. Berndt, D., Boleda, G., Gehrke, B., McNally, L.: Semantic factors in the choice between ethnic adjectives and PP counterparts: Quantitative evidence. Paper presented at QITL-4, Berlin (March 2011)
6. Bosque, I., Picallo, C.: Postnominal adjectives in Spanish DPs. Journal of Linguistics 32, 349–386 (1996)
7. Bresnan, J., Cueni, A., Nikitina, T., Baayen, R.H.: Predicting the dative alternation. In: Bouma, G., Krämer, I., Zwarts, J. (eds.) Cognitive Foundations of Interpretation, pp. 69–94. Royal Netherlands Academy of Science, Amsterdam (2007)
8. Carlson, G.N.: Reference to Kinds in English. Ph.D. thesis, University of Massachusetts at Amherst (1977)
9. Evert, S.: The Statistics of Word Cooccurrences: Word Pairs and Collocations. Dissertation, Institut für maschinelle Sprachverarbeitung, University of Stuttgart (2004); (published in 2005), urn:nbn:de:bsz:93-opus-23714
10. Fábregas, A.: The internal syntactic structure of relational adjectives. Probus 19(1), 135–170 (2007)
11. Gehrke, B., McNally, L.: Frequency adjectives and assertions about event types. In: Proceedings of SALT 19, pp. 180–197. CLC Publications, Ithaca (2011)
12. Grimshaw, J.: Argument Structure. MIT Press, Cambridge (1990)
13. Harrell, F.: Regression Modeling Strategies. Springer, Berlin (2001)
14. Harrell, F.: rms: Regression Modeling Strategies (2011), `http://CRAN.R-project.org/package=rms`, r package version 3.3-2
15. Johnson, M.: Trading recall for precision with confidence sets (2001) (unpublished technical report)
16. Kayne, R.: On certain differences between French and English. Linguistic Inquiry 12, 349–371 (1981)
17. McNally, L., Boleda, G.: Relational adjectives as properties of kinds. In: Bonami, O., Hofherr, P.C. (eds.) Empirical Issues in Syntax and Semantics, vol. 5, pp. 179–196 (2004), `http://www.cssp.cnrs.fr/eiss5`
18. Postal, P.: Anaphoric islands. In: Binnick, R.I. (ed.) Proceedings of the Fifth Regional Meeting of the Chicago Linguistics Society, pp. 205–239. University of Chicago, Department of Linguistics, Chicago (1969)

Licensing Sentence-Internal Readings in English
An Experimental Study

Adrian Brasoveanu and Jakub Dotlačil*

Linguistics, UCSC, 1156 High St., Santa Cruz, CA 95064
{abrsvn,j.dotlacil}@gmail.com

Abstract. Adjectives of comparison (AOCs) like *same*, *different* and *similar* can compare two elements sentence-internally, i.e., without referring to any previously introduced element. This reading is licensed only if a semantically plural NP is present. We argue in this paper that it is incorrect to describe a particular NP as either licensing or not licensing the sentence-internal reading of a specific AOC: licensing is more fine-grained. We use experimental methods to establish which NPs license which AOCs and to what extent and we show how the results can be interpreted against the background of a formal semantic analysis of AOCs. Finally, we argue that using Bayesian methods to analyze this kind of data has an advantage over the more traditional, frequentist approach.

Keywords: adjectives of comparison, Bayesian statistics, distributivity, acceptability judgments, pluralities.

1 The Phenomena

Most, if not all, languages have lexical means to compare two elements and express identity / difference / similarity between them. English uses adjectives of comparison (henceforth AOCs) like *same*, *different* and *similar* for this purpose. Often, the comparison is between an element in the current sentence, e.g., the italicized NP *the same movie* in (1b) below, and a sentence-external element mentioned in the previous discourse, e.g., the underlined NP 'Waltz with Bashir' in (1a). AOCs can also compare sentence-internally, that is, without referring to any previously introduced element, as shown in (2). In this kind of cases, the sentence itself, as it were, provides the context for the comparison, hence the label of **sentence-internal** reading.

(1) a. Arnold saw 'Waltz with Bashir'.
    b. Heloise saw *the same movie* / *a different movie.*

* We would like to thank Lucas Champollion, Irene Heim, John Kruschke and two anonymous reviewers for their extensive comments. The first author was supported by an SRG grant from the UCSC Committee on Research. The second author was supported by a Rubicon grant from the Netherlands Organization for Scientific Research.

M. Aloni et al. (Eds.): Amsterdam Colloquium 2011, LNCS 7218, pp. 122–132, 2012.
© Springer-Verlag Berlin Heidelberg 2012

(2) Each of the students saw <u>*the same movie*</u> / <u>*a different movie*</u>.

The sentence-internal reading is available only if the sentence in which the AOC occurs also contains a semantically (but not necessarily morphologically) plural noun. Importantly, not all semantically plural NPs can license sentence-internal readings of AOCs. This has already been observed in the previous literature (see [1], [2], [3], [4], [5], [7], [10] a.o.). The previous literature also noted that many NPs license sentence-internal readings of only some AOCs (see [3] for a recent detailed discussion and summary of the previous literature).

However, it is much less known that the majority of semantically plural NPs cannot be described as either licensing or not licensing the sentence-internal reading of a specific AOC. Licensing is more fine-grained. The gradient nature of AOC licensing has not been systematically studied, with the exception of [5] for Dutch *different*. In this paper, we report one experiment that begins to address this issue by establishing which NPs license which AOCs in English and to what extent. We also argue that using Bayesian methods to analyze the resulting data has an advantage over the more traditional, frequentist approach. We conclude with a discussion of the consequences of the experimental results for the semantic analysis of AOCs.

2 Experiment

2.1 Method

We used questionnaires to test people's intuitions about sentence-internal readings of three AOCs – *same*, *different* and *similar* – with four licensors – NPs headed by *each*, *all*, *none* and *the* – for a total of $3 \times 4 = 12$ conditions. Each condition was tested four times, twice in a scenario in which the condition was most likely judged as true and twice in a scenario in which the condition was most likely judged as false. There were 32 fillers.

An example of a scenario and three items testing the sentence-internal reading of *similar*, *same* and *different* are given below. In the actual setup, each scenario was followed by five items, two of which were fillers. For each scenario, each of its corresponding test items had a different AOC and a different licensor.

(3) Gustav, Ryan and Bill are three bank managers who share a passion for Volvo, Rolls Royce and Porsche automobiles. Last year, each of them bought a new car. Gustav bought a Volvo PY30, Ryan bought a Volvo XRT2000 and Bill bought a Volvo H4.

a. Each of the bank managers chose a similar car.

b. All the bank managers chose the same car brand.

c. None of the bank managers chose a different car brand.

Each item was judged with respect to (*i*) TRUTH: whether it is true, false or unknown given the accompanying scenario and (*ii*) ACCEPT(ABILITY): how acceptable it is on a 5-point scale (5=completely acceptable, 1=completely unacceptable). TRUTH was measured so that it could be distinguished from ACCEPT.

A total of 42 subjects in two undergraduate classes at UCSC completed the questionnaire for extra-credit. For each subject, we randomized both the order of the scenarios in the questionnaire and the order of the items for each scenario. We excluded two subjects because of their incorrect responses to fillers and one because only TRUTH was completed; one of the remaining 39 subjects filled in only three fourths of the questionnaire. Final number of observations: $N = 1856$.

Barplots of ACCEPT for the 12 conditions are shown in Figure 1, from the least acceptable, i.e., sentence-internal *different* when the licensor NP is headed by *none*, to the most acceptable, i.e., sentence-internal *same* when the licensor NP is headed by *all*.

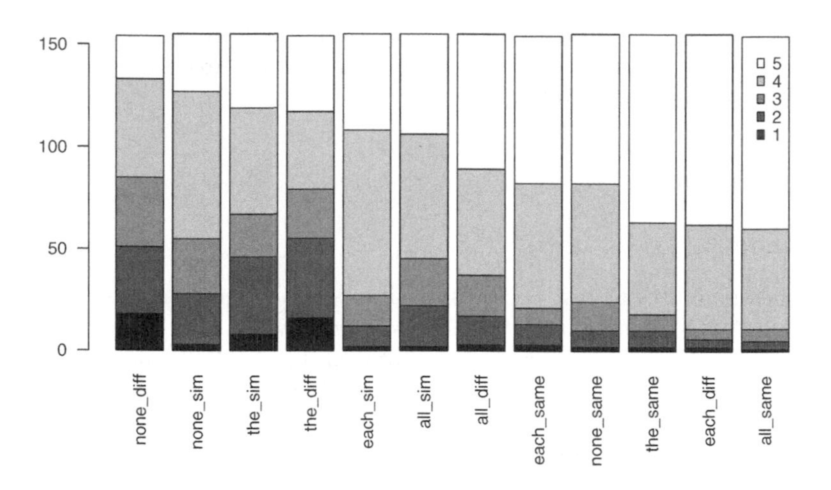

Fig. 1. Barplot of responses by quant-AOC combination

2.2 Statistical Modeling and Resulting Generalizations

The response variable ACCEPT is ordinal, so we use ordered probit regression models to analyze the data. These models are similar to linear regression models in that the predictors are linearly combined and the weights / coefficients for each predictor are estimated from the data. The linear combination of predictors provides the mean for a normal distribution with a fixed variance, set to 1^2 for simplicity. That is, the linear combination of predictors provides an 'offset' for the mean 0 of the standard normal distribution. The area under the probability density function (pdf) obtained in this way is partitioned into five regions by four thresholds, which are also estimated from the data. Each of the five regions corresponds to one value of the ordinal variable.

We have 2 fixed effects: (*i*) QUANT-AOC—factor with 12 levels since we have 12 licensor-AOC combinations, reference level: the *each-different* combination; (*ii*) TRUTH—factor with 3 levels T(rue), F(alse), U(nknown), reference level: T. Our main interest is in how QUANT-AOC affects ACCEPT, while controlling for / factoring out the influence of TRUTH on ACCEPT.

A frequentist analysis shows that adding either of the fixed effects to the null (intercept-only) model significantly decreases deviance, but the interaction of the fixed effects does not ($p = 0.31$). That is, licensor-AOC combinations and truth-value judgments significantly and additively influence acceptability judgments. Adding intercept random effects for items accounts for practically no variance, but adding random effects for subjects does (*std.dev* = 0.56). Thus, the final regression model $\mathcal{M}$ we henceforth focus on has 2 fixed effects, QUANT-AOC and TRUTH (no interaction), and intercept random effects for subjects.

Our primary interest is to establish which NPs license sentence-internal readings of which AOCs and to what extent. That is, we are interested in a wide range of pairwise comparisons between various licensor-AOC combinations. But doing this in the null-hypothesis significance testing framework would require an unfeasibly large amount of data to achieve significance given the necessary α-level correction for running all pairwise comparisons between the 12 licensor-AOC combinations (66 comparisons in total).

In contrast, any number of pairwise comparisons can be carried out in a Bayesian framework because we do not use p-values as a criterion for decision making. Instead, we simply study the multivariate posterior distribution of the parameters obtained given our prior beliefs, the data and our mixed-effects order probit regression model $\mathcal{M}$. Pairwise comparisons of various licensor-AOC combinations are just different perspectives on, i.e., different ways of marginalizing over, this posterior distribution (see [8], [9] and references therein for more discussion). To determine whether there is a credible difference between any two conditions, we check whether 0 (=no difference) is in the 95% highest posterior density interval (HDI; basically, a 95% confidence interval) of the difference: if 0 is outside the HDI, the two conditions are credibly different.

The Bayesian model we estimate has the following structure: (*i*) we assume low-information / vague priors for the non-reference levels of QUANT-AOC and TRUTH—independent normal distributions with mean 0 and variance 10^2; (*ii*) the subject random effects are assumed to come from a normal distribution with mean 0 and variance σ^2, with σ taken from a uniform distribution $Unif(0, 10)$. The function linking the linearly combined predictors and the response ordinal value is the standard normal cumulative distribution function (cdf) Φ. The support of the cdf Φ is partitioned into five intervals (since the acceptability scale was 1–5) by 4 cutoff points / thresholds. The low-information priors for the thresholds are also independent normal distributions with mean 0 and variance 10^2. We estimate the posterior distributions of the predictors QUANT-AOC and TRUTH, the standard deviation σ of the subject random effects and the 4 thresholds by sampling from them using Markov Chain Monte Carlo techniques (3 chains, 125,000 iterations per chain, we discard the first 25,000 iterations and record only every 50th one).[1]

The posterior histograms for the most relevant comparisons are shown in Figures 2–4 below, grouped by AOC. The resulting generalizations are summarized at the top of each set of plots, where > means the licensor(s) on the left is / are preferred to the licensor(s) on the right. Figure 2 shows that *each* is a better licensor of sentence-internal *different* than *all*, which in turn is better than definite plurals and negative quantifiers. But we cannot confidently distinguish between definite plurals and negative quantifiers since the HDI of the difference between them includes 0. The corresponding generalizations for *same* and *similar* are provided in Figures 3 and 4.

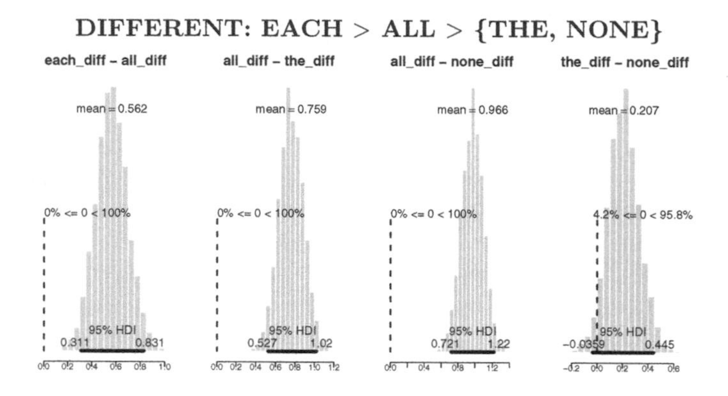

Fig. 2. Differences in acceptability between licensors of *different*

[1] Although there is no need for α-level corrections in a Bayesian framework because the posterior distribution does not depend on how many comparisons we intend to run (or any other intentions of the experimenter), we run the risk of false alarms due to sampling variability: accidental features of the collected sample can lead to spurious results in any framework for inductive inference. One way to mitigate such false alarms is to model QUANT-AOC and TRUTH as random effects, following [6]. This shrinks the estimates of distinct QUANT-AOC combinations towards the grand mean, thereby mitigating the risk of mistakenly identifying differences between any two them. We estimated the parameters of such a model and assumed two independent normal distributions with means 0 and variances τ_1^2 and τ_2^2 for the random QUANT-AOC and TRUTH effects. The hyperpriors for the standard deviations τ_1 and τ_2 were two independent folded t-distributions with means 0, variances 10^2 and 2 df. The estimates from such a model exhibited only very slight shrinkage and all the comparisons of interest remained 'significant', so for reasons of simplicity we will continue to discuss the simpler model in the main text. We are indebted to John Kruschke (p.c.) for very helpful comments about this point.

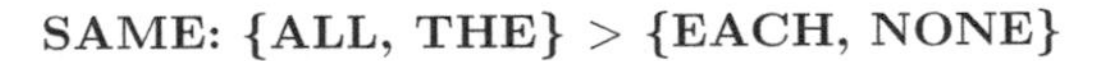

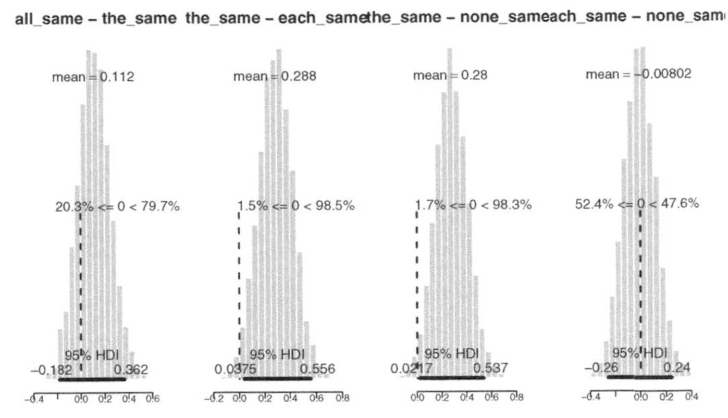

Fig. 3. Differences in acceptability between licensors of *same*

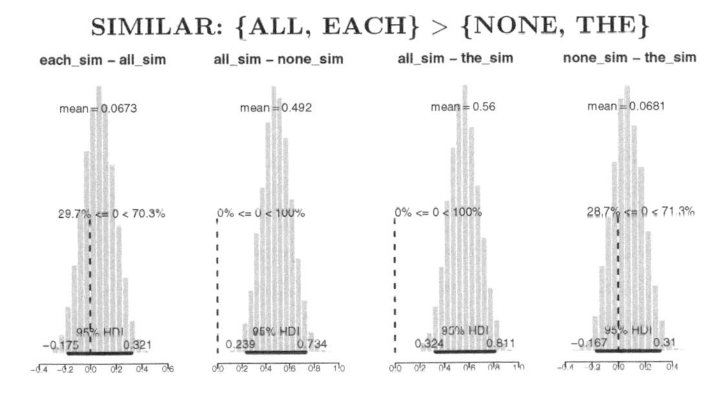

Fig. 4. Differences in acceptability between licensors of *similar*

Finally, Figure 5 below shows the posterior distributions of the 2 non-reference levels of TRUTH and the 4 thresholds. False sentences (F) and sentences whose truth values are unknown (U) because of their grammatically unclear status are less acceptable than true sentences. The rightmost plot shows the mean posterior thresholds plotted together with the standard normal pdf. This is the plot for the reference levels of QUANT-AOC and TRUTH, i.e., for true sentences exemplifying the *each-different* combination. The 4th (rightmost) threshold, for example, is the cutoff point between values 4 and 5 of the ACCEPT response variable; 5 has the highest probability of occurrence, i.e., the largest area under the pdf. Other QUANT-AOC combinations will offset the mean of the pdf, i.e., the curve moves to the left for the less acceptable QUANT-AOC combinations (the thresholds always stay put), and values other than 5 will have the highest probability.

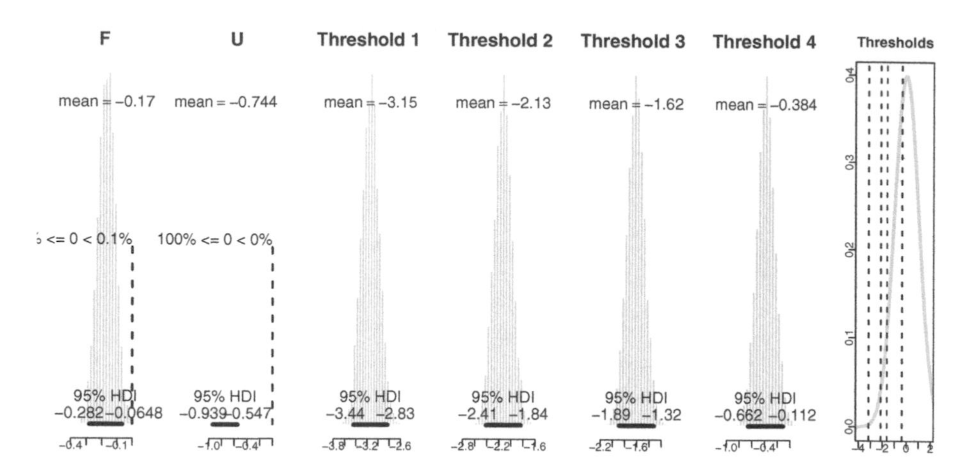

Fig. 5. Posterior distributions of TRUTH and thresholds

3 Consequences for the Semantic Analysis of AOCs

There is a long tradition of connecting the sentence-internal reading of at least some AOCs with distributivity. Here, we follow [3], who analyzes AOCs in a dynamic semantics system that provides semantic values for natural language expressions in terms of sets of variable assignments.

Consider (4) below and the sequence of figures in (5a)–(5c) depicting the sequence of dynamic updates contributed by (4). The update contributed by *each boy* stores all the boys as the value of some variable, u_0 in our example. This is pictorially depicted by the one-column table following the leftmost arrow in (5a). The interpretation of the distributive operator **dist** contributed by *each boy* and the subsequent interpretation of sentence-internal *different* are depicted in (5b). The **dist** operator provides a temporary context inside of which the interpretation proceeds in three steps, namely (*i*) pick two distinct boys, (*ii*) check that each of the two boys recited a poem and (*iii*) check that the two poems are different. In (5b-i), this sequence of steps is depicted for boy_1 and boy_2 and their corresponding poems. But **dist** requires these three steps to be repeated for any pair of boys in the set u_0, as shown in (5b-i) through (5b-v). For more details and the exact logical formulas, see [3].

(4) Each^{u_0} boy recited a^{u_1} different$^2_{u_1}$ poem.

(5) a. $\emptyset \xrightarrow{\text{Each}^{u_0}\text{boy}}$

u_0
boy_1
boy_2
boy_3

$\xrightarrow{\mathbf{dist}_{u_0}(\text{recited a}^{u_1}\text{different}^2_{u_1}\text{poem})}$

b.
i. $\begin{array}{cc} u_0 & u_1 \\ \hline boy_1 & poem_1 \end{array} * \begin{array}{cc} u_0 & u_1 \\ \hline boy_2 & poem_2 \end{array}$ & $poem_1 \neq poem_2$

ii. $\begin{array}{cc} u_0 & u_1 \\ \hline boy_1 & poem_1 \end{array} * \begin{array}{cc} u_0 & u_1 \\ \hline boy_3 & poem_3 \end{array}$ & $poem_1 \neq poem_3$

iii. $\begin{array}{cc} u_0 & u_1 \\ \hline boy_2 & poem_2 \end{array} * \begin{array}{cc} u_0 & u_1 \\ \hline boy_1 & poem_1 \end{array}$ & $poem_2 \neq poem_1$

iv. $\begin{array}{cc} u_0 & u_1 \\ \hline boy_2 & poem_2 \end{array} * \begin{array}{cc} u_0 & u_1 \\ \hline boy_3 & poem_3 \end{array}$ & $poem_2 \neq poem_3$

v. etc.

c. $\xrightarrow{\text{sum all updates}}$ $\begin{array}{cc} u_0 & u_1 \\ \hline boy_1 & poem_1 \\ boy_2 & poem_2 \\ boy_3 & poem_3 \end{array}$ where boy_1 recited $poem_1$, boy_2 recited $poem_2$, boy_3 recited $poem_3$

Thus, in this account, the **dist** operator distributes over *pairs* of individuals and is necessary to license sentence-internal *different*. Besides pairwise distributivity, [3] postulates another operator, **dist-Comp**, which creates a temporary context in which an individual is paired with all entities in the domain of quantification different from that individual. In (4), the **dist-Comp** operator would create contexts comparing, in turn, each boy and all the other boys.

Both the individual-paired-with-individual **dist** operator and the individual-paired-with-complement-set **dist-Comp** operator can capture distributive interpretations (hence their label **dist**), and both of them can account for sentence-internal readings of *different* and *same*. However, sentence-internal readings of *similar* seem to be compatible only with **dist-Comp**: similarity is computed over the entire domain of quantification, which **dist-Comp** provides, and not simply over the individual pairs contributed by **dist**.

Consider the example in (6) below. Suppose there are three managers and two of them bought the same car brand, say, Volvo. The third manager bought a BMW, the color of which is similar to one Volvo and the design of which is similar to the other Volvo. In that case, it is true that for each pair of cars, the paired cars are similar (in some respect)—but (6) is intuitively false. We capture this if sentence-internal *similar* is licensed by **dist-Comp** as opposed to **dist** and requires similarity for the full domain of quantification.

(6) Each manager bought a similar car.

Finally, in addition to being licensed by **dist** or **dist-Comp**, *same* (and plural *different*, which we do not discuss here) has another interpretation that gives rise to sentence-internal readings. If there is no distributivity in the clause, *same* has the option to check that only one entity (possibly plural) was introduced by its NP. Table 1 summarizes which operator can license sentence-internal AOCs.

Table 1. Distributivity and sentence-internal readings in [3]

	dist	dist-Comp	no distributivity
sing. *different*	✓	✓	*
sing. *same*	✓	✓	✓
sing. *similar*	*	✓	*

We are now going to indicate how this analysis, along with other accounts of sentence-internal readings, can account for the data from our experiment.

It has been observed in [5] that the distributive interpretation of predicates like *build a snowman* depends on the type of subject. The availability of this interpretation for different NP types is summarized in (7) below, where > means the NPs on the left are more readily distributive than the NPs on the right.

(7) Distributive interpretation: **EACH** > **ALL** > **THE**

The parallelism between the gradience of distributivity 'strength' associated with these determiners / quantifiers and the gradience of acceptability associated with sentence-internal readings of *different* listed in (8) below provides support for accounts in which sentence-internal *different* requires distributivity to be licensed, e.g., the account in [3] discussed above, as well as [2], [4], [5] and [10]. This is true regardless of the explanation for the gradient nature of distributivity 'strength' (but see [5] for one such explanation).

(8) Different: **EACH** > **ALL** > **{THE, NONE}**

At the same time, the results are problematic for accounts like [1], in which sentence-internal readings are incompatible with distributively interpreted licensors. From the perspective of [1], we would incorrectly expect *all* and *the* to be better licensors than *each*.

Finally, none of the current accounts can explain why negative quantifiers are dispreferred licensors for *different*. These points have already been made in [5] with respect to the Dutch data. This paper extends them to English.

Regarding *same*, we have seen the following ordering for the licensors:

(9) Same: **{ALL, THE}** > **{EACH, NONE}**

This separation into two classes of licensors supports the account of *same* in [1]. Under that analysis, *same* should not give rise to sentence-internal readings with distributive quantifiers, which squares well with the degraded status of **each** and **none**. The remaining question is why **each** and **none** are only slightly degraded, not uninterpretable, as the account in [1] would predict.

One possibility is that *same* is ambiguous, as discussed above and as assumed in [3] and [10]. One of the two meanings for *same* needs to appear in the scope of **dist** to have a sentence-internal reading, while the other meaning is compatible with a non-distributive plural licensor (see Table 1). Given the ordering in (9), the former meaning must be dispreferred / less accessible. Thus, our experiment seems to provide evidence for an ambiguity account of *same*, even though we still need to explain why one meaning of *same* should be preferred over the other. It might be that the meaning harder to evaluate is dispreferred. Consider (10a) below: under the account in [3], **dist** creates temporary contexts storing pairs of non-identical boys and *same* needs to check that within each pair, the recited poems are identical. In contrast, the meaning of *same* in (10b) only needs to check that exactly one poem was introduced in discourse by the direct object. This second meaning of *same* is simpler in that we do not need to repeatedly examine pairs of boys and their corresponding poems, we simply contribute a cardinality requirement on a set of witnesses that is easier to evaluate / verify. The investigation of the hypothesis that processing / evaluation complexity can explain the licensing gradience in (9) is left for future research.

(10) a. Each boy recited the same poem.
b. All the boys/The boys recited the same poem.

Finally, sentence-internal *similar* is associated with the following ordering:

(11) Similar: **{ALL, EACH}** > **{NONE, THE}**

The scale in (11) indicates that *similar* is close to *different*. The only difference between the two is that *similar* does not distinguish between **all** and **each**. These fine-grained contrasts between *similar* and *different* (or *same*) have not been previously noticed, as far as we know. As indicated above, the account in [3] generalizes to *similar* if we stipulate that NPs have another way of introducing distributivity, namely **dist-Comp**. But singular *similar* is overall much less acceptable than singular *same* or *different*: as the barplot in Figure 1 above shows, all 4 conditions with *similar* are among the 6 worst conditions out of the 12 QUANT-AOC combinations. We think that the strong overall infelicity of sentence-internal readings of singular *similar* overwhelms the finer grained distinction in acceptability between *each* and *all* that is observable with the much more acceptable singular *different*.

4 Conclusion

We have discussed experimental evidence showing that licensing sentence-internal readings of AOCs is gradient in nature. We have argued that this gradience supports an analysis of sentence-internal readings that connects them with distributivity. Furthermore, the particular ordering of licensors for *same* vs. *different* vs. *similar* provides evidence for an ambiguity account of *same*, as well as for two different distributivity operators. Some issues, like the particular status of negative quantifiers as licensors of *different* and *similar* or the overall infelicity of singular *similar* when compared to singular *different* or *same*, remain unclear and are left for future research.

References

1. Barker, C.: Parasitic scope. Linguistics and Philosophy 30, 407–444 (2007)
2. Beck, S.: The semantics of Different: Comparison operator and relational adjective. Linguistics and Philosphy 23, 101–139 (2000)
3. Brasoveanu, A.: Sentence-internal Different as quantifier-internal anaphora. Linguistics and Philosophy 34, 93–168 (2011)
4. Carlson, G.: Same and Different: some consequences for syntax and semantics. Linguistics and Philosphy 10, 531–565 (1987)
5. Dotlačil, J.: Anaphora and Distributivity. A study of same, different, reciprocals and others. Ph.D. thesis, Utrecht University, Utrecht (2010)
6. Gelman, A., Hill, J., Yajima, M.: Why we (usually) don't have to worry about multiple comparisons (2009), http://www.stat.columbia.edu/ gelman/research/published/multiple2f.pdf , ms.
7. Heim, I.: Notes on comparatives and related matters. University of Texas, Austin (1985)
8. Kruschke, J.K.: Bayesian data analysis. WIREs Cognitive Science 1, 658–676 (2010)
9. Kruschke, J.K.: Doing Bayesian Data Analysis: A Tutorial with R and BUGS. Academic Press/Elsevier, Oxford (2011)
10. Moltmann, F.: Reciprocals and same/different: Towards a semantic analysis. Linguistics and Philosophy 15(4), 411–462 (1992)

Evaluative Adjectives, Scale Structure, and Ways of Being Polite

Lisa Bylinina[1] and Stas Zadorozhny[2]

[1] Institute for Linguistics OTS, Utrecht
e.g.bylinina@uu.nl
[2] Yandex LLC
zador@yandex-team.ru

1 Introduction

This study is an attempt to add new facts and conclusions to draw from them to an old topic – semantics of evaluative adjectives (Bierwisch 1989). The evidence we will be using is based on a corpus study. The more general goal of this paper is to develop and describe a pipeline that would help solve quantitative semantic tasks in an easier and more systematic way, this particular study being an example.

The topic we investigate here is the semantics of evaluative adjectives (EAs) – both positive (*charming*, *industrious* etc.) and negative (*lazy*, *ugly* etc.). (Bierwisch 1989) describes EAs as a subclass of gradable adjectives along with 'dimensional adjectives' (DAs) like *tall*, *short*, *big*, *small* etc.

There are several differences between dimensional and evaluative gradable adjectives that (Bierwisch 1989) discusses (to be discussed in Section 2), the distinction basically boiling down to DAs and EAs referring to their underlying scales in different ways. The questions Biewrwisch addresses include the following: 1) What is the scale structure of EAs? Do they make use of lower-closed scales, or open scales? 2) Why the observed differences between DAs and EAs?

We will concentrate mainly on the first question, and add one more question that was not formulated by Bierwisch: are there any asymmetries within the class of EAs, more specifically, between negative and positive EAs, and if yes, how should they be accounted for?

With the support of the results of a quantitative study, we conclude that 1) there is no difference in scale structure between negative and positive EAs; all EAs make use of lower closed scales (in accordance with what Bierwisch suggests); 2) there is still significant asymmetry between positive and negative EAs which cannot be attributed to scale structure differences; we provide a pragmatic analysis for this asymmetry in terms of face-threatening acts and the theory of politeness (Brown and Levinson 1987, Leech 1983, van Rooij 2003, Mills 2003) a.m.o.

Section 2 sets the theoretic stage for the study, section 3 describes the corpus study. The results will be made sense of in section 4, and section 5 concludes.

M. Aloni et al. (Eds.): Amsterdam Colloquium 2011, LNCS 7218, pp. 133–142, 2012.
© Springer-Verlag Berlin Heidelberg 2012

2 Evaluative Gradable Adjectives and Scale Structure

2.1 Gradable Adjectives and Scale Structure

We will use the mainstream semantics for gradable adjectives that treats them as measure functions of type $\langle e, d \rangle$ from the domain of individuals to degrees (Bartsch and Vennemann 1973, Kennedy 1999, 2007): $||$ tall $||$ $= \lambda x.\mathbf{tall}(x)$, where $\mathbf{tall}(x)$ is x's height. Measure functions are converted into properties of individuals by degree morphology (comparative morphemes, intensifiers etc.). For the unmarked positive form (as in *John is tall*) a null POS morpheme is introduced, where $\mathbf{d}_s$ is a contextually appropriate standard of comparison (Kennedy 2007):

(1) $||$ POS tall $||$ $= \lambda c \in D_{\langle e,t \rangle} \lambda x.\mathbf{tall}(x)$ $!\succ$ norm $(\mathbf{tall})(c)$
c = comparison class, $!\succ$ = significantly exceed

Of particular importance is the 'significance' component that is sometimes introduced into the semantics of a positive form of adjectives like *tall* (Kennedy 2007). The basic motivation for it comes from the Sorites paradox that the positive form of *tall* is subject to: if the standard is just something like an average, we wouldn't expect the positive form to create instances of the Sorites paradox, since the average represents a crisp cut-off point. Importantly for us, this is only applicable to a subclass of gradable adjectives, namely, relative gradable adjectives. There is a substantial literature on the distinction between relative adjectives and absolute adjectives such as *closed* and the fact that absolute adjectives yield crisp judgments in the positive form, while relative ones do not (Rotstein and Winter 2004, Kennedy and McNally 2005, Kennedy 2007, Sassoon and Toledo 2011, McNally 2011). Arguably, this difference manifests itself linguistically in degree modifier distribution (Rotstein and Winter 2004, Kennedy and McNally 2005):

(2) a. ??perfectly/??slightly {tall, deep, expensive, likely}
b. ??perfectly/??slightly {short, shallow, inexpensive, unlikely}

(3) a. ??perfectly/slightly {bent, bumpy, dirty, worried}
b. perfectly/??slightly {straight, flat, clean, unworried}

(4) a. perfectly/??slightly {certain, safe, pure, accurate}
b. ??perfectly/slightly {uncertain, dangerous, impure, inaccurate}

(5) a. perfectly/slightly {full, open, opaque}
b. perfectly/slightly {empty, closed, transparent}

The patterns of degree modification have been given an explanation in terms of the differences in the structure of scales encoded by gradable adjectives (Kennedy and McNally 2005, Kennedy 2007):

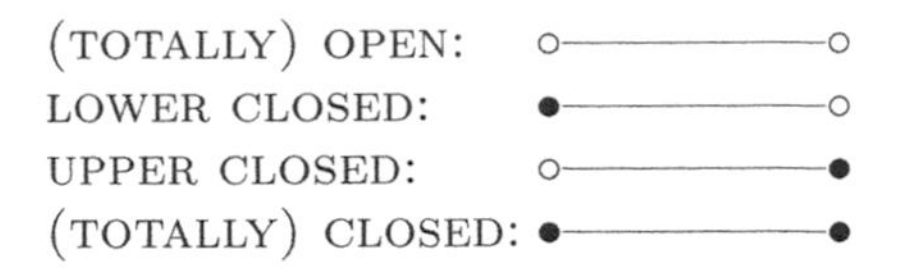

Keeping the single semantic type $\langle e, d\rangle$ for all gradable adjectives, Kennedy (2007) derives the observable differences between absolute and relative ones from a combination of conventional properties of the adjective (scale structure) and a more general principle of interpretation that would ensure that the conventionally determined standard (maximal or minimal) will be preferred over a contextually determined standard when an adjective's scale is closed (Kennedy 2007, 36).

Thus, under this view, absolute adjectives are the ones that make use of a scale that is closed at least at one end, and relative adjectives have totally open ones. We will use degree modifiers as tests for scale structure, though there are complications to this picture (McNally 2011, Solt 2011).

2.2 Evaluative Adjectives

The distinction between evaluative (EAs) and dimensional adjectives (DAs) in (Bierwisch 1989) is motivated by two observations: 1) the antonymous pairs of EAs have a less obvious relation to each other than DA antonymous pairs; sometimes it is hard to tell whether the two given EA items form a pair or not; 2) inference judgements for EAs vary a great deal inter- and intra-individually.

The intuition behind observation 1 is the following: '*Hans ist klein* 'Hans is short' assigns to Hans a certain degree of height, while *Hans ist faul* (Hans is lazy) does not mean that Hans has a certain degree of industriousness. Put somewhat differently, even a negative DA always specifies a positive value on the scale of its antonym, whereas this does not apply to a negative EA: even a short person has height, but a lazy person cannot be to any extent industrious' (Bierwisch 1989, 88).

Thus it is not obvious that every EA has a single antonym, at least in the sense that DAs do:

(6) DAs:
  a. tall ↔ short
  b. heavy ↔ light

(7) EAs:
  a. brave, bold, courageous ↔ cowardly, timid, fearful
  b. clever, bright, shrewd, intelligent, brilliant ↔ stupid, idiotic, foolish

The second observation concerns the following kinds of inference patterns and the fact that speakers do not have a consensus on whether these inferences hold:

(8) a. How hard-working/lazy is Hans?
→ Hans is hard-working/lazy.
b. Hans is as hard-working/lazy as Eva
→ Hans is hard-working/lazy.
c. Hans is more hard-working/lazier than Eva
→ Hans is hard-working/lazy.

This motivates a conclusion made in (Bierwisch 1989) that EAs are underlyingly not gradable at all, though they **can** used as gradable with the help of a type shift or coercion that results in a lower-closed scale. Our main goal is not to evaluate the idea of coercion from non-gradable to gradable per se, but to check whether there is evidence for the scale structure that Bierwisch assigns to these derived gradable predicates. Importantly, positive and negative EAs are treated the same and are ultimately assigned the same scale structure under this account, too.

However, contrasts between positive and negative EAs have been reported that look like differences in scale structure, as they involve low degree modifiers:

(9) a. Clyde is slightly stupid.
b. Clyde is slightly lazy.

(10) a. ?Clyde is slightly smart.
b. ?Clyde is slightly industrious.

There is a clear need for a more wide-ranged empirical study, to which we now turn.

3 Corpus Study

To get clearer facts about the distribution of degree modifiers with EAs, we conducted a corpus study. We did not use any of the existing corpora, but started from scratch by collecting our own corpus. The motivation for collecting a new corpus rather than using an existing one is twofold: first, there are practical issues of (free) accessibility of raw text data of existing corpora; second, and more importantly, we conducted this study as an attempt to set up a pipeline for quantitative semantic research, so that various parameters could be controlled for at all stages of the process, including corpus collection.

3.1 Corpus Collection and Processing

To collect the corpus, we used BootCaT toolkit (Baroni and Bernardini 2004) – software allowing to bootstrap corpora from web with a small initial set of seed words as its input. BootCat uses the seeds to query the specified search engines (Google, Yahoo) and to download urls from search results page to build initial corpora. These initial corpora are then used for expansion of seed words. The procedure repeats iteratively until the desired number of documents is reached.

Manipulating initial seed words is a way to control various parameters of the corpus, such as the language, genre or topic of downloaded documents. This is an extremely useful feature, though we did not make the best of it in the current study. In our work we used the general list of most frequent English words to obtain general corpora without bias towards specific genre or topic.

All the documents were removed from the boilerplate (html design elements, menus, advertising etc.). Then all the texts were split into sentences with the help of a simple heuristics – we took the end of a sentence to be marked with an exclamation mark, a question mark, or a dot followed by space and then a word beginning with an upper-case letter. The number of sentences we got as an output of this procedure is 9156630.

The resulting chunks were lower-cased, clitics were replaced with their full forms, and punctuation marks were removed. As a result, we obtained a corpus that we used in our research. Then we retrieved the bigrams (all pairs of consecutive word tokens), 631597548 total, and collected basic statistics for them – bigram frequency and marginal frequencies for both tokens in each bigram (the token frequency among all pairs).

We were interested in particular events – namely, in the co-occurrence of certain degree modifiers with certain adjectives. The modifiers we study are low degree modifiers *a bit*, *a little bit*, *slightly* and *somewhat* (=LOW). The adjectives we are interested in are, first of all, positive EAs (=POS) and negative EAs (=NEG). We also used data from relative adjectives (=REL) and lower-bound adjectives (=MIN). We drew the lists of adjectives from various sources – theoretic papers discussing gradability and scale structure (Kennedy and McNally 2005, Kennedy 1999, 2007, Rotstein and Winter 2004, Sassoon and Toledo 2011, McNally 2011) as well as quantitative studies (Solt 2011, Sassoon 2011), and also various descriptive grammars of English that introduce semantic divisions within adjectives.[1] The total number of EAs that we used is 246 (112 negative and 134 positive). Our main concern when collecting these lists was completeness that would help reduce the impact of mistakes and uncertainties in classification, as there is no conventional list of EAs discussed in the literature. The lists of relative adjectives and lower-bound adjectives, on the other hand, are much shorter (20 relative adjectives and 19 lower-bound adjectives), but these lists are rather conventional and there isn't much disagreement as to whether these adjectives belong to this or the other class. Crucially for us, the results of the test we are going to use will not depend on the number of the adjectives.

Thus, only the following kinds of events were collected:

$$\langle x, y \mid x = \text{LOW}, y = \text{POS} \rangle \tag{1}$$

$$\langle x, y \mid x = \text{LOW}, y = \text{NEG} \rangle \tag{2}$$

$$\langle x, y \mid x = \text{LOW}, y = \text{REL} \rangle \tag{3}$$

$$\langle x, y \mid x = \text{LOW}, y = \text{MIN} \rangle \tag{4}$$

[1] We thank Stephanie Solt, Galit Sassoon and Chris Kennedy for sharing their lists and thoughts on possible ways to extend the lists we are using.

For each of the events, the following frequencies were calculated: $f(x, y)$, $f(x, *)$, $f(*, y)$, $f(*, *)$, where $*$ stands for any word.

We also filtered out the events with $f(x, y)$ below 3 due to unreliability of probability estimates for low-frequency events. 3 is suggested as 'a frequency threshold, which should at least exclude the hapax and dis legomena ($f \geq 3$)' (Evert 2005, 120).

3.2 The Tests

First of all, we need to establish a measure of relatedness between a modifier and an adjective and then compare the resulting distributions for different classes of adjectives. We use pointwise mutual information (PMI) as a measure of association between two random variables (Fano 1961, Church and Hanks 1990):

$$PMI(x, y) = log \frac{P(x, y)}{P(x)P(y)} \tag{5}$$

In our case, $P(x, y)$ is the probability of a bigram, $P(x)$ is the probability of a modifier preceding any word and $P(y)$ is the probability of an adjective following any word.

Using the collected frequencies, we can calculate $\overline{PMI}$ – a maximum-likelihood estimation (MLE) of PMI (5), where MLE for probailities is $P(x, y) = \frac{f(x,y)}{f(*,*)}$, $P(x) = \frac{f(x,*)}{f(*,*)}, P(y) = \frac{f(*,y)}{f(*,*)}$:

$$\overline{PMI}(x, y) = log \frac{f(x, y)f(*, *)}{f(x, *)f(*, y)} \tag{6}$$

PMI (5) is a popular metric for collocation extraction (Evert 2005, Church and Hanks 1990), but it has a more general meaning – it quanties extra-information about possible occurrence of y when we know that first word is x and vice versa and compares the probability of observing x and y together (the joint probability) with the probabilities of observing x and y independently (by chance). If there is a genuine association between x and y, the joint probability $P(x, y)$ will be significantly higher than chance $P(x)P(y)$. Thus, the high positive values of PMI indicate stronger association or positive correlation between events; if it takes zero value, then x and y are independent; negative values indicate that x and y are in complementary distribution or negatively correlated.

Using $\overline{PMI}$ (6), we calculate PMI_{pos}, PMI_{neg}, PMI_{rel}, PMI_{min} (for each of the events (1)-(4)).

First, in order to find out whether there is indeed an asymmetry between positive and negative EAs with respect to their combination with low degree modifiers, we compare the following pair of samples:

$$PMI_{neg} \text{ vs. } PMI_{pos} \tag{7}$$

Second, we check the assumptions that we will be using about low degree modifiers as a test for absolute vs. relative distinction, more specifically, we will see whether there is a significant difference between lower-bound DAs and relative DAs in their association with low degree modifiers:

$$PMI_{min} \text{ vs. } PMI_{rel} \tag{8}$$

Finally, we check whether there are reasons to group positive or negative EAs with lower-bound or relative DAs based on co-association with low degree modifiers:

$$PMI_{neg} \text{ vs. } PMI_{min} \tag{9}$$

$$PMI_{pos} \text{ vs. } PMI_{rel} \tag{10}$$

$$PMI_{neg} \text{ vs. } PMI_{rel} \tag{11}$$

$$PMI_{min} \text{ vs. } PMI_{pos} \tag{12}$$

For each pair of samples, we formulate the following null hypothesis and an alternative hypothesis:

(11) a. H_0: There is no significant difference in association score between two compared classes of adjectives with respect to a given modifier.
b. H_1: There is a significant difference in association score between two given classes of adjectives with respect to a given modifier.

Based on our expectations from what theories of gradable adjectives predict, for some pairs we also formulate a different alternative hypothesis:

(12) H_1: The association score tends to be higher for one class then for the other.

To asses the differences between the pairs of samples, we performed series of statistical tests. We used Wilcoxon rank-sum test (also known as MannWhitney U test and MannWhitneyWilcoxon test) (MWW) (Hollander and Wolfe 1999), which allows to assess whether the two groups have a tendency to have different values (two-sided test) or whether one of the two groups tends to have greater values then the other group (one-sided test). The choice of this test was motivated by unknown distribution of samples, as MWW is a non-parametric test.

The following table represents results of the MWW test for distributions in with significance level 0.01 and frequency threshold 3:

H_1	accept/decline	p-value	W
$NEG \neq POS$	accept	3.62e-05	1924
$NEG > POS$	accept	1.81e-05	1924
$MIN \neq REL$	accept	1.35e-06	707
$MIN > REL$	accept	6.8e-07	707
$NEG \neq MIN$	decline	0.10503787	1194
$NEG > MIN$	decline	0.05251893	1194
$POS \neq REL$	accept	5.25e-06	879
$POS > REL$	accept	2.62e-06	879
$NEG \neq REL$	accept	0	2863
$NEG > REL$	accept	0	2863
$MIN \neq POS$	decline	0.15485217	425
$MIN > POS$	decline	0.07742609	425

Thus, negative EAs co-occur with low degree modifiers significantly more often than positive EAs; the assumptions about lower-bound adjectives co-occurring with low degree modifiers better than relative adjectives get the empirical support from our study; at the same time, both negative and positive EAs pattern with lower-bound rather than with relative DAs with respect do low degree modifier distribution. We will interpret this result in the next section.

4 Interpreting the Results

The result reported in the previous section is quite surprising, because, on the one hand, it shows that low degree modifier distribution groups together all EAs and lower-bound absolute DAs, quite like Bierwisch predicted; and on the other hand, there still are significant differences between negative and positive EAs that are not predicted or explained by Bierwisch's account.

The analysis we offer is pragmatic in nature. It makes crucial use of theory of politeness and the notion of FACE (public self-image) (Brown and Levinson 1987, Leech 1983, van Rooij 2003, Mills 2003). Politeness defined as expression of speakers' intention to mitigate threats carried by certain face-threatening acts (FTAs) toward another (Mills 2003, 6) describes several ways of such mitigation.

Not committing ourselves to any particular realization of this theory, we assume that the laws of politeness in conversation can be part of Gricean-like picture, and that they can function in a way similar to Gricean maxims (Grice 1975). In particular, we rely on existence of a politeness maxim of the following kind, as proposed in (Leech 1983):

(13) **The Approbation maxim**
Minimize the expression of beliefs which express dispraise of other; [maximize the expression of beliefs which express approval of other.]

The claim that we want to make is the following: assertions that involve a negative EA are FTAs. That is, a sentence like *John is lazy* damages John's face and thus is subject to politeness maxims.

One way to soften the FTA is to understate the degree to which the negative property holds of the individual. Keeping the assertion that John is lazy and at the same time politely lowering the degree of his laziness forces the speaker to refer to the 'best' part of the bad scale, which is picked precisely by low degree modifiers like *slightly*. As far as the hearer is concerned, the reasoning would be the following: interpreting the literal meaning of the speaker's utterance and being aware that this utterance is a case of FTA, the hearer can infer that the speaker's utterance lowered the degree to which the negative property holds of its subject, thus the actual degree must be higher than that (so, for example, *John is slightly lazy* is just a polite version of *John is lazy*).

This analysis is problematic for politeness theory as it stands. In the literature, FTAs are defined as essentially hearer-oriented, and it is the hearer's face that is at stake (Brown and Levinson 1987, Leech 1983). Some extensions and modifications of politeness theory are speaker-oriented (e.g. van Rooij 2003). But our claim makes a move of generalizing the notion of face that can be affected in the course of conversation to any individual that is mentioned in the discourse.

We also propose that famous cases of litote in sentences like *John is not very smart* (meaning John is stupid) or even *John is not very tall* (meaning John is short) (Horn 1989) can be explained along the same lines of generalized FTAs. This would account for the intuition that for the litote to go through for *tall* there has to be a shared assumption between a speaker and a hearer that being short is bad.

5 Conclusion

We explored negative and positive evaluative adjectives like *ugly*, *stupid* and *beautiful*, *smart*, respectively. The only available analysis of evaluative adjectives (Bierwisch 1989) groups them together with absolute dimensional adjectives with lower closed scales, like bent or dirty. The questions that we tried to answer were the following: 1) Is there empirical support for this analysis? 2) Should negative and positive EAs be analysed in the same way, or do they show significant differences that need to be reflected in the analysis?

To answer these questions, we conducted a corpus study that showed puzzling results: on the one hand, Bierwisch's idea of grouping all EAs together with lower-bound DAs was supported by low degree modifier distribution; at the same time, significant differences between positive and negative EAs were observed. We argue that there is an additional factor at play here, namely, the pragmatics of politeness (Brown and Levinson 1987, Leech 1983, van Rooij 2003, Mills 2003). We propose a pragmatic explanation of positive / negative asymmetry – namely, we treat utterances with negative EAs as face-threatening acts, and a general tendency to attenuate this kind of judgement is a face-saving technique that results in a noticeably more natural use of low degree modifiers with negative than with positive EAs. For this analysis to work, we need to extend the notion of FACE used in the theory of politeness to be applicable not only to the addressee and the speaker, but to any individual mentioned in the discourse.

References

Baroni, M., Bernardini, S.: Bootcat: Bootstrapping corpora and terms from the web. In: Proceedings of LREC 2004, pp. 1313–1316 (2004)
Bartsch, R., Vennemann, T.: Semantic Structures: A Study in the Relation between Syntax and Semantics, Frankfurt (1973)
Bierwisch, M.: The semantics of gradation. In: Bierwisch, M., Lang, E. (eds.) Dimensional Adjectives: Grammatical Structure and Conceptual Interpretation, pp. 71–261. Springer, Berlin (1989)
Brown, P., Levinson, S.C.: Politeness: Some universals in language usage. Cambridge University Press, Cambridge (1987)
Church, K.W., Hanks, P.: Word association norms, mutual information, and lexicography. Computational Linguistics 16, 22–29 (1990)
Evert, S.: The statistics of word cooccurrences: word pairs and collocations, PhD thesis. Universität Stuttgart, Stuttgart (2005)
Fano, R.: Transmission of Information: A Statistical Theory of Communications. The MIT Press, Cambridge (1961)
Grice, H.P.: Logic and conversation. Syntax and Semantics 3, 41–58 (1975)
Hollander, M., Wolfe, D.: Nonparametric statistical methods. Wiley series in probability and statistics: Texts and references section. Wiley (1999)
Horn, L.: A natural history of negation. University of Chicago Press, Chicago (1989)
Kennedy, C.: Projecting the adjective: the syntax and semantics of gradability and comparison, PhD thesis, UCSD (1999)
Kennedy, C.: Vagueness and grammar: The semantics of relative and absolute gradable predicates. Linguistics and Philosophy 30, 1–45 (2007)
Kennedy, C., McNally, L.: Scale structure, degree modification, and the semantics of gradable predicates. Language 81(2), 345–381 (2005)
Leech, G.: Principles of pragmatics. Longman, London (1983)
McNally, L.: The Relative Role of Property Type and Scale Structure in Explaining the Behavior of Gradable Adjectives. In: Nouwen, R., van Rooij, R., Sauerland, U., Schmitz, H.-C. (eds.) ViC 2009. LNCS, vol. 6517, pp. 151–168. Springer, Heidelberg (2011)
Mills, S.: Gender and Politeness. Cambridge University Press, Cambridge (2003)
Rotstein, S., Winter, Y.: Total adjectives vs. partial adjectives: Scale structure and higher-order modifiers. Natural Language Semantics 12, 259–288 (2004)
Sassoon, G.: A slightly modified economy principle: Stable properties have non-stable standards. In: Sinn und Bedeutung 16, Utrecht (2011)
Sassoon, G., Toledo, A.: Absolute vs. relative adjectives: Variance within vs. between individuals. In: Proceedings of SALT21 (2011)
Solt, S.: Comparison to fuzzy standards. In: Sinn und Bedeutung 16, Utrecht (2011)
van Rooij, R.: Being polite is a handicap: Towards a game theoretical analysis of polite linguistic behaviour. In: Tennenholtz, M. (ed.) TARK (9) (2003)

Processing: Free Choice at No Cost*

Emmanuel Chemla[1] and Lewis Bott[2]

[1] LSCP, CNRS/EHESS/ENS, Paris
[2] Cardiff University, UK

To appear in the proceedings of the Amsterdam Colloquium 2011

Abstract. We provide processing evidence that free choice inferences are derived without delay. This result is in conflict with the parsimonious view that free choice inferences are a kind of scalar implicatures (which do come at a processing cost in similar circumstances).

Keywords: Free choice permission, scalar implicatures, processing.

1 Free Choice (FC)

A disjunctive sentence such as (1) standardly carries the conjunctive inference that (2)a and (2)b are true.

(1) John is allowed to eat an apple or a banana.

(2) a. John is allowed to eat an apple.
b. John is allowed to eat a banana.

This phenomenon is known as Free Choice (FC) permission (Kamp, 1973). Current formal models tend to treat FC inferences as a *special* type of scalar implicature (mostly building on Kratzer and Shimoyama's 2002 insight, see, e.g., Schulz, 2003; Alonso-Ovalle, 2005, 2006; Klinedinst, 2006; Fox, 2007; Chemla, 2008, 2009; Franke, 2011). We present the first processing study of FC. Our results go against the expectations of recent formal analyses, and show that, unlike scalar implicatures, FC inferences come at no processing cost.

2 Scalar Implicatures

A sentence such as (3) standardly conveys that its sister sentence (4) is false. Here is a derivation: The *alternative* sentence (4) is stronger than (3); (3) is thus not the best sentence to utter, unless (4) is false.

(3) Some elephants are mammals.

(4) All elephants are mammals.

* We would like to thank Philippe Schlenker, Benjamin Spector as well as the organizers and two series of two anonymous reviewers of the Amsterdam Colloquium 2011.

M. Aloni et al. (Eds.): Amsterdam Colloquium 2011, LNCS 7218, pp. 143–149, 2012.
© Springer-Verlag Berlin Heidelberg 2012

Processing

Since Bott and Noveck (2004) (at least), a variety of experimental studies have investigated how this inference is derived in real time. They consistently found that the verification of a sentence is more demanding when its scalar implicature is taken into account (see Grodner, 2009 for discussion). To prove so, these studies relied on situations in which the target sentence would be (a) true without its scalar implicature, but (b) false with its scalar implicature. (3) is such an example: (a) it is true that there are elephants that are mammals, but (b) it is false that some *but not all* elephants are mammals. True/false judgments thus covary with the derivation of the scalar implicature. In a Truth Value Judgment Task, false answers, that correspond to the derivation of the scalar implicature, are found to be slower than true answers ("logical" answers, without the scalar implicature).

3 Free Choice as a Kind of Scalar Implicatures

It has been shown that FC could be explained in a similar fashion. One implementation of this insight relies on the following hypotheses:[1]

(H1) The sentences in (2) are alternatives to (1),
in the same sense that (4) is an alternative to (3);

(H2) Sentences (2)a/b with their scalar implicatures become (5)a/b;

(H3) The enriched versions (5) are the sentences that end up negated when (1) is uttered.

In such a view, FC is analyzed as a second order kind of scalar implicature: the scalar implicatures of the alternatives are first derived, and then the alternatives, *enriched with their own scalar implicatures*, enter in the competition process to give rise to further scalar implicatures. The result is obtained because the conjunction of (1) and of the negations of each of the enriched alternatives in (5) entails FC, i.e. the conjunction of (2)a and (2)b.

(5) a. John is allowed to eat an apple, but not a banana.
b. John is allowed to eat a banana, but not an apple.

Such a view leads to the following prediction:

> *Processing prediction.* According to the view sketched above, FC is a second order scalar implicature. Since first order scalar implicatures come with a visible processing cost, we should be able to detect a similar or higher cost for FC inferences.

[1] These hypotheses do not *follow* from standard mechanisms developed for scalar implicatures. For simplicity, we will not offer a full description of the consequences of making these hypotheses for various conceptions of what scalar implicatures are.

4 Design

We capitalized on the seminal idea that was used to test scalar implicatures: true/false answers were used as an indicator of whether FC inferences were derived. The target sentences were constructed with the help of a cover story in which the destruction of the planet was described as imminent, but that certain people were allowed to save certain types of objects. Specifically, zoologists were allowed to save living creatures, and engineers were allowed to save artificial objects. (One object per person at most, the rule says, to avoid issues about exclusive readings associated to disjunctions). We then tested sentences of the type described in Table 1.

Table 1. Schematic description of the sentences used in the experiment. Sentences were presented in small pieces: the first bit ('Mary-the-engineer') lasted on screen for 750ms, the next four words were presented for 250ms each, and the final, crucial bit remained until participants provided their true/false answers.

Identical, leading part:		Mary-the-engineer is allowed to save ...
Target	(True = no FC / False = FC)	... a monkey or a computer.
Double	(True)	... a TV or a computer.
Double	(False)	... a monkey or a lion.
Single	(True)	... a monkey.
Single	(False)	... a computer.

We presented the first bit of these sentences for 750ms, the next four words for 250ms each, and the last crucial bit remained until participants provided their true/false answers. The key example is the first one: "**Target**". According to the cover story, a FC interpretation would result in a false response (engineers are not allowed to save monkeys) whereas a logical interpretation would result in a true response (engineers are allowed to save computers). We compared response times of false responses (FC interpretations) to true responses (logical interpretations). The other conditions were included to control for various possible response biases, e.g., true responses may be faster than false responses, independently of any of the processes we are interested in.

5 Experiment 1

46 native speakers of English completed a verification task (two were excluded because they failed to answer appropriately to the control conditions). Control sentences were answered very accurately overall ($M = .93, SD = .039$), illustrating that participants understood the task and the cover story. The proportion

of free choice (false) responses to the target sentences was $M = .66, SD = .33$, and there was significantly greater variability in the target sentences than the control sentences. Overall then, the response choice data indicates that multiple interpretations were available for the target sentences.

Response Times

Figure 1 shows the pattern of RTs for the correct responses to all five types of sentences. The **Target** sentences are broken down into FC responses (false) and logical responses (true). For both types of control sentences there is a bias towards true sentences ($t_1s(43) > 3.4, t_2s(19) > 2.0$, all $ps < .05$). For the **Target** sentences, however, FC responses (false) are faster than logical (true) responses. While the simple comparison between FC and logical responses failed to reach significance, a repeated measures ANOVA with sentence type (single, double or target) and response (true or false) as factors revealed a significant interaction between sentence type and response type ($F_1(2, 76) = 3.1, F_2(2, 38) = 6.5, ps < .05$). The difference between the FC and logical interpretations is therefore significantly smaller than the difference between the true and false responses for the different control sentences.

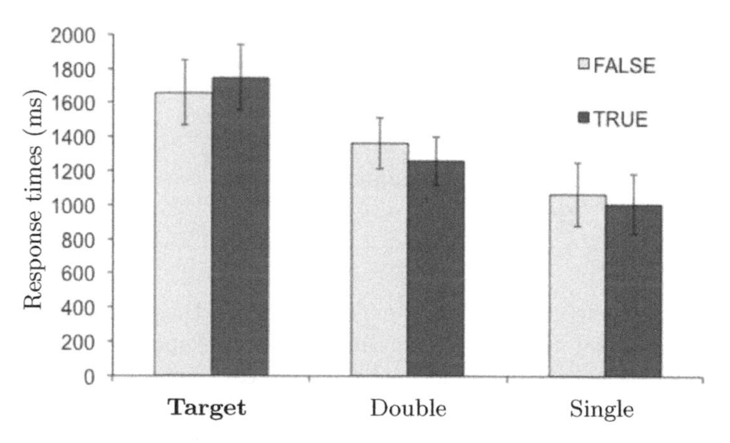

Fig. 1. Response times (ms) for Experiment 1

One explanation for our pattern of results is that FC interpretations are fast, but that a bias against false responding slowed down the FC responses. We therefore conducted an analysis in which the response bias was removed from latencies. For each participant, we computed the difference between the FC interpretations (false) and the false control sentences, and compared these scores against the difference between the logical responses (true) and the true control sentences. Any bias towards fast true responding in the **Target** condition should be removed by subtracting away the relevant control responses. This analysis revealed faster FC responses than logical responses (although only marginally so in the participants analysis, $p = .069$).

6 Experiment 2

One of the difficulties of allowing participants to choose which interpretation they made to the **Target** sentences is that it is not possible to determine which responses are errors and which responses are correct interpretations. This means that the RT analysis contains noise and consequently has low power. Experiment 2 was similar to Experiment 1, except that the 67 participants (7 excluded) were trained on sentences of the **Target** type with corrective feedback in a preliminary phase: half of them were trained to answer true (logical) and the other half to answer false (FC). Accuracy was high and approximately equal across conditions, both for the control sentences and for the **Target** sentences, in both groups ($M > .92, SD < .05$).

Response Times

Figure 2 shows the pattern of RTs. For control sentences, true responses were significantly faster than false responses (t_1s(59) > 4.0, t_2s(19) > 2.8, $ps < .01$). But for the **Target** sentences the reverse was true: FC responses (false) were derived marginally faster than logical (true) responses ($t_1(58) = 1.7, p = .10, t_2(19) = 5.0, p < .0005$). This replicates the pattern observed in Experiment 1. As an extra test of the difference between FC and logical responses we used a mixed model regression analysis that combined participants and items. This analysis demonstrated significantly faster FC responses ($p < .05$).[2] We also controlled for response bias, just as we did in Experiment 1, by removing true control RTs from logical responses to the **Target** sentences, and false control RTs from FC responses. This revealed significantly faster FC responses when using single or double sentences as controls ($ts(58) > 2.5, ps < .05$).

7 Conclusion

We provide data from two studies modeled after the original experiments that detected a cost associated to the derivation of scalar implicatures. Our results show that, contrary to arguments coming from the theoretical literature, free choice inferences are different from scalar implicatures: they come with no processing cost, if not at a *negative* cost. In fact, this pattern of result is closer to the behavior of presuppositions (see Chemla and Bott, 2012).

These results call for a greater differentiation between FC and scalar implicatures. Such a differentiation may be found in approaches that do not see a strong motivation for treating free choice inferences and scalar implicatures alike and propose to account for FC inferences as a truth-conditional component of meaning (see, e.g., Zimmermann, 2000; Geurts, 2005; Aloni, 2007). The current results could thus lead to a reevaluation of the theoretical status of free choice and offer

[2] We were unable to conduct this analysis for Experiment 1 because of the large number of missing cells for many participants, who were free to decide whether they would derive FC on every occasion.

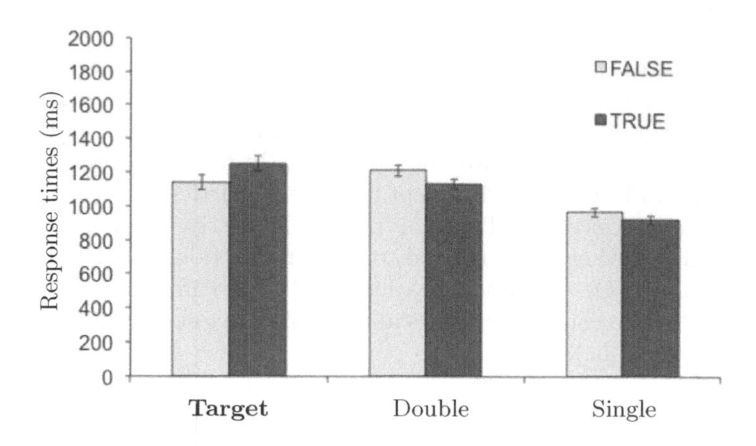

Fig. 2. Response times (ms) for Experiment 2

new arguments in favor of a truth-conditional approach of FC. Note however that experiment 1 showed that FC is a defeasible inference. This may look like a simpler and more classical result, but it also needs to be accounted for.

The present processing results may also be used to make progress in a different direction. They may inform us about the processing cost we observe for scalar implicatures. It could be that free choice really is a kind of scalar implicatures, but superficial differences between the two phenomena may distinguish them from a processing perspective. For instance, notice that the alternatives involved in the two cases are of a different nature for scalar implicatures and for free choice (replacement vs. deletion, see Katzir, 2007). If processing cost for scalar implicature is mostly associated with alternatives (alternative derivation or comparison), then it could save the parsimonious scalar implicature accounts of free choice.

References

Aloni, M.: Free choice, modals, and imperatives. Natural Language Semantics 15(1), 65–94 (2007)

Alonso-Ovalle, L.: Distributing the disjuncts over the modal space. In: Bateman, L., Ussery, C. (eds.) North East Linguistics Society (GLSA), Amherst, MA, vol. 35 (2005)

Alonso-Ovalle, L.: Disjunction in alternative semantics. Ph. D. thesis, University of Massachusetts at Amherst, Amherst, MA (2006)

Bott, L., Noveck, I.A.: Some utterances are underinformative: The onset and time course of scalar inferences. Journal of Memory and Language 51(3), 437–457 (2004)

Chemla, E.: Présuppositions et implicatures scalaires: études formelles et expérimentales. Ph. D. thesis, EHESS (2008)

Chemla, E.: Similarity: towards a unified account of scalar implicatures, free choice permission and presupposition projection. Under revision for Semantics and Pragmatics (2009)

Chemla, E., Bott, L.: Processing presuppositions: dynamic semantics vs pragmatic enrichment. Language and Cognitive Processes (to appear, 2012)

Fox, D.: Free Choice and the theory of Scalar Implicatures. In: Sauerland, U., Stateva, P. (eds.) Presupposition and Implicature in Compositional Semantics, pp. 537–586. Palgrave Macmillan, New York (2007)

Franke, M.: Quantity implicatures, exhaustive interpretation, and rational conversation. Semantics and Pragmatics 4(1), 1–82 (2011)

Geurts, B.: Entertaining Alternatives: Disjunctions as Modals. Natural Language Semantics 13(4), 383–410 (2005)

Grodner, D.: Speaker-specific knowledge in contrastive implicatures! In: MayFest in Maryland (2009)

Kamp, H.: Free choice permission. Proceedings of the Aristotelian Society 74, 57–74 (1973)

Katzir, R.: Structurally-defined alternatives. Linguistics and Philosophy 30(6), 669–690 (2007)

Klinedinst, N.: Plurality and Possibility. Ph. D. thesis, UCLA (2006)

Kratzer, A., Shimoyama, J.: Indeterminate pronouns: The view from Japanese. In: The Proceedings of the Third Tokyo Conference on Psycholinguistics, pp. 1–25 (2002)

Schulz, K.: You may read it now or later. A case study on the paradox of free choice permission. Master's thesis, University of Amsterdam (2003)

Zimmermann, T.E.: Free Choice Disjunction and Epistemic Possibility. Natural Language Semantics 8(4), 255–290 (2000)

Corpus Evidence for Preference-Driven Interpretation*

Alex Djalali, Sven Lauer, and Christopher Potts

Stanford University

Abstract. We present the Cards corpus of task-oriented dialogues and show how it can inform study of the ways in which discourse is goal- and preference-driven. We report on three experimental studies involving underspecified referential expressions and quantifier domain restriction.

1 Introduction

There is growing interest in the notion that both production and interpretation are shaped by the goals and preferences of the discourse participants. This is a guiding idea behind the question-driven models of Ginzburg (1996), Roberts (1996), Groenendijk (1999), and Büring (2003), as well as the related broadly decision-theoretic approaches of van Rooy (2003), Malamud (2006), Dekker (2007), and Davis (2011). These models, which we refer to generically as *goal-driven discourse models*, are intuitively well-motivated, but they have so far been tested against only a limited number of mostly hand-crafted examples and highly specific phenomena (Schoubye 2009; Toosarvandani 2010), with relatively little quantitative or corpus evaluation that we know of (but see Cooper and Larsson 2001; DeVault and Stone 2009; Ginzburg and Fernandez 2010).

The central goal of this short paper is to introduce a new publicly-available resource, the Cards corpus, and show how it can be used to explore and evaluate goal-driven discourse models (see also Djalali et al. 2011). The Cards corpus is built from a two-person online video game in which players collaboratively refine a general task description and then complete that task together. The game engine records everything that happens during play, making it possible to study precise connections between the players' utterances, the context, and their general strategies. Because the corpus is large (744 transcripts, ≈23,500 utterances) and its domain simple, it can be used to quantitatively evaluate specific pragmatic theories.

Here, we focus on the ways in which the players' conception of their task drives their understanding of referential and quantified noun phrases. After describing the corpus in more detail, we develop a simple goal-driven model that encodes and tracks certain important aspects of the players' preferences, and then we present three experiments designed to show how this model can be used to accurately resolve underspecified definites and quantifiers.

* We are indebted to Karl Schultz for designing the game engine underlying the Cards corpus. This research was supported in part by ONR grant No. N00014-10-1-0109 and ARO grant No. W911NF-07-1-0216.

M. Aloni et al. (Eds.): Amsterdam Colloquium 2011, LNCS 7218, pp. 150–159, 2012.
© Springer-Verlag Berlin Heidelberg 2012

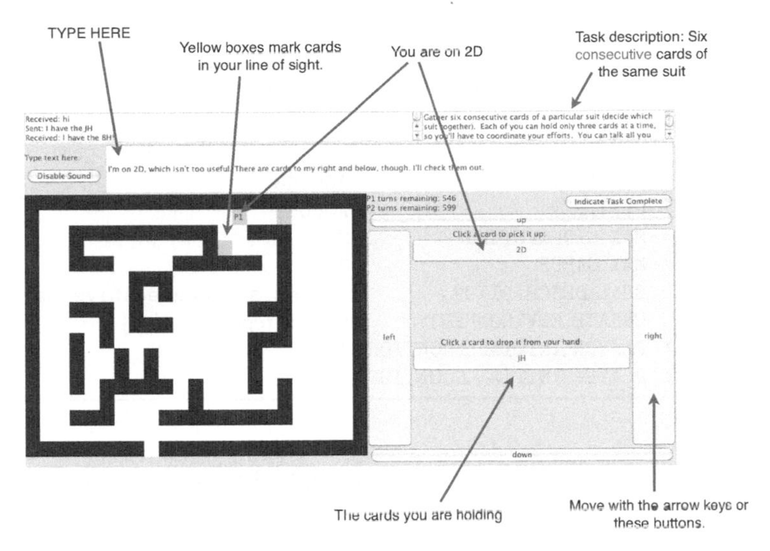

Fig. 1. An annotated version of the Cards gameboard

2 The Cards Corpus

The Cards corpus is built around a Web-based, two-person collaborative search task, partly inspired by similar efforts (Thompson et al. 1993; Allen et al. 1996; Stoia et al. 2008). We recruited players using Amazon's Mechanical Turk. The game-world consists of a maze-like environment in which a deck of cards has been randomly distributed. The players are placed in random initial positions and explore using keyboard input. A chat window allows them to exchange information and make decisions together. Each player can see his own location, but not the location of the other player. The visibility of locations of the cards are limited by distance and line-of-sight. Players can pick up and drop cards, but they can hold at most three cards at a time. In addition, while most of the walls are visible, some appear to a player only when within that player's line-of-sight.

When participants enter the game, they are presented with a description, some guidelines, and the annotated gameboard in figure 1. Before starting play, they are given the following task description (which remains visible in the upper-right of the gameboard):

> Gather six consecutive cards of a particular suit (decide which suit together), or determine that this is impossible. Each of you can hold only three cards at a time, so you'll have to coordinate your efforts. You can talk all you want, but you can make only a limited number of moves.

This task is intentionally underspecified. The players are thus forced to negotiate a specific goal and then achieve it together. In general, they begin by wandering around reporting on what they see, exchanging information as they go until a viable strategy begins to emerge. Dialogue (1) is typical. (Between utterances,

Table 1. Environment metadata in the corpus format

Agent	Time	Action type	Contents
Server	0	COLLECTION_SITE	Amazon Mechanical Turk
Server	0	TASK_COMPLETED	2010-06-17 10:10:53 EDT
Server	0	PLAYER_1	A00048
Server	0	PLAYER_2	A00069
Server	2	MAX_LINEOFSIGHT	3
Server	2	MAX_CARDS	3
Server	2	GOAL_DESCRIPTION	Gather six consecutive cards ...
Server	2	CREATE_ENVIRONMENT	[ASCII representation]
Player 1	2092	PLAYER_INITIAL_LOCATION	16,15
Player 2	2732	PLAYER_INITIAL_LOCATION	9,10

the players explore the environment and manipulate cards; this dialogue spans a total of 56 moves.)

(1) P1: i am top right
P2: im bottom left
P1: ok
P2: i have 3, 7h
P2: also found 6 and 7s...what should we go with?
P1: moving down to the bottom on that long corridor not seen a heart yet
P1: you pick
P2: u have the 9s right?
P1: yep
P2: ok lets go spades i have the 7 and 6

And then they pursue a specific solution. Because they can hold only three cards at a time, they are compelled to share information about the locations of cards, and their solutions are necessarily collaborative.

The current release (version 1) consists of 744 transcripts. Each transcript records not only the chat history, but also the initial state of the environment and all the players' actions (with timing information) throughout the game, which permits us to replay the games with perfect fidelity. In all, the corpus contains 23,532 utterances (mean length: 5.84 words), totaling 137,323 words, with a vocabulary size around 3,500. Most actions are not utterances, though: there are 255,734 movements, 11,027 card pick-ups, and 7,202 card drops. The median game-length is 414 actions, though this is extremely variable (standard deviation: 261 actions).

The transcripts are in CSV format. Table 1 is an example of the high-level environmental information included in the files, and table 2 is a snippet of gameplay. Computationally, one can update the initial state to reflect the players' actions, thereby deriving from each transcript a sequence of ⟨context, event⟩ pairs. This makes it easy to study players' movements, to make inferences about what

Table 2. A snippet of gameplay in the corpus format

Agent	Time	Action type	Contents
Player 1	566650	PLAYER_MOVE	7,11
Player 2	567771	CHAT_MESSAGE_PREFIX	which c's do you have again?
Player 1	576500	CHAT_MESSAGE_PREFIX	i have a 5c and an 8c
Player 2	577907	CHAT_MESSAGE_PREFIX	i jsut found a 4 of clubs
Player 1	581474	PLAYER_PICKUP_CARD	7,11:8C
Player 1	586098	PLAYER_MOVE	7,10

guides their decision-making and, most importantly for pragmatics, to study their language in context.

The Cards corpus is available at `http://cardscorpus.christopherpotts.net/`. The distribution includes the transcripts, starter code for working with them in Python and R, and a slideshow containing documentation. We think that the corpus fills an important niche; while there are a number of excellent task-oriented corpora available, the Cards corpus stands out for being large enough to support quantitative work and structured enough to permit researchers to isolate very specific phenomena and make confident inferences about the participants' intentions.

3 Relevance and the Evolving Task

The strategic aspects of interactions in the Cards corpus revolve around sequences of cards. In Djalali et al. 2011, we developed a hierarchy of abstract questions concerning the game, cards, and strategies, and showed that expert players and novice players negotiate this hierarchy differently: novices work systematically through it, whereas experts strategically presuppose resolutions of general issues so that they can immediately engage low-level task-oriented ones.

Here, we focus on how the relevance of particular cards changes as the players' strategies change. To this end, we define the *value* of a hand H, $\mathit{Value}(H)$, to be the minimum number of pick-up and drop moves from H to a solution to the game. For example, $\mathit{Value}(\emptyset) = 6$, $\mathit{Value}(\{5\text{H}\}) = 5$, and $\mathit{Value}(\{5\text{H}, 8\text{H}\}) = 4$. Because dropping is costly, $\mathit{Value}(\{5\text{H}, 2\text{S}, 3\text{D}\}) = 7$: at least two cards have to be dropped before forward progress towards a solution.

The *Value* function gives rise to a measure of how relevant a card c is given a hand H:

(2) $$\mathit{Relevance}(c, H) \stackrel{def}{=} V(H) - V(H \cup \{c\})$$

Where c is intuitively relevant to H, this value is $+1$, else it is -1 (or 0 if $c \in H$). Figure 2 illustrates the relevance sphere for a particular hand $H = \{2\text{H}, 4\text{H}, 5\text{H}\}$.

Our overarching hypothesis is that the players will seek cards that are relevant given their current holdings, and that this will be a driving force in how they resolve linguistic underspecification. The experiments in the next section seek to refine and support this basic idea. We should say, though, that the current notion of relevance is just an approximation of the players' underlying policy. We know, for example, that they often pick up irrelevant cards that might be relevant later — the cost of refinding cards is greater than the cost of having to drop those that turn out to be irrelevant. Our measure also does not take into account the costs of communication: the players have a slight bias for hearts, probably because this is the most iconic of the suits; they are reluctant to change suits once they have settled on one (even if this means extra exploration); and they favor solutions that don't span the Ace, since there is indeterminacy about whether such solutions are legitimate. We are confident that our results will only get stronger once these considerations are brought into the modeling. We mention them largely to emphasize that the corpus can support sophisticated investigations into decision making and pragmatics together.

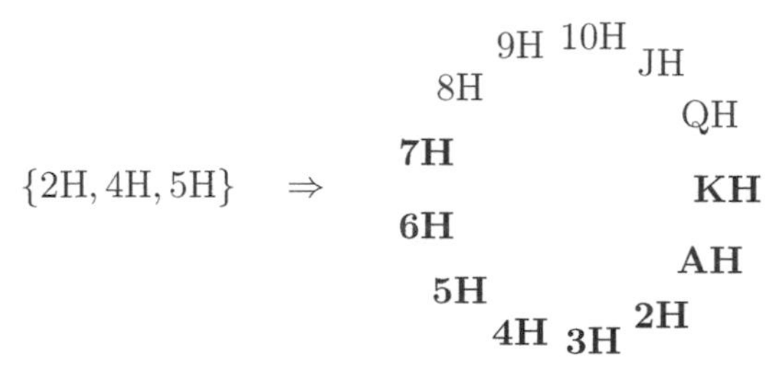

Fig. 2. Given the hand at left, the cards in bold are relevant in the sense that obtaining any one of them would move the players closer to a solution. The others would take the players farther from a solution, as measured by pick-up and drop moves, and are thus irrelevant given this hand.

4 Experiments

4.1 Experiment 1: Underspecified Card References

It is a testament to the importance of context that the majority of singular card references in the corpus (see table 4(a)) are like those in the following dialogue:

(3) P2: Look for 2.
P1: and the 3?

Absent knowledge of the context, we cannot know which cards these players are referring to. However, if we know that they are collectively holding 4H and KH, then their intentions become clear: P2 refers to 2H and P1 to 3H. Intuitively, such resolutions maximize relevance. More specifically, we hypothesized that, for any nominal referring expression N and stage of play t, the intended referent will be in the set

(4) $$\text{argmax}_{t,c \in Res(N)}\ Value(H_t) - Value(H_t \cup \{c\})$$

Table 3. Resolving underspecified card references via relevance

Phrase type	Count
Fully specified	103 (37%)
Underspecified	172 (63%)
Total	275

(a) Singular definite card references

Inference	Count
Correct	164 (95%)
Incorrect	8 (5%)
Total	172

(b) Results.

where H_t is the set of cards the players are currently holding at t, and $Res(N)$ is the set of cards consistent with the descriptive content of N. (For example, $Res(4) = \{4H, 4D, 4S, 4C\}$ and $Res(H)$ = the set of all hearts.)

To evaluate this hypothesis, we annotated the underspecified singular card references in 10 transcripts for what we took to be the intended referents and then wrote a computer program that chose the maximally relevant interpretation according to (4). (Where there was more than one such maximum, the program chose one at random.) We count a prediction as correct iff it matches the human annotation. The accuracy of this algorithm is extremely high (table 4(b)). What's more, we find that its mistakes tend to be clustered together near the start of transcripts, where even the interpreting player might have felt unsure about the speaker's intentions.

4.2 Experiment 2: Unrestricted Quantification?

From a strict logical perspective, quantifiers like *everything*, *nothing*, and *anything* carry universal force: the truth of a sentence involving them requires checking that every entity in the domain of discourse has (or does not have) the property they apply to. In this sense, P1 speaks falsely in (5) when he says, "I see nothing". We know this because the rich meta-data of the Cards corpus allows us to calculate exactly which cards P1 saw prior to this utterance. (He happened to have seen 5C and 10S.)

(5) P1: ok–i'll look at D and H, u look at C and S?
P2: ok
P1: i see nothing

It is clear why P1 is not perceived as speaking falsely, though: at the time of utterance, he had seen no cards that were relevant to his initial proposal, that is, no diamonds or hearts. In (6), the implicit restrictions are even more refined:

(6) P1: lets do spades
P1: I have the as, qs, and ks
[...]
P1: ok, i found js
P2: Ok. I haven't found anything...lol

Table 4. Experiment 2 results

Quantifier	Literally true	Literally false
anything	2	6
nothing	0	6
Total	2 (14%)	12 (86%)

Here, P2's *anything* seems to range just over the cards that are relevant to the hand {AS, QS, KS}, in the sense of (2). That is, P2 saw other cards, just not in this contextually privileged set.

The highly constrained nature of the Cards world means that we can precisely define the domain of discourse, which in turn permits us to identify exactly which contextual factors shape the domain in these cases. The first question we sought ask in this area is a seemingly simple one: what percentage of universally quantified claims are literally true, that is, true for an unrestricted interpretation of their quantifiers? To make our experiments manageable, we first extracted all utterances matching the regular expression in (7):

(7) `(find|found|see|saw) (any|no)thing`

Such phrases include simple declaratives like *I see nothing* as well as interrogatives like *Did you find anything?* Although there are 35 matches for (7), we disregard ones like *I didn't see anything around here*, as they overtly restrict the domain of discourse and often involve indexical terms, whose semantics are more difficult to define.

We define one of these quantified phrases as *literally true* just in case it is true on an unrestricted interpretation of the quantifier, that is, a quantifier that ranges over the full deck of 52 cards. At any point t in the game, each player P will have walked over, and hence seen/found ('seeing' and 'finding' amount to the same thing in the Cards world, as players can identify only the cards they are currently standing on), a subset of the full deck of cards. Call this subset $S_{P,t}$. Suppose that, at time t, player P says, "I found nothing". Then player P's claim is literally true just in case $S_{P,t} = \emptyset$ and literally false just in case $S_{P,t} \neq \emptyset$. If, at time t, a player P asks "Did you find anything" and player P' responds "Yes", then P''s response is literally true just in case $S_{P,t} \neq \emptyset$, and literally false just in case $S_{P,t} = \emptyset$. (Similarly if P' responds with 'No'.)

These definitions mean that we can quantitatively assess the percentage of universal claims that are literally true. The procedure is as follows: for each universally quantified claim made by a player P at time t, build $S_{P,t}$ by following the full path taken by P up to t and adding to $S_{P,t}$ all and only those cards walked over by P up to t. Table 4 summarizes the results of this experiment. As is evident, effectively no quantified phrases are literally true. Indeed, the only time players speak literally is during the initial stages of the game when they are trying to establish what sequence they should pursue, as in (8), in which P2 has found and is holding only the QC.

(8) P1: ok so what suit
P2: Whatever I find first.
[. . .]
P1: have you found anything yet
P2: I have QC.

4.3 Experiment 3: Goal-Based Domain Restriction

The results of experiment 2 indicate that implicit quantifier restrictions are the norm. We turn now to the task of identifying which contextual factors the players use to provide these restrictions. It turns out the players' decisions about which suit to pursue are reliable indicators of how the domain of discourse is restricted. In the following dialogue, the players agree to pursue clubs:

(9) P1: lets go clubs
[. . .]
P2: ok I finished right side and middle box did you find anything?
P1: no

Literally speaking, P1 has spoken falsely, as (s)he found 11 cards prior to uttering "no". Although P1 in fact passes over the 3C, we take P1 to regard this card as irrelevant because P2 is holding the 9C, 10 and KC and has located 7C and QC – no winning hand can be gotten by these players without renegotiating their target sequence, which we take to be a subsequence of the 7C, 8C, 9C, 10C, JC, QC and KC. Thus, P1 speaks truthfully if we regard the earlier negotiation as restricting the domain for *anything* in P2's utterance.

In the terms of section 3, it looks like P1's proposal to limit to clubs makes all and only those cards relevant, which provides the basis for interpreting the quantifier. To test this idea, we hand-annotated all the transcripts involving the stimuli in experiment 2 for the players' mutually agreed-upon suit. The annotations mark the span of text in which the negotiation occurred with the suit they agreed upon. Phrases that indicated the start of such a negotiation included, but were not limited to, "let's go (for) X", "look for X", and simple suit mentions like "hearts?".

We now define one of the quantified phrases in (7) as *restrictedly true* just in case it is true on a restricted interpretation of the quantifier, that is to say, a quantifier that ranges only over cards with the agreed-upon suit. (A prominent edge case involved players who never overtly agreed upon any suit in particular, but rather deployed a strategy of stacking cards in a large pile and looking for any winning sequence. In such cases, we made the simplifying assumption that the domain of discourse included all and only the winning sequences.)

We were able to annotate the transcripts of 12 of the 14 quantified phrases considered in experiment 1. We disregarded the two phrases that involved literally true uses of the quantifiers, as they were used when players are trying to establish what sequence they should agree upon, and reran the same experiment as above. The results are given in table 5. They are essentially the opposite of

Table 5. Experiment 3 results

Quantifier	Literally true	Literally false
anything	6	0
nothing	6	0
Total	12 (100%)	0 (0%)

those in table 4: when we restrict the domain of discourse to the players' agreed upon suit, utterances that were false become true. This is precisely the result one would expect on a model of discourse in which interpretation is governed by high-level factors relating to the discourse participants' understanding of the goals and issues at hand.

5 Conclusion

This paper introduced the Cards corpus, a highly-structured resource for doing corpus-driven computational pragmatics. In a series of experiments, we showed how the transcripts can be used to precisely define the domain, to ground denotations for quantified terms, and to pinpoint ways in which the context influences utterance understanding.

These experiments show that the Cards corpus can support quantitative evaluation of hypotheses in pragmatics. Going forward, we hope to expand our theoretical reach by pursuing the following inter-related goals:

- Increase the size and power of the corpus by collecting additional transcripts and altering the parameters of the game, e.g., number of moves, line of sight, and number of cards each player is able to hold.
- Extend our experimental techniques to a wider range of phenomena already present in the corpus. For example, there are many utterances in the corpus of the form "4H" or even just "4". In context, it is clear what the speaker intends: "Found the 4H", "Can't find the 4H", "Look for the 4H", etc. We conjecture that, just as the context can be used effectively to resolve the underspecification of "4" (section 4.1), so too can it be used to resolve which predicate the speaker intended.
- Situate the above results in a fuller question-driven model of the sort employed by Djalali et al. 2011. Both implicit and explicit questions shape players' actions (where to move, what to pick up, when to speak, and so forth). For example, the players' negotiations about which suit to pursue (experiment 3) fit neatly into a goal-driven question hierarchy of the sort envisioned by Roberts (1996). Using a question model, we can study the players' linguistic behavior, and we can pursue questions in pragmatics and decision theory simultaneously, finding new ways in which language shapes, and is shaped by, the goals and preferences of the discourse participants.

References

Allen, J.F., Miller, B.W., Ringger, E.K., Sikorski, T.: A robust system for natural spoken dialogue. In: Proceedings of ACL, pp. 62–70 (1996)

Büring, D.: On D-trees, beans, and B-accents. Linguistics and Philosophy 26(5), 511–545 (2003)

Cooper, R., Larsson, S.: Accommodation and reaccommodation in dialogue. In Göteborg Papers in Computational Linguistics. Department of Linguistics, Göteborg University (2001)

Davis, C.: Constraining Interpretation: Sentence Final Particles in Japanese. Ph.D. thesis, UMass Amherst (2011)

Dekker, P.: Optimal inquisitive discourse. In: Aloni, M., Butler, A., Dekker, P. (eds.) Questions in Dynamic Semantics, pp. 83–101. Elsevier (2007)

De Vault, D., Stone, M.: Learning to interpret utterances using dialogue history. In: Proceedings of the 12th Conference of the European Chapter of the ACL (EACL 2009), pp. 184–192. Association for Computational Linguistics, Athens (2009)

Djalali, A., Clausen, D., Lauer, S., Schultz, K., Potts, C.: Modeling expert effects and common ground using Questions Under Discussion. In: Proceedings of the AAAI Workshop on Building Representations of Common Ground with Intelligent Agents, AAAI (2011)

Ginzburg, J.: Dynamics and the semantics of dialogue. In: Seligman, J. (ed.) Language, Logic, and Computation. CSLI (1996)

Ginzburg, J., Fernandez, R.: Computational models of dialogue. In: Clark, A., Fox, C., Lappin, S. (eds.) Handbook of Computational Linguistics and Natural Language Processing. Blackwell (2010)

Groenendijk, J.: The logic of interrogation. In: Matthews, T., Strolovitch, D. (eds.) Proceedings of SALT IX, pp. 109–126. Cornell University (1999)

Malamud, S.: Semantics and Pragmatics of Arbitrariness. Ph.D. thesis, Penn. (2006)

Roberts, C.: Information structure: Towards an integrated formal theory of pragmatics. In: Yoon, J.H., Kathol, A. (eds.) OSU Working Papers in Linguistics, pp. 91–136. The Ohio State University Department of Linguistics (1996) (revised 1998)

van Rooy, R.: Questioning to resolve decision problems. Journal Linguistics and Philosophy 26, 727–763 (2003)

Schoubye, A.: Descriptions, truth value intuitions, and questions. Journal Linguistics and Philosophy 32(6), 583–617 (2009)

Stoia, L., Shockley, D.M., Byron, D.K., Fosler-Lussier, E.: SCARE: A situated corpus with annotated referring expressions. In: Proceedings of LREC (2008)

Thompson, H.S., Anderson, A., Bard, E.G., Doherty-Sneddon, G., Newlands, A., Sotillo, C.: The HCRC map task corpus: Natural dialogue for speech recognition. In: HLT 1993: Proceedings of the Workshop on Human Language Technology, pp. 25–30. ACL (1993)

Toosarvandani, M.: Association with Foci. Ph.D. thesis, UC Berkeley (2010)

Relating ERP-Effects to Theories of Belief Update and Combining Systems

Ralf Naumann

Institut für Sprache und Information
Universität Düsseldorf
Germany

Abstract. A major challenge for any attempt at combining linguistics with empirical neurophysiological data from brain research, like ERPs for example, is to give an answer to the question of how the gap between psycholinguistic and formal models of specific aspects of language on the one hand and the neural architecture underlying neurophysiological measures on the other can be bridged (Baggio et al. 2010). An interesting research program based on the extension of the event calculus in Van Lambalgen & Hamm 2005 was launched in Baggio et al (2007, 2008, 2010) where it was shown how a correlation between two known ERP-effects, the N400 and the LAN, and semantic phenomena like the progressive can be established. In this paper we will present an alternative formal theory which is based on the technique of combining systems and in which the dynamics of information change is separated from the more static aspects of knowledge representation. Using this multi-layered architecture, we hypothesize that the LAN is related to the process of updating a discourse model in the light of new information about changes in the world.[1]

Keywords: ERP, situation models, combining systems, aspect, belief revision.

1 Introduction

A major obstacle for combining neuroscience and linguistics are the diverging ways in which the meaning and understanding of linguistic expressions (words) are defined. In theories of grounded cognition the meaning of a word is linked to (or even defined in terms of) past experiences and encounters with objects of the given kind. For example, the meaning of verbs like *kick* or *lick*, which refer to actions involving leg or mouth movements, is linked or identified with somatotopic activation patterns in corresponding motor cortices. Defining meaning in this way has the advantage of directly grounding the understanding and comprehension of language in the empirical evidence available to speakers. In

[1] The research was supported by the German Science Foundation (DFG) funding the Collaborative Research Center 991 at the University of Düsseldorf. I would like to thank Wiebke Petersen and two anonymous reviewers for stimulating comments.

M. Aloni et al. (Eds.): Amsterdam Colloquium 2011, LNCS 7218, pp. 160–169, 2012.
© Springer-Verlag Berlin Heidelberg 2012

addition to past experiences with objects of the given kind a second source of learning the meaning are distributional properties abstracted from the linguistic contexts in which the word occurs. This way of defining the meaning of words fits well into the conception of viewing the primary function of language as a set of processing instructions on how to arrive at a situation model for the state of affairs referred to by a sentence or a discourse (Zwaan & Radvansky 1998).

This way of analyzing the meaning of verbs is almost orthogonal to the way they are analyzed in formal semantic theories. There verbs are classified e.g. as activities (*run* or *lick*), accomplishments (*write*) or semelfactives (*kick*). These classifications are based on properties of thematic relations like Theme and/or the notion of a nucleus structure. E.g. distinguishing *incremental* from *constant* Themes makes it possible to define accomplishment (*write*) and activity (*push*) verbs. The notion of a nucleus structure was introduced in Moens & Steedman (1988). These structures are built in terms of three different parts: *preparatory process*, *culmination* and *consequent state*. Verbs differ with respect to the type of nucleus structure assigned to them. For example, accomplishments verbs like *build* have nuclei structures consisting of all three possible parts, whereas the nucleus structure of an activity verb like *run* consists only of a preparatory process.

One way of combining the two paradigms is the following. As emphasized in Zwaan & Radvansky (1998:162), a situation model can be seen as a *multidimensional* system consisting of various dimensions like space, time, causation and objects. The various dimensions are linked to each other by particular relations guiding the flow of information between those dimensions. The different formal semantic theories can be taken as modeling a particular dimension of such a multidimensional system. For example, a nucleus structure defines the dynamic meaning component of a verb, i.e. the way an event of a given type (say a writing) evolves (or occurs) in space and time, and therefore the dimensions space and time in a situation model. The object dimension and its relation to the dynamic dimension is defined in terms of thematic relations like Actor or Theme. However, there are at least two missing links. First, the aspect of combining the different dimensions with each other (the flow of information) is missing in those semantic theories. Here I am going to propose using the framework of combining systems to fill this gap. Second, semantic theories must be related to neurophysiological and behavioural data from brain and cognitive science. In particular, the decompositions used in those theories (properties of thematic relations, the various constituents of a nucleus structure) don't seem to be directly related to the empirical evidence used by speakers to grasp the meaning of lexical items. A first step to bridge this gap has been undertaken in Van Lambalgen & Hamm (2005) where the meaning of a sentence is seen as an algorithm (or a procedure) to construct a scenario (or, formally, a logic program) from the sentence. However, even in this approach it remains unclear how the proposed decomposition clauses are related to the empirical evidence about meanings available to comprehenders. The strategy proposed here to overcome this problem is based on the following empirical considerations. Events occur in time. This occurrence can be subdivided into different stages as described by a nucleus structure. Each stage of the development is characterized by (i) a subset of the objects involved in the event and (ii) particular activities and/or properties of those objects that are undertaken or affected during that stage. Given information about the type of event (verb), the stage the event is in (aspectual and temporal markers) and the role (thematic relation) of the object in the event, it is possible to infer (based on empirical knowledge of events of that type) the corresponding activities and/or properties and to built up the temporal dimension of the situation model in terms of them.

Two important questions that arise are: (i) How are situation models computed in the brain? and (ii) What is the relation between the various ERP components and the various dimensions of a situation model? Given that a situation model is multidimensional, certain effects may be related to particular dimensions and not to others. In the following section the experimental results of Baggio et al. (2008) are presented. In the third section an alternative framework is developed and compared to that of Baggio et al.

2 The Progressive and Left Anterior Negativity

Using the formal semantic theory developed in Van Lambalgen & Hamm (2005) and Hamm & Van Lambalgen (2003) and based on ERP experiments, Baggio et al. (2008, 2010) present a unified account of the N400 and the LAN (left anterior negativity) by relating it to different semantic phenomena like the progressive and coercion. Similar to other approaches, Baggio et al. link the N400 to the processing of semantic phenomena at the level of semantic (word) composition like priming (semantic relatedness) and the unification of variables. According to the analysis of the progressive in Hamm & Van Lambalgen (2003), the meaning of a telic VP like *write a letter* is an algorithm (called a 'scenario'), implemented as a particular type of logic program from which the expression's reference, a minimal model of the scenario, can be derived via resolution (unification). For telic VPs such a scenario can be seen as a (simplified) plan to reach a particular goal. In the case of writing a letter this plan basically consists of the goal (the finished letter) and an activity to attain this goal, here writing (as opposed to say dictating). Since such a minimal model does not contain any information about obstacles (possibly) terminating the writing, it is possible to derive the obtaining of this goal at some later time using additional general constraints like the law of inertia. This conclusion concerning the existence of a goal state is defeasible. Now consider the sentences in (1).

(1) a. The girl was writing a letter when her friend spilled coffee on the paper.
    b. The girl was writing a letter when her friend spilled coffee on the tablecloth.

First, it was correctly predicted that a higher N400 effect occurred for *tablecloth* compared to *paper*. Baggio et al. explain this by the fact that in the context of a writing event (described in the main clause) the noun *paper* is semantically more expected as referring to the location where the letter comes into existence than the noun *tablecloth*. As a result, semantic unification is easier (proceeds more smoothly) for *paper* compared to *tablecloth*. Second, they (correctly) expected a higher LAN for the VP *spill coffee on the paper* compared to the VP with 'paper' replaced by *tablecloth*. This is explained by the fact that in the former case the minimal model derived for the main clause (MC) has to be recomputed due to the fact that the goal state (the completed letter) must be suppressed since the subordinate clause (SC) contains information about a terminating event. When this part of the sentence is processed by a speaker there is therefore a higher processing load which is neurophysiologically reflected in the higher sustained anterior negativity. By contrast, for *tablecloth* the minimal model is simply extended in a monotonic way. Finally, they correctly expected a correlation between higher LAN-amplitudes and speakers asserting that the letter was not finished.

3 An Alternative Approach

Following Zwaan & Radvansky (1998), we assume that the successful understanding of a sentence consists in the ability to construct a situation model for the situation (or state of affairs) described by it. A situation model is a multidimensional system in which various dimensions are combined with each other. This multidimensional character will be modeled by using the technique of combining systems (Blackburn & de Rijke 1997, Finger & Gabbay 1992). A dimension is modeled as an ontology together with a family of n-ary relations. In the present context three such ontologies are distinguished: time points, eventualities and persistent objects, which are defined as follows: (i) Object structure $\underline{\mathbf{O}} =< O, \{A_a\}_{a \in Attr} >$ with O a (non-empty) set of persistent objects (like tables and chairs) and each A_a is an *n-ary* relation on O; (ii) Eventuality structure $\underline{\mathbf{E}} =< E, \{P_v\}_{v \in Verb} >$ where E is a (non-empty) set of event tokens (or event occurrences) and each P_v is a unary relation on E, which is an event type and its elements are event occurrences or event tokens like runnings ($v = run$) or writings ($v = write$) and (iii) $\underline{\mathbf{T}} =< T, <_T>$ is a flow of time with T a (non-empty) set of time points and $<_T$ a linear and atomic ordering.

The definition of the flow of information (i.e. the connections) between the various dimensions is based on the following considerations. Each event token e occurs on a particular interval and is related to persistent objects from $\underline{O}$ that bear a particular relation to it like Actor or Theme. The first relation between $\underline{\mathbf{T}}$ and $\underline{\mathbf{E}}$ is defined by the function τ from E to $\wp(T)$, which assigns to each $e \in E$ its *run-time* and which is required to be a convex set. In terms of τ the functions PP , Cul and CS are defined, mapping an event e to the interval corresponding to the PP (preparatory process), Cul (culmination; a singleton) and CS (consequent state), respectively. This connection is called the *temporal-causal* dimension of a situation model. The relation between $\underline{\mathbf{E}}$ and $\underline{\mathbf{O}}$ is modeled by a set $\{R_{tr}\}_{tr \in TR}$ of (functional) thematic relations and is called the *static* dimension of a situation model. Events and actions bring about changes with respect to time-dependent properties of those persistent objects participating in them (i.e. which bear one of the thematic relations defined for the event/action). By way of illustration consider an event of a girl writing a letter. Such an event involves a goal (the completed letter) and an activity (writing) to bring about this goal. Since the writing involves an actor, it will be assumed that there is a state the writer is in when doing the writing. This state will be encoded by $state_{writing}(x)$, and be called a *state-fluent.* Formulas like *completed(y)*, encoding the goal, are called *goal formulas.* During the occurrence of the event of writing the formulas are evaluated in a characteristic way. Whereas $state_{writing}(girl)$ is true during the whole event (except, possibly, at the beginning and the end point), *completed(letter)* is only true at the end point of the event and, in addition, after the event occurred. This relation is captured as follows. Let $Attr_v$ be the restriction of $Attr$ to those state-fluents and goal formulas related to event tokens from P_v. Then Z is a function that assigns to each event e and time point $t \in \tau(e)$ an object structure $\underline{O'} =< O', \{A'_{a'_e}\}_{a' \in Attr_v} >$ where O' is a subset of the objects involved in e and $\{A'_{a'_e}\}_{a' \in Attr_v}$ is the restriction of $\{A'_{a'}\}_{a' \in Attr_v}$ to O'. This structure is called the *dynamic submodel of e at t.* In terms of Z a function Z^* is defined that assigns to an event e a sequence of pairs (t, M_t) that describes the evolution of the event e.

How is the function Z computed in the brain? Events evolve (develop) in time. This evolution can be subdivided into different stages as described, e.g. by a nucleus structure. Each stage of the development is characterized by (a) a subset of the objects involved in the event and (b) particular activities and/or properties of those objects that are undertaken or affected during that stage. So given information about (i) the type of the event (supplied by the verb), (ii) the stage the event is in (supplied by aspectual and temporal markers like progressive) and (iii) the role of the object (extracted from e.g. morphological information and/or word order), it is possible to retrieve, using empirically grounded knowledge about events and objects of those kinds in that particular role, appropriate activities and/or properties of the corresponding object for that stage or those stages that have already occurred up to the stage of the event described by the sentence. In our framework the above relationship is captured by axioms like those in (2) for *write.*

(2) a. $Actor(e) = x \wedge PP(e) = i \wedge P_{write}(e) \rightarrow state_{writing}(x)(i)$
b. $Theme(e) = y \wedge PP(e) = i \wedge P_{write}(e) \rightarrow state_{of\_creation}(y)(i)$
c. $Theme(e) = y \wedge Cul(e) = \{t\} \wedge P_{write}(e) \rightarrow completed(y)(\{t\})$
d. $Theme(e) = y \wedge CS(e) = i' \wedge P_{write}(e) \rightarrow completed(y)(i')$

In these axioms information about the static dimension (thematic relation) and the temporal-causal dimension (second conjunct in the antecedent) together with type information about the event is used to infer information about the dynamic dimension. For a progressive sentence, only the first two axioms (2a) and (2b) are used, since the event is described as ongoing. By contrast, for a sentence in the simple past, the third axiom is used too and for a sentence in the (present) perfect, the antecedents of all four axioms are satisfied.

Empirical evidence for axioms of this kind comes from experiments that show that aspect and other grammatical devices influence not only the way an object is simulated but also which objects are focused (Madden & Therriault 2008, Ferretti et al. 2009, Bergen & Wheeler 2010). For example, in *John is closing the drawer* (ongoing event) the mental simulation focuses on the act of pushing the drawer undertaken by the Actor John whereas in *John has closed the drawer* the simulation is focused on the goal state and therefore on the Theme in that state, i.e. the *closed* drawer. The way an object is simulated is therefore closely related to its role in the event and to its properties affected by the event. On this view, decompositional predicates like those used in the consequents of the axioms in (2) are directly related to the way objects are simulated in the brain while computing a situation model whereas in the antecedents only information is used that can be directly extracted from the surface structure of the sentence.[2]

An important aspect that is missing so far is non-determinism. If an event token occurs, it is in general not uniquely determined how the occurrence continues. For example, if a girl is writing a letter, she may finish it or not. Formally this non-determinism is modeled in the following way. Each pair (t, M_t) is related to a set of (linearly) ordered sequences of pairs $(t', M_{t'})$ such that (i) each sequence is a possible way of how the event occurring at t evolves after t and (ii) this set is ordered by a relation $\leq_t$, which is required to be at least reflexive and transitive. When taking both conditions together, one gets, for a given t, a computation tree (with root t) where each outgoing arc is assigned a weight, giving the degree of expectancy that this transition will occur (though these weights are *not* represented in the framework sketched above). This ordering reflects the expectancy or the priming a comprehender uses when processing a sentences with that verb. One basis of such an ordering is world knowledge about events of the given type, say writings. Another source is linguistic knowledge, in particular knowledge about the contexts in which the verb preferably occurs, e.g. whether the verb can be modified by temporal adverbials like *in x time* or *for x time* ('>' means that this context is more primed).

(3) a. John wrote the letter in an hour > John wrote the letter for an hour.
b. John ran for ten minutes > John ran in ten minutes.

Verbs which prime *in*-adverbials compared to *for*-adverbials are normally related to goals and/or results. E.g. writing usually involves a particular object that is

[2] As noted by one reviewer, the simulation will in general also depend on the context and background/world knowledge, e.g. on information about the type or the colour of the drawer or past experiences with (other) drawers. However, this kind of information does *not* influence (i) the status of being focused or not and (ii) the relation to a particular action/result described by the sentence. For instance, in the case of *John has closed the drawer* the focus is on the drawer simulated as being closed. Thus, it is the Aktionsart (perfect, perfective vs. imperfective) which determines whether the drawer is simulated as closed or not.

intended to be brought into existence, like a letter or an article for instance. By contrast, verbs priming *for*-adverbials are usually used to describe events that have no intended goal like running for example. From this it does not follow that *write* cannot be used without a goal, as examples like *John wrote* show, and *run* can be used as involving a goal: *Bill ran a mile/to the station*. However, these contexts are less primed. Posible orderings (primings) for *write* and *run* (or *push*) are: (i) *write*: PP Cul CS > PP and (ii) *run*: PP > PP Cul (CS).

When a word is processed, information is added at different dimensions. This information will in general be underspecified. In particular, this will be the case if information is added to a dimension that is the combination of basic dimensions like the static or the dynamic dimension because then information from *all* dimensions involved is necessary to arrive at fully specified information. For instance, when processing *wrote* in *The girl wrote a letter*, one gets the following updates: (i) type information: the event is a writing (basic dimension $\underline{E}$); (ii) unification $Actor(e) = girl$ (static dimension) since the subject has already been processed. By contrast, information about the Theme is still underspecified: $Theme(e) = y$; (iii) 'triggering' the corresponding axioms for the nucleus structure (dynamic dimension).

According to the Baggio et al. interpretation, the higher LAN is explained as follows. A formula that is true in the minimal model after processing the MC has to be retracted when this model is updated with the information contained in the (disabling) SC (revision of the model, non-monotonic update). This is formulated in the following hypothesis.

> *The Non-Monotonicity Hypothesis (NMH)*
> A higher LAN is expected when the current model has to be recomputed (revised) if new information is added that is incompatible with information in the current model. The amplitude of the LAN depends on the extent to which the model has to be recomputed.

One very simple method of calculating the measure of change to a model consists in counting the number of formulas that have to be revised (Katsuno & Mendelzon 1992). Consider again the disabling condition (1a). The spilling event causally interferes with the ongoing event of writing referred to in the MC, bringing it (at least temporarily) to an end. Let this information be encoded by the formula $clip(e_{spill}, e_{write})$, which is added to the temporal-causal dimension of the current model. Integrating this information in the model does not effect the valuation of a formula in the dynamic submodel at t_{spill}. However, the girl is no longer simulated as writing a letter after t_{spill} so that the formula $state_{writing}(girl)$ will *not* be an element of the dynamic submodel at times $t' > t_{spill}$. Similarly, the writing event is no longer described as occurring and the letter is no longer taken as coming into existence, yielding a total of three revisions. Of course, more sophisticated methods may be necessary.

The Non-Monotonicity Hypothesis not only applies to cases involving a goal state but to any non-monotonic change affecting the current model, in particular scenarios like those in (4) where a goal state is not part of the most expected nucleus structure, but the SC contains a disabling condition.

(4) The girl was pushing a cart towards the barn when she was hit by a car.

In (4) no goal state is involved because *push a cart* is atelic and the PP does not introduce a culmination. Therefore, no higher LAN is expected to occur according to the Baggio et al. analysis. By contrast, applying the NMH, the hitting terminates the pushing, so that the girl is no longer simulated as being in a pushing state. The difference between (1a) and (4) consists in the fact that for the latter the number of revision is expected to be lower because no object is coming into existence so that there are only two revisions. Thus, the amplitude of the LAN for (1a) is expected to be higher compared to that for (4) using the simple counting criterion from above.

In our approach, the following, alternative explanation is possible. After processing the MC, the event is described as ongoing. For an accomplishment verb like *write*, this means that different ways of how the event continues are still open: the writing (PP) is continued until the culmination is reached (PP Cul CS, most primed); (ii) the PP is continued but the culmination is not reached due to an interruption, leaving the eventual attainment of the goal open (PP) and (iii) the writing is definitely terminated (PP_{fail}). Thus, no *unique* nucleus structure has been determined up to that point, in particular it is *not* determined whether the culmination or goal state (e.g. a completed letter) is eventually attained or not.

When a disabling condition is encountered, the ordering on the possible continuations is changed. Several ways of how a new ordering is constructed are possible. At one extreme, the current most expected continuation is simply discarded. This will likely be the case when the MC is continued by *when she got a heart attack* or *when she died* (case (iii) above). An alternative reordering simply switches the current ordering between the different continuations so that the currently most expected ordering is still available but only demoted. This is likely to be the case for the example (1a) and case (ii) above because it is still possible that the girl (re)continues writing the letter on another sheet of paper. Changing the ordering of the possible continuations has the effect of changing the most expected scenario. This is formulated in the hypothesis below.

> *Reordering of Nuclei Structures Hypothesis (RH)*
> A higher LAN is expected whenever the ordering on the set of continuations at a point t of the temporal-causal dimension is changed due to a disabling condition.

On the RH, a reordering is also expected for (4). Recall that a verb like *push* is associated with a nucleus structure of type PP Cul CS, though this structure is less primed. Encountering the disabling condition in the SC of (4), this continuation has to be discarded. How is the cost of reordering the set of nuclei structures calculated? A very simple way of calculating the cost of a reordering operation consists in counting the number of changes wrt. the constituents of the most preferred nucleus structure. For example, if the nucleus structure PP Cul CS is discarded and PP_{fail} is assumed, both the culmination and the consequent state are no longer expected and an interferring event has to be assumed. A precise answer to this question needs, of course, further empirical investigation.

There are the following differences between the two accounts. First, according to the NMH, the minimal model always contains all constituents of the nucleus structure, either becaue the sentence and/or the context has already provided information or as the result of an inference in the limit. By contrast, on the RH neither the Cul nor the CS of the most expected nucleus structure of an accomplishment verb like *write* are part of the situation model for a progressive sentence since no information about the attaining of the goal state is (yet) given. However, this model contains axioms like (2c) and (2d) expressing the expectation that the end state is most likely to be attained. We think that there are at least two reasons why the goal should not be part of the minimal model. It seems doubtful that during the processing of a sentence a comprehender usually draws all the consequences that follow from the contents of the sentence by applying general constraints like the law of inertia. At least, this hypothesis has to be tested. Furthermore, as the empirical studies about the simulation of objects have shown, in a progressive sentence an object coming into existence is simulated as still unfinished and not as being already in the envisaged end state. Second, on the RH a comprehender always uses some form of revision strategy. His static (global) knowledge already contains a strategy of how to adjust a situation model if new incompatible information becomes available, which is given by the ordering on the set of possible evolutions. By contrast, on the NMH there is no such strategy. For instance, instead of assuming that the spilling terminated the writing, it is also possible to assume that the spilling did not have such drastic effects. The most important, and testable, difference between the two accounts has to do with (monotone) additions to the model which trigger an aspectual shift. Examples are *The girl pushed the cart to the barn* or *John ran in ten minutes*. In the former case the directional PP *to the barn* imposes a (definite) condition on the end point of the pushing path whereas in the latter case the *in*-adverbial requires such a condition on the running path which has to be inferred from general knowledge or the context. In both cases no information has to be retracted from the situation model so that no recomputation needs to take place. Thus, on the NMH, no higher LAN is to be expected. By contrast, according to the RH, the most preferred nucleus structure is of type PP, which, by the monotone addition to the situation model, gets changed to PP Cul CS. As a consequence, a higher LAN is expected to occur. In this context it is also interesting to test resultative constructions like *John washed the dishes clean* or *The mother sang the baby asleep*. For example, in the latter sentence the singing gets bounded by the baby falling asleep so that no revision for the underlying sentence *The mother sang* is required. However, the situation models for the resultative and the underlying sentence are quite different. As a result, the RH predicts a higher LAN. Generalizing, one can say that on the RH a higher LAN is expected to occur not only in cases of non-monotonicity but also in cases where the addition to the model are monotone. On the NMH (or the Baggio et al. account), a LAN is expected to occur only in relation to non-monotonic operations on the model (at least in the present formulation). So we suggest that further empirical research in this direction is likely to yield interesting results about semantic processing in the brain.

4 Conclusion

In this article we proposed both an extension of and an alternative explanation to the theory developed in Baggio et al. (2008). According to the extension, the higher LAN in the experiments depends on the number of revisions that have to be applied to the minimal model. This extension not only applies to accomplishment verbs involving a goal, but also to other verbs which do not define any particular end state. On the alternative explanation, a non-stative verb primes an ordered set of evolutions, i.e. ways events denoted by the verb can possibly develop. The interpretation of a sentence is always done wrt. a minimal element of this ordering, representing the most expected evolution of the event described by the sentence. A higher LAN is triggered whenever this expectation about the development of the event has to be changed so that an alternative evolution has to be assumed.

References

1. Baggio, G., et al.: Computing and recomputing discourse models, an ERP study. Journal of Memory and Language 59, 36–53 (2008)
2. Baggio, G., et al.: Coercion and compositionality. Journal of Cognitive Neuroscience 22, 2131–2140 (2010)
3. Baggio, G., van Lambalgen, M.: The processing consequences of the imperfective paradox. J. of Semantics 24, 307–330 (2007)
4. Bergen, B., Wheeler, K.: Grammatical aspect and mental simulation. Brain & Language 112, 150–158 (2010)
5. Ferretti, T.R., et al.: Verb aspect, event structure, and coreferential processing. Journal of Memory and Language 61, 191–205 (2009)
6. Finger, M., Gabbay, D.: Adding a temporal dimension to a logic system. J. of Logic Language and Information 1, 203–233 (1992)
7. Hamm, F., Van Lambalgen, M.: Event calculus, nominalisation, and the progressive. Linguistics and Philosophy 26, 381–458 (2003)
8. Katsuno, H., Mendelzon, A.: On the difference between updating a knowledge database and revising it. In: KR 1991, Cambridge, pp. 387–394 (1991)
9. Madden, C., Therriault, D.: Verb aspect and perceptual simulations. The Quarterly Journal of Experimental Psychology 62(7), 1294–1303 (2009)
10. Moens, M., Steedman, M.: Temporal ontology and temporal reference. Computational Linguistics 14(2), 15–28 (1988)
11. Van Lambalgen, M., Hamm, F.: The proper treatment of events. Blackwell, Oxford (2005)
12. Zwaan, R.A., Radvansky, G.A.: Situation models in language comprehension and memory. Psychological Bulletin 123(2), 162–185 (1998)

Can Children Tell Us Something about the Semantics of Adjectives?

Francesca Panzeri and Francesca Foppolo

Università degli Studi di Milano-Bicocca, Milan, Italy
{francesca.panzeri,francesca.foppolo}@unimib.it

Abstract. In this paper we discuss some data about the acquisition of relative gradable adjectives in order to evaluate two theories that have been proposed to account for the meaning of gradable adjectives, i.e. the degree-based analysis and the partial function approach. We claim that younger children start by assigning a nominal like interpretation to relative gradable adjectives (*tall* means "with a vertical dimension"), and that only at a later stage, for informativeness reasons, they access the comparative reading (*tall* means "taller than a standard"). We present and discuss the results of an experimental study in which we aimed at "turning adults into children". We show that, when informativeness is not at stake, even adults seem to access the nominal interpretation of relative adjectives. We argue that the transition from the nominal to the comparative reading of relative adjectives might be easily accounted for by a partial function approach.

Keywords: Semantics of adjectives, language acquisition, experimental semantics and pragmatics.

1 The Semantics of Gradable Adjectives

The class of adjectives is heterogeneous, and different subclasses can be identified. Focusing on the distributional pattern, a distinction is drawn between gradable adjectives (GAs, those adjectives that can enter into a comparative construction and be modified by degree expressions, e.g., *tall*, *intelligent*, *open*), and non-gradable adjectives (e.g., *Dutch*, *wooden*, *four-legged*). Ignoring the so-called intensional adjectives (cf. [15]), non-gradable adjectives can be semantically viewed as referring to properties of individuals: the meaning of the adjective *Dutch* can be identified with the set of individuals that are Dutch. Gradable adjectives, instead, cannot be straightforwardly interpreted as functions from individuals to truth-values,[1] provided that they exhibit context dependency (the same individual, for instance Bart, who is 188 cm tall, can be

[1] This is true for the subclass of *relative* gradable adjectives. *Absolute* gradable adjectives (*full*, *open*, *straight*, etc.) could in principle be analysed as denoting sets of individuals since they do not exhibit the same context dependency (cf. [10,17]). In this paper, however, we will restrict our attention to relative gradable adjectives, and will not discuss the data obtained with absolute gradable adjectives.

M. Aloni et al. (Eds.): Amsterdam Colloquium 2011, LNCS 7218, pp. 170–179, 2012.
© Springer-Verlag Berlin Heidelberg 2012

judged as *tall* for an Italian man, and as *not tall* for a basketball player) and borderline cases (if Bart is Dutch, we wouldn't be confident in judging the sentence neither true nor false, since the average height for men in the Netherlands is 185 cm).

To account for these facts, two main approaches have been proposed. The degree-based analysis (cf. [9] and references therein) assumes that GAs differ from non-gradable adjectives in that they denote a relation between an individual and a *degree*, where degrees are conceived of as abstract entities ordered along a scale associated with the dimension referred to by the adjective. According to the degree-based semantics, the GA *tall* evokes a scale of degrees ordered with respect to the dimension of height, and a sentence containing *tall* is viewed as the assertion that an individual possesses the tallness property to a certain degree. Sentences in which a bare adjective is used (e.g. "Bart is tall") are analysed as concealed comparative sentences (e.g. "Bart has a degree of height that exceeds a standard of comparison"). In case of *relative* GAs, the standard of comparison has to be contextually retrieved, for instance by making reference to the intended comparative class (e.g., Italian men or basketball players) or by taking into account other perceptually salient individuals (e.g., "tall compared to Leo").

Ewan Klein [11-13], on the other hand, proposed a unified account for adjectives, viewing them as functions from individuals to truth values. In his approach, non-gradable (extensional) adjectives are interpreted as total functions (i.e., as sets of individuals that share a property) while GAs are interpreted as *partial* functions. The idea is that, once we identify a particular domain of individuals that fall under the extension of a GA such as *tall*,[2] the meaning of *tall* consists of a function that sets apart the individuals who count as (definitely) tall and those who are (definitely) not tall. This function, however, is partial in that it remains undefined for a whole sets of individuals that lay in-between, and to whom the function is not able to assign a value. Moreover, the restriction of the domain of application of the GA to particular subclasses (e.g., the set of Italian men or the set of basketball players) leads to different outputs of the value of the function. This accounts for the fact that the same man might count as *tall* with respect to the class of Italian men, and as *not tall* with respect to the class of basketball players. Thus, Klein accounts for the existence of borderline cases by means of the partiality of the GA function, and for the contextual variability of GAs by restricting their interpretation to different comparison classes.

The degree-based and the partial function approaches have been compared on theoretical and empirical grounds. On the theoretical side, the partial function approach appears simpler and more in line with Frege's principle of compositionality, since it does not need to posit degrees as primitive entities, and it derives the meaning of the comparative construction from the meaning of the positive form (cf. [11]). On the basis of empirical findings, in which, simplifying a bit, subjects were tested by

---

[2] Even if Klein does not explicitly address this issue, we think that, in this approach, GAs are to be conceived of as presuppositional items, inasmuch as the function *tall* can assign a value only to objects that have a (salient) vertical dimension: a sphere, for instance, would not fall under the extension of *tall*. This undefinedness, however, is more akin to a presuppositional failure than to the kind of undefinedness associated to the vagueness of borderline cases.

modulating the way individuals in the comparison classes were ordered (cf. [16]), the two theories can be seen as nearly equivalent. Our aim is to look for new evidence in favour or against these approaches by taking into account the process of language acquisition. Although the proponents of these theories did not analyse this question,[3] and thus they did not make explicit predictions about the way children ought to interpret gradable adjectives, we think that children's linguistic behaviour could open a window into the semantics of adjectives.

2 Experiment 1: What Children Can Tell Us about the Semantics of Adjectives

The process of language acquisition typically passes through different phases in which children may exhibit a linguistic behaviour that does not conform to adults'. For instance, in the first phase of the acquisition of the meaning of common nouns, children often make errors of over-extension, labelling every man as *daddy*. This phenomenon has been taken as evidence of the way children establish the mapping between words and objects. Our aim is to look at the way children understand and produce adjectives, focusing in particular on the errors they make, and then discuss how the two theories presented above might account for these errors.

Despite the fact that adjectives are harder to acquire than nouns, children start using relative GAs from age 2, and by the age of 3 they produce GAs such as *big/small(little)*, *tall/short*, *long/short*, *high/low*, *heavy/light*. As far as their interpretation is concerned, this has been tested in several studies in which children have been shown to access both the normative and the perceptual readings of GAs. For example, children could judge a mitten as *big* or *small* "as a mitten" or relative to another object physically present (cf. [4-5] a.o.). Nevertheless, it has been noticed that children make a consistent series of errors. In the first place, they make substitution errors: for example, they judge an object that is meant to be *small* as *big*, cf. [3], or they substitute *tall* with *big*, cf. [1]. Also, they exhibit "extreme labeling": when they are presented with a series of seven items that decrease along a relevant dimension (for instance, seven blocks decreasing in size), they tend to label only the extreme item of the series as *big* or *small* (cf. [19, 21]), while adults tend to put the cross-over point around the mid of the series. Since the experiments designed to test children's comprehension of relative GAs explicitly provided them with a comparative class, or some perceptual cues, it is hard to tell from these studies what kind of interpretation children access in the first place, and which framework better accounts for the facts.

In a previous study we conducted (cf. [6]), we tested a group of 20 3 year-olds, 20 5 year-olds and adults as controls. In the first part of our experiment we employed a Truth-Value-Judgment Task: we presented single objects in isolation, and asked

---

3 With the exception of Syrett [21], in which she discusses children's interpretation of relative and absolute GAs within a degree-based account, even though she explicitly admits that some of the children's unexpected behaviour would be straightforwardly explained assuming a kleinian-like semantics (cf. in particular [21], p. 41-42).

subjects whether the description provided by a puppet was correct, not correct, or whether they couldn't tell. In that study, we didn't provide any contextual standard to judge relative GAs. The objects presented were "abstract" in the sense that they could not evoke any normative class, i.e. were not recognizable as having a real function or as belonging to a certain category. Also, they were presented in isolation, without other perceptual cues. For instance, we presented a 17,5 cm tall wooden rod describing it as "This is tall". Since the context did not provide any standard of comparison, we expected a prevalence of "I don't know" answers and/or chance distribution of "yes" and "no" answers. Our predictions were confirmed in case of adults: in judging descriptions with a relative GA, 60% of the times adults answered "I don't know", 30% answered "yes" and 10% "no". Quite surprisingly, 3 year-olds provided more than 88% of yes-answers (with "no" and "I don't know" answers around 6% each), independently of the polarity of the adjective (i.e., they equally accepted both antonyms). Subjects' answers distribution of relative GAs is plotted in Figure 1, differentiating by Age of participants.

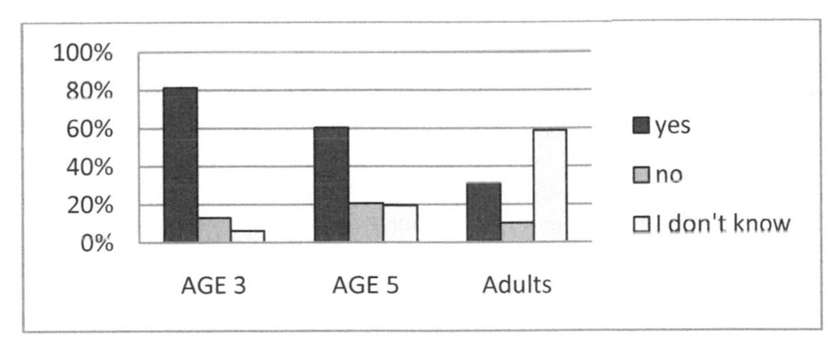

Fig. 1. Relative GAs: subjects' distribution by Age

What emerged from this study is an interesting difference between children and adults. When presented with an abstract object in isolation described by "This is relative-GA" (e.g. "This is big" or "This is small"), 3 year old children accepted the description, while adults tended to remain sceptical. Moreover, a developmental trend seemed to emerge, with the 5 year-olds answering "yes" 60% of the times, and 20% of the times opting for the answers "no" and "I don't know" respectively. Statistical comparisons by means of Fisher Exact Tests for count data revealed a significant general effect of Age in case of relative GAs ($p<.001$). A significant difference also emerged between 3 year-olds and 5 year-olds, between 3 year-olds and adults and between 5 year-olds and adults (all $ps<.001$). However, this effect might be partly due to the fact that children, in general, used the option "I don't know" much less than adults. Considering the proportion of "yes" answers over the total, the difference between the 3 and the 5 year-olds disappears ($p=.218$), while it does not comparing the 5 year-olds and the adults ($p<.01$). Summing up, younger children overwhelmingly accepted descriptions of the form "This is relative-GA" applied to abstract objects in isolation, when the context did not supply any standard of comparison.

We believe that this fact calls for an explanation. At the meantime, we think it very unlikely that this high rate of acceptance might be the result of a lack of knowledge of the meaning of the adjectives themselves combined with an alleged children's tendency to opt for a yes-answer in case of uncertainty. In fact, there's ample evidence in the literature that children know adjectives like the ones we tested from age 2 on (cf. [21]). Additional evidence against this hypothesis comes from the results of the second part of our experiment in which we tested our participants with a Scalar-Judgment-Task: we presented a series of seven objects decreasing or increasing along a relevant dimension, and asked subject whether the description "This is relative-GA" could apply to each item in the series. For example, we presented a series of seven wooden rods ranging from 20 cm to 5 cm in height, and asked for each of them "Is this tall?" (in one list) or "Is this small?" (in the other list). What we found is that the 3 year olds, though showing the well-known phenomenon of extreme labelling, were also very consistent in their answers: they consistently accepted the description for the first item of the series, and rejected it for the last item. We believe that this behaviour is evidence of the fact that children knew the meaning of the relative GAs tested. Furthermore, we exclude the hypothesis that children were simply being extremely charitable and answered "yes" whenever they did not have strong reasons to reject the description on the basis of the results that we obtained with some of the fillers, that were specifically designed to prompt an I don't know-response. For example, a toy-zebra was described as "This is obedient". As expected, more that 90% of the adults reacted to these fillers by saying "I don't know". Interestingly, though, the 3 year-old children split between "yes" and "no": around 50% accepted the description but half of them didn't, showing that it is not the case that they chose the more charitable option whenever they can.

Having excluded these simpler explanations, we propose the following explanation of the children's results. Assuming that there is a developmental trend in the interpretation of relative GAs, we propose that children start by interpreting relative GAs as if they were referring to sets of individuals sharing a certain property (just like common nouns and intersective adjectives), and only at a later stage they switch to the comparative-like reading. A similar hypothesis was proposed by H.H. Clark [2], who noticed that the positive relative GAs (*tall*, *long*, *heavy*…) are ambiguous between a nominal and a comparative use. To exemplify, let us consider *tall* in (1) and (2):

1. Lia is 156 cm tall.
2. Lia is tall.

In (1), the relative GA *tall* is interpreted nominally and is neutral: i.e., it does not imply that Lia is tall with respect to a standard of comparison, but it simply indicates that the dimension that is relevant for attributing a particular property to Lia is the vertical dimension. In (2), on the other hand, the relative GA is interpreted comparatively, and the reference to a contextual standard of comparison is necessary.

If we assume that younger children assign a nominal-like interpretation to relative GAs at first (and thus, *tall* would simply mean "that has a vertical dimension", *fast* "that has some speed", and so on) we could immediately account for the fact that younger children over-accepted the descriptions involving relative GAs in the

Truth-Value-Judgment Task. Under this nominal-like interpretation, in fact, the description used by the puppet would truthfully describe the objects presented. Only at a later stage, around 5 years of age, children might then get to realize that the nominal-like interpretation of relative GAs is highly under-informative in normal contexts, and would eventually switch to the comparative-like interpretation, which is the one that is standardly attributed to relative GAs by adults. Our hypotheses would account for the fact that (i) younger children accept "This is relative-GA" referred to objects that do possess the property predicated by the adjective even in the absence of a relevant standard of comparison; (ii) older children switch to the adult-like reading once they realise that in normal contexts the nominal-like (minimal) interpretation would not add sufficient new information, thus being underinformative.[4] This hypothesis further assumes a contiguity between the nominal-like and the comparative-like interpretations, and that the latter is derived from the first one when informativeness is at stake.

3 Experiment 2: What Adults Can Tell Us about the Semantics of Adjectives

If the hypothesis presented above is well-grounded, we should be able to detect the minimal, nominal-like, reading of relative GAs also in adults, once we suspend the standard requirements for informativeness. In a second experiment, we tested adults' interpretation of relative GAs in a context that was manipulated so as to leave informativeness aside. Our aim was that of investigating whether adults would accept relative GAs in this experimental setting, paralleling the younger children in our previous study.

We tested 73 Italian adults, randomly assigned to one of two lists. Each list comprised a total of 29 items. Of these, fifteen were control items: eight were clearly true or clearly false, while the remaining seven were cases in which the adjective used was not applicable (NA) to the object, and were meant to prompt a "no" response by the participants. For example, a typical NA-item comprised a toy plane that was described as "This is vegetarian", a description that we meant to be rejected by our participants. Given that children in our first study did not use the option "I don't know", adults were only given a dichotomic choice between responses "yes" or "no" in this case. In the second place, a training session was added, in which adults were asked to be as tolerant as they could with respect to the description provided by a puppet. The puppet was introduced as being an alien who had some hypotheses about the functioning of our language. Subjects were instructed to help him refine his hypotheses, correcting him when he was saying something that was clearly false or unacceptable, but accepting his descriptions as long as they could apply in some way to the situation or the object presented, even if they might be not optimal or not fully appropriate.

[4] In a recent paper, Katsos and Bishop [8] argue that children's performance with scalar implicatures (i.e., the fact they seem not to compute them, accepting underinformative sentences) is not imputable to the fact that they are not aware of under-informativeness, but to the fact that they more tolerant of pragmatic infelicity. Even though the question is interesting and worth pursuing, we will not explore this issue here.

During this training, for instance, we presented a situation in which a toy-boy had four coins. With respect to this situation, we recommended our subjects to reject clearly false sentences (e.g. the description "The boy has seven coins", that is blatantly false), but we also trained them to be charitable and accept the alien's under-informative sentence "The boy has three coins". To train them not to reject under-informative statements, we convinced them that they might confound the alien correcting him in cases like this: if he were told that it is false that the boy has three coins in a situation in which he actually has four, then the alien might think that the boy could not afford to buy an ice-cream that costs three coins, which is not true. After the training session, we presented abstract objects in isolation, that were described by the alien as "This is adj", and asked the subjects to accept or reject these statements.

On the basis of our hypothesis, we predicted an increase of "yes" answers in case of Relative GAs compared to our previous study, inasmuch as the adjective could truthfully (even if not optimally) describe the object under the minimal, nominal-like interpretation of the GAs. To make sure that adults were tolerant with critical cases only, we also predicted them *not* to accept the descriptions in which the adjective used did not apply to the object, as it happened for the NA-controls. For example, we expected adults to reject the description "This is vegetarian" referred to a toy-plane, on the basis that the property predicated by the adjective (i.e. "being vegetarian") cannot apply to a plane under any circumstance.

The results confirmed our predictions, as it is evident from Figure 2: adults' acceptance rate of relative GAs in this second experiment was 73%; performance with controls was at ceiling and, in particular, they did not hesitate in rejecting the description when the adjective used was not applicable to the object.

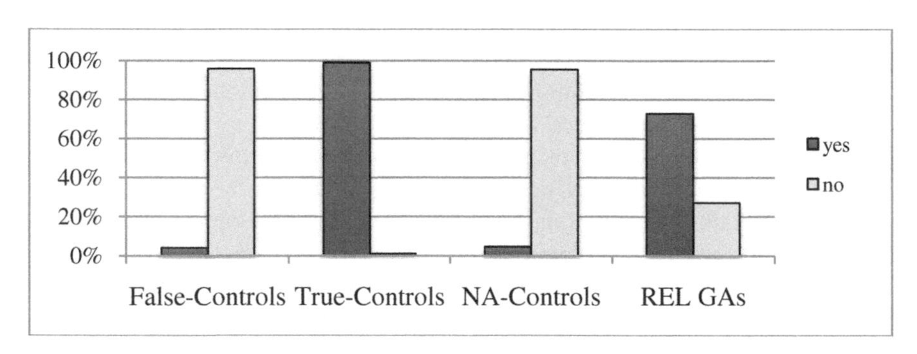

Fig. 2. Distribution of adults' responses in Experiment 2

Comparing adults' performance in the two studies in the interpretation of relative GAs, the increase of "yes" answer in this second experiment is significant, even taking into consideration the fact that in the second study the probability of answering "yes" was .50, while it was .33 in the first experiment, where three options of response were given ($\chi^2(2) = 85.35$, $p<.0001$).[5] Comparing children's and adults'

[5] The proportion of "yes" answers in the second study is also different from chance ($\chi^2(1) = 46.21$, $p<.0001$).

performance on relative GAs across the two studies, no difference was detected, as shown in Figure 3. Statistical analysis confirmed that adults did not differ from the 3 and the 5 year old children in case of relative GAs ($\chi^2(1)= 2.27$, $p = 0.13$; $\chi^2(1) = 1.77$, $p = 0.18$). The aim of turning adults into children did work, in the end.

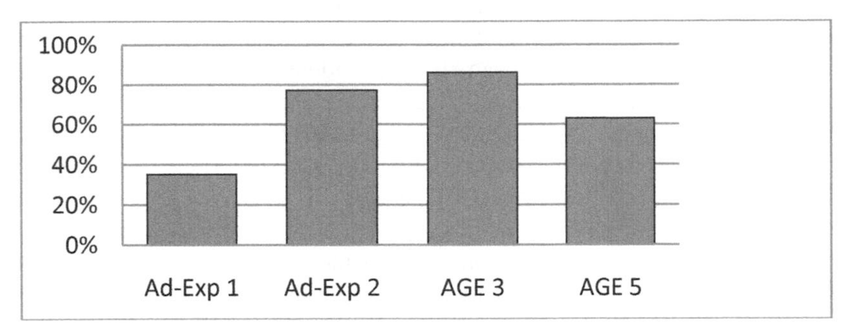

Fig. 3. Proportion of "yes" answers to relative GAs across studies (Experiment 1 and 2) and populations (adults, and children aged 3 and 5)

4 Conclusion

Summing up, we suggest that children start interpreting relative GAs as common nouns and non-gradable adjectives, i.e. as denoting sets of individuals, and by identifying their denotation with the domain of application of the GA. This hypothesis is in line with the cognitive constraints that have been assumed to guide children's first acquisition of the mapping between a word and its denotatum, in particular under the assumption that a novel word will refer to stable properties of an object, i.e., in semantic terms, it will denote a set of individuals that share a property. Only at a later stage, for informativeness reasons, children realize that a stronger meaning – one that involves a comparison – is needed. Before reaching this adult-like competence, however, we suggest that children pass through an evolutionary phase in which they switch from a nominal to a comparative interpretation.[6] Recall that the normative interpretation of relative GAs requires the identification of the intended comparison class, and the setting of an appropriate standard to identify objects in that class as having or not having (and possibly neither having nor not-having) the GA property. We believe that 3 year-olds are in fact able to take the contextual comparison class into account when this is explicitly provided (e.g., in [4], children were asked to judge a given mitten as big or small *as a mitten*), but they encounter problems in setting the appropriate standard: while adults posit it as the average (or the median) of the comparison class, younger children tend to be categorical and make crisp judgments that lead them to judge only the extreme of a series as having the GA property. As already hinted at, the evolutionary step required to reach an adult-like performance consists in the ability to retrieve a comparison class even when this is not explicit, and in setting the standard of comparison in a less categorical way.

---

[6] One could draw an interesting parallelism with the lexical proposal for Scalar Implicatures in children (cf. [7]), a topic that we won't address here.

We believe that the hypothesis of a contiguity from a nominal to a comparative interpretation of relative GAs can be easily accounted for by a partial function analysis of GAs: children's nominal interpretation would amount to the preliminary, presuppositional-like, step that requires the identification of the appropriate domain of application of relative GAs: in this sense, the GA function *tall* would apply only to objects that have a salient vertical dimension (see footnote 2). Children would then encounter problems in the subsequent steps, namely in the identification of the intended subdomain (comparison class) and in realizing that the output of the GA function is a tripartition of that domain. Children could in fact resort to the nominal-like interpretation when the objects they are asked to judge are "abstract" (i.e., not amenable to a comparison class). When the relative GA is used to modify a common noun that clearly evokes a comparison class instead (tall *child*, big *mitten*), children could persist in adopting a total function interpretation of the relative GA, interpreting it in a subsective way: *tall child* would then identify, within the set of children, the subset of those that are tall (cf. [15]).

On the contrary, we don't see any easy way to account for young children's behaviour within a degree-based approach. One could claim, as an anonymous referee suggested, that at least adults' behaviour in Experiment 2 could be explained by assuming that adults are in fact interpreting relative GAs such as *tall* as meaning "having a degree of height that exceeds a *minimum* standard of height", i.e., as if relative GAs were lower bounded absolute GAs (cf. [10]). However, we believe that such an approach would face problems in accounting for children's behaviour. As for this, the degree-based analysis should postulate that young children are indeed able to associate abstract entities like scales of degrees to gradable adjectives, since this is seen as a preliminary step to interpret GAs. In the meantime, though, children (i) would mistakenly set the standard of evaluation to a lower boundary (i.e., the minimum standard) in the case they are presented with a single "abstract" object, as documented by our Experiment 1, or with a series of objects, as emerges from their well attested extreme labelling tendency; (ii) at the same time, they should be able to use a standard that is set to the average of a comparison class when they are asked to judge a "real life" object, as shown by their adult-like performance in judging, for example, a mitten to be big or small *as a mitten*. Summing up, we believe that a partial function analysis would account for children's and adults' behaviour in a more straightforward way, whereas a degree-based approach would necessitate ancillary hypotheses to get the facts right.

References

1. Brewer, W., Stone, J.B.: Acquisition of spatial antonym pairs. Journal of Experimental Child Psychology 19, 299–307 (1975)
2. Clark, H.H.: The primitive nature of children's relational concepts. In: Hayes, J.R. (ed.) Cognition and the Development of Language, pp. 269–278. Wiley, New York (1970)
3. Donaldson, M., Wales, R.J.: On the acquisition of some relational terms. In: Hayes, J.R. (ed.) Cognition and the Development of Language, Wiley, New York (1970)

4. Ebeling, K.S., Gelman, S.A.: Coordination of size standards by young children. Child Development 59, 888–896 (1988)
5. Ebeling, K.S., Gelman, S.A.: Children's use of context in interpreting big and little. Child Development 65, 1178–1192 (1994)
6. Foppolo, F., Panzeri, F.: Do children know when their room counts as clean? In: NELS 40: Proceedings of the 40th Annual Meeting of the North East Linguistic Society. GLSA, Amherst (in press)
7. Foppolo, F., Guasti, M.T., Chierchia, G.: Scalar Implicatures in Child Language: Give Children a Chance. Language Learning & Development (in press)
8. Katsos, N., Bishop, D.V.M.: Pragmatic tolerance: Implications for the acquisition of informativeness and implicature. Cognition 120(1), 67–81 (2011)
9. Kennedy, C.: Vagueness and Grammar: The Semantics of Relative and Absolute Gradable Adjectives. Linguistics and Philosophy 30, 1–45 (2007)
10. Kennedy, C., McNally, L.: Scale structure and the semantic typology of gradable predicates. Language 81(2), 345–381 (2005)
11. Klein, E.: A semantics for positive and comparative adjectives. Linguistics and Philosophy 4, 1–45 (1980)
12. Klein, E.: The interpretation of adjectival comparatives. Journal of Linguistics 18, 113–136 (1982)
13. Klein, E.: Comparatives. In: von Stechow, A., Wunderlich, D. (eds.) Semantics: An International Handbook of Contemporary Research, pp. 673–691. de Gruyter, Berlin (1991)
14. Palermo, D.S.: More about less: A study of language comprehension. Journal of Verbal Learning and Verbal Behavior 12, 211–221 (1973)
15. Partee, B.H.: Lexical semantics and compositionality. In: Osherson, D. (series ed.), Gleitman, L., Liberman, M. (volume eds.) Invitation to Cognitive Science. Part I: Language, pp. 311–360. MIT Press, Cambridge (1995)
16. van Rooij, R.: Vagueness and linguistics. In: Ronzitti, G. (ed.) Vagueness: A Guide. Springer, Berlin (2010)
17. Rotstein, C., Winter, Y.: Total Adjectives Vs. Partial Adjectives: Scale Structure And Higher-Order Modifiers. Natural Language Semantics 12, 259–288 (2004)
18. Sera, M., Smith, L.B.: Big and Little: "Nominal" and Relative Uses. Cognitive Development 2, 89–111 (1987)
19. Smith, L.B., Cooney, N.J., McCord, C.: What is "High"? The Development of Reference Points for "High" and "Low". Child Development 57(3), 583–602 (1986)
20. von Stechow, A.: Comparing Semantic Theories of Comparison. Journal of Semantics 3, 1–77 (1984)
21. Syrett, K.: Learning about the structure of scales: Adverbial modification and the acquisition of the semantics of gradable adjectives. Doctoral dissertation. Northwestern University, Department of Linguistics (2007)

Underspecified Representations of Scope Ambiguity?

Janina Radó and Oliver Bott

SFB 833, University of Tübingen

Abstract. We tested whether quantifier scope is left underspecified until disambiguating information is encountered. We measured reading times while comprehenders read German scope ambiguous doubly quantified sentences with a configuration of quantifiers involving scope conflict. After reading the sentence a disambiguating picture was presented. Half of the pictures were only compatible with surface scope while the other half disambiguated towards inverse scope. To avoid scope reanalysis perceivers should thus delay scope interpretation and maintain an underspecified representation. Contrary to this prediction, indications of scope conflict could be observed already during reading the second quantifier, ie. well before the disambiguation. To find out whether scope computation starts even before the processor has encountered a complete predication, the experiment also included a (inverse linking) construction in which the two quantifiers appeared before the verb. In this configuration we didn't find any indication of scope conflict at the second quantifier. Taken together, our study provides evidence for immediate scope assignment, but only if the processor is dealing with a complete minimal sentence including not only the quantifiers but also the verbal predicate.

Underspecified representations are commonly used to model semantic ambiguity, and scope ambiguity in particular. They offer an elegant way to capture the interpretation possibilities in a single compact representation, thus avoiding a combinatorial explosion of readings as the number of operators increases (see [3] for a critical discussion). This property has repeatedly been claimed to make them cognitively plausible: since only one representation is constructed, scope-ambiguous sentences can be interpreted in a fast and efficient way.

There is indeed evidence that perceivers only specify the interpretation of the sentence as much as necessary for a particular task (eg. pronouns are not automatically resolved, cf. [6], [11]). A similar claim has been made with respect to quantifier interpretation: scope is only resolved when necessary, eg. when disambiguation is encountered, and may remain completely underspecified in some cases (see [11] for an example). Sanford and Sturt use the lack of scope preferences in some constructions to argue for underspecified scope representations.

It is important to note that the connection between underspecification and shallow processing is only indirect. Comprehenders often do not compute a representation that is detailed, complete, and accurate with respect to the input, but only one that is 'good enough' (see [4]). They process only as deeply as

M. Aloni et al. (Eds.): Amsterdam Colloquium 2011, LNCS 7218, pp. 180–189, 2012.
© Springer-Verlag Berlin Heidelberg 2012

necessary under the circumstances and may end up with just a partial representation if that is sufficient for the task at hand rather than constructing a single connected representation (which may or may not be fully specified) for the complete sentence. By contrast, underspecified representations are complete compact representations encoding all interpretation possibilities. They thus require deep processing but may not result in a specified representation (eg. of scope), if there is no disambiguating information.

Intuitively, some scope readings are easier to get than others. This is also supported by psycholinguistic findings; eg. [8] find a preference for the wide–scope interpretation of the second quantifier in the inverse linking construction in (1):

(1) George has a photograph of every admiral.

Such graded preferences have been captured in underspecification accounts by postulating weighted dominance constraints (eg. [7]), which make some representations cheaper and easier to derive than others. The question now is whether the different readings are computed (and ranked) automatically, or only when needed for the task, as underspecification theory would have it. As it turns out, existing evidence does not allow us to answer this question (cf. [2]). For instance, [5] examined eye movements while perceivers read sentences like (2) (see also [10]) in order to find evidence for early disambiguation of quantifier scope.

(2) a. The celebrity gave a reporter from the newspaper every in-depth interview, but the reporter(s) was/were not very interested.
b. The celebrity gave every reporter from the newspaper an in-depth interview, but the interview(s) was/were not very interesting.

They report longer *total* reading times at the second quantifier in (2-a) than in (2-b), and take it as evidence for a scope conflict arising at the second quantifier in (2-a): linear order and grammatical role favor the wide-scope existential reading, whereas inherent properties of the quantifier support the wide-scope universal interpretation. In (2-b) all factors point to the wide-scope existential reading, thus processing is easier. This is also reflected in off-line ratings: the $\exists\forall$ reading is accepted less in (2-a) than in (2-b). Crucially, however, both of these are late effects; the scope conflict in (2-a) does not have any measurable influence on first-pass reading times. The results thus do not exclude the possibility that perceivers only computed a fully specified scope interpretation when they read the disambiguating information in the second clause.

To test the cognitive plausibility of scope underspecification we need to use a task that guarantees deep processing and look for evidence for scope interaction before disambiguating information is present. We designed a combined self-paced reading/verification experiment that we think fits the bill. Participants first read doubly quantified sentences and then had to decide whether the sentence matched a scope disambiguating card display. The sentences in the reading task exhibited a scope conflict between the inherent properties of the quantifiers and the construction they occurred in. Crucially, disambiguation was not available during the reading phase, so any reading time effects could be attributed to on-line resolution of scope relations. Moreover, participants were instructed to be accommodating and accept the picture if the corresponding reading was available at all. This way we hoped to encourage them to maintain an underspecified interpretation as long as possible.

1 Methods

Materials: We tested German inverse linking constructions and manipulated the quantifier in the embedded position. One type of complex quantifiers was built of the determiners *genau ein* (*exactly one*) and *alle* (*all*) as illustrated in the sample item in (3-a), whereas the other type embedded distributive *jeder* (*each*) as in (3-b). The asterisks indicate segmentation in self-paced reading.

(3) Genau* ein Tier* auf* a) allen b) jeder* Karte(n)* ist* ein Affe.
Exactly one animal on a) all b) each card(s) is a monkey.

Condition (3-a) exemplifies a case of scope conflict between quantifiers in the same DP: the inverse linking construction strongly biases towards inverse scope, but since the non-distributive *alle* (*all*) prefers not to outscope the first quantifier we expected competition between readings in the *all* conditions. This is different in (3-b) with *jeder* (*each*). Here, both factors point in the same direction therefore *each* should be interpreted with wide scope without any difficulty. As for the final interpretation, scope conflict should lead to fewer inverse interpretations in the *all* than in the *each* conditions.

In addition, we also tested doubly quantified sentences in which the quantifiers appeared in a non-embedded configuration.

(4) Genau* ein Affe* ist* auf* a) allen b) jeder* Karte(n)* zu* finden.
Exactly one monkey is on a) all b) each card(s) to find.

Intuitively, the conditions (4) show the same contrast as (3), thus we expect the same pattern of interpretations in both construction types. In addition, comparing the DP- and the sentence conditions we can test when scope computation takes place if the initial interpretation is not underspecified. We will elaborate this point in the Predictions.

Each of the doubly quantified conditions (3) and (4) was paired with two disambiguating card displays yielding eight doubly quantified conditions in a 2 x 2 x 2 (*construction* x *quantifier* x *picture*) factorial design. The linear ∃!∀ card displays had the same object on all three cards, but the second card contained an additional object of the same category, as in Figure 1a. The inverse ∀∃! card displays had exactly one object of the relevant kind on each card, but different ones (cf. Figure 1b). Again, the second card provided disambiguation.

To control for potential differences in lexical processing between *all* and *each* we included the controls in (5). The controls were always paired with a card display that had the same object on all three cards (cf. Figure 1c).

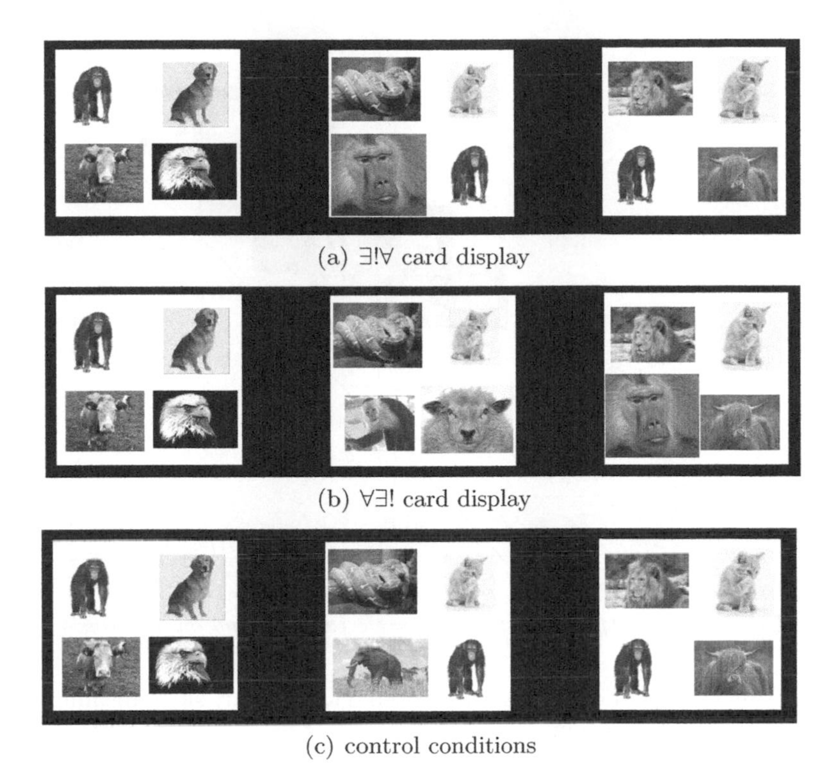

(a) ∃!∀ card display

(b) ∀∃! card display

(c) control conditions

Fig. 1. Fully uncovered card displays for the sample item. Figure 1a is only compatible with wide scope of *exactly one* (∃!). Figure 1b disambiguates towards wide scope of the universal quantifier. Figure 1c is the card display presented with the control conditions. In the experiment, card displays were uncovered card by card from left to right.

(5) Auf* a) allen b) jeder* Karte(n)* ist* ein Schimpanse.
On a) all b) each card(s) is a chimpanzee.

In total, this yielded ten conditions and we constructed 60 experimental items (ie. sentence/picture sets) in each of them. Additionally, we created 80 fillers which served several purposes. First, we made sure that all sorts of quantifiers would appear in the experiment. Second, the fillers varied in structure and in a number of cases presented crucial information at the sentence final segment (eg. *A drill can be found on all cards in their upper half*). Third, a number of fillers had pictures that required accommodating interpretations, ie. a total of three objects for *einige* (*a few*). Finally, 40 of the fillers were clearly false to ensure that on a reasonable proportion of trials participants had to reject the card display. We took the experimental items and the fillers and constructed 10 lists according to a latin square.

Predictions: Assuming underspecification with weighted constraints, the quantifier manipulation should have no effect on the reading times. We do expect

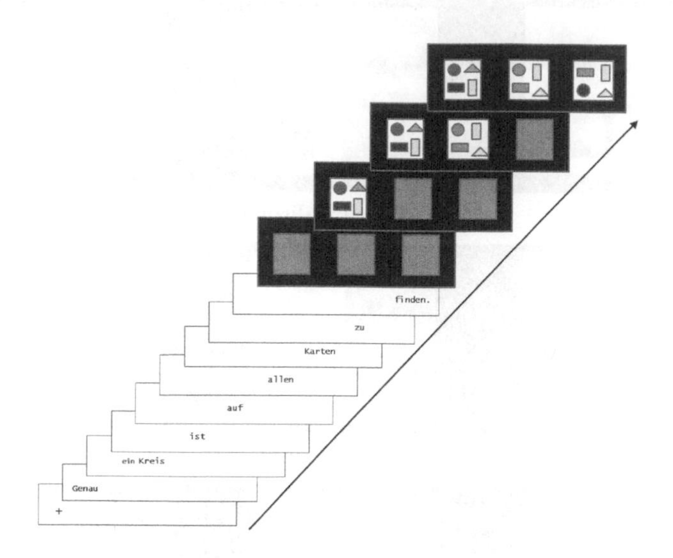

Fig. 2. Sample trial (*'Exactly one circle can be found on all cards'*)

differences at the disambiguating card (card 2), however, since disambiguation should force scope resolution.

By contrast if scope is interpreted automatically, i.e. even without disambiguation, then the scope conflict should lead to processing difficulty already before the second card. In fact, we may find processing effects at the earliest point where scope computation may take place, ie. at the second quantifier. We will dub this *immediate full interpretation.*

Finally, it is conceivable that semantic interpretation lags behind syntactic processing. At least some semantic interpretation processes require domains larger than individual words [1]. The same may hold for quantifier scope as well, especially as thematic roles seem to be one factor in determining the scope needs of a quantifier [9]. Then full scope interpretation would only be expected to take place once the verb and all of its arguments have been received. We will call this the *minimal domain* account. Under this view scope interpretation should be possible at the second quantifier in the sentence conditions in (4), but scope would remain underspecified until the lexical verb in the DP conditions in (3). Note that this still assumes the automatic computation of scope relations as soon as all relevant information is available.

To sum up, all three approaches predict a quantifier effect at the disambiguating card. In addition, under *immediate full interpretation* and the *minimal domain* approach we also expect reading time differences either at the second quantifier, or at least when the verb and all its arguments have been received. Thus the pattern of data in the inverse linking conditions will be crucial to distinguish between these two accounts.

Procedure and Participants: Participants first read a sentence phrase by phrase employing non-cumulative self-paced reading with a moving window display. Segments outside the presentation window were masked by replacing all characters including punctuation with empty spaces. This was done to ensure that participants would not know how much material was yet to come. After reading the final segment the sentence disappeared from the screen and a layout of cards was presented. Participants had to uncover the cards one by one and decide at each whether the sentence is already true or false or whether more information is needed and in that case move on to the next card. A sample trial is illustrated in Figure 2.

An experimental session took approximately 30 minutes. It started with written instructions followed by a practice session of 10 trials. In the instructions, we emphasized that whenever a picture was compatible with a sentence (even in case the interpretation is hard to get) participants should judge "yes, true". Then followed the experiment in a single block with an individually randomized presentation order for each participant. 40 participants (mean age 25.9; 31 female) from Tübingen University were paid 5 Euro for their participation.

2 Results and Discussion

Judgments: Table 1 presents the mean judgments in the ten conditions. Across the board, the linear reading was only accepted 17.2% of the time whereas the inverse conditions were accepted on average 85.1% of the time. However, acceptance in the linear conditions was 9.8% higher than in the false fillers which were only accepted 7.4% of the time indicating that linear scope, even though highly dispreferred, was still possible. Besides this very strong general preference for inverse scope, the distribution of readings showed a clear contrast between *all* and *each*. The inverse conditions were accepted 77.5% in the *all* sentence condition, 90.4% in the *each* sentence condition, 81.2% in the *all* DP condition and 91.3% in the *each* DP condition. We analyzed the judgments in the inverse linking and sentence constructions in a logit mixed effects model analysis including the fixed effects of *construction*, *quantifier* and *reading* and the random intercepts of participants and items. The analysis revealed significant fixed effects of *reading* ($z = -13.48$; $p < .01$) and of *quantifier* ($z = 3.15$; $p < .01$). Besides a marginally reliable interaction between *quantifier* and *reading* ($z = -1.67$; $p = .09$) no other effects reached significance (all $p \geq .29$). To further break down this interaction we computed separate logit mixed effects model analyses for the linear and the inverse scope conditions. While the linear scope conditions didn't differ significantly from each other (all $p \geq .23$), the analysis of the inverse conditions revealed a significant effect of *quantifier* ($z = 3.44$; $p < .01$). Taken together, the judgments provide evidence for a clear *all*/*each* contrast. Whereas *each* very strongly biases towards inverse scope, the bias is weaker with *all*. On the other hand, the complete lack of effects involving the factor *construction* suggests that the two constructions had comparable scope distributions.

Table 1. Mean proportions of "yes, true" judgments in percent (+ SE of the mean)

	Universal Quantifier	
	all	*each*
Inverse Linking ∃!∀	15.4 (2.3)	19.2 (2.5)
Inverse Linking ∀∃!	81.2 (2.5)	91.3 (1.8)
Sentence ∃!∀	16.3 (2.4)	17.5 (2.5)
Sentence ∀∃!	77.5 (2.7)	90.4 (1.9)
Control	95.8 (1.3)	97.1 (1.1)

Verification Stage: Besides the judgments we analyzed two further dependent measures in the verification stage: the decision point, ie. the particular card at which participants made their decision and the RTs of "yes, go on" button presses. The analysis of decision points revealed that participants decided at the earliest possible point. Most of the rejections in the linear card displays were launched directly from the second, disambiguating card without ever inspecting the last card. In the sentence condition with *all* and the linear display 64.7% of the rejections came from the disambiguating card, the corresponding *each* condition had 58.6% rejections, the inverse linking *all* condition had 59.6% rejections and the inverse linking *each* condition had 66.5% rejections at card 2. In the inverse card displays the situation was different. Here in 92.6% of the trials participants uncovered all three cards before they provided their judgment which in the majority of all cases was "yes, true". 6.2% out of the 7.4% early rejections came from card 2.

The analysis of RTs of "yes, go on" button presses provides us with a measure of how difficult it was to evaluate a particular card. We corrected RTs for outliers by excluding values above 5s. RTs up to the second card were analyzed with mixed effects models including the fixed effects of *construction* (three levels: *sentence* vs. *inverse linking* vs. *control*) and *determiner* (two levels: *all* vs. *each*) and the random intercepts of participants and items. The picture RTs of the fully covered card display and the first card are depicted in Figure 3[1]. While there were no reliable differences between the conditions at the fully covered card display (LME: all $t < 1$), the picture RTs of the (identical) first card clearly differed between conditions. Participants spent a mean RT of 1305ms on the first card in the doubly quantified conditions compared to only 890ms in the control conditions. Whereas the former conditions required evaluation of *exactly one*, ie. categorization of all four objects and counting, the latter only required detection of the object in question. This clear-cut difference led to a significant contrast between the controls and the doubly quantified constructions (LME: $estimate = -387.00$, $t = -7.70$, $p < .01$). This was the only effect that turned out to be reliable. The contrast between the control condition and the doubly quantified conditions indicates that participants had already interpreted the quantifiers

[1] Note that RTs of the second and third cards could not be properly analyzed because of almost no data in the linear conditions and systematic differences between cards in the different conditions.

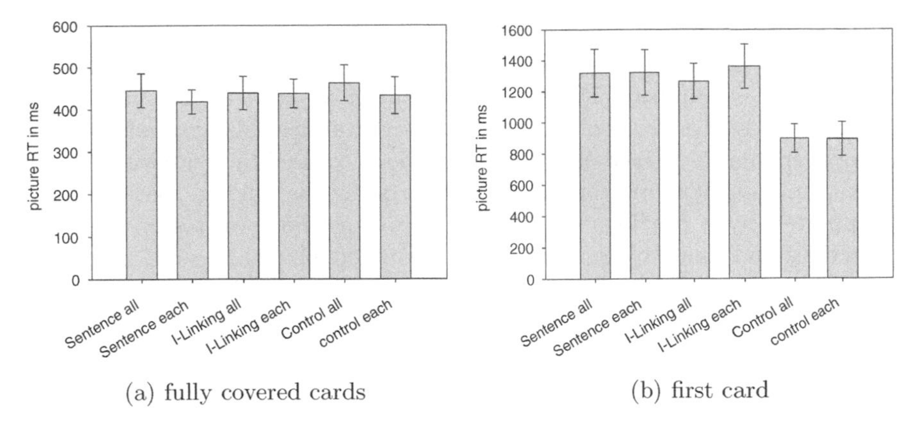

(a) fully covered cards (b) first card

Fig. 3. Mean RTs of "yes, go on" button presses in ms (+ 95% confidence intervals; computed by participants) in the six sentence conditions. Figure 3a shows the RTs of the fully covered card layout. Figure 3b shows the RTs of the first card.

before they uncovered the first card. The results do not unambiguously indicate, though, whether the scope interpretation had been computed before inspecting the first card. To determine that we need to examine the reading times.

Reading Stage: We analyzed the reading times in the following six conditions: inverse linking constructions with *each* vs. *all*, sentence constructions with *each* vs. *all* and control conditions with *each* vs. *all*. To correct for outliers, we trimmed reading times by replacing values below 200ms by a value of 200ms and values above 2000ms by a value of 2000. This correction affected 3.65% of the data.

Figure 4 shows the mean reading times in the three construction types. In the sentence conditions there was a clear contrast between *all* and *each*: the determiner *all* was read on average 23.5ms more slowly than *each*. In contrast, neither the inverse linking nor the control conditions revealed similar differences between determiners (mean difference *jeder* minus *alle* in *inverse linking*: -7.6ms; *control*: 0.7ms). We analyzed the RTs of the determiner *all* vs. *each* in the inverse linking and the sentence constructions in repeated measures ANOVAs. The observed differences led to a significant main effect of *construction* ($F_1(1,39) = 10.25, p < .01$; $F_2(1,59) = 8.14, p < .01$) and a significant interaction between *construction* and *determiner* ($F_1(1,39) = 6.17$, $p < .05$; $F_2(1,59) = 6.27$, $p < .05$). To further break down the interaction we computed pairwise comparisons in each construction type. In the sentence conditions *all* took reliably longer to read than *each* ($t_1(39) = 2.45, p < .05; t_2(59) = 2.93, p < .01$). Except for the highly predictable following segment *cards* ($t_{1/2} < 1$) this effect persisted until the end of the sentence (*to*: $t_1(39) = 3.36$, $p < .01$; $t_2(59) = 2.81$, $p < .01$; *find*: $t_1(39) = 2.20$, $p < .05$; $t_2(59) = 2.22$, $p < .05$). This was different in the inverse linking conditions where *all* and *each* didn't differ at any region from the second quantifier until the end of the sentence (all $t_{1/2} < 1.5$). The same held for the control conditions where *all* and *each* didn't differ anywhere in the sentence (all $t_{1/2} < 1.2$). The lack of difference between *all* and *each* in the control conditions shows that the difficulty in the sentence conditions cannot be attributed to lexical factors. Instead, it must be due to interference between quantifiers.

The *minimal domain* account would predict scope conflict at the point when the predication is complete. Finding no indication of scope conflict at the end of the sentence is therefore somewhat unexpected under this account. We suspect, however, that a delayed effect may have been covered by sentence wrap-up. The assumption that scope conflict is present in inverse linking, too, was actually supported by a previous eyetracking study in which we tested the same constructions and found clear indications of conflict (cf. [2]).

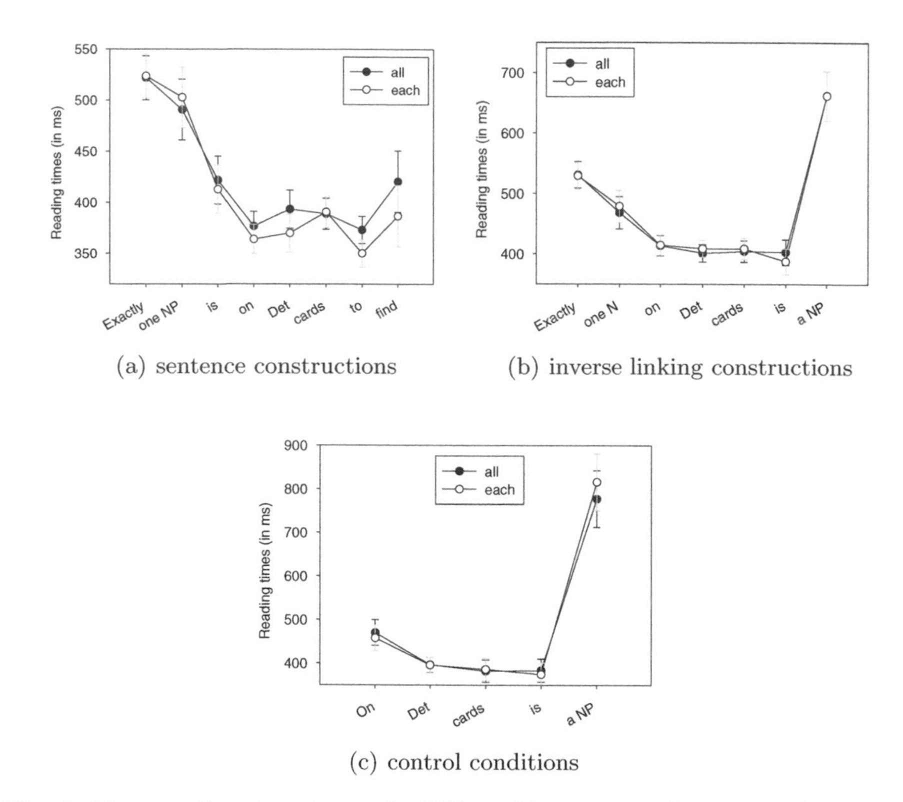

(a) sentence constructions

(b) inverse linking constructions

(c) control conditions

Fig. 4. Mean reading times in ms (+ 95% confidence intervals; computed by participants) as a function of determiner

3 Conclusions

To sum up, the experiment provides evidence for quantifier interpretation before disambiguation is encountered. Even though the participants were instructed to delay scope interpretation as long as possible, they computed scope relations already during reading. Our findings are thus clearly inconsistent with semantic underspecification accounts. The observed scope conflict might indicate that readers selected one interpretation right away. However, it is also possible that they computed both interpretations and ranked them according to the scope factors we have discussed. The results do not allow us to decide between these alternatives. We have shown, however, that a complete minimal domain is required for scope resolution. Our data thus call for introducing the notion of processing domain into cognitively realistic models of semantic interpretation. The results suggest that interpretation processes within a processing domain differ from those that take place at the domain boundary, and that the former may not be fully specified yet.

References

1. Bott, O.: The processing domain of aspectual interpretation. In: Arsenijevic, B., Gehrke, B., Marín, R. (eds.) Subatomic Semantics of Event Predicates. Springer (to appear)
2. Bott, O., Featherston, S., Radó, J., Stolterfoht, B.: The application of experimental methods in semantics. In: Maienborn, C., von Heusinger, K., Portner, P. (eds.) Semantics: An International Handbook. De Gruyter (2011)
3. Ebert, C.: Formal investigations of underspecified representations. PhD thesis. London. King's College (2005)
4. Ferreira, F., Patson, N.D.: The 'Good Enough' approach to language comprehension. Language and Linguistics Compass 1(1-2) (2007)
5. Filik, R., Paterson, K.B., Liversedge, S.P.: Processing doubly quantified sentences: Evidence from eye movements. Psychonomic Bulletin & Review 11(5) (2004)
6. Greene, S.B., McKoon, G., Ratcliff, R.: Pronoun resolution and discourse models. JEP: LMC, 18 (1992)
7. Koller, A., Regneri, M., Thater, S.: Regular tree grammars as formalism for scope underspecification. In: Proceedings of ACL 2008 (2008)
8. Kurtzman, H.S., MacDonald, M.C.: Resolution of quantifier scope ambiguities. Cognition 48 (1993)
9. Pafel, J.: Quantifier Scope in German. John Benjamins (2005)
10. Paterson, K.B., Filik, R., Liversedge, S.P.: Competition during the processing of quantifier scope ambiguities: Evidence from eye movements during reading. QJEP 61(3) (2008)
11. Sanford, A.J., Sturt, P.: Depth of processing in language comprehension: Not noticing the evidence. TCS 6(9) (2002)

Projective Behaviour of *Nur* – Quantitative Experimental Research

Agata Maria Renans*

Universität Potsdam, Department Linguistik
SFB 632 – Informationsstruktur
Karl-Liebknecht-Str. 24-25,
D-14476 Potsdam
renans@uni-potsdam.de

Abstract. The paper presents the results of the experiment on the projective meaning of *nur* (German: *only*). The data from German shows that the prejacent of *nur* projects easily out of counterfactual *if*-clauses, whereas its projective behaviour changes when it is embedded in indicative *if*-clauses. The obtained results classify projection out of counterfactuals as a reliable test for projective meanings in the cross-linguistic perspective, on the one hand, while shedding more light on the semantics of *nur* and conditionals, on the other.

Keywords: presupposition projection, counterfactual and indicative conditionals, semantics of *nur* (*only*) and *auch* (*also*).

1 Introduction

The meaning of a sentence with *only* can be fully described by two meaning components: (1) the prejacent (the meaning of the sentence without an exclusive particle), and (2) the universal (the negative, exclusive meaning of the sentence with *only*), e.g.:

(1) Only Mike ate ice-cream.
  a. Prejacent → Mike ate ice-cream.
  b. Universal → Nobody else but Mike ate ice-cream.

In the case of sentence (1), the prejacent is the proposition that *Mike ate ice-cream*, and the universal is the proposition that *Nobody else but Mike ate ice-cream*. Whereas it is commonly assumed that the universal is asserted, there are many competing theories regarding the semantic status of the prejacent of *only*,

* I would like to thank Antje Sauermann for help with statistical analysis and Mira Grubic, Anne Mucha, Radek Šimik, and Malte Zimmermann for discussion. This work was supported by the German Research Foundation DFG as part of the Collaborative Research Centre (SFB) 632 'Information Structure' at the University of Potsdam, the Humboldt-Universität zu Berlin, and the Freie Universität in Berlin, Project A5.

M. Aloni et al. (Eds.): Amsterdam Colloquium 2011, LNCS 7218, pp. 190–199, 2012.
© Springer-Verlag Berlin Heidelberg 2012

e.g., claiming that the prejacent is presupposed [3], conversationally implicated [9], [5], entailed [1], or a non-speaker-oriented implicature [7].

The question is which diagnostics we can use to check the given proposals. One of the most popular tests for defferentiating between projective meaning components (e.g., presupposition) and non-projective meaning components (e.g., assertion) is the so-called *family of sentences* test. It is claimed that the projective meaning should be interpreted outside the scope of (1) negation, (2) interrogation, (3) modals, and (4) the antecedent of a conditional.

In the experiments, I evaluate the *projection out of the antecedent of a conditional* test. The results of the experiments show that projective meaning components (including the prejacent of *nur*) behave differently depending on whether they are embedded in counterfactual or indicative *if*-clauses.

2 Projective Behaviour of *Nur*—Experiments

2.1 Experimental Set-Up

To check whether the projective behaviour of the prejacent of the sentences with *nur* changes when the given sentence is embedded in the antecedent of the counterfactual and indicative conditionals, four experiments were conducted. Since in German indicative conditionals can be expressed either with the use of the particle *wenn* ($\approx$*if*) or *falls* ($\approx$ *if, in case*), in order to eliminate the possibility that the results are modulated by the semantics of the given particle, two experiments with *wenn* and two experiments with *falls* were conducted. Moreover, experiments differed in the associated elements of *nur*: *nur* was associated either with Subject or with Direct Object of the antecedent. All conditionals in all four experiments were past-oriented[1] and in each experiment the same lexical material was used in the antecedents of the conditionals[2]. 17–26 first and second-year-students of linguistics took part in each of 4 experiments (84 participants together, 75 women, 9 men, average age: 21,91, all German native speakers). The summary of the properties of all four experiments is presented in Table 1.

2.2 Methodology

The design of the experiments is based on the methodology presented in [6]. Each task in each experiment consisted of a short context description, a target conditional sentence with *nur* (either (i) counterfactual or (ii) indicative) and a question (a) about the prejacent or (b) about the universal together with three possible answers *ja* (*yes*), *nein* (*no*), and *nicht erkennbar* (*It's not known*). The experiments were in the standard 2x2 conditions design, which is presented in Table 2. The informant's task was to answer the given questions.

[1] It is due to the fact that the participants of the experiments were asked, whether the given event took place or not. It is not possible to answer such questions in the case of the future-oriented conditionals.

[2] The consequents differed due to different association patterns with *nur*. Keeping the same lexical material in the consequent of the conditionals in all four experiments would make the conditional sentences pragmatically infelicitous.

Table 1. Experiments

	Indicative introduced by:	Element associated with *nur*
Experiment 1	*Wenn*	Subject
Experiment 2	*Wenn*	Direct Object
Experiment 3	*Falls*	Subject
Experiment 4	*Falls*	Direct Object

Table 2. 2x2 conditions design in experiments (1)–(4)

	Type of conditional:	Question about
Condition a	counterfactual	prejacent
Condition b	counterfactual	universal
Condition c	Indicative	prejacent
Condition d	Indicative	universal

Each experiment comprised 6 items per condition (24 target sentences in total) and 26 fillers (50 tasks together). All the experiments were balanced: each participant saw the same amount of conditions and no participant saw the same item twice.

The tasks looked as the following example[3]. Here is the exemplification of the counterfactual *if*-clause and *nur* associated with the Direct Object:

(2) Am Montag sollte jedes Kind seine Lieblingsspielzeuge mit in den
On Monday should each child his favourite.toy with in ART
Kindergarten bringen. **Wenn Franz nur ein Bärchen**
kindergarten bring. If Franz only ART teddy-bear
mitgebracht hätte, wären die Erzieherinnen sehr erstaunt
bring.PAST have.KONJ, be.KONJ ART teacher very surprise
gewesen.
be.PART
'On Monday every child was to bring his favourite toys to the kindergarten. If Franz had brought only a teddy-bear, the teachers would have been surprised.'

a. Hat Franz ein Bärchen mitgebracht?
have.PAST Franz ART teddy-bear bring.PART
'Did Franz bring a teddy bear?'
-ja (yes) -nein (no) -nicht erkennabar (It is not known)

b. Hat Franz noch etwas außer einem Bärchen
have.PAST Franz PART something beside ART teddy-bear
mitgebracht?
bring.PART

[3] Note that in the experiment, informants saw either question (a) or question (b).

Did Franz bring anything else than a teddy-bear?
-ja (yes) -nein (no) -nicht erkennbar (It is not known)

If the participant answers a question about the prejacent (condition *a* and *c*) with *yes*, this means that the prejacent is interpreted outside the scope of the *if*-clause and hence projects out of the respective *if*-clause type. If the participant answers a question about the universal (condition *b* and *d*) with *no*, this means that the universal projects out of the scope of the *if*-clause. The interpretation of the results is shown in Table 3. Expected answers, assuming that the prejacent of *nur* projects and the universal is asserted, are written in boldface. Note that in the case of the question about the universal, the expected answers depend on the type of the conditional. It is due to the fact that in the case of the counterfactuals, antecedent is known to be false. Hence, if the proposition is interpreted within the scope of the *if*-clause, it is known that it does not hold in the actual world. Whereas in the case of the indicatives, it is not known, whether the antecedent is false or not. Hence, if the proposition is interpreted within the scope of the *if*-clause, it is not known whether it holds in the actual world or not.

Table 3. Interpretation of the results

		No	Yes	It's not known
Counterfactual	Prejacent (*a*)	¬ projection	+ **projection**	¬ projection
	Universal (*b*)	+ projection	¬ **projection**	¬ projection
Indicative	Prejacent (*c*)	¬ projection	+ **projection**	¬ projection
	Universal (*d*)	+ projection	¬ projection	¬ **projection**

2.3 Results

The results of the experiments are shown in Tables 4, 5, 6, and 7. In all four experiments the answer patterns are similar. The results concerning the projective behaviour of the universal are as expected. They suggest that the universal does not project out of the *if*-clauses and its projective behaviour does not change with the change of the used conditionals. In all four experiments there are significantly more *yes* answers than *no* and *It's not known* answers in the case of the counterfactuals and there are significantly more *It's not known* answers than *no* and *yes* answers in the case of the indicatives, which are the expected answer patterns for the non-projective meaning components. Results concerning the projective behaviour of the prejacent are more surprising. We can observe the differences in its behaviour when it is embedded in the counterfactual and indicative *if*-clauses: prejacent of *nur* projects easily out of the counterfactual *if*-clauses, whereas it does not project so easily out of the indicative *if*-clauses.

The *chi-square* test showed that there was a significant interaction between conditions and the given answers, which means that the conditions influenced the answer patterns (for all experiments $p < 0.001$).

Table 4. Experiment1: *Nur* associated with Subject, indicatives with *wenn*

		No	Yes	It's not known
Counterfactual	Prejacent (*a*)	8 (8%)	**86 (80%)**	**12** (12%)
	Universal (*b*)	5	90 (88%)	7 (7%)
Indicative	Prejacent (*c*)	2 (2%)	**41 (40%)**	**59 (58%)**
	Universal (*d*)	16 (16%)	5 (5%)	81 (79%)

Table 5. Experiment 2: *Nur* associated with Object, indicatives with *wenn*

		No	Yes	It's not known
Counterfactual	Prejacent (*a*)	7 (6%)	**86 (71%)**	**27 (23%)**
	Universal (*b*)	0	104 (87%)	16 (13%)
Indicative	Prejacent (*c*)	1 (1%)	**41 (34%)**	**78 (65%)**
	Universal (*d*)	11 (9%)	7 (6%)	102 (85%)

Table 6. Experiment 3: *Nur* associated with Subject, indicatives with *falls*

		No	Yes	It's not known
Counterfactual	Prejacent (*a*)	7 (6%)	**111 (88%)**	**8 (6%)**
	Universal (*b*)	5 (4%)	113 (90%)	8 (6%)
Indicative	Prejacent (*c*)	5 (4%)	**48 (38%)**	**73 (58%)**
	Universal (*d*)	6 (5%)	21 (17%)	99 (78%)

Table 7. Experiment 4: *Nur* associated with Object, indicatives with *falls*

		No	Yes	It's not known	-
Counterfactual	Prejacent (*a*)	9 (6%)	**123 (79%)**	**24 (15%)**	
	Universal (*b*)	5 (3%)	133 (85%)	18 (12%)	
Indicative	Prejacent (*c*)	5 (3%)	**49 (31,5%)**	**102 (64,5%)**	
	Universal (*d*)	4 (2,4%)	11 (7%)	140 (90%)	1 (0,6%)

To check whether the different answers were influenced by the type of the used conditional *linear mixed-effects models* (*LME*) were calculated. These models correspond to (logistic) regression models that take into account the variation due to participants and items. Since *no* answers were rare, *LME* models were calculated for *yes* and *It's not known* answers[4]. *LME* models showed that in all four experiments the type of the used conditional influenced the probability of *yes* and *It's not known* answers for the questions about the prejacent (*Yes* answers: Exp. 1: $z = -6.095$, $p < 0.001$, Exp 2: $z = -6.092$, $p < 0.001$, Exp. 3: $z = -7.815$, $p < 0.001$, Exp. 4: $z = -8.786$, $p < 0.001$; *It's not known* answers: Exp. 1: $z = 6.687$, $p < 0.001$, Exp. 2: $z = 6.798$, $p < 0.001$, Exp. 3: $z = -7.781$, $p < 0.001$, Exp 4: $z = 9.069$, $p < 0.001$). It means that *Yes* and *It's not known* answers for the question about the prejacent are modulated by the type of the conditional.

To assess the difference between the probability of *Yes* and *It's not known* answers in the case of the question about the prejacent in the indicative *if*-clauses (condition *c*), *LME* models were calculated on the proportions of each answer for each participant. There were significantly higher proportions[5] of *Yes* than *It's not known* answers for Exp. 2 ($t = -4.933$) and 4 ($t = -4.046$), indicating that the prejacent of *nur* does not project out of the indicative *if*-clauses when it is associated with the Direct Object. For Exp. 3, the effect was marginal but with a trend in the same direction. However for Exp. 1 the effect was not significant. Nevertheless, the tendency towards non-projecting is also visible (40% of *Yes* answers vs. 58% of *It's not known* answers).

Summarizing, the results of the experiments show that the prejacent of *nur* behaves differently when it is embedded in counterfactual and indicative *if*-clauses: the prejacent does not project or tends not to project out of the scope of the indicative *if*-clauses. In order to check whether the obtained results are due to the semantics of *nur* specifically or due to the properties of the projective meaning components in general, an experiment on projective behaviour of *auch* (German: *too, also*) was conducted.

2.4 Projective Behaviour of *Auch* – Experiment

The experimental set-up, the methodology, and the lexical material of the *auch*-experiment was the same as in Exp. (1). The tasks looked as follows:

(3) Am Montag sollte jedes Kind seine Lieblingsspielzeuge mit in den Kindergarten bringen. Wenn **auch** Franz ein Bärchen mitgebracht hätte, wären die Erzieherinnen nicht erstaunt gewesen.
'On Monday every child was to bring his favourite toys to the kindergarten. If also Franz had brought a teddy-bear, the teachers would not have been surprised.'

a. Hat noch jemand außer Franz einen Bärchen mitgebracht?
'Did Franz bring anything else than a teddy-bear?'

---

4 Because of the lack of space, only the most important calculations are refered.

5 For these post-hoc comparisons, $|t| \geq 2.4$ was taken to be significant.

b. Hat Franz ein Bärchen mitgebracht?
'Did Franz bring a teddy bear?'

Answer *Yes* suggests that the respective meaning component projects, whereas answer *No* and *It's not known* suggest that the respective meaning component does not project. The interpretation of the results of *auch*-experiment is presented in Table 8 (expected answers, assuming that the additive meaning component projects and the core meaning is asserted, are written in boldface).

Table 8. Interpretation of the results: *auch*-experiment

		No	Yes	It's not known
Counterfactual	Additive meaning (*a*)	¬ projection	**+ projection**	¬ projection
	Core meaning (*b*)	**¬ projection**	+ projection	¬ projection
Indicative	Additive meaning (*c*)	¬ projection	**+ projection**	¬ projection
	Core meaning (*d*)	¬ projection	+ projection	**¬ projection**

13 second-year students of linguistics (11 women, 2 men, average age: 23,23, all German native-speakers) took part in the experiment. The results are presented in Table 9.

Table 9. Experiment 5: *Auch* associated with Subject, indicatives with *wenn*

		No	Yes	It's not known
Counterfactual	Additive meaning (*a*)	2 (2,5%)	**63 (80,8%)**	**13 (16,7%)**
	Core meaning (*b*)	77 (98,7%)	0 (0%)	1 (1,3%)
Indicative	Additive meaning (*c*)	1 (1,3%)	**37 (47,4%)**	**40** (51,3%)
	Core meaning (*d*)	3 (4%)	11 (14%)	64 (82%)

As in the case of Exp. (1)–(4), results for counterfactuals are as expected. They show that the additive meaning component projects easily out of the counterfactual *if*-clauses. The results for the indicative *if*-clauses are more surprising. Similar amount of *Yes* and *It's not known* answers for the question about the additivity in the case of the indicative *if*-clauses suggests that the additive meaning component can be interpreted both in and out of the scope of the *if*-clause. Moreover, as in the case of the prejacent of *nur*, we can observe the difference in the projective behaviour of the additive meaning component of *auch* depending on the type of the used *if*-clause.

The *chi-square* test showed that an interaction between the conditions and the given answers was significant ($p < 0.001$). In order to see whether the answers were modulated by the type of the conditional *LME* were calculated. They showed that the probability of *Yes* and *It's not known* answers were influenced by the type of the used conditional (*Yes* answer: $z = 4.636$, $p < 0.001$, *It's not known* answer: $z = -4.782$, $p < 0.001$). The results from Exp. (1) – (5) suggest that the described asymmetry is more systematic and is not caused by the specific semantics of *nur*. Notice, however that the observed effects are stronger in the case of the sentences with *nur* than with the sentences with *auch*.

3 Analysis

Presuppositions can be accommodated either globally and then they project or locally and then they do not project [2]. It seems that the differences in the projective behaviour of *nur* and *auch* are due to the differences in accommodation. The results of the experiments suggest that in the case of counterfactuals, presuppositions are accommodated globally, whereas in the case of indicatives, presuppositions can be interpreted both locally and globally.

To explain the differences in accommodation, I adopt the restrictor analysis of conditionals in Kratzer [4]. She claims that in both kinds of conditionals there is overt or covert modality. To understand conditional modality, we must start from Kratzer's approach to modals [4]. She captures the semantics of modals with the use of three notions: conversational background (CB), modal base, and ordering source.

The CB (a function from possible worlds to premise sets, which are sets of propositions) is an accessibility relation. By transferring the premises to the closely related worlds, a realistic CB indicates a **modal base** (a set of accessible worlds) in reference to which the truth-values of modal sentences are resolved. In addition, worlds which are in the modal base are ordered according to how closely they are related to the evaluation world taking into consideration the normal course of events in the evaluation world. The ordering is done by another function: **the ordering source**. With the use of these two notions Kratzer captures the context dependency of sentences with modals.

I argue that the presuppositions are present in the modal base. I claim, following Stalnaker [8], that the notion of presupposition should not be limited to what is known or believed, but it should be treated as the accessibility relation. Since this is the case, presuppositions should be present in the CB. Therefore they should take part in indicating the set of accessible possible worlds (the modal base).

Given the definitions of modal base and ordering source, Kratzer [4] defines conditional modality. She claims that an *if*-clause additionally restricts the modal base of the consequent.

Definition 1. *For any conversational backgrounds f and g:* $[[if \alpha \beta]]^{f,g} = [[\beta]]^{f+,g}$, *where for all* $w \in W, f^{+}(w) = f(w) \cup \{[[\alpha]]^{f,g}\}$. *(Kratzer 2011a)*

Different kinds of conditionals are defined by different settings of the modal base and ordering source. Crucially, counterfactuals (in contrast to indicatives) are interpreted with the use of an empty modal base [4]. Note that presuppositions are in the modal base. Thus, in order to accommodate them locally in the antecedent of the conditional, the initial modal base $f(w)$ cannot be empty.

Let us see what the mechanism of interpreting the counterfactuals looks like. The modal base restricts the set of the possible worlds over which the covert modal in the consequent of the conditional quantifies. Since presuppositions are in a modal base *f(w)*, they also restrict the quantification domain of the covert modal in the consequent of a conditional. The crucial thing is that in the case of counterfactuals the initial modal base *f(w)* (from Def. 1) is empty: the modal base in reference to which the consequent is interpreted ($f^+(w)$) is updated with the asserted proposition carried by the antecedent. However, since the initial modal base $f(w) = \emptyset$, there are no presuppositions in $f^+(w)$. What is more, the presuppositions cannot come to $f^+(w)$ by accommodation, because it would lead to an infelicitous discourse[6]. Hence, the presuppositions do not take part in the restriction of the quantification domain of the consequent. Thus, in the case of counterfactuals global accommodation is obligatory. Let us consider the following example:

(4) Also Mira came to the party.

a. core meaning → Mira came to the party.

b. additive meaning → Somebody else than Mira came to the party.

c. If also Mira had come to the party, Anne would have been happy.

From Def. 1 it follows that the consequent in (4-c) is interpreted in reference to the modal base $f^+(w)$, which is the union of the empty set and the proposition expressed by the antecedent: $f^+(w) = \emptyset \cup \{[[\text{Mira came to the party}]]^{\emptyset,g}\}$. We can see that the presupposition (here: the proposition that somebody else than Mira came to the party) does not restrict the possible worlds regarding to which the consequent is evaluated. In addition, an empty modal base does not allow presuppositions to be accommodated locally in the antecedent of the conditional. Hence, the additive meaning component of *auch* is interpreted globally, outside of the scope of the *if*-clause.

Indicative conditionals are interpreted with reference to a non-empty modal base. This means that the presuppositions (which are present in the initial modal base $f(w)$) restrict the modal base in reference to which the consequent is evaluated. Let us consider (4) and let us embed it in the indicative *if*-clause:

(5) If also Mira came to the party, Anne was happy.

Sentence (5) is interpreted with respect to a non-empty modal base $f(w)$ which includes the presupposition carried by the antecedent. The consequent of the *if*-clause is evaluated in reference to the modal base $f^+(w)$ which includes both the presupposition and the core meaning of the antecedent: $f^+(w) = f(w) \cup \{[[\text{Mira came to the party}]]^{f,g}\}$, where the presupposition [[Somebody else came to the party]] is included in $f(w)$. The presupposition restricts the worlds over which the consequent quantifies and therefore local accommodation is possible.

[6] For a discussion about the different kinds of accommodation and when they produce a felicitous discourse, see [2].

The mechanism of interpreting the conditionals with *nur* is the same as in the case of those with *auch*. The stronger effect of not projecting out of the scope of the indicative *if*-clauses is caused by the universal, which is an asserted part of the proposition in the antecedent and therefore obligatorily restricts the modal base. In the sentence *If only Mira came to the party, Anne was sad*, the universal is the proposition that *nobody else than Mira came to the party*. In order to indicate all the possible worlds where nobody else than Mira came to the party, we are inclined to select the worlds where Mira came to the party. That is why the results of Exp. (1)–(4) showed a tendency to interpret the prejacent of *nur* in the scope of the indicative *if*-clause (locally).

Summing up, I argued that in the case of the counterfactuals the presuppositions are accommodated globally (due to the empty modal base), whereas in the case of indicatives the presuppositions can be interpreted both locally and globally (due to the non-empty modal base).

4 Conclusion

The paper reports the results of the experiments based on the methodology presented in [6]. Exp. (1)–(5) showed that there is a significant difference in the behaviour of the projective meaning components when they are embedded in indicative and counterfactual *if*-clauses. The analysis of the results explained where this difference comes from. Therefore, the *Projection out of the indicative if-clause* test is evaluated as being inappropriate for semantic fieldwork. The results of Exp. (1)–(5) showed that quantitative experimental research can shed a new light on old (formal) semantic problems.

References

1. Atlas, J.D.: The Importance of Being 'Only': Testing the Neo-Gricean Versus Neo-Entailment Paradigms. Journal of Semantics 10, 301–308 (1993)
2. Beaver, D., Zeevat, H.: Accommodation. In: Ramchand, G., Reiss, C. (eds.) Oxford Handbook of Linguistics Interfaces. Oxford University Press (2007)
3. Horn, L.: A Presuppositional Analysis of 'Only' and 'Even'. In: Proceedings of the Annual Meeting of Chicago Linguistics Society (1969)
4. Kratzer, A.: Modals and Conditionals. Oxford University Press (2012)
5. McCawley, J.: Everything that Linguists Have Always Wanted to Know about Logic but Were Ashamed to Ask. University of Chicago Press (1981)
6. Renans, A., Zimmermann, M., Greif, M.: Questionnaire on Focus Semantics. Potsdam Universität Verlag (2011)
7. Roberts, C.: 'Only', Presupposition and Implicature. Journal of Semantics (2006) (accepted with revisions)
8. Stalnaker, R.: Common ground. Linguistics and Philosophy 25, 701–721 (2002)
9. van Rooij, R., Schulz, K.: Only: Meaning and Implicatures. In: Maier, E., Bary, C., Huitink, J. (eds.) Proceedings of Sinn und Bedeutung 9 (2005)

Presupposition Processing – The Case of German *wieder**

Florian Schwarz[1] and Sonja Tiemann[2]

[1] Department of Linguistics and IRCS, University of Pennsylvania
floriansQling.upenn.edu
[2] Eberhard-Karls Universität Tübingen
sonja.tiemannQuni-tuebingen.de

Abstract. Presuppositions are vital for language comprehension, but little remains known about how they are processed. Using eye tracking in reading, we investigated two issues based on German *wieder* ('again'). First, we looked at the time course of presupposition processing by testing for processing costs of unsupported presuppositions. Secondly, we tested whether embedding *wieder* under negation affected a potential mismatch effect. Presupposition-induced effects showed up immediately after *wieder*, but only in the unembedded context, suggesting that embedding interferes with the detection of the mismatch. However, judgments in a follow-up rating study indicate that a mismatch is perceived in both the embedded and unembedded conditions when the PSP is not supported by the context. Taken together, these results suggest that detection of the mismatch under embedding is delayed in processing.

Keywords: Presuppositions, Psycholinguistics, Presupposition Processing, Presupposition Projection, Eye Tracking.

1 Introduction

While the recent literature has seen a renewed peak in theoretical discussion of presuppositions, together with consideration of ever more intricate data (see Schlenker 2010 and Beaver and Geurts, to appear, for recent surveys), experimental approaches to presuppositional phenomena are still in their beginnings. Based on the general notion that presuppositions require some form of contextual support, previous experimental studies have found that lack of such support is reflected in various processing effects, e.g. regarding the choice of interpretation of a syntactically ambiguous structure and increase in reading times (Schwarz 2007) or the need for accommodation (Tiemann et al. 2011; see also Chemla 2009, and Chemla and Bott 2010 for other recent experimental studies). In a reading study using eye tracking, we investigated two issues concerning the processing of presuppositions. First, we test what form effects of presupposition failure, which have been found previously using Self-Paced Reading, have in

* This work was supported by EURO-XPRAG (ESF) and SFB 833 (DFG).

M. Aloni et al. (Eds.): Amsterdam Colloquium 2011, LNCS 7218, pp. 200–209, 2012.
© Springer-Verlag Berlin Heidelberg 2012

this methodology. Eye tracking is more naturalistic and faster than Self-Paced Reading, and thus provides a more precise perspective on the time course of cognitive processes during reading. Secondly, our design compares these effects for unembedded occurrences with cases where a presupposition is introduced in the (syntactic) scope of negation (but standardly interpreted globally).

2 Background

Since different expressions that introduce presuppositions seem to vary in terms of ease of accommodation and other general properties (see, for example, the distinction between soft and hard triggers in Abusch 2009), it seems most prudent to focus experimental investigations on one presupposition trigger at a time. The present experiment focuses on the German trigger *wieder* ("again").

In the theories of Stalnaker 1973 and Heim 1990, presuppositions are restrictions on appropriate contexts. This means that a sentence like (1) is only felicitous if the context entails that Sue had danced before.

(1) Sue danced again.

An unmet presupposition results in presupposition failure and thus in uninterpretability of the sentence (cf. Heim and Kratzer 1998). Within this tradition, it is generally assumed that presuppositions are lexically encoded in the meaning of the presupposition trigger. The lexical entry for *wieder* then looks as in (2):

(2) $[\![\text{wieder}]\!] = \lambda P.\lambda x.\lambda t.\lambda w\text{: } \exists t'[t'<t \;\&\; P(x)(t')(w)].\ P(x)(t)(w)$

(3) captures formally that a sentence like (1) can only update a context if the context entails that there is a time t' before t at which Sue has danced (c is Stalnaker's context set).

(3) $\lambda c\text{: } c \subseteq \{w\text{: } \exists t'<t \;\&\; \text{Sue danced at } t' \text{ in } w\}.\ c \cap \{w\text{: Sue dances at } t \text{ in } w\}$

In contrast with semantic theories of presupposition along these or similar lines, e.g. in the frameworks of dynamic semantics (Heim 1982) and Discourse Representation Theory (DRT, Kamp 1981, van der Sandt and Geurts 1991), several issues have recently given rise to a revived debate that includes various proposals for deriving presuppostions (of at least some presupposition triggers) pragmatically (Schlenker 2009, Simons 2001). In addition to the theoretical arguments that have been brought fourth in order to distinguish between these theories, experimental investigations can contribute to the debate in that the two types of accounts suggest different time courses for computing presuppositional content in online processing. The reasoning here is very much parallel to that presented in the experimental literature on scalar implicature processing. Increases in processing time have been taken to argue in favor of accounts of implicature generation where Gricean reasoning is carried out after the core literal content is computed (Bott and Noveck 2004, and much subsequent work). To the extent that pragmatic accounts of presuppositions also appeal to Gricean reasoning, we then might expect similar delays in effects related to the interpretation of presuppositions. If there is no delay in processing presupposed content, on the other hand, that would seem to fit more squarely with a view where presuppositions are encoded conventionally as part of the lexical content of the triggers (though it may not necessarily rule out certain versions of pragmatic accounts). Our experiments vary whether the context that *again* appears in supports its presupposition or not, which allows us to evaluate the time course of presupposition processing using the high temporal resolution of eye tracking during reading and thus contributes new empirical evidence to this debate.

A second experimental manipulation relates to one of the key properties of presupposed content, namely the fact that it is not affected by various embedding operators (including negation and various attitude verbs). For example, in (4),

(4) Sue [did NOT [dance again]].

even when assuming the syntactic structure indicated by the bracketing, the presupposition escapes the scope of negation, as it were, so that the entire sentence still presupposes that Sue danced before (rather than that she didn't dance before), just like the original version without negation. From the perspective of processing, keeping apart asserted and presupposed content and taking care to interpret these distinct aspects of meaning appropriately with respect to operators like negation constitutes a fairly delicate and complex task. *Again* appears syntactically in the scope of negation and has to combine with the verb in order to derive the appropriate presupposition, but the result of this then has to be interpreted globally, rather than in the scope of negation (unlike the asserted content contributed by, e.g., the verb). Investigating the online processing of presupposition triggers in the scope of operators like negation thus has the potential to provide important insights to our understanding of the underlying processes by which the global interpretation of presupposed content is derived. Our experiment is a first attempt to shed light on this issue by directly comparing processing effects based on presuppositional content both in configurations where it appears in the scope of negation as well as in global ones.

3 Experiment

3.1 Methods and Material

Design & Stimuli. Our design makes use of a feature of German syntax, where *wieder* (again) and *nicht* (not) can appear in adjacent positions in either order. This makes it possible to construct target sentences which are minimally different with respect to whether *again* appears inside or outside the scope of negation. We presented such sentences in two different contexts, each of which supported the presupposition of one of the orders of *wieder* and *nicht* and contradicted the other. In the sample item from our materials below, the context sentence (5) supports the presupposition of (7a) (that Tina went ice-skating before), while (6) contradicts it (if not strictly speaking logically, then at least pragmatically).

Conversely, (6) supports the presupposition of (7b) (that there was a preceding occasion where Tina did not go ice-skating), while (5) is inconsistent with it.[1]

(5) *Tina **went ice** skating for the first time last week with Karl. The weather was beautiful, and they **had a great time**.*

(6) *Tina **wanted** to go ice skating for the first time with Karl last week. But the weather was miserable and they **gave up on their plan**.*

(7) Dieses Wochenende war Tina {(a) **nicht wieder** / (b) **wieder nicht**}
This weekend, was Tina {(a) not again / (b) again not}
Schlittschuhlaufen, weil das Wetter so schlecht war.
ice skating because the weather so bad was

The pairing of sentences and contexts yielded a fully counterbalanced 2×2 interaction design with two factors: **Firstword** (whether *wieder* or *nicht* appeared first) and **Felicity** (whether the context supports the presupposition or not).

Procedure & Participants. 24 sentences with versions for each of the four conditions were created. In addition to the experimental items, there were 48 unrelated filler items. Subjects read the sentences on a computer screen while there eyes were being tracked by an EyeLink 1000 eye tracker from SR Research. For half of the items (of both the fillers and experimental sets), participants had to answer yes/no questions, which followed directly after the sentence, to ensure full comprehension of the materials. 32 native speakers of German from the University of Tübingen community participated in the experiment. Subjects were split into 4 groups, where each subject saw 6 of the sentences per condition.

3.2 Results

The primary focus in our analysis were the reading times on the verb following the {*wieder nicht*} sequence. Since the presupposition of *wieder* crucially relies on the verb of its clause, it is only at the point of the verb that it becomes recoverable from explicitly given materials. Reading times were also examined for {*wieder nicht*} itself. Standard reading measures were calculated for purposes of analysis. Based on prior self-paced reading experiments using the same general approach (Schwarz 2007, Tiemann et al. 2011), we expect increases in reading time when sentences are presented in contexts that are inconsistent with the presupposition. The time point at which such increases arise is indicative of the relevant presupposition having been computed at this point.

Data Analysis. All analyses used mixed-effect models with subjects and items as random effects, using the *lmer* function of the *lme4* package in *R* (Bates 2005), together with MCMC estimates for significance (Baayen et al. 2008). All effects significant at the $p < .05$ threshold are reported. Five reading time measures were computed (Rayner 1998): first fixation duration, which measures the length of

[1] At least on a global interpretation. See below for local interpretations.

the very first fixation on the region of interest (here the verb); go-past time, which here is taken to measure the sum of all fixations on the region of interest prior to any fixations to the right of this region (but not including the time of regressive fixations); first pass time, which includes all fixations on the region when it is looked at the first time, up until leaving the region (to either the left or right); total duration, which sums all the fixations on the region of interest, no matter when they occur; regression path duration, which measures all fixations from first entering the region to first leaving it to the right (including all potential regressive fixations; this is sometimes also referred to as go past time); first pass regression proportion, which is the proportion of regressive eye movements following the first time of entering the region.

Means for the reading time measures on the verb are presented in table 1.[2] The primary result is an interaction betweeen **Firstword** and **Felicity**: when *wieder* was first (i.e., not embedded under negation), reading times on the verb were significantly higher in the infelicitous condition. When *nicht* was first, on the other hand (resulting in *wieder* being embedded under negation), there was no such slow-down (and except in total reading time and first pass time, no significant difference between the felicitous and infelicitous context conditions).

Table 1. Reading time measures (in ms) and First Pass Regression Proportion (in %) on the verb

	wieder nicht		*nicht wieder*	
Reading Measure	felicitous	infelicitous	felicitous	infelicitous
First Fixation	194	210	199	192
Go-Past	292	359	324	285
First Pass	270	281	275	247
Total	309	405	370	307
Reg. Duration	395	619	438	479
Reg. Proportion	17.0%	33.5%	17.4%	20.6%

There was a significant interaction for first fixation duration ($p < .05$), go-past ($p < .01$), and total ($p < .001$) reading times, as well as a marginal interaction for first pass duration ($p = 0.067$), regression path duration ($p = 0.056$), and and for first pass regression proportion ($p = .058$). There was a main effect of **Firstword** on all measures ($p < .05$) with faster reading times (and lower regression proportions) on the verb in the *nicht wieder* conditions, which is not generally interpretable on its own given the cross-over interaction. The interaction was primarily driven by a significant simple effect of **Felicity** for the *wieder nicht* conditions, with increases in reading measures for the infelicitous context (go-past time: $p < .05$, total time: $p < .01$, first fixation: $p < .05$, regression path: $p < .01$, regression proportion: $p < .001$). For the *nicht wieder* conditions, the

[2] The numbers and statistics below differ slightly from the pre-proceedings version, due to the discovery of minor errors in data treatment and reading time computation.

only simple effect of **Felicity** appeared in the total reading time and first pass time (both $p < .05$), and it was in the opposite direction, with a decreased reading time in the infelicitous condition. Regarding simple effects of **Firstword**, the only effect for the felicitous conditions was for total reading time ($p < .05$), where reading times were faster in the *wieder nicht* condition than in the *nicht wieder* condition. In the infelicitous conditions, the *wieder nicht* conditions displayed faster reading times on the verb than the *nicht wieder* conditions (total: $p < .01$, first fixation: $p < .05$, first pass: $p < .01$, go-past: $p < .0.1$, regression path duration: $p < .1$, regression proportion: $p < .01$). With respect to reading times on {*wieder nicht*} (taken as a unit, in either order), a parallel interaction effect showed up in the total reading times, with corresponding simple effects of **Felicity** for the *wieder nicht* condition and of **Firstword** for the infelicitous condition. There were no other significant effects in this region.

Given the lack of an effect of **Felicity** in the *nicht wieder* condition on the verb, follow-up analyses on later and larger regions were carried out. No increases in reading times were found for regions consisting of the verb plus 2 following words, the 3 words following the verb, the entire section of the sentence from *nicht wieder* to the end, or, for that matter, for the entire trial duration (i.e., total reading time for the entire paragraph).

3.3 Discussion

There are two main points to discuss with respect to the experimental results. The interaction shows that the effect of encountering a presupposition in a context that is inconsistent with it differs based on whether we are dealing with an embedded or an unembedded trigger. Furthermore, the presupposition of unembedded *wieder* gives rise to fairly immediate effects of inconsistency that are reflected throughout a variety of reading time measures. Of particular interest with regards to the latter point are the simple effects for first fixation duration and first pass regression proportion. Already during the first fixation of the verb (which last less than 200 ms), the beginning of which arguably is the logically earliest point possible to fully compute the presupposition of *wieder* based on what has been explicitly provided, a 12 ms effect emerges. Based on the experimental design, the delay can be attributed to the inconsistency between the expressed presupposition and the provided context. But for such an inconsistency to arise, the relevant presupposition must of course have been computed. Similarly, the increase in first pass regression proportions indicates that upon first looking at the verb, there is an increased likelihood of returning to look at the preceding context, which is presumably triggered by noticing the same inconsistency. The experiment thus provides evidence that the presupposition of *again* is computed rapidly online. As mentioned above, this seems most consistent with theoretical proposals that assume it to be conventionally encoded, rather than derived by some type of pragmatic reasoning, which - based on what we know about scalar implicature processing - would seem to require some extra processing time.[3]

[3] The extent to which this generalizes to other presupposition triggers remains to be explored. Triggers very well may vary precisely in this respect (cf. Simons 2001).

Returning to the first point, the picture is rather different for cases where *wieder* is embedded under negation. Given the standard global interpretation, the two contexts vary in precisely the same way as was the case for the unembedded occurrence of *wieder* (albeit their roles are reversed), with one context supporting the presupposition, while the other is inconsistent with it. If the global interpretation of the presupposition were available while reading the verb, we would expect to see an effect on reading times similar to the unembedded condition. However, on none of the reading measures was there a significant increase for the infelicitous condition. In fact, the only significant simple effect (for total reading times) went in the opposite direction (which is something that we do not yet have a clear explanation for). The lack of such an increase thus can be taken as an indication that the global interpretation is not available while the verb is being read.

In principle, there are two possible explanations for why this might be the case. First, it could be that more is involved in deriving a global interpretation in the context of an embedding operator like negation, compared to simply recognizing the presupposition of an unembedded trigger. Thus, the lack of an effect might be due to a lag in generating the appropriate presupposition in this more complex sentential context. An alternative exists, however, based on the possibility (ignored in our discussion so far) of local interpretations of presuppositions in the scope of negation. Perhaps the most well-known case of this concerns the existence presupposition of the definite article, as in (8), where a global interpretation of the presuppostion is inconsistent with the continuation.

(8) The King of France is not bald - because there is no King of France!

Similarly, in (9) it seems possible to negate the presupposition that Tina had been ice-skating before, rather than the asserted content.

(9) Tina didn't go ice-skating again last weekend - this was the first time!

A simple way of modeling this local interpretation is to simply assume that both the presupposed and the asserted content remain in the scope of negation, so that the overall interpretation of the sentence can be paraphrased as follows:

(10) NOT [Tina went ice-skating before AND went ice-skating this weekend]

While in principle, the falsity of either conjunct in the scope of negation would suffice to make this true, the fact that one could express the negation of the second conjunct more straightforwardly (by simply leaving out the presupposition trigger altogether) might bias this towards an interpretation where it is indeed the falsity of the conjunct contributed by the presuppostion trigger that is conveyed by an utterance of this sentence.

In any case, given a paraphrase along the lines of (10), if a local interpretation were available for the target sentence in the experimental materials, the *nicht wieder* sentences have interpretations that are perfectly consistent with either context. If the context states that Tina had been ice-skating some time recently, then the regular global presupposition of course remains consistent with that (and the local interpretation is not strictly speaking inconsistent with this either, if the paraphrase above is correct). And if the context states that she did not go ice-skating (and had never done it before, either), then the local interpretation (which is generally taken to convey that she had not been ice-skating before) is perfectly consistent with that. Thus, if both global and local interpretations for the presupposition of *wieder* in the scope of negation are available, we would not expect to see any reflexes of inconsistency in the reading times, since at least one of the readings always is consistent with the given contexts. In order to test whether local interpretations are indeed available for the experimental materials, a follow-up rating experiment was carried out.

3.4 Follow-Up Rating Experiment

If local interpretations are indeed available for the presupposition of *wieder* when it appears in the scope of negation, we would expect this to affect speakers' acceptability judgments of these sentences in the two contexts. In particular, the type of interaction that we saw in the reading times should also be present in the judgments. If the local interpretation is not available (or only to a very limited extent), on the other hand, the *nicht wieder* sentences in what we have labeled as the infelicitous context above should be judged to be less acceptable than in the felicitous context. A rating questionnaire was conducted via the web using the WebExp2 software (`http://www.webexp.info`). The materials were exactly the same as those used in the eye tracking experiment, including all the fillers. Subjects were asked to rate the appropriateness of a given discourse on a scale from 1 (least appropriate) to 5 (most appropriate). The results are summarized in table 2.

Table 2. Results of the rating experiment

	wieder nicht		*nicht wieder*	
	felicitous	infelicitous	felicitous	infelicitous
Mean Rating	3.94	2.63	3.23	2.34

While there was a marginally significant interaction between **Firstword** and **Felicity** ($p = .059$), as well as a marginally significant main effect of **Firstword** ($p = .059$), more importantly there was a clearly significant main effect of **Felicity** ($p < 0.001$), with items containing felicitous contexts getting higher (= better) ratings than those containing infelicitous contexts. While this effect was slightly more pronounced in the *wieder nicht* items (giving rise to the marginal interaction), there nonetheless is a significant simple effect for *nicht wieder* in the same direction ($p < .001$), just as there is for *wieder nicht* ($p < .001$). Thus the rating results clearly show that for both embedded and unembedded *wieder*, the **Felicity** manipulation had a clear effect and resulted in decreased acceptability when the context sentence was inconsistent with the (global) presupposition of *wieder*. This would be unexpected if the local interpretation of *wieder* under negation were readily available. The explanation of the reading time results in terms of the availability of such an interpretation thus is undermined by the rating results.

4 Conclusion

The results from the eye tracking experiment showed that reading times on the verb following {*wieder nicht*} were affected differently based on the order (and corresponding scope) of negation and *wieder*, with clear effects of infelicity in the unembedded *wieder* condition and no (or opposite) effects in the embedded *wieder* condition. The immediate presence of presupposition-based effects arguably is more consistent with *semantic* accounts of presupposition, which assume that the presupposed content is conventionally encoded in the lexical entries for the triggers.

With respect to the absence of reading time effects of **Felicity** in the embedded *wieder* condition, the results from the rating study show that this cannot be attributed to the general availability of a local interpretation of the presupposition of *wieder*, as this would predict the same interaction to show up in the ratings. An alternative explanation for the lack of reading time effects in this condition is that computing the global interpretation in the syntactic context of negation is more complex in terms of processing, and that this interpretation therefore is not immediately available. What remains somewhat mysterious at this point is that no slow-downs in reading are to be found on any of the subsequent and larger regions that we analyzed. Characterizing the result as involving a delayed computation of the presupposition might lead us to expect to find the same type of increase in reading times, but on a later region. Nonetheless, the rating study clearly shows that the target sentences are perceived to be infelicitous in the infelicitous context, and it's hard to explain the absence of a reading time effect for the *nicht wieder* conditions if we assume that this infelicity becomes apparent immediately. Furthermore, there was a suggestive numerical increase in response times for the ratings in the infelicitous *nicht wieder* condition. While this did not give rise to a statistically significant interaction, there was a potential hint of a marginal simple effect of **Felicity** for this order ($p = .12$). If it were possible to substantiate such an increase in a study that is more directly targeted at capturing the time course of the acceptability judgment, that could lend further and even more direct support to the hypothesis that computing global presupposition interpretations in the context of negation is more costly.

While limitations of space as well as the experimental focus of the present research have kept us from evaluating the impact on theoretical discussions of the interaction in detail, it would be of high theoretical significance if embedded presuppositions indeed involve more processing effort. In particular, this would seem very much consistent with theories that posit explicit and complex operations on levels of representation in the computation of global interpretations, such as the DRT analysis by van der Sandt and Geurts 1991 and van der Sandt 1992 (though this is certainly not the only possible account consistent with the data). But a more thorough exploration of such theoretical implications must await future occasion.

References

Abusch, D.: Presupposition Triggering from Alternatives. Journal of Semantics 27, 37–80 (2009)

Bates, D.M., Sarkar, D.: lme4: Linear mixed-effects models using S4 classes, R package version 0.99875-6 (2007)

Bates, D.M.: Fitting linear mixed models in R. R News 5, 27–30 (2005)

Baayen, R.H., Davidson, D.J., Bates, D.M.: Mixed-effects modeling with crossed random effects for subjects and items. Journal of Memory and Language 59, 390–412 (2008)

Beaver, D.I., Geurts, B.: Presupposition. In: Zalta, E.N. (ed.) The Stanford Encyclopedia of Philosophy (2011), `http://plato.stanford.edu/archives/sum2011/entries/presupposition/`

Bott, L., Noveck, I.A.: Some utterances are underinformative: The onset and time course of scalar inferences. Journal of Memory and Language 51, 437–457 (2004)

Chemla, E., Bott, L.: Processing Presuppositions: Dynamic Semantics vs. Pragmatic Enrichment. Cardiff University, IJN, LSCP (2010) (manuscript)

Chemla, E.: Presuppositions of Quantified Sentences: Experimental Data. Natural Language Semantics 17, 299–340 (2009)

Heim, I., Kratzer, A.: Semantics in Generative Grammar (1998)

Heim, I.: Presupposition Projection. Reader for the Nijmegen Workshop on Presupposition, Lexical Meaning, and Discourse Processes (1990)

Heim, I.: The Semantics of Definite and Indefinite Noun Phrases, Doctoral Dissertation, University of Massachusetts at Amherst (1982)

Kamp, H.: A Theory of Truth and Semantic Representation. Formal Methods in the Study of Language, 277–322 (1981)

Rayner, K.: Eye Movements in Reading and Information Processing: 20 Years of Research. Psychological 124, 372–422 (1998)

van der Sandt, R.: Presupposition Projection as Anaphora Resolution. Journal of Semantics 9, 333–377 (1992)

van der Sandt, R., Geurts, B.: Presupposition, Anaphora, and Lexical Content. In: Herzog, O., Rollinger, C.-R. (eds.) LILOG 1991. LNCS, vol. 546, pp. 259–296. Springer, Heidelberg (1991)

Schwarz, F.: Processing Presupposed Content. Journal of Semantics 24, 373–416 (2007)

Schlenker, P.: Presuppositions and Local Contexts. Mind 119(474), 377–391 (2010)

Schlenker, P.: Local Contexts. Semantics & Pragmatics 2, 1–78 (2009)

Simons, M.: On the Conversational Basis of Some Presuppositions. In: Proceedings of Semantics and Linguistics Theory 11 (2001)

Stalnaker, R.: Presuppositions. Journal of Phillosophical Logic 2, 447–457 (1973)

Tiemann, S., Schmid, M., Bade, N., Rolke, B., Hertrich, I., Ackermann, H., Knapp, J., Beck, S.: Psycholinguistic Evidence for Presuppositions: On-line and Off-line Data. Sinn und Bedeutung 15. In: Proceedings of the Annual Conference of the Gesellschaft für Semantik, pp. 581–597 (2011)

Presupposition Projection Out of Quantified Sentences: Strengthening, Local Accommodation and Inter-speaker Variation

Yasutada Sudo[1], Jacopo Romoli[2], Martin Hackl[1], and Danny Fox[1,3]

[1] Massachusetts Institute of Technology
[2] Harvard University
[3] Hebrew University of Jerusalem
{ysudo,fox,hackl}@mit.edu, jromoli@fas.harvard.edu

Abstract. Presupposition projection in quantified sentences is at the center of debates in the presupposition literature. This paper reports on a survey revealing inter-speaker variation regarding which quantifier yields universal inferences—which Q in $Q(B)(\lambda x.C(x)_{p(x)})$ supports the inference $\forall x \in B$: $p(x)$. We observe an implication that if *some* yields a universal inference for a speaker, *no*, and *any* in a polar question do as well. We propose an account of this implication based on a trivalent theory of presupposition projection together with auxiliary assumptions suggested by [8].

Keywords: Presupposition Projection, Quantified Sentences, Trivalent Logic, Inter-speaker Variation.

1 Introduction

The judgments regarding the presuppositions associated with quantificational sentences are often delicate, and different judgments are reported in the literature ([1, 4, 12]). In this paper we focus on three particular types of quantificational sentences that are illustrated in (1).

(1) a. Some of the students drive their car to school
  b. None of the students drive their car to school
  c. Do any of the students drive their car to school?

We will refer to these three types of quantified sentence, an existential sentence, a negative existential sentence, and an existential polar question, respectively.

It is generally taken for granted that existential sentences like (1-a) do not have a universal inference, $\forall x$: $p(x)$ (but see [3, 5]). According to some theories such as [16] (see also [12]), however, all quantifiers, including existential quantifiers, are predicted to give rise to a universal presupposition.

For negative existential sentences, [6] claims that it has a universal presupposition, while [1] contends that its presupposition is always existential, i.e. $\exists x$: $p(x)$. [1] goes one step further, and proposes a theory where the presupposition is existential for all quantificational determiners, not just *none*.

M. Aloni et al. (Eds.): Amsterdam Colloquium 2011, LNCS 7218, pp. 210–219, 2012.
© Springer-Verlag Berlin Heidelberg 2012

More recently [4] conducted experiments whose results suggest that there are subtle differences among quantifiers. In particular, [4] provides evidence that existential sentences with modified numerals indeed lack a universal inference, while negative existential sentences tend to have a universal inference. However, [4]'s evidence is based on data pooled from all subjects and thus is not informative on possible variations in judgment among speakers. As we will show, such variation exists and it is an important aspect of the phenomenon that calls for an explanation.

This paper presents the results of an online survey that aims to investigate the possibility of inter-speaker variation in the distribution of universal inferences across the three types of quantified sentences mentioned above. In particular, our results indicate the following implication: if an existential sentence yields a universal inference for a speaker, a negative existential sentence and an existential polar question do as well. We offer an account of this implication framed in a trivalent theory of presupposition projection ([2, 7–10], among others). Trivalent theories predict a disjunctive presupposition for the three types of quantified sentences. Following [8], we assume that this presupposition is pragmatically marked and that two strategies can be used to mitigate this markedness: (i) pragmatic strengthening and (ii) insertion of the A-operator (defined below). We will see that the first strategy always yields a universal inference, whereas the second strategy never yields a universal inference for existential sentences, but could yield a universal inference for the other two types of quantificational sentences. We will propose that speakers vary in the strategy they prefer to use, and demonstrate how this can account for the implication found in the survey.

The organization of the paper is as follows. The survey is presented in Section 2. Our theoretical assumptions are introduced in Section 3, and our theory regarding inter-speaker variation is proposed in Section 4.

2 Survey

We conducted an on-line survey on Amazon Mechanical Turk (MTurk)[1] whose main purpose was to investigate inter-speaker variation on which quantificational determiner yields a universal inference. We focus on the three types of sentences mentioned above, i.e. existential and negative existential sentences, and existential polar questions.

2.1 Design

We employed the 'covered box' method of [13]. The covered box method is a variant of the picture selection method. In each trial of our survey, participants saw a sentence and a pair of pictures, and were asked to pick the picture that the sentence was about. One of the pictures was covered and invisible, while the other picture was overtly displayed. Participants were instructed to choose

[1] `https://www.mturk.com/mturk/welcome`

the covered picture *only if the overt picture was not a possible match for the sentence.*

The survey consists of 3 target trials, 3 control trials and 18 filler trials. The sentences used in the target trials are given below. They all contain *both*, which presupposes that there are exactly two entities satisfying the restriction.

(2) a. Some of these three triangles have the same color as both of the circles in their own cell
b. None of these three circles have the same color as both of the squares in their own cell
c. Do any of these three squares have the same color as both of the triangles in their own cell?

The overt picture in each of the target trials was designed in such a way that the universal inference is not satisfied in it. In addition for the trials with declarative sentences, the overt picture satisfies what is asserted (and implicated), e.g. some but not all of the three triangles have the same color as all of the circles in their own cell for (2-a).[2] Therefore, the prediction is that the covered picture will be chosen if and only if the speaker gets a universal inference.

The overt pictures for the target trials are given in Fig. 1.[3] Each picture contains three cells, each of which in turn contains exactly one restrictor figure (e.g. a triangle for (2-a)). Crucially, only two of the cells have exactly two nuclear scope figures (e.g. circles for (2-a)), and the remaining one has only one. For trials with a polar question such as (2-c), the overt picture is colorless, and participants were instructed to imagine that somebody who is incapable of distinguishing colors is asking the question, and guess which picture they are asking about.

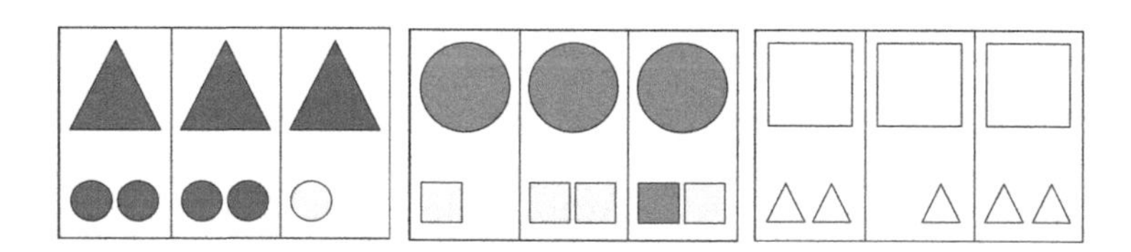

Fig. 1. Overt pictures in the target items

The three control items are identical in structure to the target items except for the following two points: the sentence mentions different restrictor and nuclear figures from the corresponding target item, and the overt picture satisfies the universal inference (i.e. all of the cells contain exactly two nuclear figures). Therefore all the participants are expected to choose the overt picture.

---

[2] The sentences had additional presupposition triggers besides *both*, namely the definite DPs *the* ... and *their own* ..., which are satisfied universally.

[3] The pictures are converted to gray scale here. In the original pictures, (3-a') contains blue triangles and blue and yellow circles, and (3-b') contains red circles and red and yellow squares.

The eighteen filler items involve non-ambiguous quantificational sentences, eight of which are polar questions. None of them contain *both* taking scope below another quantifier. For half of the filler trials the overt picture matches the sentence, and for the other half, the overt picture does not satisfy the assertion and/or the presupposition of the sentence.

2.2 Results

274 participants were employed on MTurk, among which 15 non-native speaker participants and 73 other participants whose accuracy rate for the filler items was less than 75% (i.e. 5 or more mistakes) are excluded from the analysis. All participants were paid $0.20 for their participation, and 59 of them are paid additional $0.25 for answering all of the filler items correctly.

As there are two possible answers for each of the three target trials, there are eight possible answer patterns. The data from 186 native speakers of English is summarized in Table 1, where $\forall$ stands for the covered picture, and $\exists$ stands for the overt picture.[4]

Table 1. Results of the survey

	'Some'	'None'	'?any'	# of Participants		'Some'	'None'	'?any'	# of Participants
1	$\exists$	$\exists$	$\exists$	60	5	$\forall$	$\exists$	$\exists$	2
2	$\exists$	$\exists$	$\forall$	49	6	$\forall$	$\exists$	$\forall$	1
3	$\exists$	$\forall$	$\exists$	21	7	$\forall$	$\forall$	$\exists$	2
4	$\exists$	$\forall$	$\forall$	47	8	$\forall$	$\forall$	$\forall$	19

The distribution of participants across the answer patterns is clearly non-uniform. In particular, the patterns 5-7 are very small in number, compared to the others. From this observation, we draw the following generalizations.

(3) For a given speaker,

a. if the existential sentence has a universal inference, then the negative existential sentence and the existential polar question do too (i.e. 8 vs. 5-7);

b. if the existential sentence does not have a universal inference, then the negative existential sentence and the existential polar question can but need not have a universal inference (i.e. 1-4)

It should also be remarked that 24 out of 186 participants chose the covered picture for the existential sentence, indicating that they obtained a universal

[4] The error rate for the filler items is rather high, but inclusion of more subjects by lowering the cutoff accuracy rate does not undermine our results. Specifically, by lowering the cutoff accuracy to 70% (5 or less mistakes are allowed), 221 among the 259 native speaker participants, and by lowering it to 65% (6 or less mistakes are allowed), 234 subjects remain. In both cases, all the patterns but 5-7 show an increase in number roughly proportionate to the number of the additional subjects, while 5-7 do not exceed 2.

inference for it. As remarked at the outset, it is generally considered that existential sentences have non-universal inferences, and our data shows a clear tendency in line with this intuition (see also [4]). Nonetheless, the existence of speakers preferring a universal reading is theoretically interesting, and as we will demonstrate, our theory accounts for both types of speakers.

The results of the control items are as follows. Recall that the overt picture satisfies the universal inference and hence is expected to be chosen uniformly. In fact only 3 participants chose the covered picture for the existential sentence, and 15 subjects did so for the existential polar question. However, contrary to our expectation, the covered picture was chosen for the negative existential sentence by 43 out of 186 subjects. Although we cannot offer an explanation for this unexpectedly high figure for the negative existential sentence, it does not undermine the above observations, since excluding these subjects still indicates the same implicational patterns (the numbers are omitted for reasons of space).[5]

3 Trivalent Theory of Presupposition Projection

3.1 Three Truth Values and the Felicity Condition

In the rest of this paper, we will offer a theoretical explanation of the implicational generalizations in (3) framed in a trivalent theory of presupposition projection.[6] Trivalent theories ([2, 7–10, 15]) postulate three, rather than two, truth values, denoted here by 0, 1 and #. The projection property of a given sentence is predicted via a pragmatic principle that requires a sentence to denote either 0 or 1 in each of the possible worlds in the current context set (in the sense of [19]). This pragmatic principle is stated as in (4).

(4) **Felicity Condition**
A (declarative) sentence S can be felicitously used given a context set C only if for all $w \in C$, $[\![S]\!](w) \neq \#$

The Felicity Condition can be given a pragmatic motivation. It can be thought of as a consequence of a principle of conversation demanding that an utterance of a declarative sentence tell the conversational participants which worlds in the context to retain and which ones to discard. This demand will not be met if in

[5] An anonymous reviewer of the Amsterdam Colloquium worried that this could indicate that there is a stronger bias toward the covered picture for negative existential sentences than for the other types of sentences. This suggests a possibility that for the negative existential sentence, a choice of the covered picture does not necessarily imply the universal inference, and hence the figures for the patterns 3, 4, 7 and 8 are overestimated to some extent. Crucially, however, the asymmetry between the patterns 1-4 and the patterns 5-8 will remain even if we correct for a tendency to choose the covered box for negative existentials.

[6] A reviewer suggested the possibility that the distribution of facts follows directly from a probabilistic theory of presupposition projection. We think that this is a potentially interesting avenue to investigate. However, we were unable to come up with a predictive a general theory of projection that would derive the needed probabilities.

any of the worlds in the context set the declarative sentence is neither true nor false (cf. [19]).

In this system, the presupposition of a declarative sentence is the proposition that needs to be true for the sentence to denote either 0 or 1. As an illustration, consider the simple example below, and imagine that the existence presupposition of the possessive is the only presupposition.

(5) John drives his car to school

The denotation of this sentence is as in (6).

(6) $$\lambda w. \begin{cases} 1 \text{ if John has a car and drives it to school in } w \\ 0 \text{ if John has a car and does not drive it to school in } w \\ \# \text{ otherwise} \end{cases}$$

According to the Felicity Condition in (4), for an utterance of (5) to be felicitous, it is required that John have a car in all of the possible worlds in the context set. Therefore (5) presupposes that John has a car.

3.2 Extension to Polar Questions

We now extend the above theory to polar questions. Following [14], we assume that questions denote sets of propositions. Notice that the Felicity Condition in (4) does not apply to question denotations, and therefore we postulate a separate pragmatic condition for the use of questions.[7] We hypothesize the weakest possible condition in (7).[8]

(7) **Felicity Condition for Questions**
A question Q can be felicitously used given a context set C only if for all $w \in C$, there is $q \in [\![Q]\!]$ such that $q(w) \neq \#$

This is evidently not meant to be the only condition on a felicitous use of a question. Other conditions include, for example, that the answer is not known yet, and that all propositions in the denotation are not known to be false, which amounts to requiring that for all $q \in [\![Q]\!]$, there is $w \in C$ such that $q(w) = 1$. But as it turns out, only (7) matters for our purposes at hand.

In (8) we give a simple example for illustration.

(8) Does John drive his car to school?

Suppose that the denotation of this question is in (9).

[7] [8] suggests a bivalent reformulation of the theory using the notion of relevance with the aim of giving declarative sentences and questions a uniform treatment. We will not pursue this alternative in this paper.

[8] Note that if we strengthen the condition to a universal requirement, this won't affect our results since the two are equivalent for polar questions, given that the two members of the question denotation have the same presupposition.

(9) $\{[\![\text{John drives his car to school}]\!], \neg[\![\text{John drives his car to school}]\!]\}$
$= \{[\![(5)]\!], \neg[\![(5)]\!]\}$

The condition in (7) demands that in each world w in the context set, either $[\![(5)]\!](w) = 1$ or $[\![(5)]\!](w) = 0$. Therefore, (8) has the same presupposition as (5), namely that John owns a car.

3.3 Disjunctive Presuppositions for Quantified Sentences

As we will demonstrate in this subsection, trivalent theories assign a disjunctive presupposition to all of the three types of quantified sentences that we are interested in in the present paper.[9] For ease of exposition, we schematically represent the meanings of quantificational sentences as in (10).

(10) $Q(B)(\lambda x.C(x)_{p(x)})$ where
 a. Q is a determiner denotation
 b. B is the restrictor of Q
 c. $\lambda x.C(x)_{p(x)}$ is the nuclear scope of Q with the presupposition $p(x)$

The predicted presupposition of the three types of quantificational sentences that we are after is $[\exists x \in B\colon p(x) \wedge C(x)] \vee [\forall x \in B\colon p(x)]$. Let us look at the three cases in turn.

Firstly, the truth conditions of an existential sentence are given in (11).

(11) $[\![\text{some}]\!](B)(\lambda x.C(x)_{p(x)})$

$$= \lambda w. \begin{cases} 1 \text{ if } \exists x \in B\colon p(x) \wedge C(x) \text{ in } w \\ 0 \text{ if } [\forall x \in B\colon p(x)] \wedge [\neg\exists x \in B\colon p(x) \wedge C(x)] \text{ in } w \\ \# \text{ otherwise} \end{cases}$$

The Felicity Condition requires that for each possible world w in the context set, either $\exists x \in B\colon p(x) \wedge C(x)$ or $[\forall x \in B\colon p(x)] \wedge [\neg\exists x \in B\colon p(x) \wedge C(x)]$. Thus the presupposition of (11) is the disjunction of these two propositions, i.e. $[\exists x \in B\colon p(x) \wedge C(x)] \vee [\forall x \in B\colon p(x)]$.

Given that a negative existential sentence is the negation of the corresponding existential sentence, i.e. (12), the exact same presupposition is predicted for negative existential sentences.

(12) $[\![\text{none}]\!](B)(\lambda x.C(x)_{p(x)})$

$$= \lambda w. \begin{cases} 1 \text{ if } [\![\text{some}]\!](B)(\lambda x.C(x)_{p(x)})(w) = 0 \\ 0 \text{ if } [\![\text{some}]\!](B)(\lambda x.C(x)_{p(x)})(w) = 1 \\ \# \text{ otherwise} \end{cases}$$

Furthermore, an existential polar question has the same disjunctive presupposition, if we assume the denotation of $[\![?]\!][\![\text{any}]\!](B)(\lambda x.C(x)_{p(x)})$, where '?' is the question operator, to be the set $\{[\![\text{some}]\!](B)(\lambda x.C(x)_{p(x)}), [\![\text{none}]\!](B)$

[9] As demonstrated in [2, 7] and in particular in [9, 10], a trivalent theory makes predictions for the entire language, including various kinds of connectives, but as they are not our central concern, we will ignore them here.

$(\lambda x.C(x)_{p(x)})\}$. The Felicity Condition demands that for each possible world in the context set, either of these propositions is true, i.e. $[\exists x \in B\colon p(x) \wedge C(x)] \vee [\forall x \in B\colon p(x)]$ must be true in each world. Therefore the same disjunctive presupposition $[\exists x \in B\colon p(x) \wedge C(x)] \vee [\forall x \in B\colon p(x)]$ is predicted.

3.4 Two Strategies

Following [8], we assume that the disjunctive presupposition $[\exists x \in B\colon p(x) \wedge C(x)] \vee [\forall x \in B\colon p(x)]$ is pragmatically marked and triggers one of two repair strategies: (i) pragmatic strengthening or (ii) insertion of the operator that turns a trivalent proposition into a bivalent one, which we call the A-operator following [2].

Again following [8], we furthemore assume that when applied to the disjunctive presupposition, pragmatic strengthening yields the universal inference $\forall x \in B\colon p(x)$.[10] Since three types of quantified sentences we are interested in here have the same disjunctive presupposition as shown in the previous subsection, pragmatic strengthening invariably yields the universal inference for all of them.

On the other hand, the A-operator can result in a universal or weaker inference depending on its scope and the quantifier. The denotation of the operator is given in (13).

(13) $$[\![\mathrm{A}]\!] = \lambda p_{\langle s,t\rangle}.\lambda w. \begin{cases} 1 \text{ if } p(w) = 1 \\ 0 \text{ if } p(w) = 0 \text{ or } p(w) = \# \end{cases}$$

We assume that the A-operator is a phonologically null operator that may appear in any syntactic position where the sister constituent is of the propositional type $\langle s,t\rangle$.[11] Let us now examine what the A-operator yields for the three types of quantified sentences.

Consider first an existential sentence with the A-operator above the quantifier, i.e. $[\![\mathrm{A}]\!]([\![\text{some}]\!](B)(\lambda x.C(x)_{p(x)}))$.

(14) $$\lambda w. \begin{cases} 1 \text{ if } [\![\text{some}]\!](B)(\lambda x.C(x)_{p(x)})(w) = 1 \\ 0 \text{ otherwise} \end{cases}$$

This does not have a universal inference, and only entails an existential inference, $\exists x \in B\colon p(x)$. Also it is easy to see that the predicted denotation is the same, when the A-operator is applied below the quantifier, $[\![\text{some}]\!](B)(\lambda x.[\![\mathrm{A}]\!](C(x)_{p(x)}))$. Therefore if the A-operator is present, the existential sentence has only an existential entailment.

For a negative existential sentence, on the other hand, insertion of the A-operator still can result in a universal inference. Consider $[\![\mathrm{A}]\!]([\![\text{none}]\!](B)(\lambda x.C(x)_{p(x)}))$ whose denotation is given in (15).

[10] See [8] for discussion of the way issues pertaining to the proviso problem ([11, 17, 18]) arise in this context and how they might be dealt with.

[11] Unlike [2], we do not impose any other condition on the use of the A-operator here, although ultimately some notion of preference among different scopes might be necessary.

(15) $$\lambda w.\begin{cases} 1 \text{ if } [\![\text{none}]\!](B)(\lambda x.C(x)_{p(x)})(w) = 1 \\ 0 \text{ otherwise} \end{cases}$$

This entails the universal statement $\forall x \in B$: $p(x)$. In addition, there is a second use of the A-operator which gives rise to a weaker inference: when it applies below the quantifier, i.e. $[\![\text{none}]\!](B)(\lambda x.[\![\text{A}]\!](C(x)_{p(x)}))$, the sentence is true in w iff $\neg\exists x \in B$: $[p(x) \wedge C(x)]$ is true in w, from which neither an existential nor a universal inference is derived.

Finally the A-operator can yield a universal inference for an existential polar question too. When it is applied above the question operator, the predicted meaning is $\{[\![\text{A}]\!]([\![\text{any}]\!](B)(\lambda x.C(x)_{p(x)})), [\![\text{A}]\!](\neg[\![\text{any}]\!](B)(\lambda x.C(x)_{p(x)}))\}$. Since the presupposition of a polar question is that either of the answers is true in each world in the context set, the presupposition is the disjunction $[\exists x \in B$: $p(x) \wedge C(X)] \vee [\forall x \in B$: $p(x)]$. Thus the A-operator used in this way still results in the disjunctive presupposition. Because the disjunctive presupposition is marked and needs to be remedied, and pragmatic strengthening is still available, a universal inference ensues in this case. Alternatively, the A-operator may take lower scope, yielding $\{[\![\text{A}]\!]([\![\text{any}]\!](B)(\lambda x.C(x)_{p(x)})), \neg[\![\text{A}]\!]([\![\text{any}]\!](B)(\lambda x.C(x)_{p(x)}))\}$. Since this is a bipartition of any set of possible worlds, the presupposition is trivial. Similarly, when the A-operator takes scope below the quantifier, the resulting set, $\{[\![\text{any}]\!](B)(\lambda x.[\![\text{A}]\!](C(x)_{p(x)})), \neg[\![\text{any}]\!](B)(\lambda x.[\![\text{A}]\!](C(x)_{p(x)}))\}$, is a bipartition of any set of possible worlds and only a trivial presupposition is predicted.

To sum up, pragmatic strengthening invariably yields a universal inference for all of the three types of quantified sentences. The A-operator, on the other hand, may give rise to weaker inferences depending on its scope and the quantifier. More specifically, existential sentences are never associated with a universal inference regardless of the scope of the A-operator, while negative existential sentences and existential polar questions can but do not have to have a universal inference.

4 Account of the Generalizations

The generalizations in (3) found in the survey are repeated here.

(3) For a given speaker,

a. if the existential sentence has a universal inference, then the negative existential sentence and the existential polar question do too

b. if the existential sentence does not have a universal inference, then the negative existential sentence and the existential polar question can but need not have a universal inference

In order to account for these generalizations, we propose that there are two types of speakers: (i) those who use the A-operator in one way or another for all three sentences, and (ii) those who never do. This explains the implication (3) as follows. Those who do not use the A-operator always resort to pragmatic strengthening, obtaining a universal inference for all of the three cases. This accounts for the implicational generalization in (3-a). On the other hand, those who use the A-operator never get a universal inference for existential sentences, but may get a universal inference for negative existential sentences and existential polar questions, depending on where it is inserted. This accounts for the generalization in (3-b).

References

1. Beaver, D.: Presupposition and Assertion in Dynamic Semantics. CSLI, Stanford (2001)
2. Beaver, D., Krahmer, E.: A partial account of presupposition projection. Journal of Logic, Language and Information 10, 147–182 (2001)
3. Charlow, S.: "Strong" predicative presuppositional objects. In: Klinedinst, N., Rothschild, D. (eds.) Proceedings of ESSLLI 2009 Workshop: New Directions in the Theory of Presupposition (2009)
4. Chemla, E.: Presuppositions of quantified sentences: experimental data. Natural Language Semantics 17, 299–340 (2009)
5. Chemla, E.: (ms) Similarity: towards a unified account of scalar implicatures, free choice permission and presupposition projection. Ms., ENS
6. Cooper, R.: Quantification and Semantic Theory. Reidel, Dordrecht (1983)
7. Fox, D.: Two short notes on Schlenker's theory of presupposition projection. Theoretical Linguistics 34, 237–252 (2008)
8. Fox, D.: Presupposition projection, trivalent and relevance. Handout of the talk at University of Connecticut, Storrs (2010)
9. George, B.: A new predictive theory of presupposition projection. In: Proceedings of SALT 18, pp. 358–375 (2008a)
10. George, B.: Presupposition Repairs: a Static, Trivalent Approach to Predicting Projection. MA thesis, University of California, Los Angeles (2008b)
11. Geurts, B.: Presuppositions and Pronouns. Elsevier, Amsterdam (1999)
12. Heim, I.: On the projection problem of presuppositions. In: Proceedings of WCCFL 2, pp. 114–125 (1983)
13. Huang, Y.T., Spelke, E., Snedeker, J.: (ms.) What exactly do numbers mean? Ms., Harvard University
14. Karttunen, L.: Syntax and semantics of questions. Linguistics and Philosophy 1, 3–44 (1977)
15. Peters, S.: A truth-conditional formulation of Karttunen's account of presupposition. Synthese 40, 301–316 (1979)
16. Schlenker, P.: Local contexts. Semantics and Pragmatics 3 (2009)
17. Schlenker, P.: The proviso problem: a note. To appear in Natural Language Semantics (to appear)
18. Singh, R.: Modularity and Locality in Interpretation. Ph.D. dissertation, Massachusetts Institute of Technology (2008)
19. Stalnaker, R.: Assertion. In: Cole, P. (ed.) Syntax and Semantics 9: Pragmatics, pp. 315–332. Academic Press, NY (1978)

Focus, Evidentiality and Soft Triggers

Márta Abrusán

Lichtenberg Kolleg, Georg–August Universität Göttingen, Germany

Abstract. Soft triggers are fairly easily suspendable in context. Two main environments in which this happens have been identified: (a) The presupposition of soft triggers can be suspended by focus (cf. Beaver 2004) (b) Simons (2007) has observed that many soft triggers such as *hear, see, believe, discover, know, etc.* have semantically parenthetical uses which are not presuppositional. This paper offers a way of predicting these facts in the context of a theory applicable to soft triggers.

1 Introduction

Soft triggers are fairly easily suspendable in context, as was pointed out by Karttunen [1971b], Stalnaker [1974], Chierchia and McConnell-Ginet [2000], Simons [2001], Beaver [2004], Abbott [2006], Klinedinst [2009], Romoli [2011], among others. Observe a classic example:

(1) As far as I know, everything I've said is correct. But if I realize later that I have not told the truth, I will confess it to everyone. (Karttunen 1971b)

Such examples of suspendability have been taken to suggest by many of the above authors that—at least in the case of soft triggers—a pragmatic explanation of why the presupposition arises is desirable.

Yet how presuppositions could be predicted has been an elusive and rarely addressed question. While the few attempts in the literature to explain presuppositions of at least certain items provided valuable insights (cf. Sperber and Wilson 1979, Simons 2001, Abusch 2010, Simons et al. 2010), they either did not make correct predictions or failed to be sufficiently explanatory. Stalnaker [1974] and Schlenker (2010) laid out a blueprint for a triggering mechanism, but did not provide a theory themselves.

Abrusán [2011] proposed a mechanism for predicting the presuppositions of soft triggers according to which entailments of a sentence S that are independent from the main point of S are presupposed. The main point of a sentence S is defined by grammar, and is given by those entailments that are by nature about the event time of the matrix predicate of S. Entailments of S that describe events that are not necessarily about the event time of the matrix predicate of S are independent from the main point, hence presupposed. However this proposal was not context-sensitive and was therefore not able to handle cases of presupposition suspension.

This paper argues that besides the grammatically defined main point, a secondary, pragmatic main point can be derived as well by markers such as focus

M. Aloni et al. (Eds.): Amsterdam Colloquium 2011, LNCS 7218, pp. 220–229, 2012.
© Springer-Verlag Berlin Heidelberg 2012

and evidential expressions. In these cases, sentences have two main points that are relevant for presupposition triggering: the default (grammatical) and a secondary (pragmatic) one. Entailments that are to be presupposed have to be independent from both of these. This predicts that in the examples of presupposition suspension no presupposition is triggered to begin with. A more detailed exposition of the theory presented here can be found in Abrusán [2012].

2 Attention and Aboutness

The central idea behind this paper is that presuppositions of soft triggers arise from the way our attention structures the informational content of a sentence. Some aspects of the information conveyed are such that we pay attention to them by default, even in the absence of contextual information. On the other hand, contextual cues or conversational goals can divert attention to types of information that we would not pay attention to by default. Either way, whatever we do not pay attention to, be it by default, or in context, is what ends up presupposed.

This two-tier view of how attention structures information is familiar from studies of vision. (see e.g. Goldstein 2009, Itti and Koch 2001, Navalpakkam and Itti 2005 among many others). A flickering light, a red dot on gray background, or seeing our own name written on a screen attracts our attention immediately in any context. These are examples that manifest the default, bottom-up processes of attention. Interestingly, when looking at a scene, visual attention is also influenced by what the scene is about: semantically more relevant cues attract attention. At the same time, when looking at a photograph we might easily overlook somebody's shadow or reflection in the water, or the lack of these, even if these occupy a comparatively large portion of the photograph. However, in a context in which shadows or reflections are made salient and relevant, suddenly we pay attention to such cues as well. These are examples of top-down, goal oriented mechanisms of attention. Further, we also tend to assume that aspects of familiar scenes that we do not pay attention to (for example shadows in a neutral context) are nevertheless there.

The grammatical analogy with vision is the following. We instinctively pay attention to information that is about the main event described by the sentence. This corresponds to the default, bottom-up process of attention. Any information that is also conveyed by the sentence but is not about the main event described is presupposed, unless there is some contextual factor that directs attention to this information as well. Such contextual factors correspond to top-down processes of attention. In these cases what would normally be presupposed is not presupposed any more: i.e we have presupposition suspension. Note that what happens is not that attention is completely diverted, rather, extra information is brought under the spotlight of attention as well. This corresponds to the observation that there are no cases reported in the literature (at least to my knowledge) where context would swap the presupposed and the asserted aspects of the meaning of a soft trigger. Instead, contextual effects tend to amount to the removal of presuppositions.[1]

2.1 The Default Main Point

The question now is, what is the default point described by a sentence and among all the myriad propositions that are entailed by a sentence, how do we find the ones that are *about* the main point described by the sentence?

The intuition that we want to capture is that presuppositional assertions describe complex states of events, some parts of which are independent from the main events. So what we want to achieve is to tell independent events apart: Select the main event described by the sentence, and decide what other information conveyed by the sentence describe independent events from the main one. But this is a very difficult task and cannot be easily accomplished just by looking at events themselves because of the very complex mereological structure of events. For example, is the event of raining part of the complex event of John's knowing it? If not, why not?

To simplify matters, I will map events to their event times. The idea of looking at event times instead of events themselves serves the purpose of making independence more tractable: Events that happen at different times are clearly different events. Further, in some cases, e.g. sentences involving mathematical truths (*John knows that 2+2=4*), the possibility of invoking events is not obvious at all. However, since event times are more abstract than events, evoking event times is still possible even in these cases.

Thus I will assume that the default main point of a sentence is given by those entailments that are by nature about the event time of the matrix predicate. Propositions that describe events that are not (or do not have to be—in the sense to be introduced in subsection 2.3) about the event time of the matrix predicate of S are independent, and hence presupposed. Let's illustrate the idea with a simple example. Consider (2), in which t_1 denotes the event time interval of the matrix predicate, and t_2 is some interval before t_1, given by the context. Let's look at the sentence S and two its (many) entailments, φ and ψ:

(2) S=John knows (at t_1) that it was raining (at t_2).
  a. φ= John believes (at t_1) that it was raining (at t_2).
  b. ψ= It was raining (at t_2).

In an intuitive sense, φ is about the time denoted by t_1, but ψ is not: changing the properties of the world at t_1 will not affect the truth value of ψ but it might affect the truth value of φ.

For concreteness, I will assume that event times denote salient intervals whose value is assigned by the context. As such, they are rather like pronouns (cf. Partee 1973)

[1] A different case of context dependency is when presuppositions are added to otherwise non-presuppositional expressions, as discussed in Schlenker (2010). These are discussed in Abrusan (2012).

In this system, predicates have an extra argument slot for time, thus what are usually assumed to be one place predicates such as intransitive verbs are going to be two place predicates, taking an individual and a time argument. Tense morphemes introduce time variables that saturate the time argument slot of predicates in the syntax. The denotation of this variable is given by the interpretation function *i* supplied by the context, which assigns it an element from the domain of time intervals. E.g. the sentence in (3) is true iff John is tired at the time assigned to t_2 by *i*.

(3) John is tired at t_2

Now we might ask what does it mean for a sentence to be about the entity denoted by one of its arguments, in our case about its time argument? Exactly this notion is captured by Demolombe and Fariñas del Cerro's (2000) definition of aboutness, introduced below.

2.2 Being about an Entity

The intuitive idea of Demolombe and Fariñas del Cerro [2000] is that the truth value of a sentence that is not about an entity should not change if we change the truth value of the facts about that entity. To capture this intuition, they give a proposal that has two parts. The first is the definition of variants of an interpretation with respect to an object. Given this notion, the property of a sentence being about an object can be defined.

Definition 1. The syntax of the language L_c Let L_c be a first order predicate calculus language, where c is some constant symbol. The equality predicate is not allowed in the language.

The primitive vocabulary of L_c consists of the following:

1. A set of constants designated in the metalanguage by letters *a,b,c,...*
2. A denumerably infinite set of variables v_1, v_2, v_3,... The constants and variables together constitute the *terms*.
3. A set of predicates, *p, q, r*, each with fixed arity
4. The logical connectives $\neg, \vee, \wedge, \rightarrow, \leftrightarrow$
5. The quantifier symbols $\exists, \forall$
6. The parentheses (,)

The set F of formulas of L_c is defined by the following rules:

1. If *p* is an n-ary predicate and *t* an n-tuple of terms, then $p(t) \in L_c$
2. If $F \in L_c$, and $G \in L_c$, then $\neg F \in L_c$, $(F \vee G) \in L_c$, $(F \wedge G) \in L_c$, $(F \rightarrow G) \in L_c$, $(F \leftrightarrow G) \in L_c$
3. If $F \in L_c$, then $(\exists v F) \in L_c$ and $(\forall v F) \in L_c$
4. All the sentences in L_c are defined by the rule 1-3.

Definition 2. Interpretation An interpretation M of L_c is a tuple M=<D, i> such that

1. D is a non-empty set of individuals and time intervals
2. i is a function that assigns
   (a) to each predicate symbol of arity n a subset of D^n,
   (b) to each variable symbol an element of D (As a notational convention, I will use *t* for variables over tense intervals, and *x* for variables over individuals)
   (c) to each constant symbol an element of D

The domain of M will be denoted by D_M and the interpretation of M by i_M.

NB: As indicated in the previous section, verbal predicates are assumed to have a tense argument on top of any individual arguments. I will also make use of the following simplifying assumptions: (a) definite descriptions denote individuals (b) the denotation of indexicals is given by the interpretation function and their indexical content is presupposed. E.g. *you* has the presupposition that it can only be used felicitously if i_M(you) denotes the addressee in the context. (c) For the purposes of calculating aboutness I will also assume that the complement of attitude predicates is absorbed into the attitude verb, so that *know(x,t,p)* is in fact a 2-place predicate *know-p(x,t)*. The reason why this is not harmful is because we are not attempting to derive the entailments of verbs from their lexical semantics–we are treating the origin of these entailments as a black box. This allows turning these expressions into simple extensional predicates and ignoring the intricate semantics of attitude verbs. Note that none of the above assumptions (nor L_c in general) are meant as an adequate theory of language: they are only simplifications that allow us to abstract away from complex aspects of language that are not relevant for our purposes when calculating aboutness.

Definiton 3. Satisfiability conditions Let M be an interpretation of the language L_c. The truth of a formula F in M is denoted by M $\models$F, and is inductively defined as follows:

1. If F is an atomic sentence of the form *p(k)*, where *k* is a tuple of constant or variable symbols, M $\models$F iff $i_M(k) \in i_M(p)$.
2. M $\models$F and M $\models$F $\vee$ G are defined from M $\models$F and M $\models$G as usual.
3. M $\models\exists$vF iff there exists an interpretation $M_{v/d}$ that only differs from M by the interpretation of the variable symbol v, st. $i_{Mv/d}(v)$ is the element d of $D_{Mv/d}$ and $M_{v/d}$ $\models$F.

Definition 4. Variants of an interpretation with regard to an object Roughly speaking the notion of variants of an interpretation with regard to an object denoted by constant symbol c is the set of interpretations M^c that only differ from M by the truth assignment of atomic sentences where c appears as an argument.

Let L_c be a first order predicate calculus language that contains the constant c and does not contain the identity predicate. M' is a c-variant of a model M iff it meets the constraints listed below:

1. $D_{M'}=D_M$
2. $i_{M'}=i_M$, for every variable symbol and constant symbol
3. $i_{M'}$ is defined from i_M for each predicate symbol as follows: if p is a predicate symbol of arity n
   (a) if k is an n-tuple of terms of language L_c that contain no occurrence of the constant symbol c, then $i_{M'}(k) \in i_{M'}(p)$ iff $i_M(k) \in i_M(p)$.
   (b) if an element $\langle d_1, ..., d_n \rangle$ of D^n is such that for every j in [1,n], $d_j \neq i_M(c)$, then $\langle d_1, ..., d_n \rangle \in i_{M'}(p)$ iff $\langle d_1, ..., d_n \rangle \in i_M(p)$.

M^c will be used to denote the set of c-variant interpretations *M'* defined from *M*.

An example: Let L_c be a language with a unique unary predicate symbol p, and the constant symbols *a, b, c*. Let M be an interpretation of L_c defined by: $D = \{d_1, d_2, d_3, d_4\}, i_M(a) = d_1, i_M(b) = d_2, i_M(c) = d_3$ and $i_M(p) = \{d_1, d_3, d_4\}$. For every variant *M'* in M^c, $i_{M'}(p)$ contains d_1, because $i_M(p)$ contains d_1 and d_1 is the interpretation of the constant symbol *a*, which is different from constant symbol *c*. Therefore the sentence *p(a)* is true in every variant *M'*. At the other extreme, there are variants *M'* of *M* such that d_3 is not in $i_{M'}(p)$, because d_3 is the interpretation of *c*. In these variants *p(c)* is false, although it is true in *M*.

Aboutness. Let *S* be a sentence of language L_c. *S* is not about an object named by the constant symbol *c* iff for every interpretation *M*, $M \models S$ iff for every interpretation *M'* in M^c $M' \models S$:

(4) $NA(S, c)$ holds iff $\forall M (M \models S$ iff $\forall M' \in M^c\ M' \models S)$

A formula F is about an object named by c if it is not the case that $NA(F, c)$:

(5) $A(S, c)$ holds iff $\exists M (\exists M' \in M^c (M \models S$ and $M' \not\models S))$

2.3 Triggering Mechanism—Default Version

The definition above defines what it means for a sentence to be about an object. In principle however we are interested in whether propositions (entailments of S) are about an object. The relationship between the two is somewhat indirect: we need to check whether sentences that can be used to express a proposition are about an object. If yes, I will assume that the proposition expressed by the sentence is about that object as well.

We are now in the position to give the first version of the default triggering mechanism for soft presuppositions:

(6) *Presupposition triggering (1st version, to be revised)*
Entailments of a sentence S that can be expressed by sentences that are not about the event time of the matrix predicate of S are presupposed.

Being *about* is to be understood as defined in the preceding subsection. This predicts (2): the (sentence expressing the) entailment that John believes at t_1

that it is raining at t_2 is about the matrix event time t_1, hence not presupposed, while (the sentence expressing) the entailment that it is raining at t_2 is not about the matrix event time t_1, hence it is presupposed. (The reasoning that predicts this as well as further examples are spelled out in Abrusan (2012)).

Yet with this simple approach the obvious question arises: what about sentences such as (7)? The proposal in (6) predicts that the embedded proposition in (7) is not independent from the main assertion, and therefore, not presupposed, contrary to fact:

(7) John knows (at t_1) that it is raining (at t_1).

a. φ= John believes (at t_1) that it is raining (at t_1).

b. ψ= It is raining (at t_1).

We need to distinguish accidental co-temporaneity from non-accidental one. In the above example, though it so happens that the embedded proposition and the matrix proposition are true at the same time, this is only an accident, it could be otherwise. But the co-temporaneity of the matrix time of φ with the matrix time of S is not an accident, but follows from the lexical interpretation of *know*.

To remedy this, I will assume that the default presupposition triggering mechanism looks beyond the actual sentence and assesses the properties of alternative sentences that I call temporal-alternatives (or just T-alternatives for short). T-alternatives are obtained by replacing the temporal arguments of the matrix and embedded predicates with different ones. More precisely, we replace the temporal variables with ones which the assignment function maps to different intervals than the original time of the matrix predicate. E.g:

(8) John knows (at time t_1) that it was raining (at time t_1)
T-alternative: John knows (at time t_1) that it was raining (at time t_2)

(9) John managed (at time t_1) to solve the exercise (at t_1)
T-alternative: *John managed (at time t_1) to solve the exercise (at t_2)

Let's say that p and p' are corresponding entailments if they can be expressed by sentences that only differ in their temporal arguments. Take an entailment p of S. If there is a well formed alternative S' to S such that the corresponding entailment to p (namely p' of S') can be expressed by a sentence that is not about the event time of the matrix clause of S', then I will say that p is only accidentally about the matrix event time of S. In (8), the entailment that it was raining (at time t_1) of the original sentence is only accidentally about t_1, because there is a T-alternative (*John knows (at time t_1) that it was raining (at time t_2)*) whose corresponding entailment (that it was raining at t_2) is not about the matrix tense of the T-alternative. On the other hand, (9) does not have a well formed T-alternative where the two temporal arguments differ (cf. Karttunen 1971a on temporal restrictions of implicatives): for this reason the entailment of the original sentence in (9) that John solved the exercise at t_1 is non-accidentally about the matrix event time.

A revised version of the default triggering mechanism is as follows:

(10) *Presupposition triggering (2nd version, to be revised)*
Entailments of a sentence S that can be expressed by sentences that are not accidentally-about the event time of the matrix predicate of S are presupposed.

We now make correct predictions about (7-b) as well: this entailment is predicted to be presupposed because although it is about t_1, it is only accidentally so.

2.4 Triggering Mechanism—Context Sensitive Version

Besides the default, grammatically defined main point, it is possible that the context or the intentions of the participants of the conversation raise interest in aspects of the entailed meaning of the sentence that would otherwise "pass under the radar", and be presupposed. There are two ways in which this might happen: by evidential verbs and by focus.(See Abrusan (2012) for more discussion) This is not intended as an exhaustive list, indeed it is likely other factors will turn out to be relevant as well in the future.

One factor that can bring extra elements under the spotlight of attention is focus. As Beaver [2004] observes, (11) does not suggest that the student has plagiarized his work:

(11) If the TA discovers that [your work is plagiarized]$_F$, I will be [forced to notify the Dean] (Beaver 2004, slightly modified)

Focus is usually taken to be the part of a sentence that conveys the new or highlighted information, thus the information that directly answers a background question. In this sense, focus grammatically signals the presence of a background question. I will propose that grammatically marked background questions can introduce a secondary (or pragmatic) main point. Secondary main points concern the event time of the sentence expressing the most direct proposition that answers the background question. The presupposition triggering mechanism looks both at the default (grammatical) and the secondary (pragmatic) main points and requires the presupposition to be independent from both of these. This derives the above data in the present framework.

(12) *Presupposition triggering (3rd, final version)*
Entailments of a sentence S that can be expressed by sentences that are neither accidentally about the event time of the matrix predicate of S nor about the event time of the sentence expressing the most direct answer to the (grammatically signaled) background question are presupposed.

Put more simply, the proposal above requires that presuppositions be independent from both the default and the secondary (pragmatic) main points. Secondary main points can be introduced by grammatical markers such as focus and evidential verbs (and presumably others). In (11), focusing the embedded clause indicates that the background question is *What will I discover?*

The direct answer to this question is a proposition, namely the proposition denoted by the embedded clause *your work is plagiarized*. The pragmatic, secondary main point therefore concerns the information that is about the tense argument of the sentence expressing this proposition, i.e. the tense argument of the embedded clause. Thus the information conveyed by the embedded clause is not independent from the secondary main point, and is not predicted to be presupposed.

Let's look now at evidential uses of verbs. Simons [2007] observes that certain clause embedding verbs such as *hear, see, believe, discover, know, etc.* have semantically evidential uses. In these cases the embedded clause carries the main point of the utterance, while the matrix clause serves an evidential function of identifying information source, emotional attitude, etc. Cf. the example below:

(13) A: Why didn't Louise come to the meeting yesterday?
B: I heard that she's out of town.

Simons notes that when used in this evidential manner, these verbs loose their presuppositionality. Simons' (2007) idea has interesting consequences for the present proposal. In cases where the matrix verb is used in an evidential way, there is also a second (pragmatic) main point besides the grammatically defined default one, which is derived contextually. The second main point concerns the clause "modified" by the evidential, the syntactically embedded clause. Technically the secondary main point is the information that is about the time of the syntactically embedded verb. Since presupposed material has to be independent from the main point (i.e. not be about the event time of the main point), once the embedded clause becomes the main point as well, the content of the embedded clause is not predicted to be presupposed any more, despite being entailed.

Focus and the evidential uses of verbs are two cases where the presuppositions of verbs can be suspended by grammatically signaled contextual factors. In most cases however there is either no grammatically signaled secondary main point, or this is the same as the default main point, and thus no contextual suspension of presupposition is observed.

Acknowledgments. Many thanks to the audience at SEP in Göttingen and the Amsterdam Colloquium for their helpful questions about this material, all the people who have commented on previous, related papers and the Mellon Foundation and the DFG in the framework of the Lichtenberg-Kolleg of the Georg-August-Universität Göttingen for financial support.

References

Abbott, B.: Where have some of the presuppositions gone? In: Birner, B.J., Ward, G. (eds.) Drawing the Boundaries of Meaning: Neo-Gricean Studies in Pragmatics and Semantics in Honor of Laurence R. Horn, pp. 1–20. John Benjamins, Philadelphia (2006)

Abrusán, M.: Triggering verbal presuppositions. In: Li, N., Lutz, D. (eds.) Semantics and Linguistic Theory (SALT) 20, pp. 684–701. eLanguage (2011)

Abrusán, M.: Predicting the Presuppositions of Soft Triggers. Linguistics and Philosophy (to appear, 2012)

Abusch, D.: Presupposition triggering from alternatives. Journal of Semantics 27(1), 37–80 (2010)

Beaver, D.: Have you noticed that your belly button lint colour is related to the colour of your clothing? In: Bauerle, R., Reyle, U., Zimmerman, T.E. (eds.) Presupposition: Papers in Honor of Hans Kamp (2004)

Chierchia, G., McConnell-Ginet, S.: Meaning and grammar: an introduction to semantics. MIT Press, Cambridge (2000)

Demolombe, R., Fariñas del Cerro, L.: Towards a logical characterization of sentences of the kind "sentence p is about object c". In: Holdobler, S. (ed.) Intellectics and Computational Logic: Papers in Honor of Wolfgang Bibel. Kluwer Academic Press (2000)

Goldstein, E.: Sensation and perception. Wadsworth Pub. Co. (2009)

Itti, L., Koch, C.: Computational modeling of visual attention. Nature Reviews Neuroscience 2(3), 194–203 (2001)

Karttunen, L.: Implicative verbs. Language 47(2), 340–358 (1971a)

Karttunen, L.: Some Observations on Factivity. Papers in Linguistics 5, 55–69 (1971b)

Keshet, E.: Infinitival Complements. In: Gronn, A. (ed.) Proceedings of Sinn und Bedeutung 12 (2008)

Klinedinst, N.: Totally Hardcore Semantic Presuppositions. ms. UCL (2009)

Kusumoto, K.: On the quantification over times in natural language. Natural Language Semantics 13(4), 317–357 (2005)

Navalpakkam, V., Itti, L.: Modeling the influence of task on attention. Vision Research 45(2), 205–231 (2005)

Partee, B.H.: Some structural analogies between tenses and pronouns in English. The Journal of Philosophy 70(18), 601–609 (1973)

Romoli, J.: The presuppositions of soft triggers are not presuppositions. Talk presented at SALT 21 (2011)

Simons, M.: On the conversational basis of some presuppositions. In: Hastings, R., Jackson, B., Zvolensky, Z. (eds.) Semantics and Linguistic Theory, 11 (2001)

Simons, M.: Observations on embedding verbs, evidentiality, and presupposition. Lingua 117(6), 1034–1056 (2007)

Simons, M., Tonhauser, J., Beaver, D., Roberts, C.: What projects and why. In: Li, N., Lutz, D. (eds.) Proceedings of Semantics and Linguistic Theory 20, pp. 309–327. Ithaca, Cornell (2010)

Sperber, D., Wilson, D.: Ordered entailments: An alternative to presuppositional theories. In: Syntax and Semantics XI: Presupposition, pp. 299–323 (1979)

Stalnaker, R.C.: Pragmatic presuppositions. Semantics and Philosophy, 197–214 (1974)

Polarities in Logic and Semantics

Arno Bastenhof

Utrecht University

Abstract. We ask to what extent a satisfactory categorial analysis of non-local scope construal can already be realized while restricting to a non-commutative, non-associative setting, and without relaxing compositionality to a relation. In response, we show a variety of data on the topic can be dealt with through an adaptation of classical non-associative Lambek calculus (**CNL**, [3]) that incorporates Girard's concept of polarity ([5]). The latter notion originated in the constructivization of classical logic, recognizable by computer scientists as the definition of continuation-passing style (CPS) translations.

Keywords: Classical non-associative Lambek calculus, formal semantics, scopal ambiguities, continuations, type-shifting.

1 Introduction

Categorial type logics (CTL, [10]) attempt the reduction of natural language grammar to proof theory. Thus, formulas replace syntactic categories, and a sentence's derivability amounts to the provability of its grammaticality. CTL realizes Montague's Universal Grammar program ([9]), in that it offers a mathematical framework for studying both the syntax and semantics of natural and formal languages, in accordance with compositionality. In particular, CTL benefits from the close correspondence between the logician's formulas and the computer scientist's types, as cemented in the Curry-Howard isomorphism.

In practice, CTL acts as an umbrella term covering a plethora of logical calculi, each more or less agreeing on a fixed context-free core, but diverging in their explorations into the expressivity realms beyond. Still, one typically observes a shared asymmetry inherited from Lambek's founding work [7]. Roughly, derivability relates a possible multitude of hypotheses (the categories, or formulas, assigned to the individual words) with a unique conclusion (the formula categorizing the phrase made up from the hypotheses).

Proposals for restoring symmetry are of more recent times, inspired by Girard's classical linear logic. Despite mathematical elegance, little improvement has been claimed towards strengthening linguistic coverage.[1] We challenge this situation, treating scopal ambiguities. Typically, their categorial analyses involve either non-context-free mechanisms, realized by controlled use of structural rules, or relax compositionality to a relation, mapping derivations into non-singleton

[1] See [11] for an exception, and §5 for a comparison with the approach discussed here.

M. Aloni et al. (Eds.): Amsterdam Colloquium 2011, LNCS 7218, pp. 230–239, 2012.
© Springer-Verlag Berlin Heidelberg 2012

sets of readings. In contrast, we show that once symmetry is embraced, neither compromise need be made to still account for all combinatorially possible scopal readings. We use a variation on *Classical Non-associative Lambek calculus* (**CNL**, [3]), dubbed 'polarized **CNL**', or simply **CNL**^{pol}. While ensured equivalent provability-wise with **CNL**, **CNL**^{pol} makes a finer distinction between proofs, mapped in turn to separate readings. The concept of polarity, our main ingredient, originates in Girard's constructivization of classical logic ([5]), recognizable by computer scientists as a continuation-passing style (CPS) translation.

We proceed as follows. §2 introduces **CNL**^{pol}, with §3 motivating its linguistic applicability by example. To gain better understanding of the flexibility of our approach, §4 shows the derivability of Hendriks' type-shifting rules ([6]) within **CNL**^{pol}. §5 evaluates, including discussion on the limitations of our work.

2 Polarized Classical Non-associative Lambek Calculus

De Groote and Lamarche ([3]) introduced **CNL** as a classical conservative extension of Lambek's non-associative syntactic calculus (**NL**, [8]). Here, 'classicality' refers to the existence of a linear negation $A^{\perp}$ on formulas A satisfying involutivity ($A^{\perp\perp} = A$). Besides the usual multiplicative conjunctions, or *tensors* ($A \otimes B$), their dual disjunctions, or *pars* ($A \oplus B$) now arise through the De Morgan laws as $(B^{\perp} \otimes A^{\perp})^{\perp}$. Consequently, direction-sensitive implications (A/B) and ($B\backslash A$) need no longer be considered primitive, rather being defined by ($A \oplus B^{\perp}$) and ($B^{\perp} \oplus A$) respectively. **CNL** faithfully embeds **NL**, as shown in [3].

We adapt **CNL** according to Girard's constructivization of classical logic ([5]), dubbing the result *polarized* **CNL** (**CNL**^{pol}). Roughly, formulas are taken as being of either *positive* or *negative polarity*, with two additional connectives, the *shifts* $\downarrow$ and $\uparrow$ mediating between the two. Polarity interchanges through $\cdot^{\perp}$, lending proof-theoretic support. Intuitively, shifts amount to 'semantic annotations', recording places for inserting negations inside the semantics. The following definition details the formula language, explicating the concept of polarity.

Definition 1. Positive formulas ($P, Q, \ldots$) and their negative duals ($M, N, \ldots$) are defined by mutual induction, starting from a countable set of atoms $p, q, \ldots$.

$$\begin{array}{rcll} P, Q & ::= & p \mid (P \otimes Q) \mid (\downarrow N) & \text{(Positive(ly polar) formulas)} \\ M, N & ::= & \bar{p} \mid (M \oplus N) \mid (\uparrow P) & \text{(Negative(ly polar) formulas)} \end{array}$$

Linear negation $\cdot^{\perp}$ acts as primitive on atoms only (written $\bar{\cdot}$), while extending to complex formulas through De Morgan's laws: $p^{\perp} = \bar{p}$, $\bar{p}^{\perp} = p$, $(P \otimes Q)^{\perp} = Q^{\perp} \oplus P^{\perp}$, $(M \oplus N)^{\perp} = N^{\perp} \otimes M^{\perp}$, $(\downarrow N)^{\perp} = \uparrow N^{\perp}$ and $(\uparrow P)^{\perp} = \downarrow P^{\perp}$.

In practice, we stick to atomic formulas s (categorizing sentences), np (noun phrases) and n (common nouns). For the target language of semantic interpretation, we take simply-typed λ-calculus with base types e (for individuals) and t (truth values). Besides the familiar *function types* ($\sigma \to \tau$) (written $\langle \sigma, \tau \rangle$ in [4]), semantic types σ, τ may include *products* ($\sigma \times \tau$). In mapping formulas to types, function types are restricted to ($\sigma \to t$), abbreviated $\neg\sigma$.

$$\frac{\Delta \vdash s' : P \quad \Gamma, x : P \vdash s}{\Gamma, \Delta \vdash s[s'/x]}\ Cut \qquad \frac{}{x : P \vdash x : P}\ Ax$$

$$\frac{\Gamma, \Delta \vdash s}{\Delta, \Gamma \vdash s}\ dp^1 \qquad \frac{\Gamma \bullet \Delta, \Theta \vdash s}{\Gamma, \Delta \bullet \Theta \vdash s}\ dp^2$$

$$\frac{\Gamma \vdash s : N^{\perp}}{\Gamma, \downarrow N^x \vdash (x\ s)}\ \downarrow L \qquad \frac{\Gamma, N^{\perp x} \vdash s}{\Gamma \vdash \lambda x s : \downarrow N}\ \downarrow R$$

$$\frac{\Gamma, P^y \bullet Q^z \vdash s}{\Gamma, P \otimes Q^x \vdash s[\pi^1(x)/y, \pi^2(x)/z]}\ \otimes L \qquad \frac{\Gamma \vdash s : P \quad \Delta \vdash s' : Q}{\Gamma \bullet \Delta \vdash \langle s, s' \rangle : P \otimes Q}\ \otimes R$$

Fig. 1. Polarized classical non-associative Lambek calculus

Definition 2. We associate, by mutual induction, formulas P, N with types $[\![P]\!]^+, [\![N]\!]^-$. At the level of atoms, we set $[\![s]\!]^+ = [\![\bar{s}]\!]^- = t$ (sentences interpret by truth values), $[\![np]\!]^+ = [\![\bar{np}]\!]^- = e$ (noun phrases by entities) and $[\![n]\!]^+ = [\![\bar{n}]\!]^- = \neg e$ (common nouns by first-order properties), while for complex types,

$$[\![P \otimes Q]\!]^+ = [\![P]\!]^+ \times [\![Q]\!]^+ \quad [\![M \oplus N]\!]^- = [\![N]\!]^- \times [\![M]\!]^-$$
$$[\![\downarrow N]\!]^+ = \neg [\![N]\!]^- \qquad\qquad [\![\uparrow P]\!]^- = \neg [\![P]\!]^+$$

Note $[\![P]\!]^+ = [\![P^{\perp}]\!]^-$, by a routine induction. Proofs establish pairings of binary-branching trees Γ, Δ of formulas, referred to as *structures*. Their leaves carry positive formulas, negative N appearing only when $\downarrow$-prefixed. Anticipating the term assignment to derivations, formulas are labeled by (λ-)variables $x, y, \ldots$

Definition 3. Structures Γ, Δ collect pairings P^x into binary-branching trees:

$$\Gamma, \Delta ::= P^x \mid (\Gamma \bullet \Delta)$$

Proofs involve derivability judgements $\Gamma, \Delta \vdash s$ and $\Gamma \vdash s : P$, with s a λ-term of type t, respectively $[\![P]\!]^+$, while containing free variables x of type $[\![Q]\!]^+$ for each Q^x in $\Gamma(, \Delta)$. Compared to [3], we unconventionally write sequents left-sided, better reflecting the intuitionistic nature of the semantic target language.

Definition 4. Figure 1 mutually defines derivability judgements $\Gamma, \Delta \vdash s$ and $\Gamma \vdash s : P$. Double inference lines indicate derivability goes in both directions.

We make the following observations regarding the above definition.

1. The rules (dp^1) and (dp^2) bear resemblance to Belnap's display postulates ([2]) in allowing one to isolate the main formula of a logical inference as the whole of one of the sequent's components by turning its context 'inside out'. They are not to be mistaken for associativity and commutativity postulates.
2. Compared to **CNL**, **CNL**^{pol} uses both sides of the turnstile $\vdash$, though treats them asymmetrically to restore the tight connection with λ-calculus. The newly added shifts serve solely to control traffic across $\vdash$, applying linear negation. Provability-wise, these are harmless extensions over **CNL**, as argued in Theorems 6 and 7. At the level of the actual proofs, however, shifts,

together with (dp^1) and (dp^2), allow abstraction over any of the variables found in a sequent without compromising structure-sensitivity. In contrast, **NL** severely restricts the abstraction mechanism, accounting for its limited semantic expressivity.

Theorem 5. Cut is admissible, as are non-atomic instances of axioms.

Theorem 6. Under the forgetful translation, removing shifts and writing all sequents one-sided, derivations in $\mathbf{CNL}^{pol}$ map into those of **CNL**.

Theorem 7. There exists a decoration of **CNL**'s formulas by shifts relative to which derivations can be translated into $\mathbf{CNL}^{pol}$.

The previous theorems, proven in [1], ensure equivalence, provability-wise, of **CNL** and $\mathbf{CNL}^{pol}$. The proof of Theorem 7 in particular bears resemblance to a continuation-passing style (CPS) translation, as further elaborated upon in [1].

3 Case Analyses

We illustrate our approach through several classical examples. The underlying generalization is that any combinatorially possible scopal reading can be accounted for through generous use of shifts inside the lexicon. Conversely, parsimonious use allows for certain readings to be blocked (§3.4), although this approach has its limitations (cf. §5). We first provide a sample lexicon (§3.1), allowing to explain object-wide scope readings (§3.2, as a warm-up), embedded scope (§3.3), scope sieves (§3.4), and intensionality and coordination (§3.5).

3.1 Lexicon

A lexicon associates words w with formulas P and semantic interpretation a term s of type $[\![P]\!]^+$, and is written informally as a collection of sequents $w \vdash s : P$. If this notation is taken seriously, allowing in particular words at the leaves of structures, lexical insertion proceeds through Cut. Lexical denotations s may contain predicate-logical connectives, considered constants $\wedge$, $\vee$, $\supset$ $(t \to (t \to t)$, written in infix style), $\forall$ and $\exists$ (of types $(e \to t) \to t$). Here, $\supset$ refers to implication, its notation meant to prevent confusion with function types $(\sigma \to \tau)$.

Figure 2 presents a sample lexicon, using constants ALICE (of type e), PERSON, UNICORN, YAWN $(e \to t)$, FIND $(e \to (e \to t))$, SEEK $(((e \to t) \to t) \to (e \to t))$, THINK and HEAR $(t \to (e \to t))$. We write (N/P) and $(P\backslash N)$ for $(N \oplus P^{\perp})$ and $(P^{\perp} \oplus N)$ respectively, facilitating comparison with traditional CTL. Finally, we use paired abstractions $\lambda\langle x, y\rangle s$ as abbreviations for $\lambda z s[\pi^1(z)/x, \pi^2(z)/y]$.

3.2 Object-Wide Scope

As a warm-up, consider the familiar linear and inverse scope readings (1a), (1b)

1. Everyone found a unicorn.
1a. For every person x, there exists some unicorn y s.t. x found y.
1b. There exists some unicorn y s.t. for every person x, x found y.

$$\begin{array}{rl}
\text{Alice} \vdash & \textsc{alice} : np \\
\text{everyone} \vdash & \lambda Q \forall x((\textsc{person}\ x) \supset (Q\ x)) : \downarrow\uparrow np \\
\text{a} \vdash & \lambda\langle P, Q\rangle \exists x((P\ x) \wedge (Q\ x)) : \downarrow(\uparrow np/n) \\
\text{unicorn} \vdash & \textsc{unicorn} : n \\
\text{yawn(ed)} \vdash & \lambda\langle q, x\rangle (q\ (\textsc{yawn}\ x)) : \downarrow(np\backslash\uparrow s) \\
\text{found} \vdash & \lambda\langle y, \langle q, x\rangle\rangle (q\ ((\textsc{find}\ y)\ x)) : \downarrow((np\backslash\uparrow s)/np) \\
\text{sought} \vdash & \lambda\langle Y, \langle q, x\rangle\rangle (q\ ((\textsc{seek}\ Y)\ x)) : \downarrow((np\backslash\uparrow s)/\downarrow\uparrow np) \\
\text{thinks} \vdash & \lambda\langle Q, \langle q, x\rangle\rangle (q\ ((\textsc{think}\ (Q\ \lambda pp))\ x)) : \downarrow((np\backslash\uparrow s)/\downarrow\uparrow s) \\
\text{heard} \vdash & \lambda\langle p, \langle q, x\rangle\rangle (q\ ((\textsc{hear}\ p)\ x)) : \downarrow((np\backslash\uparrow s)/s) \\
\text{and} \vdash & \lambda\langle R, \langle\langle Y, \langle q, x\rangle\rangle, S\rangle\rangle ((S\ \langle Y, \langle q, x\rangle\rangle) \wedge (R\ \langle Y, \langle q, x\rangle\rangle)) : \downarrow((\downarrow tv\backslash tv)/\downarrow tv)
\end{array}$$

Fig. 2. Sample lexicon, where tv abbreviates $(np\backslash(\uparrow s/\downarrow\uparrow np))$

The formulas assigned to the quantified noun phrases involved reflect their interpretation as second-order properties through use of double shifts, inserting double negations at the level of types (i.e., implications with result type t). We construct the following derivations, using the entries found in Fig.2 (cf. Fig.3)

$$\downarrow\uparrow np^x \bullet (\downarrow((np\backslash\uparrow s)/np)^y \bullet (\downarrow(\uparrow np/n)^u \bullet n^v)), \downarrow \bar{s}^z \vdash \begin{cases} (x\ \lambda a(u\ \langle v, \lambda b(y\ \langle b, \langle z, a\rangle\rangle)\rangle)) \\ (u\ \langle v, \lambda b(x\ \lambda a(y\ \langle b, \langle z, a\rangle\rangle))\rangle) \end{cases}$$

with lexical insertion (through Cut) deriving, after β-conversion, the desired

$$\text{everyone} \bullet (\text{found} \bullet (\text{a} \bullet \text{unicorn})), \downarrow \bar{s}^z \vdash \begin{cases} \forall x((\textsc{person}\ x) \supset \exists y((\textsc{unicorn}\ y) \wedge (z\ ((\textsc{find}\ y)\ x)))) & \text{(1a)} \\ \exists y((\textsc{unicorn}\ y) \wedge \forall x((\textsc{person}\ x) \supset (z\ ((\textsc{find}\ y)\ x)))) & \text{(1b)} \end{cases}$$

noting the presence of a free variable z for the resulting category. If desired, one may apply $(\uparrow)$ to create an abstraction, deriving $\uparrow\downarrow\bar{s}$.

3.3 Embedded Scope

We illustrate non-local scope construal from inside a complement clause:

2. Alice thinks a unicorn yawned.
2a. Alice thinks there exists a unicorn y s.t. y yawned.
2b. For some unicorn y, Alice thinks y yawned.

By having the verb select for a clausal complement of category $\downarrow\uparrow s$ instead of s, we provide the deductive machinery with enough freedom to construct both of the desired readings. Using the formulas found in Fig.2, we derive three terms,

$$np^a \bullet (\downarrow((np\backslash\uparrow s)/\downarrow\uparrow s)^t \bullet ((\downarrow(\uparrow np/n)^u \bullet n^v) \bullet \downarrow(np\backslash\uparrow s)^w)), \downarrow \bar{s}^z \vdash \begin{cases} (t\ \langle \lambda q(u\ \langle v, \lambda y(w\ \langle q, y\rangle)\rangle), \langle z, a\rangle\rangle) \\ (u\ \langle v, \lambda y(t\ \langle \lambda q(w\ \langle q, y\rangle), \langle z, a\rangle\rangle)\rangle) \\ (u\ \langle v, \lambda y(w\ \langle \lambda p(t\ \langle \lambda q(q\ p), \langle z, a\rangle\rangle), y\rangle)\rangle) \end{cases}$$

collapsed into two readings after lexical insertion:

$$\text{Alice} \bullet (\text{thinks} \bullet ((\text{a} \bullet \text{unicorn}) \bullet \text{yawned})), \downarrow \bar{s}^z \vdash \begin{cases} (z\ ((\textsc{think}\ \exists y((\textsc{unicorn}\ y) \wedge (\textsc{yawn}\ y)))\ \textsc{alice})) & (2a) \\ \exists y((\textsc{unicorn}\ y) \wedge (z\ ((\textsc{think}\ (\textsc{yawn}\ y))\ \textsc{alice}))) & (2b) \end{cases}$$

$$\downarrow\bar{s}^z \vdash z : \downarrow\bar{s} \ (Ax) \qquad np^a \vdash a : np \ (Ax)$$
$$np^b \vdash b : np \ (Ax) \qquad \downarrow\bar{s}^z \bullet np^a \vdash \langle z, a\rangle : \downarrow\bar{s} \otimes np \quad \otimes R$$
$$np^b \bullet (\downarrow\bar{s}^z \bullet np^a) \vdash \langle b, \langle z, a\rangle\rangle : np \otimes (\downarrow\bar{s} \otimes np) \quad \otimes R$$
$$np^b \bullet (\downarrow\bar{s}^z \bullet np^a), \downarrow((np\backslash\uparrow s)/np)^y \vdash (y\ \langle b, \langle z, a\rangle\rangle) \quad \downarrow L$$
$$(\downarrow\bar{s}^z \bullet np^a) \bullet \downarrow((np\backslash\uparrow s)/np)^y, np^b \vdash (y\ \langle b, \langle z, a\rangle\rangle) \quad Dp$$
$$n^v \vdash v : n \ (Ax) \qquad (\downarrow\bar{s}^z \bullet np^a) \bullet \downarrow((np\backslash\uparrow s)/np)^y \vdash \lambda b(y\ \langle b, \langle z, a\rangle\rangle) : \downarrow\bar{np} \quad \downarrow R$$
$$n^v \bullet ((\downarrow\bar{s}^z \bullet np^a) \bullet \downarrow((np\backslash\uparrow s)/np)^y) \vdash \langle v, \lambda b(y\ \langle b, \langle z, a\rangle\rangle)\rangle : n \otimes \downarrow\bar{np} \quad \otimes R$$
$$n^v \bullet ((\downarrow\bar{s}^z \bullet np^a) \bullet \downarrow((np\backslash\uparrow s)/np)^y), \downarrow(\uparrow np/n)^u \vdash (u\ \langle v, \lambda b(y\ \langle b, \langle z, a\rangle\rangle)\rangle) \quad \downarrow L$$
$$(\downarrow((np\backslash\uparrow s)/np)^y \bullet (\downarrow(\uparrow np/n)^u \bullet n^v)) \bullet \downarrow\bar{s}^z, np^a \vdash (u\ \langle v, \lambda b(y\ \langle b, \langle z, a\rangle\rangle)\rangle) \quad Dp$$
$$(\downarrow((np\backslash\uparrow s)/np)^y \bullet (\downarrow(\uparrow np/n)^u \bullet n^v)) \bullet \downarrow\bar{s}^z \vdash \lambda a(u\ \langle v, \lambda b(y\ \langle b, \langle z, a\rangle\rangle)\rangle) : \downarrow\bar{np} \quad \downarrow R$$
$$(\downarrow((np\backslash\uparrow s)/np)^y \bullet (\downarrow(\uparrow np/n)^u \bullet n^v)) \bullet \downarrow\bar{s}^z, \downarrow\uparrow np^x \vdash (x\ \lambda a(u\ \langle v, \lambda b(y\ \langle b, \langle z, a\rangle\rangle)\rangle)) \quad \downarrow L$$
$$\downarrow\uparrow np^x \bullet (\downarrow((np\backslash\uparrow s)/np)^y \bullet (\downarrow(\uparrow np/n)^u \bullet n^v)), \downarrow\bar{s}^z \vdash (x\ \lambda a(u\ \langle v, \lambda b(y\ \langle b, \langle z, a\rangle\rangle)\rangle)) \quad Dp$$

$$\downarrow\bar{s}^z \vdash z : \downarrow s \ (Ax) \qquad np^a \vdash a : np \ (Ax)$$
$$np^b \vdash b : np \ (Ax) \qquad \downarrow\bar{s}^z \bullet np^a \vdash \langle z, a\rangle : \downarrow\bar{s} \otimes np \quad \otimes R$$
$$np^b \bullet (\downarrow\bar{s}^z \bullet np^a) \vdash \langle b, \langle z, a\rangle\rangle : np \otimes (\downarrow\bar{s} \otimes np) \quad \otimes R$$
$$np^b \bullet (\downarrow\bar{s}^z \bullet np^a), \downarrow((np\backslash\uparrow s)/np)^y \vdash (y\ \langle b, \langle z, a\rangle\rangle) \quad \downarrow L$$
$$(\downarrow((np\backslash\uparrow s)/np)^y \bullet np^b) \bullet \downarrow\bar{s}^z, np^a \vdash (y\ \langle b, \langle z, a\rangle\rangle) \quad Dp$$
$$(\downarrow((np\backslash\uparrow s)/np)^y \bullet np^b) \bullet \downarrow\bar{s}^z \vdash \lambda a(y\ \langle b, \langle z, a\rangle\rangle) : \downarrow\bar{np} \quad \downarrow R$$
$$(\downarrow((np\backslash\uparrow s)/np)^y \bullet np^b) \bullet \downarrow\bar{s}^z, \downarrow\uparrow np^x \vdash (x\ \lambda a(y\ \langle b, \langle z, a\rangle\rangle)) \quad \downarrow L$$
$$(\downarrow\bar{s}^z \bullet \downarrow\uparrow np^x) \bullet \downarrow((np\backslash\uparrow s)/np)^y, np^b \vdash (x\ \lambda a(y\ \langle b, \langle z, a\rangle\rangle)) \quad Dp$$
$$n^v \vdash v : n \ (Ax) \qquad (\downarrow\bar{s}^z \bullet \downarrow\uparrow np^x) \bullet \downarrow((np\backslash\uparrow s)/np)^y \vdash \lambda b(x\ \lambda a(y\ \langle b, \langle z, a\rangle\rangle)) : \downarrow\bar{np} \quad \downarrow R$$
$$n^v \bullet ((\downarrow\bar{s}^z \bullet \downarrow\uparrow np^x) \bullet \downarrow((np\backslash\uparrow s)/np)^y) \vdash \langle v, \lambda b(x\ \lambda a(y\ \langle b, \langle z, a\rangle\rangle))\rangle : n \otimes \downarrow\bar{np} \quad \oplus$$
$$n^v \bullet ((\downarrow\bar{s}^z \bullet \downarrow\uparrow np^x) \bullet \downarrow((np\backslash\uparrow s)/np)^y), \downarrow(\uparrow np/n)^u \vdash (u\ \langle v, \lambda b(x\ \lambda a(y\ \langle b, \langle z, a\rangle\rangle))\rangle) \quad \downarrow L$$
$$\downarrow\uparrow np^x \bullet (\downarrow((np\backslash\uparrow s)/np)^y \bullet (\downarrow(\uparrow np/n)^u \bullet n^v)), \downarrow\bar{s}^z \vdash (u\ \langle v, \lambda b(x\ \lambda a(y\ \langle b, \langle z, a\rangle\rangle))\rangle) \quad Dp$$

Fig. 3. Deriving *everyone found a unicorn.* (Dp) abbreviates multiple (dp^2, dp^1)-steps.

3.4 Scope Sieves

Further insight into the categorization of verbs selecting for clausal complements is provided by the following example, illustrating scope sieves. Here, non-local scope is enforced, as typically observed with perception verbs (cf. [6], p.108):

3. Alice heard a unicorn yawn.

The desired result obtains by having the matrix verb select for s instead of $\downarrow\uparrow s$. For reasons of space, we only show the end result obtained after lexical insertion.

$$\text{Alice} \bullet (\text{heard} \bullet ((\text{a} \bullet \text{unicorn}) \bullet \text{yawn})), \downarrow\bar{s}^z$$
$$\vdash \exists y((\textsc{unicorn}\ y) \wedge (z\ ((\textsc{hear}\ y)\ \textsc{alice})))$$

On the other hand, it does not seem clear how to enforce the local reading (e.g., as needed for scope islands). See §5 for further discussion.

3.5 Coordination and Intensionality

We conclude with intensionality and coordination, treated together to show their interplay. Our treatment of intensionality is simplified w.r.t. our choice of semantic type for interpreting sentences, although this can be easily remedied if the reader so desires. Furthermore, *found* is treated extensionally, in line with [4].

4. Alice sought and found a unicorn.
4a. Alice sought a unicorn, and there exists a unicorn y s.t. Alice found y.
4b. There exists a unicorn y s.t. Alice sought y and Alice found y.

Again, showing only the end results after lexical insertion for reasons of space:

$$\text{Alice} \bullet ((\text{sought} \bullet (\text{and} \bullet \text{found})) \bullet (\text{a} \bullet \text{unicorn})), \downarrow \bar{s}^z$$
$$\vdash \begin{cases} (z\ ((\textsc{seek}\ \lambda P \exists y((\textsc{unicorn}\ y) \wedge (P\ y)))\ \textsc{alice})) \wedge \exists y((\textsc{unicorn}\ y) \wedge (z\ ((\textsc{find}\ y)\ \textsc{alice}))) & (4a) \\ \exists y((\textsc{unicorn}\ y) \wedge (z\ ((\textsc{seek}\ \lambda P(P\ y))\ \textsc{alice})) \wedge (z\ ((\textsc{find}\ y)\ \textsc{alice}))) & (4b) \end{cases}$$

4 Type-Shifting

The previous section described $\mathbf{CNL}^{pol}$'s semantic expressivity by empirical illustration. For a more precise characterization, we now compare our treatment of scopal ambiguities with a more well-understood solution. Hendriks ([6]), generalizing [12], proposed a flexible mapping between syntax and semantics by closing the readings associated with a given syntactic derivation under the following derived semantic inference rules, referred to collectively by *type shifting*:

1. **Value Raising.** Assuming right-associative bracketing, terms of type $\sigma_1 \to \ldots \to \sigma_n \to \sigma$ may be lifted to the type $\sigma_1 \to \ldots \to (\sigma \to \tau') \to \tau'$, for any τ'. E.g., a proper name 'Alice' with interpretation of type e undergoes value raising to derive a term of the type $(e \to t) \to t$ for generalized quantifiers.
2. **Argument Raising.** Inhabitants of $\sigma_1 \to \ldots \to \sigma_n \to \tau$ may transition into $\sigma_1 \to \ldots \to ((\sigma_i \to \tau') \to \tau') \to \tau$. Typically applies to the subject and object positions of transitive verbs. E.g., $e \to (e \to t)$ undergoes argument raising twice to derive the type of a binary relation on generalized quantifiers. Depending on which of the arguments is raised first, we get different readings.
3. **Argument Lowering.** Inhabitants of $\sigma_1 \to \ldots \to \sigma_i \to \tau$ derive from terms in $\sigma_1 \to \ldots \to ((\sigma_i \to \tau') \to \tau') \to \tau$. E.g., lowering and subsequent raising (for $\tau' = t$) of an intensional verb's object position derives a de re reading.

By applying the right combinations of these rules, one can derive all combinatorially available scopal readings for a given sentence, as proved by Hendriks. What is important to note is that all rules involved are derivable within λ-calculus. In contrast, analogous rules for the traditional incarnations of CTL necessitate full associativity and commutativity, leading to serious overgeneration. Consequently, type-shifting seemed exclusive to the realm of semantics, necessitating the relaxation of compositionality if it was to be made any use of.

We claim the existence of derivable rules of inference within $\mathbf{CNL}^{pol}$ mapping to Hendriks' type-shifting schemas under semantic interpretation. Consequently, the strict correspondence between syntax and semantics can be restored, while preserving semantic expressivity. One restriction, however, applies: the types τ' in the above explanation of type-shifting always are to be t. In practice, this still leaves us with sufficient generality to account for scoping within clausal domains.

Theorem 8. The following rules of *Value Raising* (VR), *Argument Raising* (AR) and *Argument Lowering* (AL) are derivable within $\mathbf{CNL}^{pol}$:

$$\begin{array}{rr}
\Gamma \vdash s : P \Rightarrow \Gamma \vdash \lambda x(x\ s) : {\downarrow}{\uparrow} P & (VR)\\
\Gamma \vdash s : {\downarrow}(N/P) \Rightarrow \Gamma \vdash \lambda\langle x, v\rangle(x\ \lambda u(s\ \langle u, v\rangle)) : {\downarrow}(N/{\downarrow}{\uparrow} P) & (AR^r)\\
\Gamma \vdash s : {\downarrow}(P\backslash N) \Rightarrow \Gamma \vdash \lambda\langle v, x\rangle(x\ \lambda u(s\ \langle v, u\rangle)) : {\downarrow}({\downarrow}{\uparrow} P\backslash N) & (AR^l)\\
\Gamma \vdash s : {\downarrow}(N/{\downarrow}{\uparrow} P) \Rightarrow \Gamma \vdash \lambda\langle u, v\rangle(s\ \langle\lambda x(x\ u), v\rangle) : {\downarrow}(N/P) & (AL^r)\\
\Gamma \vdash s : {\downarrow}({\downarrow}{\uparrow} P\backslash N) \Rightarrow \Gamma \vdash \lambda\langle v, u\rangle(s\ \langle v, \lambda x(x\ u)\rangle) : {\downarrow}(P\backslash N) & (AL^l)
\end{array}$$

Example 9. We illustrate the above result with a lexical entry from Figure 2:

$$\text{found} \vdash \lambda\langle y, \langle q, x\rangle\rangle(q\ ((\textsc{find}\ y)\ x)) : {\downarrow}((np\backslash{\uparrow} s)/np)$$

Note the term involved is of type $\neg(e \times (\neg t \times e))$, which by uncurrying is isomorphic to $e \to (e \to \neg\neg t))$. Applying Argument Raising (i.e., (AR^r)), we get

$$\text{found} \vdash \lambda\langle Y, \langle q, x\rangle\rangle(Y\ \lambda y(q\ ((\textsc{find}\ y)\ x))) : {\downarrow}(({\downarrow}{\uparrow} np\backslash{\uparrow} s)/np)$$

which is of type $\neg(\neg\neg e \times (\neg t \times e))$, i.e., $\neg\neg e \to (e \to \neg\neg t)$.

Still lacking is the means to target other positions besides the direct object for Argument Raising (or -Lowering), a possibility rendered available in Hendriks' presentation by allowing for any number of arguments preceding the one to which the operation was applied. In our case, the following result allows any intervening arguments to be stripped off one by one, and to be added back after the desired type-shifting rule has been applied.

Lemma 10. The following are derivable rules of inference:

$$\frac{\Gamma \bullet P^x \vdash s : {\downarrow} N}{\Gamma \vdash \lambda\langle x, y\rangle(s\ y) : {\downarrow}(N/P)} > \qquad \frac{P^x \bullet \Gamma \vdash s : {\downarrow} N}{\Gamma \vdash \lambda\langle y, x\rangle(s\ y) : {\downarrow}(P\backslash N)} <$$

$$\frac{\Gamma \vdash s : {\downarrow}(N/P)}{\Gamma \bullet P^x \vdash \lambda y(s\ \langle x, y\rangle) : {\downarrow} N} >' \qquad \frac{\Gamma \vdash s : {\downarrow}(P\backslash N)}{P^x \bullet \Gamma \vdash \lambda y(s\ \langle y, x\rangle) : {\downarrow} N} <'$$

Example 11. Continuing where we left off in Example 9, (AR^l) combines with $(>)$ and $(>')$ to allow for Argument Raising of the subject position, resulting in

$$\text{found} \vdash \lambda\langle Y, \langle q, X\rangle\rangle(X\ \lambda x(Y\ \lambda y(q\ ((\textsc{find}\ y)\ x)))) : {\downarrow}(({\downarrow}{\uparrow} np\backslash{\uparrow} s)/{\downarrow}{\uparrow} np)$$

being of type $\neg\neg e \to (\neg\neg e \to \neg\neg t)$ after uncurrying. Note we could also have applied (AR^l) first, followed by (AR^r) to derive

$$\text{found} \vdash \lambda\langle Y, \langle q, X\rangle\rangle(Y\ \lambda y(X\ \lambda x(q\ ((\textsc{find}\ y)\ x)))) : {\downarrow}(({\downarrow}{\uparrow} np\backslash{\uparrow} s)/{\downarrow}{\uparrow} np)$$

The latter allows for the derivation of object-wide scope readings, whereas raising the subject after the object favors subject-wide scope.

5 Evaluation

We have presented polarized classical non-associative Lambek calculus, arguing for its applicability to the analysis of scopal ambiguities despite being both non-commutative and non-associative, and keeping to a strict correspondence between syntax and semantics. In contrast, earlier categorial accounts typically involved non-context-free mechanisms by allowing for controlled associativity and commutativity, or relaxed compositionality to a relation. We have compared our approach to one such traditional account, to wit, Hendriks' use of type-shifting in associating syntactic derivations with sets of readings. Below, we briefly discuss, first, a point of critique concerning our capacity to block scopal readings that, while combinatorially possible, are unrealized by linguistic reality. Second, we draw attention to a curious discrepancy with Hendriks' account of verbs taking clausal complements, and, finally, make a brief comparison with a proposal for the categorial analysis of scope closely related to ours.

5.1 Blocking Scopal Readings

While we could ensure derivability of all combinatorially available scopal readings for the various examples we considered, this same property also constitutes the main limitation of our approach: situations where linguistic reality excludes certain readings are not so easily accounted for. While we had a small success with the analysis of scope sieves, scope islands fall outside our coverage. The reasons for this go deep: through the display postulates, any variable-labeled formula may be isolated as the whole of one of the sequents' components, which may subsequently be abstracted over through ($\uparrow$). In other words, accounting for scope islands means to restrict the display postulates, which constitute an essential ingredient for the notion of structure adopted by **CNL**, as exemplified graphically by De Groote and Lamarche's proof nets ([3]).

5.2 On Verbs Taking Clausal Complements

Our analysis of verbs taking clausal complements is host to an interesting curiosity, setting it apart from Hendriks' account. One might first expect the category $\downarrow((np\backslash\uparrow s)/s)$ assigned to *heard* in §3.4 to provide at least the local reading, with type-shifting required for non-local scope construal. Such a situation would parallel Hendriks' [6], which starts from a minimal semantic type assignment. Instead, we find that only the non-local reading is derived, whereas a local reading necessitates the prefixing of the s argument by a double shift, cf. the category $\downarrow((np\backslash\uparrow s)/\downarrow\uparrow s)$ assigned to *thinks* in §3.3.

5.3 Lambek-Grishin Calculus

Another account of scopal ambiguities similar to ours was recently put forward by Moortgat ([11]), working within the closely related *Lambek-Grishin calculus* (**LG**). There, scopal ambiguities were accounted for by adding minimal (co)negations to the logical vocabulary (referred to by (co-)Galois connections), used similarly to our shifts. We claim polarization allows for a better understanding of this proposal. Like **CNL**, **LG** turns out amendable to a polarized reformulation, as demonstrated in [1]. Moortgat's compositional semantics may then be recast as a translation of **LG** into $\mathbf{LG}^{pol}$, revealing that his minimal negations simply serve to enforce the appearance of shifts in the target.

Acknowledgements. I thank Michael Moortgat and Vincent van Oostrom for helpful discussions, as well as an anonymous referee for valuable comments. All remaining errors are my own.

References

1. Bastenhof, A.: Polarized Classical Non-associative Lambek Calculus and Formal Semantics. In: Pogodalla, S., Prost, J.-P. (eds.) LACL 2011. LNCS, vol. 6736, pp. 33–48. Springer, Heidelberg (2011)
2. Belnap, N.D.: Display logic. Journal of Philosophical Logic 11(4), 375–417 (1982)
3. De Groote, P., Lamarche, F.: Classical non associative Lambek calculus. Studia Logica 71, 355–388 (2002)
4. Gamut, L.T.F.: Logic, Language and Meaning. Intensional Logic and Logical Grammar, vol. 2. University of Chicago Press (1991)
5. Girard, J.-Y.: A new constructive logic: classical logic. Mathematical Structures in Computer Science 1(3), 255–296 (1991)
6. Hendriks, H.: Studied flexibility. Categories and types in syntax and semantics. PhD thesis, ILLC Amsterdam (1993)
7. Lambek, J.: The mathematics of sentence structure. American Mathematical Monthly 65, 154–169 (1958)
8. Lambek, J.: On the calculus of syntactic types. In: Jakobson, R. (ed.) Proceedings of the Twelfth Symposium in Applied Mathematics, Structure of Language and its Mathematical Aspects (1961)
9. Montague, R.: Universal grammar. Theoria 36(3), 373–398 (1970)
10. Moortgat, M.: Categorial type logics. In: Handbook of Logic and Language, pp. 93–177. Elsevier (1997)
11. Moortgat, M.: Symmetric categorial grammar: residuation and Galois connections. CoRR, abs/1008.0170 (2010)
12. Partee, B., Rooth, M.: Generalized conjunction and type ambiguity. In: Bäuerle, R., Schwarze, C., von Stechow, A. (eds.) Meaning, Use, and Interpretation of Language, pp. 361–383. Gruyter, Berlin (1983)

Implicit Arguments: Event Modification or Option Type Categories?

Chris Blom[1], Philippe de Groote[2], Yoad Winter[3], and Joost Zwarts[3]

[1] CAI Master Program, Utrecht University
[2] LORIA/INRIA
[3] UiL OTS, Utrecht University

Abstract. We propose a unified syntactic-semantic account of passive sentences and sentences with an unspecified object (*John read*). For both constructions, we employ *option types* for introducing implicit arguments into the syntactic-semantic categorial mechanism. We show the advantages of this approach over previous proposals in the domains of scope and unaccusatives. Unlike pure syntactic treatments, option types immediately derive the obligatory narrow scope of existential quantification over an implicit argument's slot. Unlike purely semantic, event-based treatments, our proposal naturally accounts for syntactic contrasts between passives and unaccusatives, as in *the door *(was) opened by John*.

1 Introduction

Many verbs allow adjacent constituents to be optional, as illustrated in (1).

(1) a. John read (the book).
b. John introduced Paul (to Mary).
c. The vase was broken (by John).

Unspecified objects (UOs, Levin [12]) as in (1a) are licensed with verbs like *eat, drink, bake* etc. In semantic terms, the optionality of object NPs with such verbs is similar to the optionality of subcategorized PPs as in (1b) and *by* phrases in passive sentences like (1c). When these constituents are missing the sentence still makes an existence claim with respect to the unfilled predicate slot [5]. This is illustrated by the equivalences in (2).

(2) John read ⇔ John read something.
John introduced Paul ⇔ John introduced Paul to someone.
The vase was broken ⇔ The vase was broken by someone or something.

M. Aloni et al. (Eds.): Amsterdam Colloquium 2011, LNCS 7218, pp. 240–250, 2012.
© Springer-Verlag Berlin Heidelberg 2012

Because of this 'semantic implicitness' of the verbal slots in (2), we refer to the underlined constituents in (1) as optional *arguments.*[1]

Our aim in this paper is to give a basic syntactic-semantic account of optional arguments in categorial grammar and explain their semantic properties. The main two problems we deal with are therefore:

P1 How can an argument be missing while the sentence still makes an existential import involving the unfilled slot (2)?

P2 How can verbs with an optional argument compose with this argument when it is present, neutralizing the purely existential effect?

After reviewing some previous accounts and their limitations, we propose a new account of these two questions using *option types* in Abstract Categorial Grammar (ACG [6,13]).

2 Meaning Postulates and Event Modification

Existential implicit arguments must be interpreted with narrow scope (Fodor & Fodor, [5]). Consider the following examples.

(3) a. John did not read yesterday.
    b. John did not read something yesterday.

(4) a. The door was not opened yesterday.
    b. The door was not opened by something/someone yesterday.

The overt indefinites in (3b) and (4b) are scopally ambiguous. By contrast, in (3a) and (4a) the only reading is existential narrow scope: "John did not read anything" and "the door was not opened by anything/anyone", respectively. As Fodor & Fodor observe, obligatory narrow scope of quantifiers over implicit arguments challenges pure syntactic theories that introduce an existential argument at some covert syntactic level. F&F account for contrasts as in (3) using a meaning postulate connecting a transitive entry of UO verbs with another, intransitive entry. As F&F show, this immediately accounts for narrow scope effects in cases like (3a). A similar idea governs ambiguity-based derivations of passives, with or without a *by* phrase (Bach

[1] We will not try to give here a full syntactic analysis of the distinction between optional arguments and adjuncts. First, the disability to iterate (optional and obligatory) arguments (cf. **John ate/bought a sandwich a hamburger*) extends to other constituents that syntacticians may consider as adjuncts: **John went to Paul to Mary*, **John baked a cake for Mary for Sue*, **John cut the rope with a knife with an axe*, etc. Another challenge for defining the distribution of optional arguments is that some typical adjuncts may also have an existence entailment when missing. For instance, *John will sing* entails *John will sing sometime.* Note however that many adjuncts are clearly distinguished from optional arguments in having no existential entailment: *Mary worked* does not entail *Mary worked for someone*, *John baked a cake* does not entail *John baked a cake for someone*, *John went* does not entail *John went somewhere* (he may have evaporated in space), and *John cut the rope* does not entail *John cut the rope with something* (he may have had super-natural powers).

[1], Dowty [4]). However, this 'classical' approach to passives is quite cumbersome (Landman [9]). For instance, to account for the optionality of the two underlined constituents in a passive sentence such as *Paul was introduced by John to Mary*, classical accounts would have to assign four different meanings to the passive form of the verb *introduce*, connected to each other using meaning postulates.

A more promising semantic approach to passives and UOs was suggested by Carlson [2], who treats both *by* phrases in passives and objects with UO verbs as neo-Davidsonian event modifiers. Carlson treats the object in *John read 'Lolita'* as an adjunct that modifies reading events. When the object is missing (*John read*), Carlson treats the existence entailment as an implied property of events: when there is a reading, something must be read (cf. Larson [10], Lasersohn [11]). Similarly, Carlson treats the understood agent in a passive sentence like *Mary was praised* as a property of 'praising events': when there is a praising there is a praiser. These intuitions can be implemented as a single meaning postulate on predicates over events, which nicely improves F&F's approach. A similar event-based account of passives is proposed by Landman in [9].

However, Carlson's approach becomes less appealing when considering *unaccusative verbs*. A sentence like *the door opened* is not equivalent to *someone/something opened the door*. This shows that 'opening events' do not require an agent. Hence, the understood agent in passives like *the door was opened* is not explained.[2] Furthermore, treating *by* phrases as adjuncts does not account for their ungrammaticality with unaccusatives: if *by* phrases simply modify events by specifying an agent, why are they systematically disallowed in cases like **the door opened by John*? To block a meaning like "John opened the door" for such strings, Carlson (p.268) must resort to ad hoc syntactic assumptions.

3 Option Types in Abstract Categorial Grammar

The considerations in section 2 support a semantic approach that is more consistent with current syntactic theories about implicit arguments (Landau [8]). We propose the following principle (cf. Dekker [3]).[3]

[2] This is no problem for Landman's proposal, which uses the passive morphology for deriving the semantic requirement that an agent exists in events described by passive sentences. However for this reason, Landman's approach does not naturally extend to UOs, where intransitive forms are homonymic to transitives. Hence, in such situations overt morphology alone cannot account for the requirement that a patient/theme exists in all events described by UO verbs.

[3] Dekker proposes treating some optionality phenomena using dynamic binding. Extending Dekker's approach to passives and UOs may have comparable advantages to the present account. It would require augmenting Dekker's proposal with a clear distinction between verb modifiers (e.g. many adverbials), which can iterate, and optional arguments that cannot iterate (e.g. optional objects and *by* phrases in passives). In Dekker's system the former must be treated as dynamic argument modifiers, whereas the latter are treated as static argument saturators. Once this distinction is properly treated, we believe that Dekker's approach can be extended to an approach that is very similar to our account of optional arguments – see especially Dekker's remarks on this issue on [3, p.573].

Optional Argument Principle (OAP): *Verbal forms may specify optional arguments in their syntactic-semantic type. The existential semantic interpretation of an implicit argument is lexically introduced.*

The OAP allows us to account for the two problems mentioned above for Carlson's approach. First, the understood agent in passives like *the door was opened* is analyzed as a result of an optional argument. This argument can be materialized as a *by* phrase, but it also leads to the requirement that an agent exists when the *by* phrase is missing. By contrast, an unaccusative sentence like *the door opened* is analyzed, following Carlson, as a simple intransitive statement. The semantic (uni-directional) entailment from *the door was opened* to *the door opened* can be handled in event semantics, as in Carlson's account.[4] However, in our approach, unlike Carlson's proposal, there is no general strategy of *by* phrase adjunction. Thus, unaccusative sentences like *the door opened*, which are analyzed as intransitive sentences, do not license *by* phrases.

To implement OAP-based optionality in categorial grammar, we mark optionality by adding the symbol 'o' on the optional argument's type. For instance:

Intransitive verbs (*smile*, unaccusative *open*):
$\mathsf{np} \to \mathsf{s}$
Transitive verbs with obligatory object (*praise*, unergative *open*):
$\mathsf{np} \to (\mathsf{np} \to \mathsf{s})$
Transitive verbs with optional object (*read*):
$\mathsf{np}^{\circ} \to (\mathsf{np} \to \mathsf{s})$
Passive forms of transitive verbs (*was read, was praised, was opened*):[5]
$\mathsf{np}^{\circ}_{\mathsf{by}} \to (\mathsf{np} \to \mathsf{s})$

Option types are simple cases of *sum types* in functional programming [7]. There are two ways of filling in an optional slot of a function: by providing an argument of the appropriate type, or by using a universal filler, marked '$*$'. Let F be a function of type $\mathsf{a} \to \mathsf{b}$, with an obligatory argument slot of abstract type a. To make this slot into an optional one, an *option operator* distinguishes the case where the argument is present from usages of the filler. For the second case, we use a *default result.* Formally, an object x of type a° is either of type a or the $*$-filler (disjoint from a's domain). Let d be a default result of type b. The **option** operator of type $(\mathsf{a}^{\circ} \times (\mathsf{a} \to \mathsf{b}) \times \mathsf{b}) \to \mathsf{b}$ is defined by:

$$\mathbf{option}(x, F, d) = d \text{ if } x = *; \text{ otherwise } \mathbf{option}(x, F, d) = F(x).$$

The function $\lambda x.\mathbf{option}(x, F, d)$ is therefore of type $\mathsf{a}^{\circ} \to \mathsf{b}$.

In ACG, the connection between the morpho-syntactic level and the semantic level is handled by mapping each simple *abstract type* (np, s etc.) to a morpho-syntactic type and a semantic type. We assume the simple morpho-syntactic type

[4] For space considerations we do not develop here this event-based account but refer the reader to [14], where an event semantics is handled within an ACG framework compatible with the current proposal.

[5] The abstract type $\mathsf{np}_{\mathsf{by}}$ is used for *by* phrases in passive constructions, and is needed for morpho-syntactic reasons alone (see section 5). The semantics of this type for the purposes of this paper is treated as equivalent to the semantics of np types.

f for *strings* and the simple semantic types e and t for *entities* and *truth-values* respectively. We assume the following mappings of the simple abstract type np and s to morpho-syntactic and semantic types:

$\mathsf{np} \mapsto \langle f, e\rangle$: a noun phrase surfaces as a string and denotes an entity

$\mathsf{s} \mapsto \langle f, t\rangle$: a sentence surfaces as a string and denotes a truth-value

Complex types like $\mathsf{np} \rightarrow \mathsf{s}$ (for intransitive verbs) or $\mathsf{np} \rightarrow (\mathsf{np} \rightarrow \mathsf{s})$ (for transitive verbs) are inductively mapped to the corresponding morpho-syntactic and semantic types according to the following rule.

Let a *be an abstract type that is mapped to the morpho-syntactic type* a_1 *and the semantic type* a_2*. Let* b *be an abstract type that is mapped to the morpho-syntactic type* b_1 *and the semantic type* b_2*. Then* $\mathsf{a} \rightarrow \mathsf{b}$ *is an abstract type that is mapped to the morpho-syntactic type* $a_1 \rightarrow b_1$ *and the semantic type* $a_2 \rightarrow b_2$*. Likewise,* a° *is an abstract type that is mapped to the morpho-syntactic type* a_1^{o} *and the semantic type* a_2^{o}*.*

In short we denote:

$\mathsf{a} \mapsto \langle a_1, a_2\rangle$
$\mathsf{b} \mapsto \langle b_1, b_2\rangle$
$(\mathsf{a} \rightarrow \mathsf{b}) \mapsto \langle a_1 b_1, a_2 b_2\rangle$
$\mathsf{a}^{\circ} \mapsto \langle a_1^{o}, a_2^{o}\rangle$

For instance:

$(\mathsf{np} \rightarrow \mathsf{s}) \mapsto \langle ff, et\rangle$
: an intransitive verb surfaces as a function from strings to strings, and denotes a function from entities to truth-values

$(\mathsf{np} \rightarrow (\mathsf{np} \rightarrow \mathsf{s})) \mapsto \langle f(ff), e(et)\rangle$
: a transitive verb surfaces as a function from strings to functions from strings to strings, and denotes a function from entities to functions from entities to truth-values

Specifically, we assume the entry SMILE for the intransitive verb *smile*:

$\mathsf{SMILE} = \lambda s_f.s \bullet \mathit{smiled}_f \ : \ \mathbf{smile}_{et}$

In words: the morpho-syntactic entry for the verb *smile* is the function sending every subject string s to the corresponding sentence string composed of s and the string *smiled*; the corresponding semantic function is the function **smile** from entities to truth-values.

Similarly, we assume the following ACG entry PRAISE for the transitive verb *praise*:

$\mathsf{PRAISE} = \lambda o_f.\lambda s_f.s \bullet \mathit{praised}_f \bullet o \ : \ \mathbf{praise}_{e(et)}$

In words: the morpho-syntactic entry for the verb *praise* is the function sending every object string o to the function mapping every subject string s to the corresponding sentence string composed of o, s and the string *praised*; the corresponding semantic function is the function **praise** from entities to functions from entities to truth-values.

Simple sentences like *John smiled* and *John praised Lolita* are analyzed at the abstract level as SMILE(JOHN) and PRAISE(LOLITA)(JOHN), respectively. Using the verbal entries above and the treatment of names as entity-denoting strings, these sentences are associated with the following strings and truth-values:

$\mathsf{SMILE}(\mathsf{JOHN}) = \textit{John} \bullet \textit{smiled} \;:\; \mathbf{smile}(\mathbf{john})$

$\mathsf{PRAISE}(\mathsf{LOLITA})(\mathsf{JOHN}) = \textit{John} \bullet \textit{praised} \bullet \textit{Lolita} \;:\; \mathbf{praise}(\mathbf{lolita})(\mathbf{john})$

Using option types in ACG we extend this simple treatment to transitive verbs with an optional object. For instance, we assume the following ACG entry READ for the transitive verb *read*, of abstract type $\mathsf{np}^{\circ} \to (\mathsf{np} \to \mathsf{s})$:

$$\mathsf{READ} = \lambda o_f.\lambda s_f.\mathbf{option}(o\;,\; \lambda u_f.s \bullet \textit{read}_f \bullet u\;,\; s \bullet \textit{read}) \;: \\ \lambda o_e.\lambda s_e.\mathbf{option}(o\;,\; \lambda u_e.\mathbf{read}_{e(et)}(u)(s)\;,\; \exists x.\mathbf{read}(x)(s))$$

In words: the morpho-syntactic entry for the verb *read* uses the object string o in the object position if it is given, and leaves the object position null if the object argument is not given; the corresponding semantic function uses the object entity o as the first argument of the binary function **read**, and existentially saturates the position if the object argument is not given.

In this way, the sentences *John read* and *John read Lolita* are associated with the following strings and truth-values:

$\mathsf{READ}(\mathsf{JOHN}) = \textit{John} \bullet \textit{read} \;:\; \exists x.\mathbf{read}(x)(\mathbf{john})$

$\mathsf{READ}(\mathsf{LOLITA})(\mathsf{JOHN}) = \textit{John} \bullet \textit{read} \bullet \textit{Lolita} \;:\; \mathbf{read}(\mathbf{lolita})(\mathbf{john})$

The treatment of passives is analogous. For instance, the passive form *was praised* (*by*) is analyzed as follows.

$$\mathsf{PRAISE}_{pass} = \lambda o_f.\lambda s_f.\mathbf{option}(o\;,\; \lambda u_f.s \bullet \textit{was-praised}_f \bullet u\;,\; s \bullet \textit{was-praised}_f) \;: \\ \lambda o_e.\lambda s_e.\mathbf{option}(o\;,\; \lambda u_e.\mathbf{praise}_{e(et)}(s)(u)\;,\; \exists x.\mathbf{praise}(s)(x))$$

The sentences *Lolita was praised* and *Lolita was praised by John* are associated with the following strings and truth-values (for the definition of BY see Table 1):

$\mathsf{PRAISE}_{pass}(\mathsf{LOLITA}) = \textit{Lolita} \bullet \textit{was-praised} \;:\; \exists x.\mathbf{praise}(\mathbf{lolita})(x)$

$\mathsf{PRAISE}_{pass}(\mathsf{BY}(\mathsf{JOHN}))(\mathsf{LOLITA}) = \textit{Lolita} \bullet \textit{was-praised} \bullet \textit{by} \bullet \textit{John} \;:\; \mathbf{praise}(\mathbf{lolita})(\mathbf{john})$

4 Some Exceptional Cases

There are some exceptional kinds of UO verbs and passive verbs that do not give rise to existential entailments when the object or *by* phrase is missing. Consider the following examples.

(5) a. John kicked $\overset{?}{\Rightarrow}$ John kicked something.

b. John bit $\overset{?}{\Rightarrow}$ John bit something.

(6) a. Mary was left alone $\overset{?}{\Rightarrow}$ Someone left Mary alone. [2].

b. The traps were avoided $\overset{?}{\Rightarrow}$ Someone avoided the traps. [11]

Similar cases were used by Carlson [2] to argue for event-oriented modification by objects and *by* phrases. Carlson assumes that since existence entailments do not appear with some such cases, this motivates an ontology-based account of the existence entailments that as a rule appear with these constructions. In addition to the reasons we gave in section 2 for doubting Carlson's general approach to UOs and passives, it should be remarked that lexical UOs like *kick* and *bite* are rather rare.[6] Because of their rarity, the UO verb alternations that such transitive verbs show in English may be a result of an accidental lexical ambiguity/polysemy, rather than a general process, as Carlson assumes. With passives as in (6), it is even more unclear whether there is any need to assume a general non-existential strategy for passives. Note that both sentences in (6) involve a negative component for the verbal element (*alone* = with no one, *avoid* = not get close to). Arguably, the non-existential effect is due to an existential reading of the agent argument which is in the scope of these negative components. Thus, examples such as (6) do not provide evidence for "non-existential passives".

5 A General Strategy of 'Optionalization'

So far in this paper we have exemplified specific entries of various option types and the way they are interpreted. We would now like to define a general procedure of 'optionalization'. Thus, given an ACG lexicon $\mathcal{L}$ without option types, we would like to transform $\mathcal{L}$ into an lexicon $OPT(\mathcal{L})$ with option types. This is done using a specification of some of the arguments of entries in $\mathcal{L}$ as optional arguments and using this specification for systematically optionalizing $\mathcal{L}$.

Consider the optionality-free toy ACG verbal lexicon in Table 1.[7] This lexicon contains the verbal forms discussed in section 3, but without any treatment of optionality: all arguments are treated as obligatory, against common linguistic judgements about the verbal arguments that are underlined in Table 1. Our general method transforms such a description into a proper ACG lexicon with types for optional arguments and the **option** operator in entry values as required by their types. This general method guarantees an economical representation for the two features common to all the optional arguments treated: no effect of the presence or lack of an optional argument on the *form* of the verb; and narrow scope existential quantification over the semantic slot in the verb's *meaning*.

Our general procedure for mapping such an option-free lexicon $\mathcal{L}$ to the corresponding optionalized lexicon uses option-free entries of abstract type $\underline{\mathsf{a}} \to \mathsf{b}$, where the a argument is targeted as requiring optionality. Our general procedure of 'optionalization' maps such an entry to a lexicon entry of abstract type $\mathsf{a}^{\circ} \to \mathsf{b}$, with the proper morpho-syntactic and semantic treatment. For instance, the entry READ in Table 1 does not respect the optionality of the verb's object argument. However, the procedure below will guarantee that the underlined argument in

[6] In her extensive typology of verb alternations in English, Levin [12] characterizes verbs such as *kick* or *bite* as showing "characteristic property alternations". This is because of sentences like *our horse kicks* or *our dog bites regularly*, which indicate a tendency towards kicking or biting. Levin does not mention simple past tense sentences as in (5), where transitive verbs without objects give rise to episodic readings.

[7] To achieve a more conspicuous notation we avoid parentheses in this table, reading the type notation $\alpha \to \beta \to \gamma$ with right-association as in $\alpha \to (\beta \to \gamma)$.

Table 1. Toy ACG lexicon without option types

Entry name	Type	Value
BY	$\mathsf{np}\to\mathsf{np}_{\mathsf{by}}$	$\lambda x_f.by \bullet x \ :\ \lambda x_e.x$
TO	$\mathsf{np}\to\mathsf{np}_{\mathsf{to}}$	$\lambda x_f.to \bullet x \ :\ \lambda x_e.x$
SMILE	$\mathsf{np}\to\mathsf{s}$	$\lambda s_f.s \bullet smiled_f \ :\ \mathbf{smile}_{et}$
PRAISE	$\mathsf{np}\to\mathsf{np}\to\mathsf{s}$	$\lambda o_f.\lambda s_f.s \bullet praised_f \bullet o : \mathbf{praise}_{eet}$
PRAISE$_{pass}$	$\underline{\mathsf{np}_{\mathsf{by}}}\to\mathsf{np}\to\mathsf{s}$	$\lambda b_f.\lambda s_f.s \bullet was\text{-}praised_f \bullet b \ :\ \lambda x_e.\lambda y_e.\mathbf{praise}(y)(x)$
READ	$\underline{\mathsf{np}}\to\mathsf{np}\to\mathsf{s}$	$\lambda o_f.\lambda s_f.s \bullet read_f \bullet o \ :\ \mathbf{read}_{eet}$
READ$_{pass}$	$\underline{\mathsf{np}_{\mathsf{by}}}\to\mathsf{np}\to\mathsf{s}$	$\lambda b_f.\lambda s_f.s \bullet was\text{-}read_f \bullet b \ :\ \lambda x_e.\lambda y_e.\mathbf{read}_{eet}(y)(x)$
INTRODUCE	$\underline{\mathsf{np}_{\mathsf{to}}}\to\mathsf{np}\to\mathsf{np}\to\mathsf{s}$	$\lambda t_f.\lambda o_f.\lambda s_f.s \bullet introduced_f \bullet o \bullet t : \mathbf{introduce}_{eeet}$
INTRODUCE$_{pass}$	$\underline{\mathsf{np}_{\mathsf{to}}}\to\underline{\mathsf{np}_{\mathsf{by}}}\to\mathsf{np}\to\mathsf{s}$	$\lambda t_f.\lambda b_f.\lambda s_f.s \bullet was\text{-}introduced_f \bullet b \bullet t \ :$ $\lambda r_e.\lambda a_e.\lambda p_e.\mathbf{introduce}(r)(p)(a)$

the type $\underline{\mathsf{np}}\to\mathsf{np}\to\mathsf{s}$ will be properly treated as optional in the resulting lexicon, with an entry for the same verb of type $\mathsf{np}^{\mathsf{o}}\to\mathsf{np}\to\mathsf{s}$, and the appropriate values in its morpho-syntactic and semantic treatment.

Let $\varphi = \varphi_1 : \varphi_2$ be a lexical entry of abstract type $\mathsf{a}\to\mathsf{b}$, where the argument type a is defined as desirably optional. Such entries are marked $\underline{\mathsf{a}}\to\mathsf{b}$ in Table 1. In order to define optionalization inductively, we assume that the abstract types a and b have no underlined sub-parts. To capture correctly the behavior of optional arguments, we assume further that these types satisfy the following restrictions:

- The concrete morpho-syntactic type of a is string:
  $\mathsf{a}\mapsto\langle f, \tau_{\mathsf{a}}\rangle$.
- The concrete semantic type of b is boolean:
  $\mathsf{b}\mapsto\langle\sigma_{\mathsf{b}}, \tau_{\mathsf{b}}\rangle$, where τ_{b} is a boolean type.[8]

The 'string' requirement is needed to allow filling in optional slots by the empty string; the 'boolean' requirement is needed to allow existential quantification over morpho-syntactically empty slots.

Given these assumptions on the type of the entry φ, we define the corresponding optionalized entry $opt(\varphi) = \varphi_1' : \varphi_2'$ of type $\mathsf{a}^{\mathsf{o}}\to\mathsf{b}$ as follows:

$$\varphi_1' = \lambda x_{f^o}.\mathbf{option}(x, \varphi_1, \varphi_1(\epsilon))$$

$$\varphi_2' = \lambda x_{\tau_{\mathsf{a}}{}^o}.\mathbf{option}(x, \varphi_2, CLOS(\varphi_2))$$

The operator $CLOS$ existentially saturates the first argument of φ_2, which is inductively defined as follows:

If $\tau_{\mathsf{b}} = t$ then $CLOS(\varphi_2) = \exists x_{\tau_{\mathsf{a}}}.\varphi_2(x)$.

Otherwise, let $\tau_{\mathsf{b}} = \tau_1\tau_2$, where τ_2 is boolean by induction. We define:

$$CLOS(\varphi_2) = \lambda y_{\tau_1}.CLOS(\lambda x_{\tau_{\mathsf{a}}}.\varphi_2(x)(y)).$$

For instance, for the constant **read** of type $e(et)$, we have:

$$CLOS(\mathbf{read}) = \lambda y_e.CLOS(\lambda x_e.\mathbf{read}(x)(y)) = \lambda y.\exists x.\mathbf{read}(x)(y)$$

[8] Standardly, we define the set of 'boolean' semantic types as the smallest set of semantic types that contains t and any type $\tau_1\tau_2$ where τ_2 is boolean.

As a result, the optionalized version of the entry for READ in Table 1 is $opt(\text{READ}) = \varphi'_1 : \varphi'_2$ s.t.:

$$\begin{aligned} \varphi'_1 &= \lambda x_{f^o}.\mathbf{option}(x,\ \lambda o_f.\lambda s_f.s \bullet \mathit{read}_f \bullet o,\ \lambda s_f.s \bullet \mathit{read}_f \bullet \epsilon) \\ \varphi'_2 &= \lambda x_{e^o}.\mathbf{option}(x,\ \mathbf{read}_{eet},\ \mathit{CLOS}(\mathbf{read}_{eet})) \\ &= \lambda x_{e^o}.\mathbf{option}(x,\ \mathbf{read}_{eet},\ \lambda y.\exists x'.\mathbf{read}(x')(y)) \end{aligned}$$

The definition of the *opt* operator is not yet sufficient in order to optionalize entries of abstract type $\mathsf{a} \to \mathsf{b}$ where b has underlined types in it. This is necessary, as illustrated by the entry INTRODUCE_{pass} in Table 1, representing the form and meaning of the verbal passive *be introduced* (*by X to Y*), which has two optional arguments. To allow optionalization of such entries as well, let $\varphi = \varphi_1 : \varphi_2$ be a lexical entry of an abstract type X. The type X may be primitive. In case $\mathsf{X} = \mathsf{a} \to \mathsf{b}$, we adopt the same assumptions above on a (string morpho-syntactic type) and b (boolean semantic type), but now b possibly has underlined types in it. Given these assumptions, we inductively define the optionalized entry $OPT(\varphi)$:

If X is primitive, then the entry $OPT(\varphi) = \varphi$.
If $\mathsf{X} = \mathsf{a} \to \mathsf{b}$ (the argument a should not be optionalized) , then the optionalized entry is inductively defined by $OPT(\varphi) = \lambda x_{\mathsf{a}}.OPT(\varphi(x))$.
If $\mathsf{X} = \underline{\mathsf{a}} \to \mathsf{b}$ (the argument a should be optionalized) , then the optionalized entry is inductively defined by $opt(\lambda x_{\mathsf{a}}.OPT(\varphi(x)))$.

For instance, applying OPT to the entry INTRODUCE_{pass} in Table 1, of type $\underline{\mathsf{np_{to}}} \to \underline{\mathsf{np_{by}}} \to \mathsf{np} \to \mathsf{s}$ (first two arguments are optional), results in:

$$\begin{aligned} & OPT(\text{INTRODUCE}_{pass}) \\ =\ & opt(\lambda x_{np_{to}}.OPT(\text{INTRODUCE}_{pass}(x))) \\ =\ & opt(\lambda x_{np_{to}}.opt(\lambda y_{np_{by}}.OPT(\text{INTRODUCE}_{pass}(x)(y)))) \\ =\ & opt(\lambda x_{np_{to}}.opt(\lambda y_{np_{by}}.\text{INTRODUCE}_{pass}(x)(y))) \end{aligned}$$

In the morpho-syntactic level, this amounts to:

$$\begin{aligned} & opt(\lambda x_f.opt(\lambda y_f.(\lambda t_f.\lambda b_f.\lambda s_f.s \bullet \textit{was-introduced}_f \bullet b \bullet t)(x)(y))) \\ =\ & opt(\lambda x_f.opt(\lambda y_f.\lambda s_f.s \bullet \textit{was-introduced}_f \bullet y \bullet x)) \\ =\ & \lambda w_f.\mathbf{option}(w\ ,\lambda x_f.\ \lambda z_f.\mathbf{option}(z, \lambda y_f.\ \lambda s_f.s \bullet \textit{was-introduced}_f \bullet y \bullet x \\ & \qquad\qquad\qquad\qquad\qquad\qquad ,\ \lambda s_f.s \bullet \textit{was-introduced}_f \bullet \epsilon \bullet x\ ) \\ & \qquad\qquad ,\ \lambda z_f.\mathbf{option}(z, \lambda y_f.\ \lambda s_f.s \bullet \textit{was-introduced}_f \bullet y \bullet \epsilon \\ & \qquad\qquad\qquad\qquad\qquad\qquad ,\ \lambda s_f.s \bullet \textit{was-introduced}_f \bullet \epsilon \bullet \epsilon\ )) \end{aligned}$$

This covers the four combinations of present/missing morpho-syntactic arguments. Similarly, in the semantic level we have:

$$\begin{aligned} & opt(\lambda x.opt(\lambda y_e.(\lambda r_e.\lambda a_e.\lambda p_e.\mathbf{introduce}_{e(e(et))}(r)(p)(a))(x)(y))) \\ =\ & opt(\lambda x_e.\lambda z.\mathbf{option}(z\ , \lambda y_e.\lambda p_e.\mathbf{introduce}_{e(e(et))}(x)(p)(y) \\ & \qquad\qquad ,\ CLOS(\lambda y_e.\lambda p_e.\mathbf{introduce}_{e(e(et))}(x)(p)(y)))) \end{aligned}$$

And applying $CLOS$ gives:

$$\begin{aligned} & opt(\lambda x_e.\lambda z.\mathbf{option}(z\ , \lambda y_e.\lambda p_e.\mathbf{introduce}_{e(e(et))}(x)(p)(y) \\ & \qquad\qquad ,\ \lambda p_e.\exists y_e.\mathbf{introduce}_{e(e(et))}(x)(p)(y))) \end{aligned}$$

$$
\begin{aligned}
= \lambda w_{e^o}.\mathbf{option}(w\,,\lambda x_e.\lambda z_{e^o}.\mathbf{option}(z\,,\lambda y_e.\lambda p_e.&\mathbf{introduce}_{e(e(et))}(x)(p)(y)\\
&,\lambda p_e.\exists y_e.\mathbf{introduce}_{e(e(et))}(x)(p)(y))\\
,\lambda z_{e^o}.\exists x_e.\mathbf{option}(z\,,\lambda y_e.\lambda p_e.&\mathbf{introduce}_{e(e(et))}(x)(p)(y)\\
&,\lambda p_e.\exists y_e.\mathbf{introduce}_{e(e(et))}(x)(p)(y)))
\end{aligned}
$$

This covers the four combinations of the two present/existentially-closed semantic arguments. The same procedure also correctly derives the other specific entries discussed in section 3.

6 Conclusions

Optionality with verbal arguments is a well-known phenomenon that is treated in one way or another by most syntactic theories. We started out by pointing out the narrow-scope behavior of existential quantifiers over empty argument slots, which, although familiar, is not treated systematically by syntactic frameworks we are aware of. The behavior of unaccusative verbs in contrast to their passive forms has led us to conclude that *by* phrases in passives should be treated similarly to optional arguments, and not using event-modifiers. The Optional Argument Principle aims to describe the relations between the syntactic optionality of arguments and their semantic properties. Within the framework of Abstract Categorial Grammar (ACG), we introduced a standard interpretation of option types, which led to a general transformation of grammars without argument optionality to grammars that encode it in a uniform way.

Acknowledgements. We thank Marijana Marelj for helpful discussions. The work was partially supported by a Van Gogh grant to the 2nd and 3rd author by the French-Dutch Academy. The work of the 3rd author was partially supported by a VICI grant number 277-80-002 by the Netherlands Organisation for Scientific Research (NWO). The work of the 4th author was supported by an NWO grant number 360-70-340.

References

1. Bach, E.: In defense of passive. Linguistics and Philosophy 3, 297–341 (1980)
2. Carlson, G.: Thematic roles and their role in semantic interpretation. Linguistics 22, 259–279 (1984)
3. Dekker, P.: Existential disclosure. Linguistics and Philosophy 16, 561–588 (1993)
4. Dowty, D.: Grammatical relations and Montague Grammar. In: Jacobson, P., Pullum, G. (eds.) The Nature of Syntactic Representation. Kluwer, Dordrecht (1982)
5. Fodor, J.A., Fodor, J.D.: Functional structure, quantifiers, and meaning postulates. Linguistic Inquiry 11, 759–770 (1980)
6. de Groote, P.: Towards abstract categorial grammars. In: Proceedings of the 39th Annual Meeting of the Association for Computational Linguistics, ACL (2001)
7. Gunter, C.: Semantics of programming languages: structures and techniques. The MIT Press (1992)
8. Landau, I.: The explicit syntax of implicit arguments. Linguistic Inquiry 41, 357–388 (2010)

9. Landman, F.: Events and Plurality: the Jerusalem lectures. Kluwer, Dordrecht (2000)
10. Larson, R.K.: Events and modification in nominals. In: Proceedings of Semantics and Linguistic Theory, SALT8, pp. 145–168 (1998)
11. Lasersohn, P.: Lexical distributivity and implicit arguments. In: Proceedings of Semantics and Linguistic Theory, SALT3 (1993)
12. Levin, B.: English Verb Classes and Alternations. The University of Chicago Press, Chicago (1993)
13. Muskens, R.: Language, Lambdas, and Logic. In: Kruijff, G.J., Oehrle, R. (eds.) Resource Sensitivity in Binding and Anaphora, pp. 23–54. Kluwer (2003)
14. Winter, Y., Zwarts, J.: Event Semantics and Abstract Categorial Grammar. In: Kanazawa, M., Kornai, A., Kracht, M., Seki, H. (eds.) MOL 12. LNCS, vol. 6878, pp. 174–191. Springer, Heidelberg (2011)

Each vs. *Jeweils*: A Cover-Based View on Distance Distributivity

Lucas Champollion

New York University
10 Washington Place, New York, NY 10003
champollion@nyu.edu

Abstract. Zimmermann [2002] identifies two kinds of distance-distributive items across languages. The first kind (e.g. *each*) is restricted to distribution over individuals; the second kind (e.g. German *jeweils*) can also be interpreted as distributing over salient occasions. I explain this behavior by formally relating this split to the two distributivity operators proposed in the work of Link (atomic operator) and Schwarzschild (cover-based operator), which I reformulate in a Neo-Davidsonian framework.

Keywords: distributivity, crosslinguistic semantics, algebraic semantics.

1 Introduction

Across languages, distributive items have different syntactic uses and different meanings. In English, *each* can be used in three essentially synonymous ways:

(1) a. **Adnominal:** The children saw two monkeys *each*.
b. **Adverbial:** The children *each* saw two monkeys.
c. **Determiner:** *Each* child saw two monkeys.

There are many terms for these three uses. Adnominal *each* is also called binominal or shifted; adverbial *each* is also called floated; and determiner *each* is also called prenominal. Following Zimmermann [2002], I will refer to adnominal and adverbial *each* as distance-distributive items, or DD items for short.

In German, adnominal and adverbial *each* are translated by one word, *jeweils*. Determiner *each*, however, is translated by another one, *jed-*. I gloss DD items as DIST since, as we will see, they have a wider range of readings than *each*.

(2) a. **Adnominal:** Die Kinder haben [*jeweils* [zwei Affen]] gesehen.
The children have DIST two monkeys seen.
b. **Adverbial:** Die Kinder haben [*jeweils* [zwei Affen gesehen]].
The children have DIST two monkeys seen.
c. **Determiner:** *Jedes/*Jeweils* Kind hat zwei Affen gesehen.
Each.sg.n/DIST child has two monkeys seen.

Though adverbial and adnominal *jeweils* take the same surface position in (2a) and (2b), they can be teased apart syntactically, as shown in Zimmermann [2002]. However, this distinction will play no role in this paper.

M. Aloni et al. (Eds.): Amsterdam Colloquium 2011, LNCS 7218, pp. 251–260, 2012.
© Springer-Verlag Berlin Heidelberg 2012

2 Crosslinguistic Variation

Zimmermann [2002] classifies about a dozen languages depending on whether the DD item can also function as a distributive determiner, as in English, or not, as in German. Across these languages, he observes that DD items which can also be used as determiners (e.g. *each*) always distribute over individuals, as determiners do. In contrast, many of those DD items which are formally distinct from determiners (e.g. *jeweils*) can also distribute over salient occasions, that is, over chunks of time or space. See also Moltmann [1997] for an earlier discussion of *each* vs. *jeweils*, and Zimmermann [2002] for a critique of Moltmann's analysis.

The best way to illustrate Zimmermann's observation is to start by considering German *jeweils*, a DD item which cannot double as a distributive determiner. *Jeweils* can distribute over individuals like English *each*, but also over spatial or temporal occasions, as long as context provides a salient set of such occasions. I call this the occasion reading.[1] The following examples illustrate this. Sentence (3) is ambiguous between a reading that distributes over individuals – the ones of which their plural subject consists, (3a) – and one that distributes over occasions (3b).

(3) Die Kinder haben jeweils zwei Affen gesehen. *(German)*
The children have DIST two monkeys seen.
a. 'Each of the children has seen two monkeys.'
b. 'The children have seen two monkeys each time.'

While the former reading is always available, the latter requires a supporting context. That is, when (3) is uttered out of the blue, it only has the reading (3a). The reading (3b), by contrast, is only available in contexts where there is a previously mentioned or otherwise salient set of occasions, such as contexts in which the children have been to the zoo on several previous occasions.

Unlike *each*, *jeweils* can also occur with a singular subject, as in (4), which only has an occasion reading, 'Hans has seen two monkeys on each occasion.'

(4) Hans hat jeweils zwei Affen gesehen. *(German)*
Hans has DIST two monkeys seen.

[1] The occasion reading corresponds to what Balusu [2005] calls the *spatial key* and *temporal key* readings. I leave open the question of whether the spatial and temporal cases should be distinguished as two separate readings. Another, less theory-neutral term for it is *event-distributive reading* [Oh, 2001]. Zimmermann [2002] uses the term *adverbial reading* for it. This term is misleading, because it suggests that only the adverbial use of *jeweils* can give rise to this reading. But as he documents in his Chapter 5, adnominal *jeweils* can give rise to it as well. For example, in (i), *jeweils* is part of the subject DP and is therefore adnominal. However, as shown by the paraphrase, this instance of *jeweils* distributes over occasions, not over individuals.

(i) Jeweils zwei Jungen standen Wache.
DIST two boys stood watch.
'Each time, two boys kept watch.'

This sentence is odd out of the blue, and it requires supporting context in the same way as reading (3b) does. Its other potential reading would involve vacuous distribution over only one individual, Hans, and is presumably blocked through the Gricean maxim of manner "Be brief".

While *jeweils* allows distribution both over individuals and over salient occasions, this is not the case for all DD items, as Zimmermann reports. Crosslinguistically, many adnominal DD items can only distribute over individuals. For example, English adnominal *each* lacks the occasion reading:

(5) The children have seen two monkeys each.
  a. *Available:* 'Each of the children has seen two monkeys.'
  b. *Unavailable:* 'The children have seen two monkeys on each occasion.'

When adnominal *each* is used in a sentence whose subject is singular, distribution over individuals is not possible, again presumably for Gricean reasons:

(6) *John has seen two monkeys each.

Unlike (4), this sentence lacks an occasion reading, even with supporting context. Why does *each* lack the occasion reading? We have seen in Section 1 that *each* also differs from *jeweils* in that only the former can also be used as a determiner. Adnominal DD items in Dutch, Norwegian, Italian, Russian, and French [Zimmermann, 2002] and in Turkish (Tuğba Çolak, p.c.) all behave like adnominal *each* in two ways: They can also be used as distributive determiners, and they lack the occasion reading.[2] Following Zimmermann [2002], we can generalize:

(7) **Zimmermann's Generalization:** If a DD item can also be used as a distributive determiner, it lacks the occasion reading.

This generalization goes only one way, that is, the "if" cannot be strengthened to "if and only if". This is because, as Zimmermann shows, the Japanese DD item *sorezore* cannot be used as a determiner but lacks the occasion reading. But "if and only if" may still be true as a tendency. Zimmermann reports that in addition to German *jeweils*, adnominal DD items in Czech, Bulgarian, and Korean have occasion readings and cannot be used as determiners.[3]

The following requirements for a semantic analysis of distance-distributivity emerge. First, the synonymy of the determiner, adnominal and adverbial uses

[2] The French case is somewhat controversial. Adnominal *chacun* and determiner/adnominal *chaque* are not exactly identical, but Zimmermann [2002] argues (p. 44) that they are historically related and can still be considered formally identical.

[3] Many languages express adnominal distance distributivity by reduplicating a numeral [Gil, 1982]. In this category, we both find cases where reduplication does not give rise to occasion readings, such as Hungarian [Farkas, 1997; Szabolcsi, 2010], and cases where it does, such as Telugu [Balusu, 2005]. The import of these cases on Zimmermann's generalization is unclear, as reduplication is not usually thought of as a free morpheme and is therefore not expected to be able to act as a determiner.

of *each* in English should be captured, ideally by essentially identical lexical entries. Second, the fact that DD items across languages share some part of their meanings (namely their individual-distributive readings) should be represented, as well as the fact that some of them can also have occasion readings. Third, the analysis should clarify the connections between DD items and distributivity theory. Finally, there should be a way to capture Zimmermann's Generalization. I now propose an analysis that fulfills these requirements. Section 3 presents distributivity operators; Section 4 relates them to DD items. Section 5 concludes and offers a speculation on how Zimmermann's Generalization can be captured.

3 Distributivity Operators in Algebraic Event Semantics

The following analysis is placed in the context of algebraic event semantics [Krifka, 1989] and of the theory of distributivity developed by Link [1987] and Schwarzschild [1996]. Link postulates a silent operator that shifts a VP to a distributive interpretation, that is, one that holds of any individual whose atomic parts each satisfy the unshifted VP. This so-called D operator is defined as follows. Here, the variable x is resolved to a plural entity, the subject, and y ranges over its atomic parts, that is, the singular individuals of which it consists.

(8) $[\![\text{D}]\!] = \lambda P_{et} \lambda x \forall y[y \leq x \wedge \text{Atom}(y) \rightarrow P(y)]$ [Link, 1987]

The optional presence of the D operator derives the ambiguity between distributive and scopeless readings. For example, (9a) represents a scopeless reading and (9b) a distributive reading. I use the term "scopeless" to refer both to collective and cumulative readings. The distinction between these two readings does not matter for this paper. See Landman [2000] for discussion.

(9) a. The children saw two monkeys.
≈ The children between them saw two monkeys. *scopeless*
b. The children [D [saw two monkeys]].
≈ The children each saw two monkeys. *distributive*

I propose that DD items should be essentially thought of as versions of this D operator (cf. Link [1986] for a similar claim for German *je*, a short form of *jeweils* which seems to lack the occasion reading). Clearly, Link's D operator and *each* are similar, as can be seen from the paraphrase of (9b). I take adverbial *each* and related DD items in Dutch, Norwegian, Italian, Russian, French, and Turkish to be D operators. As for *jeweils* and its relatives in Czech, Bulgarian, and Korean, we have seen that they can distribute over spatial and temporal intervals – arguably nonatomic entities. Link's D operator always distributes down to individual atoms and can therefore not be extended to these cases.

However, Schwarzschild [1996] argues on independent grounds that Link's D operator should be modified to allow for "nonatomic distributive" interpretations in a limited set of circumstances, namely whenever there is a particularly salient way to divide a plural individual. A good example of what Schwarzschild has in mind is provided by Lasersohn [1998]. Shoes typically come in pairs, so a sentence like *The shoes cost $50* can be interpreted as saying that each pair of shoes costs $50, as opposed to each shoe or all the shoes together. To model this kind of example, Schwarzschild modifies D and makes it anaphoric to a salient cover (a partition of a plural individual that allows overlap). C, the "cover variable", is free and anaphoric on the context. Schwarzschild assumes that C is a cover of the entire universe of discourse, but for most purposes one can instead think of C as a cover or a partition of the sum individual in question into salient parts, which may be plural sums. In this case, C partitions the sum of shoes into pairs. Schwarzschild refers to his own version of the D operator as Part.

(10) $[\![\text{Part}_C]\!] = \lambda P_{et} \lambda x \forall y [y \leq x \wedge \text{C}(y) \rightarrow P(y)]$ [Schwarzschild, 1996]

This operator optionally applies to a VP and shifts it to a nonatomic distributive reading. For example, Lasersohn's sentence is modeled as follows:

(11) The shoes [Part [cost $50]].
≈ Each salient plurality of shoes costs $50. *nonatomic distributive*

It is of course possible to think of D as a special case of Part, namely the one that results when the variable C is resolved to the predicate *Atom*. However, I assume that both D and Part are present in the grammar. This assumption will allow us to capture the distinction between *each* and *jeweils*. The former corresponds to D and the latter corresponds to Part. This accounts for the fact that *jeweils* and its relatives across languages have a wider range of readings than *each* and its relatives do.

In count domains, distributivity over atoms is expected to be salient in almost all contexts and to obscure the presence of nonatomic distributive readings [Schwarzschild, 1996]. It is therefore useful to look for nonatomic VP-level distributivity in a noncount domain, such as time. Here we find once again that the readings in question are available given appropriate contextual information or world knowledge. Example (12) is based on observations in Moltmann [1991]. It is odd out of the blue because pills cannot be taken repeatedly, but it is acceptable in a context where the patient's daily intake is discussed. Example (13) is from Deo and Piñango [2011], and is acceptable because it is clear that snowmen are typically built in winter.

(12) The patient took two pills for a month and then went back to one pill.

(13) We built a huge snowman in our front yard for several years.

Since *for*-adverbials are otherwise not able to cause indefinites to covary [Zucchi and White, 2001], and since Part is dependent on a salient level of granularity just like (12) and (13) are, it is plausible to assume that a temporal version of Part is responsible for the distributive interpretation of these sentences. See Champollion [2010] for more discussion of this point. The contribution of this temporal version of Part can be paraphrased as *daily* in (12) and *yearly* in (13).

The original formulations of the operators in (8) and (10) can only "target" (that is, distribute over parts of) the subject. Examples like (12) and (13) motivate a reformulation of the operators that allows them to target different thematic roles, including time. I will represent the relationship between D and the thematic role it targets through coindexation. For evidence that this relationship can be nonlocal, which justifies the use of coindexation, see Champollion [2010]. Coindexation also allows us to capture the fact that DD items can also target different thematic roles [Zimmermann, 2002]. For example, (14) can either involve two stories per boy or two stories per girl, depending on which thematic role is targeted by *each*.

(14) The boys told the girls two stories each.

In the following, I assume a Neo-Davidsonian algebraic semantic system loosely based on Krifka [1989] and Champollion [2010]. Events, verbs and thematic roles are each assumed to be closed under sum formation. Verbs and their projections are all of type vt (event predicates). Here is a sample entry of a verb.

(15) $[\![\text{see}]\!] = \lambda e\, [{}^*\text{see}(e)]$

This entry includes the star operator from Link [1983] as a reminder that the predicate is closed under sum formation. The star operator maps a set P to the predicate that applies to any sum of things each of which is in P. It can be easily generalized to functions such as thematic roles [Champollion, 2010].

Noun phrases are interpreted in situ (I do not consider quantifier raising in this paper). Silent theta role heads, which denote functions of type ve (event to individual), are located between noun phrases and verbal projections. I will often omit them in the LFs for clarity. The precise nature of the compositional process is not essential, but it affects the types of the lexical entries of DD items so let me make it concrete. I assume that the following type shifters apply first to the theta role head, then to the noun phrase, and finally to the verbal projection.

(16) a. Type shifter for indefinites: $\lambda\theta_{ve}\lambda P_{et}\lambda V_{vt}\lambda e[V(e) \wedge P(\theta(e))]$
b. Type shifter for definites: $\lambda\theta_{ve}\lambda x\lambda V_{vt}\lambda e[V(e) \wedge \theta(e) = x]$

Each of these type shifters combines a noun phrase with its theta role head to build an event predicate modifier of type $\langle vt, vt\rangle$. For example, after the noun phrases *the children* (definite) and *two monkeys* (indefinite) combine with the theta role heads *agent* and *theme* respectively, their denotations are as follows. Here, $\bigoplus$ child stands for the sum of all children, a plural individual of type e.

(17) $[\![[\text{agent [the children]}]]\!] = \lambda V\lambda e[V(e) \wedge {}^*\text{ag}(e) = \bigoplus\text{child}]$

(18) $[\![[\text{theme [two monkeys]}]]\!] = \lambda V\lambda e[V(e) \wedge |{}^*\text{th}(e)| = 2 \wedge {}^*\text{monkey}({}^*\text{th}(e))]$

After the verb has combined with all its arguments, the event variable is existentially bound if the sentence is uttered out of the blue. If the sentence is understood as referring to a specific event, the event variable is instead resolved to that event. If the noun phrases combine directly with the verb, we get a scopeless reading as in (19). Here and below, I write *2M* as a shorthand for $\lambda e[|{}^*\text{th}(e)| = 2 \wedge {}^*\text{monkey}({}^*\text{th}(e))]$.

(19) $[\![\text{The children saw two monkeys}]\!] = \exists e[^*\text{ag}(e) = \bigoplus \text{child} \wedge {}^*\text{see}(e) \wedge 2\text{M}(e)]$

To generate distributive readings, we use Link's D operator. Since VPs are event predicates, VP-level operators must be reformulated as event predicate modifiers. As described above, I assume that the D operator is coindexed with a thematic role θ, its target. My reformulation of Link's D operator is therefore as follows:[4]

(20) $[\![\text{D}_\theta]\!] = \lambda V_{vt} \lambda e[e \in {}^*\lambda e'[V(e') \wedge \text{Atom}(\theta(e'))]]$

As an example, the distributive reading of (19) is derived like this:

(21) $[\![\text{The children D}_{ag} \text{ [saw two monkeys]}]\!]$
$= \exists e[^*\text{ag}(e) = \bigoplus \text{child} \wedge e \in [[\![D_{ag}]\!](\lambda e'[^*\text{see}(e') \wedge 2\text{M}(e')])]]$
$= \exists e[^*\text{ag}(e) = \bigoplus \text{child} \wedge e \in {}^*\lambda e'[^*\text{see}(e') \wedge 2\text{M}(e') \wedge \text{Atom}(\text{ag}(e'))]]$

This formula is true just in case there is an event e whose agent is the children, and which consists of seeing-two-monkeys events whose agents are atomic. Remember that events and thematic roles are closed under sum, so e can be a plural event with a plural agent. The formula does not explicitly state that the seeing-two-monkeys events have children as agents. However, this fact is entailed by the assumption that thematic roles are closed under sum formation together with the assumption that the entities in the denotation of singular count nouns like *child* are atoms. Specifically, the existentially quantified event can only have the children as its agent if it consists of events whose individual agents are children.

4 *Each* and *Jeweils* as Distributivity Operators

Adverbial *each* is a VP modifier and can therefore be given the same entry as the D operator in (20). Adnominal and determiner *each* need to be type-shifted, but both are defined in terms of (20). This reflects their synonymousness:[5]

(22) $[\![\text{each}_\theta]\!]_{adverbial} = [\![D_\theta]\!] = (20)$

(23) $[\![\text{each}_\theta]\!]_{adnominal} = \lambda P_{et} \lambda \theta_{ve} \lambda V_{vt} \lambda e \, [[\![D_\theta]\!](\lambda e'[V(e') \wedge P(\theta(e'))])(e)]$

(24) $[\![\text{each}]\!]_{determiner} = \lambda P_{et} \lambda \theta_{ve} \lambda V_{vt} \lambda e \, [\theta(e) = \bigoplus P \wedge [\![D_\theta]\!](V)(e)]$

Adnominal *each* combines with an indefinite noun phrase and then with a theta head. Determiner *each* combines first with a nominal and then with a theta head. It is not coindexed with anything because it is not a DD item. Since both entries happen to have the same type, I assume that the syntax is responsible for restricting their distribution (syntactically speaking, one is an adverb and

[4] This is not the only way to reformulate the D operator. See Lasersohn [1998] and Dotlačil [2011] for other proposals. This particular definition is taken from Champollion [2010], except that *PureAtom* has been changed to *Atom*. This change is immaterial because we do not distinguish between pure and impure atoms here.

[5] For other semantic analyses of the DD items *each* and *jeweils*, see for example Moltmann [1997], pp. 205ff., and Zimmermann [2002]. For a recent compositional analysis of *each* that uses plural compositional DRT, see Dotlačil [to appear].

the other one is a determiner). In both cases, the result is a phrase of VP modifier type $\langle vt, vt\rangle$, which is also the type of D_θ. Some intermediate steps of the derivations of (1) are shown in (25) and (26).

(25) $[\![[[\text{two monkeys}]\ \text{each}_{ag}]\ \text{theme}]]\!]$
$= \lambda V_{vt} \lambda e[[\![D_{ag}]\!](\lambda e'[V(e')\ \wedge\ 2\text{M}(e')])(e)]$
$= \lambda V_{vt} \lambda e[e \in {}^*\lambda e'[{}^*\text{see}(e')\ \wedge\ 2\text{M}(e')\ \wedge\ \text{Atom}(\text{ag}(e')]]$

(26) $[\![[[\text{Each child}]\ \text{agent}]]\!]$
$= \lambda V_{vt} \lambda e[{}^*\text{ag}(e) = \bigoplus \text{child} \wedge [\![D_{ag}]\!](V)(e)]$
$= \lambda V_{vt} \lambda e[{}^*\text{ag}(e) = \bigoplus \text{child} \wedge e \in {}^*\lambda e'[V(e') \wedge \text{Atom}(\text{ag}(e'))]]$

The result of these derivations is always the same, which reflects their synonymy:

(27) $[\![\text{The children each}_{ag}\text{ saw two monkeys}]\!]$
$= [\![\text{The children saw two monkeys each}_{ag}]\!]$
$= [\![\text{Each child saw two monkeys}]\!]$
$= (21) = [\![\text{The children D}_{ag}\text{ saw two monkeys}]\!]$

We now come to the event-based reformulation of Part. We obtain it by replacing *Atom* in (20) with a free variable C, which is assumed to be anaphoric on the context. This minimal change reflects the close connection between D and Part.

(28) $[\![\text{Part}_{\theta,C}]\!] = \lambda P_{vt} \lambda e[e \in {}^*\lambda e'[P(e')\ \wedge\ C(\theta(e'))]]$

Part takes an event predicate P and returns a predicate that holds of any event e which can be divided into events that are in P and whose θs satisfy the contextually salient predicate C. Note that the definition of (28) entails that C is a cover of $\theta(e)$. The operator (28) is also the lexical entry of adverbial *jeweils*. The same type shift as in (23) brings us from (28) to adnominal *jeweils*:

(29) $[\![\text{jeweils}_{\theta,C}]\!]_{adverbial} = [\![\text{Part}_{\theta,C}]\!] = (28)$

(30) $[\![\text{jeweils}_{\theta,C}]\!]_{adnominal} = \lambda P \lambda \theta \lambda V \lambda e[[\![\text{Part}_{\theta,C}]\!](\lambda e'[V(e') \wedge P(\theta(e'))])(e)]$

As in the case of the Part operator, the C parameter of *jeweils* can be set to *Atom* so long as θ is set to a function which points into a count domain, such as *ag*. In that case, *jeweils* distributes over individuals and is equivalent to *each*. The following example illustrates this with sentence (2a); sentence (2b) is equivalent.

(31) Die Kinder haben jeweils$_{ag,Atom}$ zwei Affen gesehen.
The children have DIST two monkeys seen.
"The children have each seen two monkeys."

If – and only if – there is a supporting context, the anaphoric predicate C can be set to a salient antecedent other than *Atom*, and in that case θ is free to adopt values like τ (runtime). This leads to occasion readings. Suppose for example that it is in the common ground that the children have been to the zoo to see animals last Monday, last Wednesday and last Friday, and that (2a) is uttered with reference to that state of affairs, or sum event. It is interpreted as follows.

(32) ⟦Die Kinder haben jeweils$_{\tau,\text{zoovisit}}$ zwei Affen gesehen.⟧ =
$^*\text{ag}(e_0) = \bigoplus \text{child} \wedge e_0 \in {}^*\lambda e'[^*\text{see}(e') \wedge 2\text{M}(e') \wedge \text{zoovisit}(\tau(e'))]$
"The children have seen two monkeys on each occasion."

Since the sentence refers specifically to the sum e_0 of the three events in question, the event variable in (32) is resolved to e_0 rather than being existentially bound. The predicate that is true of any time interval at which a zoo visit takes place, call it *zoovisit*, is also salient in this context. So C can be resolved to *zoovisit* rather than to *Atom*. Since there are no atoms in time, it is only now that θ can be set to τ, rather than to *ag* as in (31). What (32) asserts is that e_0 has the children as its agents; that it can be divided into subevents, each of whose runtimes is the time of a zoo visit; and that each of these subevents is a seeing-two-monkeys event. Runtime is closed under sum just like other thematic roles ($\tau = {}^*\tau$), or in other words, it is a sum homomorphism [Krifka, 1989]. This means that any way of dividing e_0 must result in parts whose runtimes sum up to $\tau(e_0)$. Assuming that $\tau(e_0)$ is the (discontinuous) sum of the times of the three zoo visits in question, this entails that each of these zoo visits is the runtime of one of the seeing-two-monkeys events. This is the occasion reading.

5 Summary and Discussion

This analysis has captured the semantic similarities between DD items across languages, as well as their variation, by relating them to distributivity operators. DD items can be given the same lexical entry up to type shifting and parameter settings. The parameters provided by the reformulation of the D and Part operators capture the semantic variation: DD items like English *each* are hard-coded for distribution over atoms, which blocks distributivity over a noncount domain like time. DD items like German *jeweils* can distribute over noncount domains, but only if they can pick up salient nonatomic covers from context.

The remaining question is how to capture the correlation expressed in Zimmermann's Generalization (7). That is to say, why does a DD item which can also be used as a distributive determiner lack the occasion reading? Zimmermann himself proposes a syntactic explanation: Determiners must agree with their complement; DD *each* also has a complement, a proform that must acquire its agreement features from its antecedent, the target of *each*; only overt targets have agreement features. Alternatively, a semantic explanation seems plausible: Distributive determiners like English *each* are only compatible with count nominals (*each boy*, **each mud*). Formally, this amounts to an atomicity requirement of the kind the D operator provides. This requirement can be seen as independent evidence of the atomic distributivity hard-coded in the entry (24) via the D operator (20). In other words, the DD item inherits the atomicity requirement of the determiner. Both explanations are compatible with the present framework.

Acknowledgments. Thanks to Anna Szabolcsi, Robert Henderson and to the reviewers for helpful comments. I am also grateful to audiences at the 2011 Stuttgart workshop on quantification and at the 2011 Amsterdam Colloquium.

References

Balusu, R.: Distributive reduplication in Telugu. In: Davis, C., Deal, A.R., Zabbal, Y. (eds.) NELS 36, pp. 39–52. GLSA, Amherst (2005)

Champollion, L.: Parts of a whole: Distributivity as a bridge between aspect and measurement. PhD thesis, University of Pennsylvania, Philadelphia (2010)

Deo, A., Piñango, M.M.: Quantification and context in measure adverbs. In: Ashton, N., Chereches, A., Lutz, D. (eds.) SALT 21, pp. 295–312 (2011)

Dotlačil, J.: Fastidious distributivity. In: Ashton, N., Chereches, A., Lutz, D. (eds.) SALT 21, pp. 313–332 (2011)

Dotlačil, J.: Binominal each as an anaphoric determiner: Compositional analysis. In: Aguilar, A., Chernilovskaya, A., Nouwen, R. (eds.) Sinn und Bedeutung 16, MITWPL (to appear, 2012)

Farkas, D.: Dependent indefinites. In: Corblin, F., Godard, D., Marandin, J.-M. (eds.) Empirical Issues in Formal Syntax and Semantics, pp. 243–268 (1997)

Gil, D.: Distributive numerals. PhD thesis, University of California, Los Angeles (1982)

Krifka, M.: Nominal reference, temporal constitution and quantification in event semantics. In: Bartsch, R., van Benthem, J., van Emde Boas, P. (eds.) Semantics and Contextual Expression, pp. 75–115. Foris, Dordrecht (1989)

Landman, F.: Events and plurality: The Jerusalem lectures. Kluwer, Dordrecht (2000)

Lasersohn, P.: Generalized distributivity operators. Linguistics and Philosophy 21(1), 83–93 (1998)

Link, G.: The logical analysis of plurals and mass terms: A lattice-theoretical approach. In: Link (1998), pp. 11–34 (1983)

Link, G.: Je drei Äpfel - three apples each: Quantification and the German 'je'. In: Link (1998), pp. 117–132 (1986)

Link, G.: Generalized quantifiers and plurals. In: Link (1998), pp. 89–116 (1987)

Link, G.: Algebraic semantics in language and philosophy. CSLI (1998)

Moltmann, F.: Measure adverbials. Linguistics and Philosophy 14, 629–660 (1991)

Moltmann, F.: Parts and wholes in semantics. Oxford University Press (1997)

Oh, S.-R.: Distributivity in an event semantics. In: Hastings, R., Jackson, B., Zvolensky, Z. (eds.) SALT 11, pp. 326–345. CLC Publications, Ithaca (2001)

Schwarzschild, R.: Pluralities. Kluwer, Dordrecht (1996)

Szabolcsi, A.: Quantification. Research Surveys in Linguistics. Cambridge University Press (2010)

Zimmermann, M.: Boys buying two sausages each: On the syntax and semantics of distance-distributivity. PhD thesis, University of Amsterdam (2002)

Zucchi, S., White, M.: Twigs, sequences and the temporal constitution of predicates. Linguistics and Philosophy 24, 187–222 (2001)

Cross-Categorial Donkeys*

Simon Charlow

Department of Linguistics, New York University
simon.charlow@nyu.edu

Abstract. Data from surprising sloppy readings of verb phrase ellipsis constructions argue that ellipsis sites can partially or totally consist of dynamically bound pro-forms. I give an account, integrating Muskens' CDRT with a focus-based theory of ellipsis and deaccenting.

1 Surprisingly Sloppy

An influential theory of verb phrase (VP) ellipsis has it that elided VPs ('ε') and their antecedents ('α') must share an interpretation/LF (Keenan [8]; Sag [17]; Williams [25]). If pronouns have bound and referential uses, this correctly predicts that (1a) is ambiguous between a *strict* reading (Chris thinks Simon is smart: $[\![\alpha]\!]^g, [\![\varepsilon]\!]^g \equiv \lambda x.\, x$ thinks Simon is smart) and a *sloppy* reading (Chris thinks Chris is smart: $[\![\alpha]\!]^g, [\![\varepsilon]\!]^g \equiv \lambda x.\, x$ thinks x is smart). But requiring α and ε to mean the same thing, though appealing, turns out to be too restrictive. Sloppy pronouns/traces are sometimes bound only *outside* of ε (cf. 1b) and sometimes lack a c-commanding antecedent altogether (cf. 1c).

(1) (a) Simon [$_\alpha$ thinks he's smart], and CHRIS$_\text{F}$ does ε too.

(b) [$_{\text{S}_\alpha}$ Bagels$_i$ I [$_\alpha$ like t_i]]
[$_{\text{S}_\varepsilon}$ DONUTS$_{\text{F},j}$ I DON'T$_\text{F}$ [$_\varepsilon$ like t_j]] (Evans [3])

(c) [$_{\text{S}_\alpha}$ the cop who arrested John$_i$ [$_\alpha$ INSULTED$_\text{F}$ him$_i$]]
[$_{\text{S}_\varepsilon}$ the cop who arrested BILL$_{\text{F},j}$ DIDN'T$_\text{F}$ [$_\varepsilon$ insult him$_j$]] (Wescoat [24])

To deal with cases like (1b), Rooth [14] proposes a two-part theory of ellipsis: (i) α and ε must be syntactically identical, *but only up to variable names (and F-marks)*. (ii) A node dominating ε must also CONTRAST with a node dominating α, in the sense of Definition 1. (CONTRAST prevents rank over-generation: "John likes him, and BILL does too" can't mean John likes Steve, and Bill likes Bill.)

Definition 1. CONTRAST(ϕ, ψ) at g iff: $[\![\phi]\!]^g \neq [\![\psi]\!]^g$, and $[\![\psi]\!]^g \in \langle\!\langle \phi \rangle\!\rangle^g$, with $\langle\!\langle \cdot \rangle\!\rangle^g$ the standard Roothian (1985) function into focus sets, as follows:

- Focus values for non-F-marked terminals: $\langle\!\langle \phi \rangle\!\rangle^g = \{[\![\phi]\!]^g\}$
- Focus values for F-marked nodes: $\langle\!\langle \phi_\text{F} \rangle\!\rangle^g = \{x : x_{\tau(\phi)}\}$
- For any non-F-marked branching node ϕ dominating γ and δ, if $[\![\gamma]\!]^g([\![\delta]\!]^g)$ is defined, $\langle\!\langle \phi \rangle\!\rangle^g = \{c(d) : c \in \langle\!\langle \gamma \rangle\!\rangle^g \wedge d \in \langle\!\langle \delta \rangle\!\rangle^g\}$

* Thanks to Mark Baltin, Chris Barker, Daniel Hardt, Irene Heim, Kyle Johnson, Salvador Mascarenhas, Philippe Schlenker, Mike Solomon, and Anna Szabolcsi. I was supported by an NSF Graduate Research Fellowship and NSF grant BCS-0902671.

M. Aloni et al. (Eds.): Amsterdam Colloquium 2011, LNCS 7218, pp. 261–270, 2012.
© Springer-Verlag Berlin Heidelberg 2012

This accounts for (1b): α and ε are structurally identical modulo indices, and since $\langle\!\langle S_\varepsilon \rangle\!\rangle^g = \{I\ f(\text{like } x) : x_e, f_{et,et}\}$, $[\![S_\alpha]\!]^g \in \langle\!\langle S_\varepsilon \rangle\!\rangle^g$. But it doesn't explain (1c). While (1c)'s α and ε are identical up to indices, no choice of any two nodes satisfies CONTRAST: since *him*$_j$ isn't c-commanded by a co-indexed expression, it must—assuming Reinhart's [12] view of the syntax-semantics interface, anyway—be interpreted referentially. This entails that, e.g., $\langle\!\langle S_\varepsilon \rangle\!\rangle^g = \{\text{the cop who arrested } x \text{ insulted Bill} : x_e\}$. $[\![S_\alpha]\!]^g$ is not in this set. CONTRAST fails.

Yet CONTRAST *must* be satisfiable! Here's why: (2) has a reading entailing that, for all x other than Bill, I didn't hear that the cop who arrested x insulted x (Tomioka [22]). Given a standard semantics for *only* (Definition 2), some LF for (2)'s S-node, call it '$\mathcal{L}$', must be such that $\langle\!\langle \mathcal{L} \rangle\!\rangle^g = \{\text{the cop who arrested } x \text{ insulted } x : x_e\}$. But $\mathcal{L}$ must also be available as an LF for (1c)'s S_ε, the elliptical variant of S. Since (1c)'s $[\![S_\alpha]\!] \in \langle\!\langle \mathcal{L} \rangle\!\rangle^g$, CONTRAST must be satisfiable, after all.

(2) I only heard that [$_S$ the cop who arrested BILL$_{F,i}$ insulted him$_i$]!

Definition 2. $[\![\text{only VP}]\!]^g = \lambda x \colon [\![\text{VP}]\!]^g(x).\, \forall Q \in \langle\!\langle \text{VP} \rangle\!\rangle^g.\, Q(x) \rightarrow Q = [\![\text{VP}]\!]^g$

Sloppy Elliptical VPs. Surprising sloppiness is cross-categorial. Sentence (3) can mean that when John has to clean, he doesn't want to clean (Hardt [5]; Schwarz [19]). But treating ellipsis as simple non-pronunciation of LF material yields the LFs in (3b), where α_2 and ε_2 aren't even identical up to indices! Nor is CONTRAST satisfiable; (3b)'s S_ε, for example, is associated with the focus set $\langle\!\langle S_\varepsilon \rangle\!\rangle^g = \{\text{if John has to } P\text{, he doesn't want to clean}) : P_{et}\}$.

(3) (a) If John has to cook, he doesn't WANT to. If he has to CLEAN, he doesn't either.

(b) [$_{S_\alpha}$ if John has to [$_{\alpha_1}$ cook] he doesn't [$_{\alpha_2}$ [WANT to]$_F$ [$_{\varepsilon_1}$ cook]]]
[$_{S_\varepsilon}$ if he has to [$_{\alpha_3}$ CLEAN$_F$] he doesn't [$_{\varepsilon_2}$ want to [$_{\varepsilon_3}$ clean]]]

Yet, like (2), "I only heard that if John has to CLEAN he doesn't want to" has a covarying reading—such that for all P_{et} other than clean, I didn't hear that if John has to P, he doesn't want to P (cf. related claims in Kratzer [9]). Again there's reason to believe that, potentially, (3b)'s $[\![S_\alpha]\!]^g \in \langle\!\langle S_\varepsilon \rangle\!\rangle^g$. But, again, how?

The Scoping Theory. Schwarz [19] argues that (a) *syntactic binding* (i.e. with LF c-command) underlies all sloppy readings, and (b) elided VPs are sometimes (but not always) null variable pro-VPs ('P_n'). Moreover, he suggests, VPs can QR to positions of sentential scope. So, according to Schwarz, LFs like (4) underlie the sloppy reading of (3) (analogous LFs can be mooted for (1c) and (3)).

(4) [$_{S_\alpha}$ cook λ_1 [if John has to t_1 he doesn't [$_\alpha$ [WANT to]$_F$ P_1]]]
[$_{S_\varepsilon}$ CLEAN$_F$ λ_2 [if he has to t_2 he doesn't [$_\varepsilon$ want to P_2]]]

Here, α and ε are identical up to indices and F-marks. Moreover, $[\![S_\alpha]\!] \in \langle\!\langle S_\varepsilon \rangle\!\rangle = \{\text{if John has to } P\text{, he doesn't want to } P : P_{et}\}$. So the sloppy reading is generated.

But there are issues. (i) The proforms in (4) get bound from an $\bar{\text{A}}$-position, something Reinhart [12] deems possible only for *traces*. (ii) The account requires covert movement out of scope/extraction islands—including asymmetric QR out of conjunctions (against the Coordinate Structure Constraint), cf. (5). (iii) To explain (5b), NPs—not subject to overt movement—must QR, again across potentially unbounded distances (Elbourne [2]). (iv) Pro-VPs, if instantiated as variables, lack internal syntax; so it should be impossible to extract out of them, inconsistent with the grammatical sloppy reading of (5c) (Tomioka [23]).

(5) (a) If I'm stressed and John says something awful, I get mad at him.
If I'm stressed and BILL does, I DON'T.

(b) If you lose your visa, you get another.
If you lose your PASSPORT, you DON'T.

(c) I bought everything I was SUPPOSED to and SOLD everything I WASN'T.

But most troubling for the scoping theory is that it fails to generate the *correspondence reading* of constructions like (6)—the one entailing that Sue waves to whoever Mary does (Rooth & Partee [16]; Stone [20]). Just as no amount of QR gives donkey truth conditions for sentences like "if someone$_i$ knocked, she$_i$ left", no amount of QR yields the correspondence reading of sentences like (6).

(6) If Mary waves to John or Bill, then SUE does too.

Hardt's Dynamic Theory. Hardt [5] gives a dynamic account of surprising sloppy readings using Muskens's [11] Compositional DRT (CDRT). I'll postpone the details of CDRT until the following section. For now, it suffices to note that Hardt assigns (3) the LF in (7), with P_n, as before, a phonologically null pro-VP.

(7) if John^1 has to cook^{*,2} he doesn't [want to P_*]^3
if he$_1$ has to clean^{*,4} he doesn't P_3

Superscripted indices correspond to the introduction of a discourse referent (dref); subscripted items denote previously introduced drefs. There is a dedicated index '$*$' which Hardt allows to be overwritten and dubs the "center". Informally, in (7) both α and ε denote the property of wanting to σ, with σ the *current value of the center*. Since *clean*^{*,4} overwrites $*$ with clean (roughly), P_3 evaluates to the property of wanting to clean, and the sloppy reading is derived.

Like the theory we began with, Hardt requires semantic identity of α and ε. This is why destructive update is crucial: for surprising sloppy configurations, it seems like the only way, in a dynamic theory, for α and ε to denote identical properties! But this creates problems. For one, Hardt is forced to posit *two* indices on items U that update the center. The reason: though $*$ may subsequently be overwritten, this shouldn't preclude subsequent "ellipsis" of U. But even with this complication, problems remain. As Sauerland [18] points out, there can be multiple surprising sloppy things of a single type (cf. 8, after Sauerland's ex. 10; NB: the indexing here merely indicates the intended reading). So Hardt's theory actually needs, in principle, an infinity of rewritable indices $*_1, *_2$, etc.

(8) When a woman$_i$ buys a blouse$_j$ we [$_\alpha$ ask that she$_i$ try it$_j$ on]
When a MAN$_k$ buys a SHIRT$_l$ we DON'T [$_\varepsilon$ ask that he$_k$ try it$_l$ on]

Hardt's account also makes heavy use of structure-less pro-VPs (again, this is difficult to square with extraction cases like (5c)) and lacks an account of correspondence readings (though one could be added). But the biggest issue with the theory is that focus is not implicated (remember, Hardt simply requires

α and ε to mean the same thing). There's at least two problems with this: (i) A story about surprising sloppy readings should also have something to say about surprising covarying association-with-focus readings. (ii) Sentences like *John likes his mom, and Bill does too, but Sam doesn't* utterly lack a reading on which Bill likes John's mom, and Sam likes Sam's mom ('strict–sloppy') (Fiengo & May [4]). But Hardt generates that reading straightaway with the LF in (9a). Similarly, *Mary's dad thinks she's smart, and Sue does too* lacks a sloppy reading (Bos [1]). And again, Hardt over-generates with the LF in (9b).

(9) (a) John^{*,1} [likes his$_*$ mom]^2. Bill^3 does P_2 too. Sam^{*,4} doesn't P_2.

(b) Mary^{*,1}'s dad [thinks she$_*$ is smart]^2, and Sue^{*,3} does$_2$ too.

The unavailability of these readings falls out of a CONTRAST–based theory. In the first case, the strict reading of the *Bill*-clause corresponds to a proposition (viz. that Bill likes John's mom) not in the focus set associated with the sloppy reading of the *Sam*-clause—viz. $\{x \text{ likes } x\text{'s mom} : x_e\}$. So if CONTRAST is operative here, the strict–sloppy reading is predicted bad. Likewise for the second example: the proposition that Mary's dad thinks she's smart isn't in $\{x \text{ thinks } x \text{ is smart} : x_e\}$ (cf. also Bos [1]). So CONTRAST rules out that sloppy reading, as well.

Summing Up. Schwarz's account of surprising sloppy readings incorporates focus but relies on an ad hoc variant of QR and fails to explain correspondence readings. Hardt's solution is dynamic and avoids these worries. But his reliance on semantic identity and (thus) destructive update in lieu of a focus-based theory means his account needs an infinity of rewritable indices and struggles with over-generation. Both theories have a paucity of structure at or inside ellipsis sites, making it difficult to see how extraction happens. What we need is a theory that references CONTRAST (or something like it), achieves covariation across focus alternatives despite a lack of syntactic binding, and allows "extraction out of" surprisingly sloppy items. I sketch such a theory in the next section.

2 A Theory

The Fragment. Following Hardt [5], I adopt a higher-order variant of Muskens' [11] Compositional DRT (CDRT). The underlying system is a classical type logic with three primitive types: e, t, and s (for 'states'). DRT boxes are syntactic sugar for λ-terms encoding the usual dynamic relations on states, as follows:

Definition 3. DRT conditions to type logic formulae ('$\rightsquigarrow$' = 'translates as'):
· $R_{\tau_1 \to \ldots \to \tau_n \to t}(\alpha_{s \to \tau_1}) \ldots (\Omega_{s \to \tau_n}) \rightsquigarrow \lambda i.\, R(\alpha(i)) \ldots (\Omega(i))$
· $K \Rightarrow K' \rightsquigarrow \lambda i.\, \forall j.\, K(i)(j) \to \exists k.\, K'(j)(k)$
· $\neg K \rightsquigarrow \lambda i.\, \neg \exists j.\, K(i)(j)$
· $\alpha = \beta \rightsquigarrow \lambda i.\, (\lambda \hat{x}.\, \alpha(\hat{x})(i) = \lambda \hat{x}.\, \beta(\hat{x})(i))$, ('$\hat{x}$' is a possibly empty sequence).

Definition 4. Box sequencing (relational composition):
· $K; K' \rightsquigarrow \lambda ij.\, \exists k.\, K(i)(k) \wedge K'(k)(j)$

Definition 5. Interpretation of boxes:
· $[\nu_1 \ldots \nu_m \mid \kappa_1, \ldots, \kappa_n] \rightsquigarrow \lambda ij.\, i[\nu_1, \ldots, \nu_m]j \wedge \kappa_1(j) \wedge \ldots \wedge \kappa_n(j)$
· $i[\nu_1, \ldots, \nu_m]j$ iff i and j differ at most in the values they assign to $1, \ldots, m$.

Definition 6. Merging lemma (ML): if $\nu'_1, \ldots, \nu'_m$ do not occur free in $\kappa_1, \ldots, \kappa_l$:
· $[\nu_1 \ldots \nu_k \mid \kappa_1, \ldots, \kappa_l]; [\nu'_1 \ldots \nu'_m \mid \kappa'_1, \ldots, \kappa'_n] = [\nu_1 \ldots \nu_k\ \nu'_1 \ldots \nu'_m \mid \kappa_1, \ldots, \kappa_l, \kappa'_1, \ldots, \kappa'_n]$

Definition 7. Truth and entailment in CDRT:
· K is *true* at i ('$\mathsf{True}_i(K)$') iff $\exists j. K(i)(j)$. K is true *simpliciter* iff $\forall i \exists j. K(i)(j)$.
· K entails K' at i ('$K \models_i K'$') iff if K is true at i, K' is true at i.

I add two pieces to Muskens' basic system (call the extension 'CDRT$^+$'). The first is the notion of a *variable dynamic property*—a box parametrized both to the usual arguments and incoming states (Hardt [5], Stone & Hardt [21]). The second is a (externally) dynamic entry for disjunction—relational union, i.e. an instance of generalized disjunction (Rooth & Partee [16]). Variable dynamic properties are an important part of the account of surprising sloppy readings. Dynamic disjunction is crucial for the account of correspondence readings (Stone [20]).

Definition 8. Variable dynamic properties:
· For any ν_n of type $s \to \tau_1 \to \ldots \to \tau_m \to s \to s \to t$, and any (possibly empty) sequence of arguments $\hat{x}$ of length m: $\nu_n(\hat{x}) := \lambda ij. \nu_n(i)(\hat{x})(i)(j)$.

Definition 9. Box disjunction (externally dynamic):
· $K \sqcup K' \rightsquigarrow \lambda ij. K(i)(j) \vee K'(i)(j)$

Table 1. CDRT$^+$ fragment

Expression(s)	Translation	Type
a^n	$\lambda PQ. [u_n \mid]; P(u_n); Q(u_n)$	(et)(et)t
the$_n$	$\lambda PQ. P(u_n); Q(u_n)$	(et)(et)t
every^n	$\lambda PQ. [\mid ([u_n \mid]; P(u_n)) \Rightarrow Q(u_n)]$	(et)(et)t
John^n	$\lambda P. [u_n \mid u_n = \mathsf{john}]; P(u_n)$	(et)t
man	$\lambda v. [\mid \mathsf{man}(v)]$	et
met	$\lambda \mathcal{Q} v. \mathcal{Q}(\lambda v'. [\mid \mathsf{met}(v')(v)])$	((et)t)et
he$_n$, t_n	$\lambda P. P(u_n)$	(et)t
P_n	P_n	s(et)
R_n	R_n	s(eet)
λ_n X	$\lambda u_n. [\![\mathsf{X}]\!]$	e(τ(X))
X$^{\uparrow n}$	$\lambda \hat{x}. [\nu_n \mid \nu_n = [\![\mathsf{X}]\!]]; \nu_n(\hat{x})$	$\tau_\mathtt{t}\tau_\mathtt{t}$
if, when	$\lambda pq. [\mid p \Rightarrow q]$	ttt
and, C$_0$	$\lambda fg\hat{x}. f(\hat{x}); g(\hat{x})$	$\tau_\mathtt{t}\tau_\mathtt{t}\tau_\mathtt{t}$
or	$\lambda fg\hat{x}. f(\hat{x}) \sqcup g(\hat{x})$	$\tau_\mathtt{t}\tau_\mathtt{t}\tau_\mathtt{t}$
want to	$\lambda P. \mathbf{want}(P)$	(et)et
doesn't	$\lambda Pv. [\mid \neg P(v)]$	(et)et

Table 1 gives the lexicon. The notational conventions are as follows: 'e' abbreviates '$s \to e$', and 't' abbreviates '$s \to s \to t$' (the type of boxes). Types associate to the right; $\tau_1\tau_2\tau_3 := \tau_1(\tau_2\tau_3)$. '$\tau$' is used both as a function into types and a variable over types; '$\tau_\mathtt{t}$' stands for any type ending in t. As before, '$\hat{x}$'

stands for a (possibly empty) sequence of arguments. Finally, subscripted terms are variable functions from states, sans serif proper names like 'john' are constant functions from states to individuals, and sans serif predicates like 'man' or 'met' are the familiar functions from individual(s) to truth values.

Much in Table 1 is as in Muskens, but there are several important add-ons (along with a couple minor embellishments like dynamic entries for *John* and *the*). Variable dynamic properties—e.g. P_n and R_n—were discussed above. Additionally, I've defined a family of $^{\uparrow n}$ operators which type-shift constituents into dynamic binders. $^{\uparrow n}$ is essentially a polymorphic dynamicizing identity function: $[\![\mathsf{X}^{\uparrow n}]\!]$ introduces a variable dynamic property ν_n, sets ν_n to $[\![\mathsf{X}]\!]$, and otherwise behaves the same as $[\![\mathsf{X}]\!]$. I've also added a *generalized* entry for dynamic disjunction which disjoins any two expressions so long as they have the same type-ending-in-t (Rooth & Partee [16]). As for the syntax: it's implicit but straightforward (cf. Muskens [11]). For now, I assume with Muskens that object QPs needn't QR (cf. our entry for transitive verbs), although they can. (I'll come back to this when I consider extraction cases.)

(10) shows how the system treats a simple donkey anaphora case. As expected, the type logic translation is true (Definition 9) iff every man who knocked left.

(10) If a^1 man knocked, he$_1$ left.
$[\![\text{a}^1 \text{ man knocked}]\!] = [u_1 \mid\,]; [\,\mid \mathsf{man}(u_1)]; [\,\mid \mathsf{knocked}(u_1)]$
$=_{\mathsf{ML}} [u_1 \mid \mathsf{man}(u_1), \mathsf{knocked}(u_1)]$
$[\![\text{he}_1 \text{ left}]\!] = [\,\mid \mathsf{left}(u_1)]$
$[\![\text{if a}^1 \text{ man knocked he}_1 \text{ left}]\!] = [\,\mid [u_1 \mid \mathsf{man}(u_1), \mathsf{knocked}(u_1)] \Rightarrow [\,\mid \mathsf{left}(u_1)]]$
$\rightsquigarrow \lambda ij.\, i[\,]j \wedge \forall k.\, (j[u_1]k \wedge \mathsf{man}(u_1(k)) \wedge \mathsf{knocked}(u_1(k))) \rightarrow (\exists l.\, k[\,]l \wedge \mathsf{left}(u_1(l)))$

Quiet VPs. I assume with Schwarz [19] that there are two ways for an XP ε to go missing. The first ('DELETION') is the usual Roothian condition: ε has a salient antecedent α with which it's syntactically identical up to indices and F-marks. The second ('BINDING') applies when ε is a phonologically null or deaccented pro-XP (on this theory, English happens to lack non-pronominal pro-XPs).

We've seen ample reason to suppose that something like CONTRAST regulates DELETION. I haven't yet discussed whether unstressed (i.e. silent or deaccented) pro-XPs need to be licensed by CONTRAST, but it's clear they do. As noted previously, CONTRAST explains why "John likes him, and BILL does too" requires coreferential pronouns. But the *exact same facts* pertain to deaccented pro-forms (Rooth [14]). Neither "John likes him, and BILL *likes him* too" nor "Simon thinks he's smart, and CHRIS *thinks he's smart* too" has more interpretations than its elliptical counterpart. This follows if CONTRAST must relate a node dominating the unstressed pro-XP with some other node in the discourse.

So CONTRAST is relevant for both BINDING and DELETION. But it needs a dynamic reformulation: the version of CONTRAST I've been working with pulls the things it wants to compare out of their contexts of evaluation, in effect unbinding any dynamically bound variables (and CONTRAST is, in any case, defined for a system without assignments in the model). Moreover, as defined, CONTRAST(ϕ, ψ) requires *exact* semantic identity between $[\![\psi]\!]$ and some $\kappa \in \langle\!\langle \phi \rangle\!\rangle$. But boxes yield extremely fine-grained denotations; two truth-conditionally equivalent boxes K

and K' can nevertheless differ in context change potential (cf. also Hardt's [5] fn. 12). But (11) indicates that CONTRAST should care only about truth-conditional import, not context change potential. So requiring semantic identity is too strict.

Together, these facts suggest a reformulation of CONTRAST as a compositionally integrated presuppositional operator $\sim_n$ (after Rooth's [15] $\sim$) which is sensitive only to truth conditions and which *itself* gets dynamically bound.

(11) John met a^1 man. Then BILL *met a^2 man.* / Then BILL did $[_\varepsilon$ meet a^2 man$]$.

Definition 10. Local, dynamic reformulation of CONTRAST:
· $[\![\mathsf{X} \sim_n]\!] = \lambda\hat{x}ij : [\exists a \in \langle\!\langle \mathsf{X} \rangle\!\rangle .\, v_n \models_i a] .\, [\![\alpha]\!](\hat{x})(i)(j)$, where $f_{\tau_t} \models_i g_{\tau_t}$ iff $\forall \hat{x}.\, f(\hat{x}) \models_i g(\hat{x})$

Somewhat less formally, $[\![\mathsf{X} \sim_n]\!]$ is only defined for incoming states i such that for some $a \in \langle\!\langle \mathsf{X} \rangle\!\rangle$, $v_n \models_i a$ (this presumes an obvious redefinition of $\langle\!\langle \cdot \rangle\!\rangle^g$ to $\langle\!\langle \cdot \rangle\!\rangle$, which I leave implicit).[1,2] Assuming definedness, $[\![\mathsf{X} \sim_n]\!]$ does not differ from $[\![\mathsf{X}]\!]$. Note that $\sim_n$ *has to be* bound (free variables are prohibited in (C)DRT) and that Definition 10 leads us to expect that it may even be *donkey* bound. We'll shortly see that correspondence readings offer instances of precisely that.

Let's see how the definition works in a simple case of pronominal deaccenting, (12a). (12b) is defined for i such that if S_2 is true at i, then some alternative in $\langle\!\langle \mathrm{S}_{\delta'} \rangle\!\rangle$ is true at i. But box sequencing (Definition 4) guarantees that the only states fed to (12b) are those output by (12c)—so they will all necessarily make u_1 a man who entered and S_2 the box $[u_1 \mid \mathsf{man}(u_1), \mathsf{entered}(u_1)]$.[3] Since presumably $\langle\!\langle \mathrm{S}_{\delta'} \rangle\!\rangle = \{[\,\mid P(u_1)] : P_{et}\}$ (or something equivalent), $[\,\mid \mathsf{entered}(u_1)] \in \langle\!\langle \mathrm{S}_{\delta'} \rangle\!\rangle$. And since all the states output by (12c) already assign u_1 to a man who entered, they will necessarily assign u_1 to someone who entered. So at all relevant i, it's impossible for $\mathsf{True}_i(S_2)$ to be true and $\mathsf{True}_i([\,\mid \mathsf{entered}(u_1)])$ to be false. The presupposition is met.

(12) (a) $[_{\mathrm{S}_\alpha}$ a^1 man entered$]^{\uparrow 2}$; $[_{\mathrm{S}_\delta}$ $[_{\mathrm{S}_{\delta'}}$ he$_1$ SAT$_\mathrm{F}]$ $\sim_2]$

(b) $[\![\mathrm{S}_\delta]\!] = \lambda i : [\exists a \in \langle\!\langle \mathrm{S}_{\delta'} \rangle\!\rangle .\, S_2 \models_i a] .\, [\![\mathrm{S}_{\delta'}]\!](i)$

(c) $[\![$[a^1 man entered$]^{\uparrow 2}]\!]$
$\rightsquigarrow \lambda ij.\, i[u_1, S_2]j \wedge \mathsf{man}(u_1(j)) \wedge \mathsf{entered}(u_1(j)) \wedge S_2(j) = [\![$a^1 man entered$]\!]$

Basic Cases. I'm now ready to give LFs for (1c) and (3). Temporarily assuming definedness, (14a) derives $[\![\mathrm{S}_{\alpha_{13b}}]\!]$, and (14b) gives the meaning of $[\![\mathrm{S}_{\varepsilon_{13b}}]\!]$.[4]

[1] Recall that for variable dynamic properties, $\nu_n(\hat{x}) := \lambda ij.\, \nu_n(i)(\hat{x})(i)(j)$.

[2] A simplification. Implicational bridging requires *contextual* entailment (Rooth [14]).

[3] More accurately, at any i, $S_2(i) = \lambda j.\, i[u_1, S_2]j \wedge \mathsf{man}(u_1(j)) \wedge \mathsf{entered}(u_1(j))$.

[4] Two notes: (i) I'm treating *has to* as vacuous. The reason: in DRT, drefs introduced under modals like *has to* are typically inaccessible outside the modal's box. But constant-y things (names, specific indefinites, VP-type meanings, etc.) should be accessible *no matter where they're introduced* (cf. Hardt's [5] fn. 13). It's possible to give a semantic definition of accessibility that handles our cases without outrageous meanings for traditionally externally static items (Stone & Hardt [21]), but this requires a fully intensional semantics—way beyond what I can discuss here. (ii) I'm assuming $\sim_n$ operators can't be nested; hence the pro-form inside ε is never responsible for any presuppositions beyond those due to ellipsis of ε. Though I'm not sure how defensible this is, I see no other way to make the account work.

(13) (a) $[_{S_\alpha}$ the$_0$ cop who [arrested John^1]$^{\uparrow 2}$ $[[_\alpha$ INSULTED$_F$ him$_1$] $\sim_2]]^{\uparrow 4}$
$[[_{S_\varepsilon}$ the$_3$ cop who arrested (BILL^5)$_F$ DIDN'T$_F$ $[_\varepsilon$ insult him$_5$]] $\sim_4]$

(b) $[_{S_\alpha}$ if John^1 [has to cook$^{\uparrow 3}$]$^{\uparrow 5}$ he$_1$ doesn't $[[_\alpha$ WANT$_F$ to P_3] $\sim_5]]^{\uparrow 6}$
$[[_{S_\varepsilon}$ if he$_1$ has to (CLEAN$_F$)$^{\uparrow 4}$ he$_1$ doesn't $[_\varepsilon$ want to P_4]] $\sim_6]$

(14) (a) $[\![\text{John}^1 \text{ has-to cook}^{\uparrow 3}]\!] = [u_1\, P_3 \mid u_1 = \mathsf{john}, P_3 = [\![\text{cook}]\!]; P_3(u_1)$
$[\![\text{he}_1 \text{ doesn't want to } P_3]\!] = [\,\mid \neg P_3(u_1)]$
$[\![\text{if John}^{\uparrow 1} \text{ has-to cook}^{\uparrow 3} \text{ he}_1 \text{ doesn't [WANT to]}_F\, P_3]\!]$
$= [\,\mid ([u_1\, P_3 \mid u_1 = \mathsf{john}, P_3 = [\![\text{cook}]\!]; P_3(u_1)) \Rightarrow [\,\mid \neg \textbf{want}(P_3)(u_1)]]$

(b) $[\![\text{if he}_1 \text{ has-to (CLEAN}_F)^{\uparrow 4} \text{ he}_1 \text{ doesn't want to } P_4]\!]$
$= [\,\mid ([P_4 \mid P_4 = [\![\text{clean}]\!]; P_4(u_1)) \Rightarrow [\,\mid \neg \textbf{want}(P_4)(u_1)]]$

I omit the translations to type logic formulae here, but it's relatively straightforward to check that the resulting boxes are true iff (a) if John (has to) cook, he doesn't want to cook, and (b) if he (has to) clean, he doesn't want to clean.

So the meanings are correct. Now, I show that the conditions on ellipsis are satisfied. First: every silent XP is either a bound pro-form or has an identical-up-to-indices antecedent, so each is an instance of DELETION or BINDING. Next: the presuppositions introduced by $\sim_5$ and $\sim_6$ are both met. The case of $\sim_5$ is like (12): in both, a node dominating a dynamically bound pro-form contrasts with a node dominating the binder. Now $\langle\!\langle \alpha_{13b} \rangle\!\rangle = \{f(P_3) : f_{(et)et}\}$. I assume **has-to**$(P_3) \in \langle\!\langle \alpha_{13b} \rangle\!\rangle$. Since $\sim_5$ is bound by a constituent denoting **has-to**$([\![\text{cook}]\!])$, and the definition of $\Rightarrow$ guarantees that in every state fed to the alternative **has-to**(P_3), P_3 denotes $[\![\text{cook}]\!]$, the presupposition must be satisfied. As for $\sim_6$, I assume $[\![\text{cook}]\!] \in \langle\!\langle \text{CLEAN}_F \rangle\!\rangle$, from which it follows that $[\![\text{cook}^{\uparrow 4}]\!] \in \langle\!\langle (\text{CLEAN}_F)^{\uparrow 4} \rangle\!\rangle$, from which it follows that $[\,\mid ([P_4 \mid P_4 = [\![\text{cook}]\!]; P_4(u_1)) \Rightarrow [\,\mid \neg \textbf{want}(P_4)(u_1)]] \in \langle\!\langle S_{\varepsilon_{13b}} \rangle\!\rangle$. Since (i) $\sim_6$ is bound to $[\![S_{\alpha_{13b}}]\!]$, (ii) sequencing entails that the u_1's free in S_ε always evaluate to john, and (iii) the two boxes in (15) have identical truth conditions, the presuppositions of $\sim_6$ must be satisfied at all possible incoming i.

(15) $[\,\mid ([u_1\, P_3 \mid u_1 = \mathsf{john}, P_3 = [\![\text{cook}]\!]; P_3(u_1)) \Rightarrow [\,\mid \neg \textbf{want}(P_3)(u_1)]]$
$[\,\mid ([P_4 \mid P_4 = [\![\text{cook}]\!]; P_4(\mathsf{john})) \Rightarrow [\,\mid \neg \textbf{want}(P_4)(\mathsf{john})]]$

Mutatis mutandis, deriving truth conditions and checking definedness for (13a) works in an exactly analogous fashion.

Extraction. (16) gives LFs generating (5c)'s sloppy reading (I've split it into two sentences). Note that I've QR'ed the object. This (standard) move is forced by the (standard) assumption that the only possible antecedents for DELETION/BINDING are XPs: since the object starts in VP, the only way to generate ACD as DELETION of or BINDING by an XP is to scope the object out of VP (Sag [17]).

(16) $[_{S_\alpha}$ I [[every^{0}thing λ_2 I was $[_\alpha$ SUPPOSED$_F$ to R_3 t_2]] [λ_1 bought t_1]$^{\uparrow 3}$]]$^{\uparrow 9}$
$[[_{S_\varepsilon}$ I [[every^{8}thing λ_5 I WASN'T$_F$ $[_\varepsilon$ supposed to R_4 t_5]] [λ_6 [SOLD$_F$ t_6]]$^{\uparrow 4}$]] $\sim_9$]

Object QR is to a position under the subject, rather than to S. This is independently motivated. Merchant [10] notes NPIs can participate in ACD, e.g. "I didn't read a damn thing you asked me to". Since the NPI needs to stay in the scope of VP-negation, ACD QR must at least potentially target VP rather than S.

How are the LFs in (16) interpreted? First, we need a way to quantify into VP. I'll adopt Hendriks' [7] Argument Raising type-shifter:

Definition 11. $\mathsf{AR}(f) := \lambda \mathcal{Q}\hat{v}.\, \mathcal{Q}(\lambda u.\, [\![\mathsf{X}]\!](u)(\hat{v}))$

$\mathsf{AR}([\![\lambda_1 \text{ bought } t_1]\!])$ is $\lambda \mathcal{Q}v.\, \mathcal{Q}(\lambda u_1.\, [\,|\, \mathsf{bought}(u_1)(v)])$. If $\uparrow^3$ applies next, we have $\lambda \mathcal{Q}v.\, [R_3 \mid R_3 = \lambda \mathcal{Q}v.\, \mathcal{Q}(\lambda u_1.\, [\,|\, \mathsf{bought}(u_1)(v)])]; R_3(\mathcal{Q})(v)$. Since R_3 is accessible inside $\mathcal{Q}$, the R_3 subsequent to "supposed to" is bound. The same goes, *mutatis mutandis*, for S_ε. So adequate interpretations can be generated. As for licensing: all silent material is either an instance of DELETION or BINDING. And $\sim_9$'s presupposition is satisfied, which the reader is invited to check.

But, wait. Doesn't CONTRAST also have to license the silent pro-form R_3? Absolutely, which brings me to a slightly uncomfortable matter: since QR targets S_α's *VP*, it's hard to see which two nodes in S_α could possibly be related by $\uparrow^n/\sim_n$: the smallest constituent in which t_2 is bound already contains the subject pronoun "I"! Note that, while troubling, this doesn't seem like an issue for my proposal *per se*: QR to VP is what's creating the difficulty here, but Merchant's NPI ACD case shows that non-sentential QR is necessary.[5]

Correspondence Readings, Donkey $\sim_n$ Binding. Here, finally, are the LF and interpretation of (6) (partially following Stone's [20] informal discussion):

(17) (a) If [[John or Bill] λ_4 [Mary [meets t_4]$^{\uparrow 3}$]$^{\uparrow 6}$] [[SUE$_\mathrm{F}$ does P_3 (too)] $\sim_6$]
(b) $[\![\text{Mary [meets } t_4]^{\uparrow 3}]\!] = [P_3 \mid P_3 = \lambda v.[\,|\, \mathsf{meets}(u_4)(v)]]; P_3(\mathsf{mary})$
$[\![[\text{Mary [meets } t_4]^{\uparrow 3}]^{\uparrow 6}]\!] = [S_6 \mid S_6 = [\![\text{Mary [meets } t_4]^{\uparrow 3}]\!]]; S_6$
$[\![\lambda_4 \text{ [Mary [meets } t_4]^{\uparrow 3}]^{\uparrow 6}]\!] = \lambda u_4.[S_6 \mid S_6 = [\![\text{Mary [meets } t_4]^{\uparrow 3}]\!]]; S_6$
$[\![\text{John or Bill}]\!] = \lambda P.\, P(\mathsf{john}) \sqcup P(\mathsf{bill})$
$[\![\text{[John or Bill] } \lambda_4 \text{ [Mary [meets } t_4]^{\uparrow 3}]^{\uparrow 6}]\!]$
$= ([S_6 \mid S_6 = [P_3 \mid P_3 = \lambda v.[\,|\, \mathsf{meets}(\mathsf{john})(v)]]; P_3(\mathsf{mary})]; S_6) \sqcup$
$([S_6 \mid S_6 = [P_3 \mid P_3 = \lambda v.[\,|\, \mathsf{meets}(\mathsf{bill})(v)]]; P_3(\mathsf{mary})]; S_6)$

So the states output by the antecedent fix P_3 either to the property of meeting John or to the property of meeting Bill (this happens in slightly convoluted fashion since the introduction of P_3 is actually tucked inside the dynamic variable S_6). This guarantees truth conditions such that if Mary meets John, Sue meets John, and if Mary meets Bill, Sue meets Bill, exactly the meaning we're after.

Lastly, I show that (17a) is defined. The interesting bit here, which I alluded to previously, is is that $\sim_6$ acts like a donkey pronoun! Specifically, the states output by (17a)'s antecedent fix S_6 to one of two boxes: either $[\![\text{Mary [meets John]}^{\uparrow 3}]\!]$ or $[\![\text{Mary [meets Bill]}^{\uparrow 3}]\!]$. Now if $[\![\text{Mary}]\!] \in \langle\!\langle \text{SUE}_\mathrm{F} \rangle\!\rangle$, then $[\,|\, P_3(\mathsf{mary})] \in \langle\!\langle \text{SUE}_\mathrm{F} \text{ does } P_3 \rangle\!\rangle$. Since at each state i output by the antecedent, $\mathsf{True}_i(S_6) \rightarrow \mathsf{True}_i([\,|\, P_3(\mathsf{mary})])$, the presupposition is satisfied.

3 Conclusion

I've argued for a theory of cross-categorial surprising sloppy readings in which ellipsis sites may consist either in part or in whole of dynamically bound pro-forms. The account has three parts: the conditions regulating the distribution of elliptical

[5] There is very little work on just how CONTRAST works in ACD cases. One promising way forward might be Heim's [6] *formulas*-based Roothian account of ACD. Unfortunately, I have to postpone a real investigation.

sites and pro-XPs, a dynamicizing type-shifter $^{\uparrow n}$, and a presuppositional, dynamically bound, alternative-sensitive CONTRAST operator $\sim_n$. Dynamic binding is a more natural option for these cases than QR, and integration with a theory of focus avoids over-generation. Extraction cases are within reach (though it is not in the end clear how ACD is licensed), while dynamic disjunction generates correspondence readings and predicts the possibility of donkey-binding $\sim_n$.

References

[1] Bos, J.: Focusing Particles & Ellipsis Resolution. Verbmobil Report 61, Universitat des Saarlandes (1994)
[2] Elbourne, P.: E-type anaphora as NP deletion. Natural Language Semantics 9(3), 241–288 (2001)
[3] Evans, F.: Binding into Anaphoric Verb Phrases. In: Proceedings of ESCOL (1988)
[4] Fiengo, R., May, R.: Indices and identity. MIT Press, Cambridge (1994)
[5] Hardt, D.: Dynamic interpretation of verb phrase ellipsis. Linguistics and Philosophy 22(2), 185–219 (1999)
[6] Heim, I.: Predicates or formulas? Evidence from ellipsis. In: Lawson, A., Cho, E. (eds.) Proceedings of SALT 7, pp. 19–221 (1997)
[7] Hendriks, H.: Studied Flexibility. ILLC Dissertation Series, Amsterdam (1993)
[8] Keenan, E.: Names, quantifiers, and the sloppy identity problem. Papers in Linguistics 4(2), 211–232 (1971)
[9] Kratzer, A.: The Representation of Focus. In: von Stechow, A., Wunderlich, D. (eds.) Semantics: An International Handbook of Contemporary Research, pp. 825–832. de Gruyter, Berlin (1991)
[10] Merchant, J.: Antecedent Contained Deletion in Negative Polarity Items. Syntax 3, 144–150 (2000)
[11] Muskens, R.: Combining Montague Semantics and Discourse Representation. Linguistics and Philosophy 19, 143–186 (1996)
[12] Reinhart, T.: Anaphora and Semantic Interpretation. U. of Chicago Press (1983)
[13] Rooth, M.: Association with Focus. UMass, Amherst dissertation (1985)
[14] Rooth, M.: Ellipsis redundancy and reduction redundancy. In: Berman, S., Hestvik, A. (eds.) Proceedings of the Stuttgart Ellipsis Workshop, Stuttgart (1992a)
[15] Rooth, M.: A Theory of Focus Interpretation. Natural Language Semantics 1, 75–116 (1992b)
[16] Rooth, M., Partee, B.: Conjunction, type ambiguity, and wide scope "or". In: Flickinger, D., et al. (eds.) Proceedings of the First West Coast Conference on Formal Linguistics, pp. 353–362. Stanford University (1982)
[17] Sag, I.: Deletion and Logical Form. MIT dissertation (1976)
[18] Sauerland, U.: Copying vs. structure sharing: a semantic argument. In: van Craenenbroeck, J. (ed.) Linguistic Variation Yearbook, vol. 7, pp. 27–51. John Benjamins Publishing Company (2007)
[19] Schwarz, B.: Topics in Ellipsis. UMass, Amherst dissertation (2000)
[20] Stone, M.: Or and anaphora. In: Proceedings of SALT 2, pp. 367–385 (1992)
[21] Stone, M., Hardt, D.: Dynamic discourse referents for tense and modals. In: Bunt, H. (ed.) Computational Semantics, pp. 287–299. Kluwer (1999)
[22] Tomioka, S.: A sloppy identity puzzle. Natural Language Semantics 7(2), 217–241 (1999)
[23] Tomioka, S.: A step-by-step guide to ellipsis resolution. In: Johnson, K. (ed.) Topics in Ellipsis, pp. 210–228. Cambridge University Press (2008)
[24] Wescoat, M.: Sloppy Readings with Embedded Antecedents. Stanford ms. (1989)
[25] Williams, E.: Discourse and logical form. Linguistic Inquiry 8, 101–139 (1977)

On *Wh*-Exclamatives and Noteworthiness*

Anna Chernilovskaya and Rick Nouwen

Utrechts Instituut voor Linguïstiek

1 Introduction

We explore a new approach to the semantics of *wh*-exclamatives, like (1).

(1) What a beautiful song John wrote!

We will aim for two things: (i) extend the empirical focus beyond English *what*- and *how*-exclamatives, to include exclamatives common in other languages that are based on other *wh*-words; (ii) counter the common assumption that exclamative semantics needs to involve some kind of scalar mechanism.

Before we motivate and present our analysis, a word of caution is in order. To simplify matters for this short paper, we will be discussing the semantics of exclamatives like (1) in terms of truth-conditions. Such a move blatantly ignores the fact that an utterance of (1) counts as a speech act that comes with its own intricate and interesting properties, properties which will be quite different from those of an assertion. For the purpose of this short paper, however, we will remain agnostic as to what role the truth-conditions play in the pragma-semantics of exclamatives. [See Rett, 2012; Zanuttini and Portner, 2003, for extensive discussion.]

2 Background: Scalarity in Exclamative Semantics

Rett 2012 proposes that the semantics of *wh*-exclamatives involves degree intensification [Cf. Castroviejo, 2006; Rett, 2008a,b, for related approaches]. On her approach, the logical form of (1) will specify the (derived) degree predicate in (2). (See Rett's paper for details of the derivation.)

(2) $\lambda d.\exists x[song(x) \wedge wrote(j, x) \wedge beautiful(x, d)]$

According to Rett, an utterance of a *wh*-exclamative involves the speaker expressing that it is noteworthy that the degree property corresponding to the

* Both authors gratefully acknowledge the NWO-VIDI grant 'Degrees under Discussion' that made this work possible. For stimulating discussion and helpful comments we would furthermore like to thank audiences in Amsterdam, Chicago, Montreal and Stanford, where earlier versions were presented. In particular, we would like to thank: Alan Bale, Cleo Condoravdi, Itamar Francez, Anastasia Gianakidou, Tim Grinsell, Chris Kennedy, Sven Lauer, Bernhard Schwarz, Elizabeth Smith, Malte Willer. Any errors or shortcomings are the sole responsibility of the authors.

M. Aloni et al. (Eds.): Amsterdam Colloquium 2011, LNCS 7218, pp. 271–280, 2012.
© Springer-Verlag Berlin Heidelberg 2012

exclamative is instantiated by some value that exceeds the relevant contextual standard (in the case of (1) the standard of beauty w.r.t. songs).

If, as in (3-a), a wh-exclamative lacks an overt gradable adjective, a measurement operator $\lambda d.\lambda x.\mu_\alpha(x) = d$ is inserted. This operator basically plays the role of a silent adjective, where the relevant measurement dimension α is determined contextually. Via this operator, (3-a), too, ends up expressing a degree property, namely how α (how *beautiful*, how *weird*, how *complex*, etc.) John's song was. This is the predicate in (3-b). An utterance in (3-a), then, means that the speaker expresses that it is noteworthy that the degree of α of John's song exceeds the standard.

(3) a. What a song John wrote!
 b. $\lambda d.\exists x[song(x) \wedge wrote(j,x) \wedge \mu_\alpha(x) = d]$

Rett's approach is to treat a *wh*-exclamative as a degree phenomenon and she argues that it could really be nothing else. In particular, she argues that the noteworthiness evaluation that is part of the *wh*-exclamation is necessarily directed at predicate of *degrees*. She does so on the basis of scenarios like (4).

(4) *Imagine that by some strange coincidence someone repeatedly picks out the same two cards (say, the* $3\diamondsuit$ *and the* $6\heartsuit$*) from a (repeatedly reshuffled) pack of cards. Mary has seen this happen and now witnesses this person pick* $3\diamondsuit$ *and* $6\heartsuit$ *yet again.*

Rett observes that it is now infelicitous for Mary to utter (5):[1]

(5) What cards he picked!

What is essential to this scenario is that no matter how noteworthy the events of picking these cards were, there is nothing particularly special about the cards that were picked. That is, there is no α such that $\mu_\alpha(3\diamondsuit \oplus 6\heartsuit)$ returns a particularly high degree. In Rett's approach, this then explains why (5) is infelicitous. Rett concludes that *wh*-exclamatives are subject to a *degree restriction*: they always express that the degree to which something holds is deemed noteworthy by the speaker.

It is worth remarking that it is not necessary to interpret Rett's observation concerning (4) and (5) as saying something about the involvement of degrees in

[1] We have encountered several native speakers of English (including an anonymous reviewer) who disagree with this judgment or who at least can come up with similar scenarios in which an exclamative *is* felicitous. Below, we will show that in languages like Dutch some exclamatives are generally felicitous in a scenario like (4). We would guess that the speakers who disagree with Rett's observation speak a dialect of English that is in this respect close to Dutch. The fact that such dialects exist is further supported by the fact that we have found speakers of English that have *who*-exclamatives in their language, while Rett assumes that such exclamatives do not exist in English since English *who* is incapable of ranging over degrees. Despite such exceptions, we believe Rett's observations are important and should be taken seriously, given the clear support they receive from the vast majority of speakers.

the semantics of *wh*-exclamatives, but that instead one could argue that it just shows that *wh*-exclamatives always involve some kind of *scalar* mechanism. For instance, we believe that the influential scalar approach of Zanuttini and Portner [2003], although compositionally less specific, is in principle suitable for dealing with examples like (5) equally well (pace Rett's own assessment of that work). In the interest of space, we refrain from a detailed discussion and comparison.

3 The Scope of Noteworthiness

Rett claims that it should be impossible to insert silent measurement operators using dimensions like *noteworthiness, unexpectedness* or *surprise.* Rett suspects that such a choice would wrongly predict that *What cards he picked!* is felicitous in the card-picking scenario of (4). This does not seem entirely accurate to us. Unlike (say) $A\heartsuit \oplus A\diamondsuit \oplus A\clubsuit \oplus A\spadesuit$, it seems to us that the pair $3\diamondsuit \oplus 6\heartsuit$ lacks any noteworthy features. Also, what is surprising is not the pair of cards itself, but rather *that they were picked.* Moreover, Rett's restriction on silent measurement operators is surprising given the possibility of having overt adjectives associated to those same dimensions in *wh*-exclamatives, as in (6).

(6) What a surprising turn of events!

This discussion points in the direction of our main argument: cases like (4)/(5) are not evidence of a degree (or scalarity) restriction, but rather point out that the noteworthiness evaluation is always directed at the referent of the *wh*-phrase. Call this the *locality restriction*: readings in which the μ applies to a structure that properly contains the *wh*-phrase, as in (7), are unavailable. Such degree predicates would incorrectly predict (5) to be felicitous in (4).

(7) $\lambda d.\exists x[\mu_{noteworthy}(^\wedge cards(x) \wedge picked(h,x)) = d]$

4 Beyond English *What*-Exclamatives

Whilst English only has *what*- and *how*-exclamatives, languages like Dutch, German and Russian allow for exclamatives based on other *wh*-words too. The example in (8) presents a Dutch *who*-exclamative, and (9) a Dutch *which*-exclamative. There are similar examples in other languages, but not in English.

(8) Wie ik net gezien heb!
who I just seen have

(9) Welke vrouw ik net gezien heb!
which woman I just seen have

The meaning of these examples shows a contrast with English *what*-exclamatives. It is infelicitous to utter (8) or (9) in response to seeing a woman with some noteworthy feature, e.g. an exceptionally tall woman. These exclamatives *can* be used when the very fact that the speaker just saw the woman in question is unexpected; for instance, as a reaction to seeing Mary, an in all senses absolutely normal woman, of whom everybody thought that she had left the country.

The approaches discussed in the previous study will have a hard time accounting for these examples.[2] This is first of all because these exclamatives lack the readings that these approaches account for in the case of *what*-exclamatives. Moreover, the available readings are close to the readings that were meant to be excluded as possible interpretations of *what*-exclamatives. For instance, (10) is felicitous in the card-picking scenario in (4).

(10) Welke kaarten hij toen (weer) trok!
which cards he then (again) pulled

As we discussed above, in the scenario there is no property such that $3\diamondsuit \oplus 6\heartsuit$ has this property to a particularly high degree. What appears to be needed to account for the fact that (10) is felicitous in the given scenario is to break with the locality restriction and assume that the exclamative expresses the degree predicate in (7). This is not unproblematic, for we need the locality restriction to avoid predicting the Dutch *what*-exclamative in (11) to be felicitous in scenarios like (4), while in fact it behaves in complete parallel to its English counterpart.

(11) Wat een kaarten hij toen (weer) trok!
what a cards he then (again) pulled

5 Proposal

We propose that exclamatives directly express a noteworthiness evaluation and that exclamatives are not a degree phenomenon in Rett's sense. We furthermore categorise all *wh*-exclamatives in two distinct classes:

Type 1: expressing the noteworthiness of a referent of the wh-word
Type 2: expressing the noteworthiness of the proposition referenced in the exclamative

In addition to this, each *wh*-word introducing a *wh*-exclamative is specified for what kind(s) of noteworthiness it can mark. For example, both English and Dutch *what a*-exclamatives are type 1. Dutch *who* and *which* introduce wh-exclamatives of type 2.

As (12)-(15) show, type 1 exclamatives can be *reduced*, whilst type 2 exclamatives can not. This would be expected since the latter but not the former are dependent on a larger propositional structure. In Dutch, moreover, word order reflects the type 1/2 distinction. Dutch has SVO for main clauses, with V2, and SOV for embedded clauses. As (16)-(19) show, type 1 exclamatives may be either V2 or verb-final, whilst type 2 exclamatives are exclusively verb-final.

[2] Rett assumes that wh-exclamatives can only be formed with wh-words ranging over degrees, i.e. *who*-exclamatives are predicted to be ungrammatical. This is a fine prediction for English, but obviously not for other languages. Even if this restriction is lifted for such languages, the desired readings for examples like (8) and (9) are not derived.

(12) What a (beautiful) book!

(13) Wat een (mooi) boek!
what a (beautiful) book

(14) *Wie!
who

(15) *Welk mooi boek!
which beautiful book!

(16) Wat maakte Jan een herrie!
what made Jan a racket
'What a racket Jan made!'

(17) Wie ik net zag!
Who I just saw
'You wouldn't believe who I just saw'

(18) Wat Jan een herrie maakte!
what Jan a racket made
'What a racket Jan made!'

(19) *Wie zag ik net!
Who saw I just

5.1 Wh-Exclamatives of Type 1

According to our proposal, the key ingredient to exclamative meaning is a noteworthiness evaluation. The meaning that we propose for (20-a) is in (20-b) (though we will refine this shortly):[3]

(20) a. What a song John wrote!
b. $\exists x[song(x) \wedge wrote(j, x) \wedge noteworthy(x)]$

How does this work? First, let us note that it is probably impossible to give a maximally precise semantic definition of the predicate *noteworthy*,[4] just like it is impossible to give maximally precise truth-conditions for, say, (21-a) or (21-b).

(21) a. The achievements of Sir Alex Ferguson are noteworthy.
b. It's noteworthy that Sir Alex Ferguson is still the manager of ManU.

Despite this semantic ineffability, we have clear intuitions on what is noteworthy. Take *blackberry, chicken liver and cauliflower cake*, which we consider almost indisputably noteworthy. Similarly, *the font used in this article* most probably counts as a clear case of something non-noteworthy. Here is our own intuition about the rough concept behind the label *noteworthy*: *an entity is* **noteworthy** *iff its intrinsic characteristics (i.e. those characteristics that are independent of the factual situation)* **stand out considerably** *with respect to a comparison class of entities.*

This definition should be taken *cum grano salis*. We believe that the concept of noteworthiness is at the core of what exclamatives express, but we also believe

[3] An anonymous reviewer wonders how the exclamative picks up the individual that is being evaluated, given the general assumption that wh-phrases are not referential. Here we agree with Rett [2012] that wh-phrases in (type 1) exclamatives are best analysed as free relatives.

[4] This is possibly a weakness of the approach, since this makes it difficult for opponents to our theory to show that something is *not* noteworthy in a situation in which an exclamative is used felicitously.

that, as with any model-theoretic semantic analysis, a full characterisation of the concepts involved is outside the scope of our theory.[5]

Given the characterisation above, (20-b) is true if and only if the song written by John stands out considerably in some sense, compared to other songs. This can be because of many things (it being particularly good, or particularly weird, or particularly beautiful, etc.), which is as it should be.

If we apply the notion of noteworthy to the card-trick scenario, then it should be clear that the combination $3\diamondsuit \oplus 6\heartsuit$ is not noteworthy. This makes type 1 exclamatives infelicitous in the card-picking scenario. (Later, we will use the fact that the proposition that these cards were picked several times in a row *is* noteworthy.)

Of course, *noteworthy* is a gradable predicate. We will take this into account by adopting a vague predicate approach to degree predicates (but nothing hinges on this). So we will write $noteworthy(x)(c)$ to express that among the members of comparison class c, x would count as noteworthy. For (20-a):

(22) $\exists x[song(x) \wedge wrote(j, x) \wedge noteworthy(x)(\lambda y.song(y))]$

In case there is a gradable predicate present in the *wh*-phrase, this predicate will become part of the comparison class, as in (23) (where c is the contextually determined comparison class for *beautiful*).

(23) a. What a beautiful song John wrote!
b. $\exists x[song(x) \wedge beautiful(x)(c) \wedge wrote(j, x) \wedge noteworthy(x)(\lambda y.beautiful(y)(c) \wedge song(y))]$

Of course, the most salient reason for John's beautiful song to be a noteworthy beautiful song is that it was particularly beautiful, which accounts for the most natural reading of (23-a). However, there also exists a reading in which this *wh*-exclamative does not express how beautiful the song was, but that the beautiful song in question was noteworthy for some other reason (e.g. because it was very weird, unusually structured, or abnormally long):

(24) Q: Did John write a beautiful song?
A: Yes he did, and whàt a beautiful song he wrote! It contained 36 verses!
A′: Yes he did, and whàt a beautiful song he wrote! It has just one chord!

5.2 Wh-Exclamatives of Type 2

According to our proposal, Dutch *who* and *which* exclamatives differ from examples like (23-a) in that they do not involve the noteworthiness of the referent

[5] Note further that the given 'definition' of noteworthiness applies to entities only, while below we apply *noteworthy* to propositions too. We refrain from spelling out a similar definition for the propositional use, but do point out that the flexible application of noteworthiness is reflected in the distribution of the natural language adjectives like *noteworthy*, as is shown by (21).

corresponding to the *wh*-phrase, but rather the noteworthiness of a proposition. For instance, for Dutch examples of the form *Who/which woman I just saw!* the interpretation in (25-a) is unavailable. Instead, the predicted reading is in (25-b), where the noteworthiness evaluation takes propositional scope.[6] (We will leave the comparison class implicit. It suffices to assume it contains the speaker's experience.)

(25) a. $\exists x[saw(I,x) \wedge noteworthy(x)]$
b. $\exists x[noteworthy(^\wedge saw(I,x))]$

We will assume *noteworthy* to be factive. The form in (25-b) is true if and only if for some person, the true proposition that I saw this person is noteworthy.[7]

6 Discussion

6.1 No Interaction with Degree Constructions

In our proposal, *wh*-exclamatives are not a degree phenomenon in the sense of containing a mechanism that targets degree arguments of gradable predicates. The high degree reading of exclamatives is due to the fact that noteworthy objects are objects that stand out by possessing some attribute to an exceptionally high degree. But this high degree is lexically accessed, not compositionally. The result is that we predict that proper degree constructions, i.e. those that target degree slots of adjectives, will not stand in the way of forming a *wh*-exclamative. For instance, the degree slot in (26) is presumably saturated by *extremely*, witness the incompatibility of having further degree modification if *extremely* is there, as in (27). On a degree approach, it would be difficult to account for why (26) is felicitous, since *extremely* and *what* would target the same degree slot. (Cf. Castroviejo-Miro [2008] for discussion.)

(26) What an extremely nice man John is!

(27) a. *John is more extremely nice than Bill.
b. *John is too extremely nice.

In our proposal, however, the degree slot of *nice* in (26) plays no role in the exclamative semantics.

[6] More needs to be said on how this analysis is compositionally realised. We omit these details here given the limited space we have.

[7] An anonymous reviewer wonders what the interaction is between noteworthiness and expectation. In particular, s/he wonders whether unexpectedness is a necessary condition for being noteworthy. This is an interesting question since expectation has played an important role in the characterisation of exclamatives in the literature (compare Rett [2012] and Zanuttini and Portner [2003]). It would be good to gain some ground regarding this issue, but we have to leave this for further research, also in the light of the general discussion concerning the concept of noteworthiness in section 5.1.

6.2 An Extension to *How*-Exclamatives

In English, questions with *how* may form degree questions, as in (28-a). Given our proposal to abandon a degree semantics for *wh*-exclamatives, the question arises how we could account for exclamatives like (28-b).

(28) a. How tall is John?
b. How tall John is!

In particular, one might ask whether English *how*-exclamatives are type 1 or type 2. Although, we do not have any clearcut arguments for either choice, there is one reason to think that they are of type 1, and that is because they can be reduced, as in *How tall!*, *How bizarre!*, etc. We will leave a more decisive argumentation for a choice between type 1 and type 2 to further research. It will be informative, however, to explore the two options a little bit.

***How*-exclamatives as type 1.** If *how*-exclamatives are indeed of type 1, we get the following semantics for (28-b).

(29) $\exists d[tall(j, d) \wedge noteworthy(d)(c)]$

There are some problems with these truth-conditions. First of all, it is not clear what the comparison class c should look like, but let us assume that this is just some set of degrees of tallness that makes contextual sense. Second, and more seriously, it is unclear when a degree is or is not noteworthy. For instance, it makes no sense to assume that $noteworthy(210cm)(c)$ is true because people who are that tall are remarkable, for that would make 210cm towers exclamation-worthy. However, it is not completely inconceivable that we can have attitudes to degrees. What we need is a notion of degree that expresses the relative positions entities take up in the relevant ordering. We could, for instance, use the universal scale of Bale [2008], where every entity is mapped to a universal degree representing its relative position in the (finite) weak order of entities under discussion. For now, however, we'll use a different mechanism originating from Klein [1980]. We'll assume that *how* in exclamatives does not range over a fine-grained scale of degrees, but rather over a coarse-grained set of so-called degree functions. Instead of interpreting adjectives as degree relations, we take them to be relations between comparison classes and entities. Degree modifiers like *very* manipulate the comparison class argument [Klein, 1980; Kennedy and McNally, 2005].

(30) $[\![\text{tall}]\!] = \lambda c \lambda x.tall(x)(c)$ "x is tall in class c"

(31) $[\![\text{very tall}]\!] = \lambda c \lambda x.tall(x)(\lambda y.tall_c(y))$

So, being *very tall* just means being tall with respect to the class of tall individuals. Functions like that expressed by *very* are called degree functions. We can create an ordering of degree functions as follows. Let D_0 be $\lambda P \lambda c \lambda x.P(c)(x)$ and D_1 correspond to "very", that is, $D_1 = \lambda P \lambda c \lambda x.P(P(c))(x)$. The next, stricter degree function is $D_2 = \lambda P \lambda c \lambda x.P(P(P(c))(x)$. Generally, for $n > 0$:

$D_n = \lambda P \lambda c \lambda x.P(D_{n-1}(P))(c)(x)$. The exclamative semantics could now be given in terms of such functions:

(32) $\exists D[(D(tall))(c)(j) \wedge noteworthy(D)]$

Assume that things that are *very very* A are noteworthy, as are things that are *very very very* A, etc. In other words, $noteworthy(D_n)$ is true for $n \geq 2$. This means that (28-b) is felicitous if John is (at least) *very very tall*. In this setup we now do not predict that towers of John's height are exclamation-worthy, since those towers will not count as *very very tall*.

***How*-exclamatives as type 2.** When we consider the option of *how*-exclamatives being of type 2, we get (33) for (28-b).

(33) $\exists d[noteworthy(^\wedge tall(j,d))]$

This is completely parallel to the semantics that Nouwen [2011] gives for (34-a), as in (34-b). (Assume that the predicate *surprising* is factive).

(34) a. John is surprisingly tall.
b. $\exists d[surprisingly(^\wedge tall(j,d))]$

The parallel is intuitively attractive. Both evaluative adverbs of degree like *surprisingly* and exclamatives seem to combine high degree with some sort of attitude. The page limit for this paper does not allow for a more thorough comparison of (28-b) and (34-a), nor for a proper evaluation of (33).

7 Conclusion

We have identified two kinds of *wh*-exclamatives: those that involve the noteworthiness of the *wh*-referent and those that involve a noteworthiness evaluation of the open proposition in the exclamative. Our approach simplifies the semantic mechanism and improves the empirical coverage over competing, scalar approaches.

An anonymous reviewer suggests a more radical approach than ours. The fact that type 2 exclamatives are not found in all languages and are syntactically distinct from type 1 exclamatives could mean that they do not form a common semantic class with type 1 exclamatives. In other words, our attempt at a unified analysis (involving noteworthiness) is misguided, given the many differences between the classes. This is an interesting thought, which we have not pursued for several reasons, most importantly because intuitively there is considerable semantic overlap between the two classes. (We characterise this overlap as the concept of noteworthiness). Furthermore, it was one of our main goals for this paper to point out an oddity in the literature: an analysis like the one in Rett [2012] is only suitable for English exclamatives, whilst the completely different analysis given in d'Avis [2002] is only suitable for *who*-exclamatives in languages like German and Dutch. We show that such papers are not in opposition to each other, but analyse different types of exclamatives. At the very least, we hope to have broadened the empirical burden of theories of *wh*-exclamatives.

References

Bale, A.: A universal scale of comparison. Linguistics and Philosophy 31, 1–55 (2008)
Castroviejo, E.: Wh-Exclamatives in Catalan. Ph.D. thesis, Universitat de Barcelona (2006)
Castroviejo-Miro, E.: Adverbs in restricted configurations. In: Cabredo Hofherr, P., Bonami, O. (eds.) Empirical Issues in Syntax and Semantics, vol. 7, pp. 53–76 (2008)
d'Avis, F.-J.: On the interpretation of wh-clauses in exclamative environments. Theoretical Linguistics 28, 5–31 (2002)
Kennedy, C., McNally, L.: Scale structure, degree modification and the semantics of gradable predicates. Language 81(2), 345–381 (2005)
Klein, E.: A semantics for positive and comparative adjectives. Linguistics and Philosophy 4, 1–45 (1980)
Nouwen, R.: Degree modifiers and monotonicity. In: Egré, P., Klinedinst, N. (eds.) Vagueness and Language Use. Palgrave McMillan (2011)
Rett, J.: A degree account of exclamatives. In: Proceedings of SALT XVII (2008a)
Rett, J.: Degree modification in Natural Language. Ph.D. thesis, Rutgers (2008b)
Rett, J.: Exclamatives, degrees and speech acts. Linguistics and Philosophy (in press, 2012)
Zanuttini, R., Portner, P.: Exclamative clauses: at the syntax-semantics interface. Language 79(1), 39–81 (2003)

Generalizing Monotonicity Inferences to Opposition Inferences

Ka-Fat Chow

The Hong Kong Polytechnic University
kfzhouy@yahoo.com

Abstract. This paper generalizes the notion of monotonicities to opposition properties (OPs). Some propositions regarding the OPs of determiners will be proposed and proved. We will also define the notion of OP-chain and deduce a condition that enables us to determine the OPs of an iterated quantifier in its predicates based on the OPs of its constituent determiners.

Keywords: monotonicity inferences, opposition inferences, opposition properties, OP-chain, Generalized Quantifier Theory, Natural Logic.

1 Introduction

Van Benthem (2008) characterized monotonicity inferences, an important type of Natural Logic inferences, as "inferences with inclusion premises" of the form "$P \leq Q \Rightarrow \varphi(P) \leq \varphi(Q)$". He also proposed the study on "inferences with exclusion premises" of the form "$P \leq \neg Q \Rightarrow \varphi(P) \leq \neg\varphi(Q)$", which we call "opposition inferences". In recent years, some scholars, e.g. MacCartney and Manning (2009), MacCartney (2009), Icard (2011), have started to study opposition inferences as a new type of Natural Logic inferences from different perspectives based on different frameworks. This paper is a study of opposition inferences as a generalization of monotonicity inferences based on the Generalized Quantifier Theory (GQT) framework.

2 Basic Definitions

First of all, let us review the definition of increasing monotonicity in the right argument for type $\langle 1, 1\rangle$ generalized quantifiers (i.e. determiners)[1].

Definition 1. Let Q be a determiner, then Q is increasing in the right argument iff for all A, B, B', $B \leq B' \Rightarrow Q(A)(B) \leq Q(A)(B')$.

[1] In what follows, we use A and B to denote the left and right arguments of a determiner, respectively. We also use the symbol "$\leq$" to denote the subset relation between sets as well as the entailment relation between propositions. Note that increasing monotonicity in the left argument can be defined analogously.

M. Aloni et al. (Eds.): Amsterdam Colloquium 2011, LNCS 7218, pp. 281–290, 2012.
© Springer-Verlag Berlin Heidelberg 2012

Note that an equivalent definition can be obtained by replacing both occurrences of "$\leq$" by "$\geq$" in the above definition. Based on Definition 1, we may denote increasing monotonicity figuratively by "$\leq\rightarrow\leq$" (or equivalently "$\geq\rightarrow\geq$"). Analogously, decreasing monotonicity may be denoted figuratively by "$\leq\rightarrow\geq$" (or equivalently "$\geq\rightarrow\leq$"). Now "$\leq$" and "$\geq$" are just two possible relations between sets / propositions. If we replace them by more general binary relations R_1, R_2 (written in prefix form), we will obtain a more general definition[2].

Definition 2. Let Q be a determiner, then Q is $R_1 \rightarrow R_2$ in the right argument iff for all A, B, B', $R_1[B, B'] \Rightarrow R_2[Q(A)(B), Q(A)(B')]$.

Hence, increasing and decreasing monotonicites are just special cases of Definition 2 with R_1 and R_2 instantiated as the "inclusion relations", i.e. $\leq$ and $\geq$.

Apart from "inclusion relations", we may also consider "exclusion relations". In this paper, we will consider two "exclusion relations" that are disjunctions of relations in the classical square of opposition: CC (standing for "contrary or contradictory") and SC (standing for "subcontrary or contradictory"). In classical logic, two propositions p and q satisfy the CC relation (denoted CC[p, q] iff they cannot be both true, and they satisfy the SC relation (denoted SC[p, q]) iff they cannot be both false. More generally, we may adopt the following definitions so that these two relations are also applicable to sets: let X and X' be sets or propositions, then

$$\mathrm{CC}[X, X'] \Leftrightarrow X \leq \neg X'; \mathrm{SC}[X, X'] \Leftrightarrow \neg X \leq X' \quad (1)$$

For example, we have CC[YOUNG, OLD] and SC[AGED-OVER-50, AGED-BELOW-51] because an individual cannot be young and old at the same time, whereas an individual must be either aged over 50 or aged below 51. According to (1), we also have

$$\mathrm{CC}[X, X'] \Leftrightarrow \mathrm{SC}[\neg X, \neg X'] \quad (2)$$

By instantiating R_1 and R_2 in Definition 2 as CC and SC, we then have 4 possible properties of determiners: CC→CC, CC→SC, SC→CC and SC→SC. These 4 properties will henceforth be called "opposition properties" (OPs).

3 OPs of Determiners

Our next task is to classify some commonly used determiners according to their OPs in the two arguments. For convenience, I will denote the sets of determiners possessing or not possessing a certain OP in a certain argument by placing a "+" or "−" sign on the left and right-hand sides of the name of the OP. For instance, −CC→CC+ denotes the set of those determiners that are CC→CC in the right but not left argument.

---

[2] Similar property in the left argument can be defined analogously.

In what follows I first state and prove four propositions:

Proposition 1. A determiner Q possesses a certain OP in its right argument iff each of its outer negation (denoted $\neg Q$), inner negation (denoted $Q\neg$) and dual (denoted Q^d)[3] possesses a different OP in its right argument according to the following table:

Q	**¬Q**	**Q¬**	**Q^d**
CC→CC	CC→SC	SC→CC	SC→SC
CC→SC	CC→CC	SC→SC	SC→CC
SC→CC	SC→SC	CC→CC	CC→SC
SC→SC	SC→CC	CC→SC	CC→CC

Proof: Here we only prove the first row of the table. The remaining rows can be proved similarly. By Definition 2 and (1), $Q \in$ CC→CC+ iff

$$\text{CC}[B, B'] \Rightarrow Q(A)(B) \leq \neg(Q(A)(B')) \tag{3}$$

Now (3) is equivalent to

$$\text{CC}[B, B'] \Rightarrow \neg(\neg Q(A)(B)) \leq \neg Q(A)(B') \tag{4}$$

Substituting the arbitrary sets B and B' by their negations and using (2) and the definitions of inner negation and dual, (3) and (4) can be rewritten as

$$\text{SC}[B, B'] \Rightarrow Q\neg(A)(B) \leq \neg(Q\neg(A)(B')) \tag{5}$$

$$\text{SC}[B, B'] \Rightarrow \neg(Q^d(A)(B)) \leq Q^d(A)(B') \tag{6}$$

From (4)-(6), we have $\neg Q \in$ CC→SC+, $Q\neg \in$ SC→CC+ and $Q^d \in$ SC→SC+. □

Proposition 2. Let Q_1 and Q_2 be determiners such that $Q_1 \leq Q_2$[4].
(a) If Q_2 is CC→CC (SC→CC) in an argument, then so is Q_1 in the same argument.
(b) If Q_1 is CC→SC (SC→SC) in an argument, then so is Q_2 in the same argument.

Proof: Here we only prove part (a). Part (b) can be proved similarly. Suppose $Q_2 \in$ CC→CC+ and CC$[B, B']$. Let $\|Q_1(A)(B)\| = 1$. Then since $Q_1 \leq Q_2$, we have $\|Q_2(A)(B)\| = 1$. But then we must have $\|Q_2(A)(B')\| = 0$. By $Q_1 \leq Q_2$ again, we have $\|Q_1(A)(B')\| = 0$. We have thus proved that CC$[B, B'] \Rightarrow$ CC$[Q_1(A)(B), Q_1(A)(B')]$, i.e. $Q_1 \in$ CC→CC+. The proofs for the cases $Q_2 \in$ +CC→CC, SC→CC+ and +SC→CC are exactly the same. □

[3] Outer negation, inner negation and dual are as defined in Peters and Westerståhl (2006).

[4] $Q_1 \leq Q_2$ iff for all A, B, $Q_1(A)(B) \leq Q_2(A)(B)$. From this we can define the following: $Q_1 = Q_2$ iff $Q_1 \leq Q_2$ and $Q_2 \leq Q_1$.

Proposition 3. Let Q_1 be a symmetric determiner and Q_2 be a contrapositive determiner[5].
(a) Q_1 possesses a certain OP in an argument iff Q_1 possesses the same OP in the other argument.
(b) Q_2 is CC→CC in an argument iff Q_2 is SC→CC in the other argument. Q_2 is CC→SC in an argument iff Q_2 is SC→SC in the other argument.

Proof: Here we only prove part (b). Part (a) can be proved similarly. Suppose $Q_2 \in$ CC→CC+ and SC$[A, A']$, which by (2) is equivalent to CC$[\neg A, \neg A']$. Let $\|Q_2(A)(B)\| = 1$. By the contrapositivity of Q_2, this is equivalent to $\|Q_2(\neg B)(\neg A)\| = 1$. But then we must have $\|Q_2(\neg B)(\neg A')\| = 0$. By the contrapositivity of Q_2 again, this is in turn equivalent to $\|Q_2(A')(B)\| = 0$. We have thus proved that SC$[A, A'] \Rightarrow$ CC$[Q_2(A)(B), Q_2(A')(B)]$, i.e. $Q_2 \in$ +SC→CC. Similarly, we can prove that if $Q_2 \in$ +SC→CC, then $Q_2 \in$ CC→CC+. The proofs for the cases $Q_2 \in$ +CC→CC, CC→SC+ and +CC→SC are exactly the same. □

Proposition 4. On condition that $A \neq \emptyset$[6], $(\textit{at least } r \textit{ of}) \in$ CC→CC+ for $1/2 < r < 1$; $(\textit{more than } r \textit{ of}) \in$ CC→CC+ for $1/2 \leq r < 1$; $(\textit{exactly } r \textit{ of}) \in$ −CC→CC for $0 \leq r \leq 1$.

Proof: Let $\|(\textit{at least } r \textit{ of})(A)(B)\| = 1$ for $1/2 < r < 1$ and CC$[B, B']$. Then by (1), $B \leq \neg B'$. Since "$(\textit{at least } r \textit{ of})$" is right increasing, we have $\|(\textit{at least } r \textit{ of})(A)(\neg B')\| = 1$, which is equivalent to $\|(\textit{at most } 1-r \textit{ of})(A)(B')\| = 1$. Since $1/2 < r < 1$, this entails $\|(\textit{less than } r \textit{ of})(A)(B')\| = 1$[7], which is equivalent to $\|(\textit{at least } r \textit{ of})(A)(B')\| = 0$. We have thus shown that CC$[B, B'] \Rightarrow$ CC$[(\textit{at least } r \textit{ of})(A)(B), (\textit{at least } r \textit{ of})(A)(B')]$, i.e. $(\textit{at least } r \textit{ of}) \in$ CC→CC+ for $1/2 < r < 1$. The fact that $(\textit{more than } r \textit{ of}) \in$ CC→CC+ for $1/2 \leq r < 1$ can be proved similarly.

To prove that $(\textit{exactly } r \textit{ of}) \in$ −CC→CC for $0 \leq r \leq 1$, we devise a method for constructing counterexample models. Choose natural numbers x and y such that $x/y = r$. Construct two sets A and A' such that $|A| = |A'| = y$ and $A \cap A' = \emptyset$. Choose a subset X of A and a subset X' of A' such that $|X| = |X'| = x$. Then set $U = A \cup A'$ and $B = X \cup X'$. It is easy to check that under this model, we have CC$[A, A']$ and $\|(\textit{exactly } r \textit{ of})(A)(B)\| = \|(\textit{exactly } r \textit{ of})(A')(B)\| = 1$. In other words, we do not have CC$[(\textit{exactly } r \textit{ of})(A)(B), (\textit{exactly } r \textit{ of})(A')(B)]$, thus completing the proof. □

Based on the above propositions, we can now determine the OPs of some commonly used determiners. For example, let $1/2 < r < 1$ and $A \neq \emptyset$, then

---

[5] Symmetry and contrapositivity are as defined in Zuber (2007).

[6] In what follows, we assume that the truth condition of a proportional determiner involves $|A|$ in the denominator, e.g. $\|(\textit{at least } r \textit{ of})(A)(B)\| = 1 \Leftrightarrow |A \cap B|/|A| \geq r$. Thus when $A = \emptyset$, these proportional determiners have no truth values.

[7] This step has made essential use of a property of numerical comparison: for $1/2 < r < 1$, if $x \leq 1 - r$, then $x < r$. Note that this property is not derivable from the monotonicity of the numerical comparative determiners "$(\textit{at least } r \textit{ of})$", etc. Thus, although the definitions of the CC / SC relations in (1) are expressed in the form of subset relations, which is a characteristic relation of the monotonicity inferences, opposition inferences are not subsumable under monotonicity inferences.

since (*exactly r of*) $\leq$ (*at least r of*), by Proposition 4 and Proposition 2(a), we immediately have (*exactly r of*), (*at least r of*) $\in$ $-$CC$\rightarrow$CC$+$. Similarly, since (*exactly r of*) $\leq$ (*more than* $r - \varepsilon$ *of*) where ε represents an infinitesimal quantity such that $1/2 \leq r - \varepsilon < 1$, by Proposition 4 and Proposition 2(a) again, we have (*more than* $r - \varepsilon$ *of*) $\in$ $-$CC$\rightarrow$CC$+$. Replacing the arbitrary variable $r - \varepsilon$ by r, we can rewrite the last result as (*more than r of*) $\in$ $-$CC$\rightarrow$CC$+$ for $1/2 \leq r < 1$. From the above, we can derive even more results.

Since the outer negation of "(*at least r of*)" is "(*less than r of*)", by Proposition 1, we have (*less than r of*) $\in$ $-$CC$\rightarrow$SC$+$ for $1/2 < r < 1$. Moreover, since the inner negation and dual of "(*at least r of*)" are "(*at most* $1 - r$ *of*)" and "(*more than* $1 - r$ *of*)", respectively, by Proposition 1 again, we have (*at most* $1 - r$ *of*) $\in$ $-$SC$\rightarrow$CC$+$ and (*more than* $1 - r$ *of*) $\in$ $-$SC$\rightarrow$SC$+$ for $0 < 1 - r < 1/2$. Replacing the arbitrary variable $1 - r$ by r, we can rewrite the last results as (*at most r of*) $\in$ $-$SC$\rightarrow$CC$+$ and (*more than r of*) $\in$ $-$SC$\rightarrow$SC$+$ for $0 < r < 1/2$.

A similar analysis for "(*more than r of*)" yields the following results: (*at most r of*) $\in$ $-$CC$\rightarrow$SC$+$ for $1/2 \leq r < 1$; (*less than r of*) $\in$ $-$SC$\rightarrow$CC$+$ and (*at least r of*) $\in$ $-$SC$\rightarrow$SC$+$ for $0 < r \leq 1/2$.

For "(*exactly r of*)", its inner negation may take two forms: "(*all except r of*)" and "(*exactly* $1 - r$ *of*)". Thus, by Proposition 1, we have (*all except r of*) $\in$ $-$SC$\rightarrow$CC$+$ for $1/2 < r < 1$ and (*exactly* $1 - r$ *of*) $\in$ $-$SC$\rightarrow$CC$+$ for $0 < 1 - r < 1/2$. Replacing the arbitrary variable $1 - r$ by r, we can rewrite the last result as (*exactly r of*) $\in$ $-$SC$\rightarrow$CC$+$ for $0 < r < 1/2$. Based on the last result and using the fact that the inner negation of "(*exactly r of*)" is "(*all except r of*)" and Proposition 1 again, we obtain (*all except r of*) $\in$ $-$CC$\rightarrow$CC$+$ for $0 < r < 1/2$.

We next consider the classical determiner "*some*". We have the relation: (*at least r of*) $\leq$ *some* for $0 < r \leq 1/2$ [8], on condition that $A \neq \emptyset$. By virtue of a previous result and Proposition 2(b), we know that *some* $\in$ SC$\rightarrow$SC$+$ on condition that $A \neq \emptyset$. Note that this condition is essential because when $A = \emptyset$, $\|some(\emptyset)(B)\| = 0$ for any B, and so we can never have SC$[B, B']$ $\Rightarrow \neg some(\emptyset)(B) \leq some(\emptyset)(B')$. In other words, *some* $\notin$ SC$\rightarrow$SC$+$ when $A = \emptyset$. As for the left argument of "*some*", by the symmetry of "*some*" and Proposition 3(a), we know that *some* $\in$ $+$SC$\rightarrow$SC subject to certain condition. One can easily find that this condition is $B \neq \emptyset$. The above facts will be represented succinctly by: *some* $\in$ $+$SC$\rightarrow$SC$+$ $(B \neq \emptyset; A \neq \emptyset)$[9]. By using a similar line of reasoning and the fact that "*no*" is the outer negation of "*some*" and is a symmetric determiner, we can find that *no* $\in$ $+$SC$\rightarrow$CC$+$ $(B \neq \emptyset; A \neq \emptyset)$.

We next consider "*every*". Since "*every*" is the dual of "*some*", by Proposition 1, we know that *every* $\in$ CC$\rightarrow$CC$+$ subject to certain condition. This

[8] Although we also have (*at least r of*) $\leq$ *some* for $1/2 < r < 1$, since (*at least r of*) $\in$ CC$\rightarrow$CC$+$ in this range, this fact cannot be used to derive the OP of "*some*" by virtue of Proposition 2.

[9] The conditions $B \neq \emptyset; A \neq \emptyset$ are ordered such that the first (second) condition corresponds to the left (right) argument of the determiner.

condition is $A \neq \emptyset$ because when $A = \emptyset$, $\|every(\emptyset)(B)\| = 1$ for any B, and so we can never have CC$[B, B'] \Rightarrow every(\emptyset)(B) \leq \neg every(\emptyset)(B')$. As for the left argument of "*every*", since "*every*" is contrapositive according to Zuber (2007), by Proposition 3(b), we know that *every* $\in$ +SC→CC subject to certain condition. This condition is $B \neq U$ because when $B = U$, $\|every(A)(U)\| = 1$ for any A, and so we can never have SC$[A, A'] \Rightarrow every(A)(U) \leq \neg every(A')(U)$. The above facts will be represented succinctly by: *every* $\in$ +SC→CC $\cap$ CC→CC+ $(B \neq U; A \neq \emptyset)$. By using a similar line of reasoning and the fact that "(*not every*)" is the outer negation of "*every*" and is a contrapositive determiner, we can find that (*not every*) $\in$ +SC→SC $\cap$ CC→SC+ $(B \neq U; A \neq \emptyset)$.

Based on the above results, we can now derive valid inferential relations between quantified statements. For example, the following are instances exemplifying the facts *every* $\in$ CC→CC+ and *some* $\in$ SC→SC+ on condition that $A \neq \emptyset$ (given that CC[YOUNG, OLD], SC[AGED-OVER-50, AGED-BELOW-51]):

Example 1. (Condition: There is some member in this club.) CC["Every member of this club is young", "Every member of this club is old"]

Example 2. (Condition: There is some member in this club.) SC["Some member of this club is aged over 50", "Some member of this club is aged below 51"]

4 OPs of Iterated Quantifiers

An adequate theory on opposition inferences should achieve the following. Given an iterated quantifier composed of n constituent determiners[10] in the form:

$$Q_1(A_1)(\{x_1 : \ldots Q_n(A_n)(\{x_n : B(x_1, \ldots x_n)\}) \ldots\}) \tag{7}$$

we hope to determine the OPs of this iterated quantifier in the predicates $A_1, \ldots A_n, B$ based on the OPs of $Q_1, \ldots Q_n$. To this end, we first define the notion of "OP-chain":

Definition 3. Let X be a predicate under an iterated quantifier. Suppose X is within the i_k argument of $Q_k (1 \leq k \leq n)$, i_{k-1} argument of $Q_{k-1}, \ldots i_1$ argument of Q_1, where each of $i_k, i_{k-1}, \ldots i_1$ is one of {left, right} and $Q_k, Q_{k-1}, \ldots Q_1$ are constituent determiners of the iterated quantifier ordered from the innermost to the outermost layers. Then X has an OP-chain $\langle R_k, R_{k-1}, \ldots R_0 \rangle$, where each of $R_k, R_{k-1}, \ldots R_0$ is one of {CC, SC}, iff Q_k is $R_k \rightarrow R_{k-1}$ in the i_k argument, Q_{k-1} is $R_{k-1} \rightarrow R_{k-2}$ in the i_{k-1} argument, $\ldots Q_1$ is $R_1 \rightarrow R_0$ in the i_1 argument.

For instance, in the following iterated quantifier:

$$(at\ most\ 1/2\ of)(A_1)(\{x_1 : no(A_2)(\{x_2 : B(x_1, x_2)\})\}) \tag{8}$$

[10] Iterated quantifiers refer to polyadic quantifiers constructed from monadic quantifiers by "iteration" (Peters and Westerståhl (2006)). In this paper we only consider iterated quantifiers constructed from determiners.

A_2 is within the left argument of "*no*" and right argument of "(*at most 1/2 of*)". Since *no* $\in$ +SC→CC on condition that its right argument is non-empty and (*at most 1/2 of*) $\in$ CC→SC+ on condition that its left argument is non-empty, A_2 has an OP-chain ⟨SC, CC, SC⟩ on condition that $A_1 \neq \emptyset \wedge \{x_2 : B(x_1, x_2)\} \neq \emptyset$. Similarly, one can easily check that B has an OP-chain ⟨SC, CC, SC⟩ on condition that $A_1 \neq \emptyset \wedge A_2 \neq \emptyset$, while A_1 has no OP-chain.

To facilitate the discussion below, we first state a proposition.

Proposition 5. Let $P(x_1, \ldots x_n)$ and $P'(x_1, \ldots x_n)$ be n-ary predicates and R be one of {CC, SC}, then $\mathrm{R}[P, P'] \Rightarrow \mathrm{R}[\{x_i : P(y_1, \ldots y_{i-1}, x_i, y_{i+1}, \ldots y_n)\}, \{x_i : P'(y_1, \ldots y_{i-1}, x_i, y_{i+1}, \ldots y_n)\}]$ for any $1 \leq i \leq n$ and any particular set of $y_1, \ldots y_{i-1}, y_{i+1}, \ldots y_n$.

Proof: Here we only prove the case in which R = CC. The case in which R = SC is similar. Suppose $\mathrm{CC}[P, P']$. By (1), this is equivalent to $P \leq \neg P'$. Then for any particular set of $y_1, \ldots y_{i-1}, y_{i+1}, \ldots y_n$ and any arbitrary x_i, we have $P(y_1, \ldots y_{i-1}, x_i, y_{i+1}, \ldots y_n) \leq \neg P'(y_1, \ldots y_{i-1}, x_i, y_{i+1}, \ldots y_n)$, and so we have $\{x_i : P(y_1, \ldots y_{i-1}, x_i, y_{i+1}, \ldots y_n)\} \leq \{x_i : \neg P'(y_1, \ldots y_{i-1}, x_i, y_{i+1}, \ldots y_n)\}$. But $\{x_i : \neg P'(y_1, \ldots y_{i-1}, x_i, y_{i+1}, \ldots y_n)\}$ can be rewritten as $\neg\{x_i : P'(y_1, \ldots y_{i-1}, x_i, y_{i+1}, \ldots y_n)\}$. Thus, by (1) again, we have $\mathrm{CC}[\{x_i : P(y_1, \ldots y_{i-1}, x_i, y_{i+1}, \ldots y_n)\}, \{x_i : P'(y_1, \ldots y_{i-1}, x_i, y_{i+1}, \ldots y_n)\}]$. □

We can now deduce a condition for determining the OPs of the iterated quantifier (7) in its predicates based on the OPs of its constituent determiners. We focus on the predicate B (the other predicates can be similarly treated). Let B have an OP-chain $\langle R_n, R_{n-1}, R_{n-2}, \ldots R_0\rangle$ and $R_n[B, B']$. By Proposition 5, we have $R_n[\{x_n : B(x_1, \ldots x_n)\}, \{x_n : B'(x_1, \ldots x_n)\}]$ for any $x_1, \ldots x_{n-1}$. Moreover, by the definition of OP-chain, Q_n is $R_n \to R_{n-1}$ in the argument $\{x_n : B(x_1, \ldots x_n)\}$, and so we have $R_{n-1}[Q_n(A_n)(\{x_n : B(x_1, \ldots x_n)\}), Q_n(A_n)(\{x_n : B'(x_1, \ldots x_n)\})]$. The above reasoning can be seen as a kind of "upward derivation": from the R_n relation at the B-level, we derive the R_{n-1} relation at the Q_n-level. Now $Q_n(A_n)(\{x_n : B(x_1, \ldots x_n)\})$ can be seen as an $(n-1)$-ary predicate (with $x_1, \ldots x_{n-1}$ as arguments). Thus, we can carry out the aforesaid upward-derivation again and derive a R_{n-2} relation at the Q_{n-1}-level. The process of determining the OPs of the iterated quantifier (7) in B is essentially a repetition of this upward derivation. After n rounds of derivation, we will finally derive the R_0 relation at the Q_1 level. The net effect is thus $R_n(B, B') \Rightarrow R_0[Q_1(A_1)(\{x_1 : \ldots Q_n(A_n)(\{x_n : B(x_1, \ldots x_n)\}) \ldots\}), Q_1(A_1)(\{x_1 : \ldots Q_n(A_n)(\{x_n : B'(x_1, \ldots x_n)\}) \ldots\})]$, showing that the iterated quantifier is $R_n \to R_0$ in B.

The above derivation relies on the condition that B has an OP-chain. This condition does not hold either when at least one of $Q_1, \ldots Q_n$ possesses none of the OPs, or when the OPs possessed by $Q_1, \ldots Q_n$ do not form a chain. In either case, the absence of the OP-chain blocks the upward derivation. Based on the above discussion, we can thus formulate the following condition: let X be a predicate under an iterated quantifier Q,

$$Q \text{ is } R_k \to R_0 \text{ in } X \text{ iff } X \text{ has an OP-chain } \langle R_k, \ldots R_0\rangle \quad (9)$$

We now use (9) to determine the OPs of (8) in its predicates. Previously we have already found that A_2 and B both have the OP-chain ⟨SC, CC, SC⟩ subject to different conditions, whereas A_1 has no OP-chain. Thus, according to (9), we know that (8) is SC→SC in A_2 on condition that $A_1 \neq \emptyset \wedge \{x_2 : B(x_1, x_2)\} \neq \emptyset$ and SC→SC in B on condition that $A_1 \neq \emptyset \wedge A_2 \neq \emptyset$. Moreover, (8) possesses none of the OPs in A_1. From the above result, we can derive the following (by letting A_1 = CLUB, A_2 = AGED-OVER-50, A'_2 = AGED-BELOW-51, B = ADMIT-AS-MEMBERS):

Example 3. (Condition: There is at least a club and every club admits somebody as members.) SC["At most 1/2 of the clubs admit nobody aged over 50 as members", "At most 1/2 of the clubs admit nobody aged below 51 as members"]

The derivation process of (9) is not exclusively valid for (7). In fact the condition in (9) can also be applied to iterated quantifiers in a form different than (7). Consider the following:

$$no(A \cap \{x : some(B)(\{y : C(x, y)\})\})(D) \tag{10}$$

The above iterated quantifier represents a quantified statement whose subject contains a relative clause which is another quantified statement. Let's determine the OP of (10) in the predicate B by using (9). Since B falls within the left arguments of "*some*" and "*no*", which are +SC→SC and +SC→CC, respectively, both on condition that their right arguments are non-empty, B has an OP-chain ⟨SC, SC, CC⟩. By (9), (10) is SC→CC in B subject to the condition that $\{y : C(x, y)\} \neq \emptyset \wedge D \neq \emptyset$. From the above result, we can derive the following (by letting A = COMPANY, B = AGED-OVER-50, B' = AGED-BELOW-51, C = EMPLOY, D = GO-BANKRUPT):

Example 4. (Condition: Every company employs somebody and some company went bankrupt.) CC["No company employing somebody aged over 50 went bankrupt", "No company employing somebody aged below 51 went bankrupt"].

Note that monotonicity inferences of iterated quantifiers are governed by the same condition as opposition inferences. We can define an analogous notion of MON-chain by replacing {CC, SC} with $\{\leq, \geq\}$ in Definition 3 and modify the condition in (9) by replacing "OP-chain" with "MON-chain". The modified condition can then be used to determine the monotonicities of iterated quantifiers in its predicates.

For illustration, consider the iterated quantifier in (8) again. Let's determine the monotonicity of (8) in the predicate A_2. Since A_2 is within the left argument of "*no*" and right argument of "(*at most 1/2 of*)", and "*no*" is left decreasing while "(*at most 1/2 of*)" is right decreasing, A_2 has a MON-chain $\langle \leq, \geq, \leq \rangle$ (or equivalently, $\langle \geq, \leq, \geq \rangle$)[11]. According to the modified version of condition (9),

[11] Note that since both increasing and decreasing monotonicities have two possible representations, the determination of MON-chains is more complicated than that of OP-chains. We may need to consider all possible representations of the monotonicities involved in order to determine whether a predicate has a MON-chain.

we know that (8) is $\leq \rightarrow \leq$ (or equivalently $\geq \rightarrow \geq$), i.e. increasing, in A_2. This result is in accord with that obtained by using van Eijck (2007)'s "monotonicity calculus".

5 Concluding Remarks

According to van Eijck (2007), monotonicity inferences are an important type of inferences in modern Natural Logic. Even syllogistics, the most important type of classical inferences, are subsumable under monotonicity inferences. By proposing the study on "inferences with exclusion premises", van Benthem (2008) has opened up a new direction of studies on Natural Logic. This paper is an implementation and generalization of van Benthem (2008)'s proposal and a contribution to the studies on Natural Logic. We have proposed a number of results by which we can determine the OPs of determiners and iterated quantifiers composed of constituent determiners, and derive valid inferential relations between quantified statements.

Nevertheless, one may criticize that the inferential relations derived from the OPs of determiners are too weak. For instance, by (1) the inferential relation in Example 1 above can be rewritten as the following entailment:

> Every member of this club is young. $\Rightarrow$ Not every member of this club is old.

Although valid, the conclusion above seems too weak because intuitively, one would expect that the proper conclusion of the above inference should be "No member of this club is old".

However, entailment is not the only type of inferential relations. In some situations, we do need to establish some other types of inferential relations (such as the CC or SC relation) between sets / propositions. These situations do not only include solving logical puzzles, but also include linguistic uses. One such use is to determine the incompatibility between two sets / propositions. For instance, from the fact that *every* $\in$ −CC→CC+[12], we know that "clubs every member of which is young" and "clubs every member of which is old" are incompatible, whereas "clubs of which all young people are members" and "clubs of which all old people are members" are not.

As incompatibility is an essential element of antonyms that feature in certain linguistic structures, such as those identified by Jones (2002), the determination of incompatibility can thus help us determine the well-formedness of certain linguistic structures. For example, "X rather than Y" is a structure where X and Y should be antonyms.

Moreover, the determination of incompatibility can also help us differentiate between entailments and implicatures, especially the "alternate-value implicatures" studied by Hirsch-berg (1975). For instance, in the following discourse, B's conclusion is a logical entailment inferred from A's utterance:

[12] Note that in Section 3, we have only established that *every* $\in$ CC→CC+. But since *every* = (*exactly 100% of*), by Proposition 4, we know that *every* $\in$ −CC→CC.

A: This is a club every member of which is young.
B: So it is not a club every member of which is old.

whereas in the following discourse, B's conclusion (inferred from A's utterance under suitable context) should be seen as an alternate-value implicature that is cancellable:

A: This is a club of which all young people are members.
B: So it is not a club of which all old people are members.

Note that the difference between the aforesaid two discourses is analogous to the difference between the following two discourses:

A: She is my enemy. B: So she is not your platonic friend.
A: She is my colleague. B: So she is not your platonic friend.

References

1. van Benthem, J.: A Brief History of Natural Logic. Technical Report PP-2008-05, Institute for Logic, Language and Computation (2008)
2. van Eijck, J.: Natural Logic for Natural Language. In: ten Cate, B.D., Zeevat, H.W. (eds.) TbiLLC 2005. LNCS (LNAI), vol. 4363, pp. 216–230. Springer, Heidelberg (2007)
3. Hirschberg, J.B.: A Theory of Scalar Implicature. PhD thesis, University of Pennsylvania (1985)
4. Icard, T.: Exclusion and Containment in Natural Language. Accepted to Studia Logica (2011)
5. Jones, S.: Antonymy: A corpus-based perspective. Routledge, London (2002)
6. MacCartney, B.: Natural Language Inference. PhD thesis, Stanford University (2009)
7. MacCartney, B., Manning, C.D.: An extended model of natural logic. In: Proceedings of the Eighth International Conference on Computational Semantics (2009)
8. Peters, S., Westerståhl, D.: Quantifiers in Language and Logic. Clarendon Press, Oxford (2006)
9. Zuber, R.: Symmetric and contrapositional quantifiers. Journal of Logic, Language and Information 16(1), 1–13 (2007)

Exclusive Updates*

Elizabeth Coppock[1] and David Beaver[2]

[1] Heinrich Heine University, Düsseldorf
[2] University of Texas at Austin

Abstract. This paper develops a type of dynamic semantics in which contexts include not only information, but also questions, whose answers are ranked by strength. The questions can be local to the restrictor of a quantifier, and the quantifier can bind into them. The proposed framework satisfies several desiderata arising from quantificational expressions involving exclusives (e.g. *only, just, mere* and *sole*), allowing: (i) presupposed questions; (ii) presuppositional constraints on the strength ranking over the answers to the question under discussion; (iii) quantificational binding into such presupposed questions; and (iv) compositional derivation of logical forms for sentences.

1 Introduction

Contemporary work on information structure commonly relates focus to questions. For example, "Pedro feeds SAM" might answer the question of who Pedro feeds. In current question-based theories of focus such as that of Roberts (1996), questions are root-level entities, in the sense that there is one question per declarative utterance. In this paper, we explore the possibility that there are local, embedded questions, which may contain bound variables. Consider, for example: "If a man owns a DONKEY, then he BEATS it." The consequent here might be taken to answer the question "What does he do to it?". That question, however, is not one that could occur in current question-based theories of focus, since it contains locally bound pronouns and it is not at the root level. In this paper, we develop a dynamic model of discourse that makes it possible to bind into local questions under discussion.

We apply this model to sentences like (1) and (2), in which focus-sensitive elements occur in the restrictor of a quantifier.

(1) *No mere child* could keep the Dark Lord from returning. [web ex.]

(2) As a bilingual person I'm always running around helping *everybody who only speaks Spanish.* [web ex.]

For reasons discussed in Beaver and Clark (2008) and Coppock and Beaver (2011), we assume that exclusives such as *mere* and *only* relate to the current Question

* We would especially like to thank Robin Cooper and Jeroen Groenendijk for feedback, and we gratefully acknowlege financial support fthrough NSF grant BCS-0952862 *Semantics and Pragmatics of Projective Meaning across Languages*, NSF BCS-0904913, DoD HHM402-10-C-0100, and the NYCT TeXIT project.

M. Aloni et al. (Eds.): Amsterdam Colloquium 2011, LNCS 7218, pp. 291–300, 2012.
© Springer-Verlag Berlin Heidelberg 2012

Under Discussion (CQ): this is what explains the focus sensitivity of such particles. We take the exclusives to presuppose that the prejacent is the weakest of the true answers to the CQ, and their ordinary semantic content is that the prejacent is the strongest of them, where strength is not necessarily determined by entailment. In (1) and (2), these questions are 'open', in the logical sense. For example, (1) relates to the question, 'What properties does x have?' where x is the variable bound by the negative quantifier. The question in (2) can be rendered as 'What does x speak?', where x is bound by *everybody.*

For (1), the answer 'x is a child' (the prejacent) is weaker than for example 'x is an adult', and *mere* contributes 'x is at least a child' as a presupposition, 'x is no more than a child' as ordinary semantic content. For (2), the answer 'x speaks Spanish' (the prejacent) is weaker than 'x speaks Spanish and English', and here the presupposition is 'x speaks at least Spanish', and the ordinary semantic contribution is 'x speaks no more than Spanish'. So exclusives place presuppositional constraints on the CQ, and make use of the CQ in the computation of their at-issue content; furthermore, these presuppositions may contain a variable that is bound externally to the question. For *only,* this problem only arises when it is inside a relative clause, as in (2), but for adjectival exclusives like *mere* and *sole,* this is the typical case.

Exclusives may also place their own constraints on the CQ. Consider:

(3) a. The mere student proved Goldbach's conjecture.
 b. The only student proved Goldbach's conjecture.

In (3a), the question that *mere* depends on is 'What are the properties of x (student, professor, etc.)?' (answer: x is only a student). In (3b), the question that the adjective *only* depends on is 'Which individuals are students?' (answer: only x). This difference stems from the choice of lexical item.

Presuppositions about the question under discussion are not like the presuppositions of factive or aspectual verbs (as they are standardly analyzed). The former do not constrain the context set, or the shared assumptions or beliefs of the interlocutors. They could be described as *interrogative presuppositions*: constraints on the Question Under Discussion. We need a framework that allows us to express such things. Indeed, this conclusion was reached independently by Jäger (1996) and Aloni et al. (2007) based on the apparent presupposition of a QUD by focus, and effects of questions on *only*'s quantificational domain.

Furthermore, we need to bind into such questions.[1] Binding into questions was addressed by Krifka (2001), but for a fully satisfactory solution to the problem at hand, we need a system that allows for: (i) presupposed questions; (ii) constraints on the strength ranking over the answers to presupposed questions;

[1] Note that similar problems could arise in alternative semantics (Rooth, 1992), as long as (a) the possibility of binding into alternatives is allowed, and (b) a strict separation is maintained between two dimensions of meaning (ordinary and alternative). The technical problems that would arise are familiar from the analogous problems observed in Karttunen and Peters's (1979) discussion of the example "Somebody managed to succeed George IV", where a quantifier appears to bind material that, for Karttunen and Peters, was in a separate dimension.

(iii) quantificational binding into such questions; and (iv) compositional derivation of meanings for sentences. We do not know of any prior system that satisfies all of these desiderata, and the present work aims to produce one that does.

2 Framework

Because Beaver's (2001) dynamic semantics deals successfully with quantified presuppositions, we use this as starting point, and introduce CQs and an answer strength ranking into the context. So a context S will determine: (i) INFO_S, a set of world-assignment pairs; (ii) CQ_S, a set of answers, where each answer is an information state; (iii) $\geq_S$, a partial order over information states. For example, suppose that we have the following three information states: $I = \{\langle g, w_3\rangle, \langle g, w_2\rangle\}$, $J = \{\langle g, w_1\rangle\}$, and $K = \{\langle g, w_2\rangle\}$. Here is a possible context:

(4) a. $\text{INFO}_S = \{\langle g, w_1\rangle, \langle g, w_2\rangle, \langle g, w_3\rangle\}$
  b. $\text{CQ}_S = \{I, J, K\}$
  c. $\geq_S = \{\langle I, I\rangle, \langle I, J\rangle, \langle J, J\rangle, \langle I, K\rangle, \langle J, K\rangle, \langle K, K\rangle\}$

These three components of context are not independent. Under the assumption that the strength ranking does not rank answers other than those in the CQ, the CQ is recoverable from the strength ranking; it is its *field*:[2]

(5) $\text{CQ}_S = \text{FIELD}(\geq_S)$
  where $\text{FIELD}(R) = \{x \mid \exists y[yRx \vee xRy]\}$

Likewise, the common ground is recoverable from the QUD.[3] As we are representing questions as sets of information states, we can recover the common ground by taking the union of all the answers (information states) in the question:

(6) $\text{INFO}_S = \bigcup \text{CQ}_S$

Because all of the information that the context must provide is contained in the strength ranking, we can *identify the context with the strength relation*: $\geq_S = S$.

3 Theory of Exclusives

With the framework just introduced, we can formulate a proper theory of exclusives. We formalize Beaver and Clark's (2008) theory of *only* as follows (where infix notation is used for relations):[4]

[2] In his analysis of *at least* (which is strikingly similar to our MIN), Krifka (1999) represents alternative semantic values as relations corresponding to 'strength' in the same sense, and derives the ordinary alternatives as the field of the ranking.

[3] Jäger (1996) represents questions and information states as equivalence relations over possible worlds, which may be partial. "Hence each state nontrivially determines a certain proposition, which can be thought of as the factual knowledge shared by the conversants."

[4] We use the variables x, y, z for individuals, D for discourse markers, f, g, h for assignments (sets of discourse referent-individual pairs), I, J, K for information states (sets of world-assignment pairs), S for contexts (ranking relations over information states), C for CCPs (relations between contexts), and P for dynamic properties (functions from discourse referents to CCPs).

(7) $\text{ONLY} = \lambda C.\{\langle S, S'\rangle \mid S' \subseteq S \wedge S[\text{MIN}(C)]S \wedge S[\text{MAX}(C)]S'\}$

In (7), ONLY is defined to take a CCP (C) and return another CCP (a relation between S and S'). The presuppositional nature of the MIN component is expressed by requiring that the input state S be a reflexive point with respect to MIN and C.[5]

MAX is defined to take a CCP C and provide another CCP relating contexts S and S', where the CQ in S' is a subset of the CQ in S containing only information states J such that J is as strong (according to $\geq_S$) as the information state corresponding to C. We formalize this as in (8); the corresponding MIN is in (9).

(8) $\text{MAX} = \lambda C.\ \{\ \langle S, S'\rangle \mid S' \subseteq S \wedge \exists S''[S[C]S''] \wedge \forall J \in \text{CQ}_{S'}[J \leq_S \text{INFO}_{S''}]\ \}$

(9) $\text{MIN} = \lambda C.\ \{\ \langle S, S'\rangle \mid S' \subseteq S \wedge \exists S''[S[C]S''] \wedge \forall J \in \text{CQ}_{S'}[\text{INFO}_{S''} \leq_S J]\ \}$

These definitions differ from Beaver and Clark's (2008) definitions of MAX and MIN insofar as the meaning of the sentence is not be used as the object whose strength is being ranked, because here we have assumed that the meanings of sentences are CCPs and the answers to questions are information states. In (8) and (9) we extract an information state from C by applying C to the input state. Because we are treating CCPs as relations rather than functions, the formalism does not guarantee a unique output state, but in general we assume that there willl be a maximum of one output state.[6] So in (8) and (9) we require that it exists and refer to it by using an existential quantifier.

Similarly to Coppock and Beaver (2011), we use a type-raised (Geached) version of (7) for VP-*only* and adjectival exclusives as in (10) (D is a variable over discourse referents, and P is a variable over dynamic properties, i.e. functions from discourse referents to CCPs):

(10) $\text{G-ONLY} = \lambda P \lambda D.\ \{\ \langle S, S'\rangle \mid S[\text{ONLY}(P(D))]S'\}$

Adjectival exclusives like *mere* instantiate G-ONLY but also impose further constraints. *Mere* requires the CQ to be about what properties the referential argument has:

(11) $\text{MERE} = \lambda P \lambda D.\ \{\ \langle S, S'\rangle \mid S[\text{ONLY}(P(D))]S' \wedge \text{CQ}_S \subseteq \{I \mid \exists P' \exists S''[S[P'(D)]S''] \wedge I = \text{INFO}_{S''}\}\}$

This extra constraint ensures that in e.g. *mere herring*, *mere* ranges over a scale of properties, e.g. herring, octopus, caviar. In *(the) only herring,* in contrast, adjectival *only* would require the question to be 'What things are herrings?', and for stronger answers to imply the existence of more herrings.

---

[5] The requirement that $S \subseteq S'$ makes this a *declarative update* in Jäger's (1996) sense, as opposed to an *interrogative update,* which would not affect the information state of the context, but would change how it is divided up into smaller information states in the CQ.

[6] At this point one might question the decision to formulate dynamic meanings relationally rather than functionally. This is the strategic choice that enables Beaver (2001) to formulate a CCP model straightforwardly in a classical two-sorted type theory.

4 Examples

4.1 Predicative Example

Before addressing exclusives in quantifiers, let us warm up with a simpler example. Consider (12), where the subscript 7 on *he* indicates that the pronoun is associated with discourse referent 7.

(12) He_7 is a mere child.

According to our analysis, this sentence denotes the CCP in (13).

(13) $\textsc{mere}(\textsc{child})(7)$
$$= \{\langle S, S'\rangle |\ S' \subseteq S \wedge S[\textsc{min}(\textsc{child}(7))]S \wedge S[\textsc{max}(\textsc{child}(7))]S' \wedge \textsc{cq}_S \subseteq \{I | \exists P' \exists S''[S[P'(7)]S''] \wedge I = \textsc{info}_{S''}\}$$

where CHILD is a function from discourse referents to CCPs which require the output context to be one where the discourse referent is mapped to a child. Formally (following Beaver 2001: 180) in many respects):

(14) $$\textsc{child} = \lambda D.\{\ \langle S, S'\rangle | D \in \textsc{t-domain}(\textsc{info}_S) \wedge \textsc{info}_{S'} = \{\langle w, f\rangle \in \textsc{info}_S | \forall x[\langle D, x\rangle \in f \rightarrow \textsc{child}'(x)(w)]\}\}$$

where (cf. Beaver 2001, p. 168, 170):

(15) $$\textsc{t-domain} = \lambda I.\{D | \forall w \forall f[\langle w, f\rangle \in I \rightarrow \exists x[\langle D, x\rangle \in f]\}$$

and $\textsc{child}'$ is a a function of type $\langle e, \langle w, t\rangle\rangle$ that returns true given an individual and a world if the individual is a child in the world.

Let us consider an input context. Suppose the following has been announced:

(16) Somebody_7 has proven Goldbach's conjecture.

Let us assume further that the domain consists of the Simpsons family: Homer and Marge (adults), Bart and Lisa (children), Maggie (a baby), Santa's Little Helper (a dog), and Snowball (a cat). The conversational participants rule out 7 being mapped to a baby, dog, or cat, because such individuals lack the necessary mathematical competence. So the world-assignment pairs in the information state for the input context are all such that 7 is mapped to a child or an adult who proved Goldbach's conjecture. Let w_{hmbl} be the world where Homer, Marge, Bart, and Lisa all proved it, w_{hmb} the world where Homer, Marge, and Bart proved it, etc. The information state of this context consists of pairs such as $\langle w_b, \langle 7, \text{Bart}\rangle\rangle$, $\langle w_l, \langle 7, \text{Lisa}\rangle\rangle$, $\langle w_h, \langle 7, \text{Homer}\rangle\rangle$, $\langle w_m, \langle 7, \text{Marge}\rangle\rangle$, $\langle w_{hm}, \langle 7, \text{Marge}\rangle\rangle$, $\langle w_{hm}, \langle 7, \text{Homer}\rangle\rangle$, etc. In order to satisfy *mere*'s requirement on the CQ, these world-assignment pairs must be organized into information states that are answers to the question, "What properties does 7 have?" If this is satisfied, then all of the states where 7 is mapped to a child will be grouped into one information state ("7 is a child"), etc. So:

(17) $\textsc{cq}_S = \{I_{child}, I_{adult}\}$
where $I_{child} = \{\langle w_b, \langle 7, \text{Bart}\rangle\rangle, \langle w_l, \langle 7, \text{Lisa}\rangle\rangle, \langle w_{hmb}, \langle 7, \text{Bart}\rangle\rangle, ...\}$, and
$I_{adult} = \{\langle w_h, \langle 7, \text{Homer}\rangle\rangle, \langle w_m, \langle 7, \text{Marge}\rangle\rangle, \langle w_{hm}, \langle 7, \text{Marge}\rangle\rangle, ...\}$

Let us assume further that I_{adult} outranks I_{child} according to $\geq_S$.

MIN requires that I_{child} be the weakest of the true answers in S, which it is, as we have assumed. So the MIN presupposition is satisfied. The further requirement that *mere* imposes is that every answer to the CQ is of the form $P(7)$ for some (appropriately contrasting) P. This is satisfied here, with P instantiated as ADULT and CHILD. MAX removes the possibility of 7 being an adult in the output state S', so the output CQ ($\text{CQ}_{S'}$) is just $\{I_{child}\}$. This is intuitively the right result; (12) means that 7 is a child.

4.2 Argument NP Example

Now let us consider an example in which *mere* occurs within the scope of a quantifier. We analyze (18a) as (18b) (as a first pass).

(18) a. Some$_7$ mere child succeeded.
b. SOME(7)(MERE(CHILD))(SUCCEED)

The quantifier SOME is defined as follows (cf. Beaver 2001 p. 185):

(19) $\text{SOME} = \lambda D.\lambda P.\lambda P'.\{\langle S, S'\rangle | \exists S_{in} \exists S_{res} S[+D]S_{in}[P(D)]S_{res}[P'(D)]S'\}$

where + requires that 7 is not in the domain of the input state and introduces it in the output state (see Beaver 2001 for details). So, expanded, (18b) is:

(20) $\{\langle S, S'\rangle | \exists S_{in} \exists S_{res}[S[+7]S_{in}[\text{MERE}(\text{CHILD})(7)]S_{res}[\text{SUCCEED}(7)]S']\}$

Imagine that (18a) is spoken in the context of a particular question under discussion, "Who kept the Dark Lord from returning?" – in homage to our naturally-occurring example, *No mere child could keep the Dark Lord from returning.* Let us assume that the domain contains three individuals: Maggie, Bart, and Marge. In every world under consideration, Maggie is a baby, Bart is a child, and Marge is an adult. It is common ground that babies cannot succeed at the task under consideration, so the possible answers to the question are (assuming that someone succeeded): Bart succeeded, Marge succeeded, and Bart and Marge succeeded. Let us introduce three possible worlds, one for each of these possibilities: w_b, w_m, and w_{bm}. Pairing each of these worlds with the empty assignment $\emptyset$, we obtain three singleton information states: $\{\langle w_b, \emptyset\rangle\}$, $\{\langle w_m, \emptyset\rangle\}$, and $\{\langle w_{bm}, \emptyset\rangle\}$. Let us refer to these as I_b, I_m, and I_{bm}, respectively. The input context is a ranking over these states corresponding to the individual-sum operation:

(21) $S = \{\langle I_b, I_b\rangle, \langle I_b, I_{bm}\rangle, \langle I_m, I_m\rangle, \langle I_m, I_{bm}\rangle, \langle I_{bm}, I_{bm}\rangle\}$

This is the context against which (20) is to be evaluated.

The first step in the evaluation of (20) is adding 7 as a new discourse referent. [+7] relates the input context S to a new context in which 7 is defined in all assignments, and assigned to an arbitrary individual:

(22) $S_{in} = \{\langle I'_b, I'_b\rangle, \langle I'_b, I'_{bm}\rangle, \langle I'_m, I'_m\rangle, \langle I'_m, I'_{bm}\rangle, \langle I'_{bm}, I'_{bm}\rangle\}$
where $I'_b = \{\langle w_b, \langle 7, \text{Maggie}\rangle\rangle, \langle w_b, \langle 7, \text{Bart}\rangle\rangle, \langle w_b, \langle 7, \text{Marge}\rangle\rangle\}$,
$I'_m = \{\langle w_m, \langle 7, \text{Maggie}\rangle\rangle, \langle w_m, \langle 7, \text{Bart}\rangle\rangle, \langle w_m, \langle 7, \text{Marge}\rangle\rangle\}$, and
$I'_{bm} = \{\langle w_{bm}, \langle 7, \text{Maggie}\rangle\rangle, \langle w_{bm}, \langle 7, \text{Bart}\rangle\rangle, \langle w_{bm}, \langle 7, \text{Marge}\rangle\rangle\}$

But now we have a problem. The discourse referent is required to be new, so nothing can be known about it. 7 could be Maggie, Bart, or Marge. But MIN requires it to be at least a child, and in normal contexts, babies are inherently ranked lower than children. So Maggie should not be assignable to 7.

To eliminate assignments of Maggie to 7, we can use domain restriction. There are at least two strategies we could employ to restrict the domain. The first approach, which is the more standard one, invokes a contextually-bound domain variable – this would be a discourse referent, say, 8 – which is to stand for the set of contextually-relevant entities. Under this approach, the logical form would become, for example:

(23) $\{\langle S, S'\rangle | \exists S_{in} \exists S_{dom} \exists S_{res}$
$S[+7]S_{in}[7 \in 8 \wedge \text{MERE}(\text{CHILD})(7)]S_{res}[\text{SUCCEED}(7)]S'\}$

Another potential solution would be to restrict the domain of a quantifier to those individuals that satisfy the scope predicate (here, *succeeded*) in any world. A possibility modal could be employed to restrict the domain, thus:

(24) $\{\langle S, S'\rangle | \exists S_{in} \exists S_{dom} \exists S_{res}$
$S[+7]S_{in}[\Diamond(\text{SUCCEED}(7))]S_{dom}[\text{MERE}(\text{CHILD})(7)]S_{res}[\text{SUCCEED}(7)]S'\}$

This approach has a certain intuitive appeal: It would seem rational to restrict quantification to entities that could conceivably satisfy the restrictor predicate. However it is formulated, its effect in this case is clear. The pair $\langle S_{in}, S_{dom}\rangle$ will be in the domain restriction CCP if S_{dom} is as follows:

(25) $S_{dom} = \{\langle I''_b, I''_b\rangle, \langle I''_b, I''_{bm}\rangle, \langle I''_m, I''_m\rangle, \langle I''_m, I''_{bm}\rangle, \langle I''_{bm}, I''_{bm}\rangle\}$
where $I''_b = \{\langle w_b, \langle 7, \text{Bart}\rangle\rangle, \langle w_b, \langle 7, \text{Marge}\rangle\rangle\}$,
$I''_m = \{\langle w_m, \langle 7, \text{Bart}\rangle\rangle, \langle w_m, \langle 7, \text{Marge}\rangle\rangle\}$, and
$I''_{bm} = \{\langle w_{bm}, \langle 7, \text{Bart}\rangle\rangle, \langle w_{bm}, \langle 7, \text{Marge}\rangle\rangle\}$

But even with domain restriction taken care of, we are *still* not ready for MERE(CHILD)(7), because this presupposes the question "What properties does 7 have?" and that is not the current CQ. We propose to introduce a new CQ that takes scope only within the restrictor of *some*, and remove it and restore the CQ to its prior state once we are "done," so to speak, with the restrictor, modulo any information that we have gained in the process of processing the restrictor. We call these processes *question accommodation* and *question resetting*, respectively. Question accommodation converts the domain-restricted state S_{dom} into a new state $S_{localQin}$ where the CQ is structured correctly. In our case:

(26) $S_{localQin} = \{\langle I_{child}, I_{child}\rangle, \langle I_{child}, I_{adult}\rangle, \langle I_{adult}, I_{adult}\rangle\}$
where $I_{child} = \{\langle w_b, \langle 7, \text{Bart}\rangle\rangle, \langle w_m, \langle 7, \text{Bart}\rangle\rangle, \langle w_{bm}, \langle 7, \text{Bart}\rangle\rangle\}$, and
$I_{adult} = \{\langle w_b, \langle 7, \text{Marge}\rangle\rangle, \langle w_m, \langle 7, \text{Marge}\rangle\rangle, \langle w_{bm}, \langle 7, \text{Marge}\rangle\rangle\}$

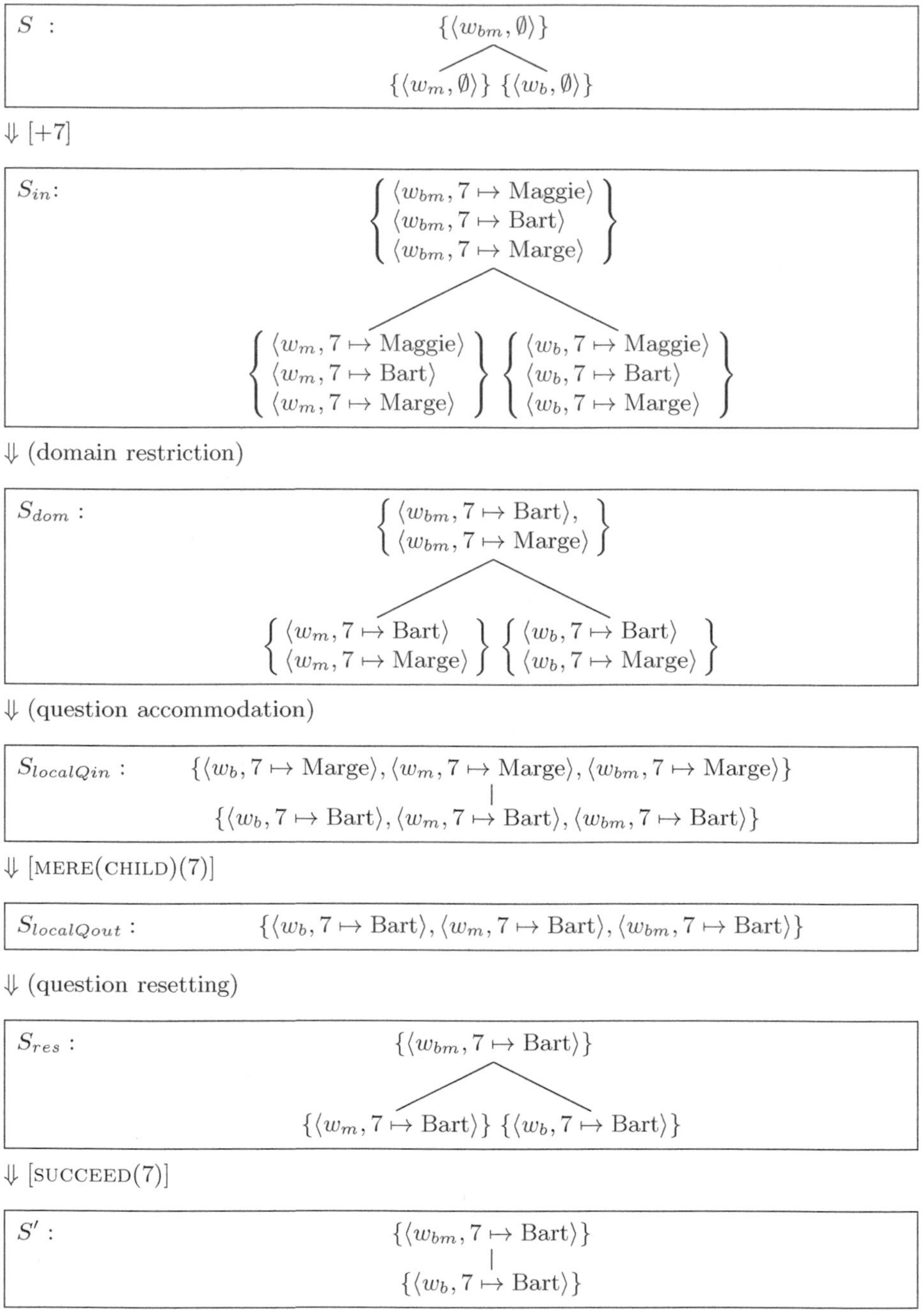

S : $\{\langle w_{bm}, \emptyset\rangle\}$ — $\{\langle w_m, \emptyset\rangle\}$ $\{\langle w_b, \emptyset\rangle\}$

⇓ [+7]

S_{in}: $\left\{\begin{array}{l}\langle w_{bm}, 7 \mapsto \text{Maggie}\rangle\\ \langle w_{bm}, 7 \mapsto \text{Bart}\rangle\\ \langle w_{bm}, 7 \mapsto \text{Marge}\rangle\end{array}\right\}$ — $\left\{\begin{array}{l}\langle w_m, 7 \mapsto \text{Maggie}\rangle\\ \langle w_m, 7 \mapsto \text{Bart}\rangle\\ \langle w_m, 7 \mapsto \text{Marge}\rangle\end{array}\right\}$ $\left\{\begin{array}{l}\langle w_b, 7 \mapsto \text{Maggie}\rangle\\ \langle w_b, 7 \mapsto \text{Bart}\rangle\\ \langle w_b, 7 \mapsto \text{Marge}\rangle\end{array}\right\}$

⇓ (domain restriction)

S_{dom} : $\left\{\begin{array}{l}\langle w_{bm}, 7 \mapsto \text{Bart}\rangle,\\ \langle w_{bm}, 7 \mapsto \text{Marge}\rangle\end{array}\right\}$ — $\left\{\begin{array}{l}\langle w_m, 7 \mapsto \text{Bart}\rangle\\ \langle w_m, 7 \mapsto \text{Marge}\rangle\end{array}\right\}$ $\left\{\begin{array}{l}\langle w_b, 7 \mapsto \text{Bart}\rangle\\ \langle w_b, 7 \mapsto \text{Marge}\rangle\end{array}\right\}$

⇓ (question accommodation)

$S_{localQin}$: $\{\langle w_b, 7 \mapsto \text{Marge}\rangle, \langle w_m, 7 \mapsto \text{Marge}\rangle, \langle w_{bm}, 7 \mapsto \text{Marge}\rangle\}$ | $\{\langle w_b, 7 \mapsto \text{Bart}\rangle, \langle w_m, 7 \mapsto \text{Bart}\rangle, \langle w_{bm}, 7 \mapsto \text{Bart}\rangle\}$

⇓ [MERE(CHILD)(7)]

$S_{localQout}$: $\{\langle w_b, 7 \mapsto \text{Bart}\rangle, \langle w_m, 7 \mapsto \text{Bart}\rangle, \langle w_{bm}, 7 \mapsto \text{Bart}\rangle\}$

⇓ (question resetting)

S_{res} : $\{\langle w_{bm}, 7 \mapsto \text{Bart}\rangle\}$ — $\{\langle w_m, 7 \mapsto \text{Bart}\rangle\}$ $\{\langle w_b, 7 \mapsto \text{Bart}\rangle\}$

⇓ [SUCCEED(7)]

S' : $\{\langle w_{bm}, 7 \mapsto \text{Bart}\rangle\}$ | $\{\langle w_b, 7 \mapsto \text{Bart}\rangle\}$

Fig. 1. Sequence of updates for *Some mere child succeeded*

We lack the space for a satisfactory characterization of question accommodation but two criteria that it should satisfy are: (i) that the information state of the pre-accommodation context should be the same as the information state of the accommodated context, and (ii) the accommodated context should be congruent to the focus alternatives of the linguistic material to be interpreted once the question has been accommodated (in this case, *mere child*). If the focus alternatives are themselves ranking relations over standard alternatives, as Krifka (1999) proposes, then this can be implemented as a constraint that the accommodated context, qua ranking relation, is a subset of the focus value of the linguistic material in question.

After question accommodation, it is possible to update with $\text{MERE}(\text{CHILD})(7)$. This update produces $S_{localQout}$, which will be $\{\langle I_{child}, I_{child}\rangle\}$ by the reasoning outlined above in the discussion of the predicative example. The final step in the processing of the restrictor is to restore the question to its original structure, taking into account the information that has been gained, yielding:

(27) $S_{res} = \{\langle I_b''', I_b'''\rangle, \langle I_b''', I_{bm}''\rangle, \langle I_m''', I_m'''\rangle, \langle I_m''', I_{bm}'''\rangle, \langle I_{bm}''', I_{bm}'''\rangle\}$

where $I_b''' = \{\langle w_b, \langle 7, \text{Bart}\rangle\rangle\}$; $I_m''' = \{\langle w_m, \langle 7, \text{Bart}\rangle\rangle\}$; $I_{bm}''' = \{\langle w_{bm}, \langle 7, \text{Bart}\rangle\rangle\}$.

Updating with $\text{SUCCEED}(7)$ eliminates world-assignment pairs where 7 did not succeed. I.e. $S_{res}[\text{SUCCEED}(7)]S'$ will hold of S' if and only if $\text{INFO}_{S'} = \{\langle w_b, \langle 7, \text{Bart}\rangle\rangle, \langle w_{bm}, \langle 7, \text{Bart}\rangle\rangle\}$. Intuitively, this is the right result; *A mere child succeeded* basically means *A child succeeded.* Retaining the structure of the CQ from the previous state, this gives the following as a final result:

(28) $S' = \{\langle I_b''', I_b'''\rangle, \langle I_b''', I_{bm}''\rangle, \langle I_{bm}''', I_{bm}'''\rangle\}$

The entire process is summarized in Figure 1.

5 Conclusion

The system we have outlined satisfies the desiderata laid out in the introduction. Because the contexts contain questions and the meanings of sentences are CCPs over such contexts, sentences may impose presuppositional constraints on the CQ, and because the answers are ranked, the presuppositions may concern the ranking over those questions. Furthermore, our system inherits properties from that of Beaver 2001 which facilitate a successful treatment of quantified presuppositions. Our system is also compositional. With this combination of properties, we can capture the meaning of sentences in which focus particles occur within the scope of a quantifier. This apparatus also has clear potential applications in further domains such as clefts where the notion of a presupposed open question is relevant.

References

Aloni, M., Beaver, D., Clark, B., van Rooij, R.: The dynamics of topics and focus. In: Aloni, M., Butler, A., Dekker, P. (eds.) Questions in Dynamic Semantics. CRiSPI. Elsevier, Oxford (2007)

Beaver, D.: Presupposition and Assertion in Dynamic Semantics. CSLI Publications, Stanford (2001)

Beaver, D.I., Clark, B.Z.: Sense and Sensitivity: How Focus Determines Meaning. Wiley-Blackwell, Chichester (2008)

Coppock, E., Beaver, D.: Sole sisters. In: Ashton, N., Chereches, A., Lutz, D. (eds.) Proceedings of SALT 21, pp. 197–217. Rutgers University, eLanguage (2011)

Jäger, G.: Only updates. In: Dekker, P., Stokhof, M. (eds.) Proceedings of the Tenth Amsterdam Colloquium. ILLC, University of Amsterdam, Amsterdam (1996)

Karttunen, L., Peters, S.: Conventional implicatures. In: Oh, C.-K., Dinneen, D.A. (eds.) Presuppositions. Syntax and Semantics, vol. 11. Academic Press, New York (1979)

Krifka, M.: At least some determiners aren't determiners. In: Turner, K. (ed.) The Semantics/Pragmatics Interface from Different Points of View, pp. 257–291. Elsevier, Oxford (1999)

Krifka, M.: Quantifying into question acts. Natural Language Semantics 9, 1–40 (2001)

Roberts, C.: Information structure in discourse: Towards an integrated formal theory of pragmatics. In: Yoon, J.-H., Kathol, A. (eds.) OSU Working Papers in Linguistics 49: Papers in Semantics. The Ohio State University, Columbus (1996)

Rooth, M.: A theory of focus interpretation. Natural Language Semantics 1, 75–116 (1992)

Steedman's Temporality Proposal and Finite Automata

Tim Fernando*

Trinity College Dublin, Ireland

Abstract. The proposal from Steedman 2005 that "the formal devices" required for temporality in linguistic semantics "are those related to representation of causality and goal-directed action" is developed using finite automata, implicit in which are notions of causality (labeled transitions) and goal-directed action (final/accepting states). A bounded granularity is fixed through a finite alphabet, the temporality of which is given special attention and applied to causal models. A notion of a string in compliance with a causal model is defined such that for a finite causal model M, the set of strings compliant with M is a regular language.

Keywords: Temporality, causal model, finite automata.

1 Introduction

In a wide-ranging study, Steedman (2005) argues that "the so-called temporal semantics of natural language is not primarily to do with time at all" (as given say, by the real line $\mathbb{R}$), and proposes rather that "the formal devices we need are those related to representation of causality and goal-directed action." Finite automata are simple candidates for such devices, with notions of causality and goal-directed action implicit in an automaton's transition table and accepting (final) states. Stepping back from automata, the strings which such automata may or may not accept are employed in Fernando 2004, 2008 to represent events of various kinds (for linguistic semantics). But does this step (back) not trivialize the notions of causality and goal-directed action offered by finite automata? The present paper is an attempt to bring these notions to the fore, and more generally, to explain why Steedman's proposal – henceforth ST – is interesting, and how it might be implemented through finite-state methods – ST_{fa}.

Motivation for ST can be found from discourse coherence, (1), down to tense and aspect, (2).

(1) Max fell. Mary pushed him.
(2) John [*has] left, but is back.

In (1), the push is most naturally understood as preceding the fall because (1) suggests a causal connection, while in (2), the difference between the simple past and the present perfect (*has left*) is that under the latter, the result of John's

* I thank my anonymous referees for suggesting changes to a previous draft.

M. Aloni et al. (Eds.): Amsterdam Colloquium 2011, LNCS 7218, pp. 301–310, 2012.
© Springer-Verlag Berlin Heidelberg 2012

departure persists through (2)'s utterance (incompatible with him back). In (1) and (2), time and its modelling by $\mathbb{R}$ are secondary to the changes or, as the case may be, non-changes that are communicated. Put crudely under ST_{fa}, time is the result of running automata. A more moderate position is that time is conceived in ST to be relative (as opposed to absolute or independent), the raison d'être of which is to place some set E of events and states in some order. Fleshed out according to Russell and Wiener (e.g. Lück 2006), this relative conception of time can be brought in line with the ST_{fa}-view of time as runtime (Fernando 2011a). Apart from their runs, however, do the automata merit a place in semantics? Consider the habitual (3a), in contrast to the episodic (3b).

(3) a. Tess eats dal.
  b. Tess is eating dal.

Not only can we assert (3a) and at the same time deny (3b), it is not entirely clear that the truth of (3a) at a world and time can be reduced to its episodic instances at that world (even if we may agree on some minimal constraints between them). Under the *rules-and-regulations* view defended in Carlson 1995, generic sentences such as (3a) are true at a world and time only if at that world and time, "structures" exist that are not "the episodic instances but rather the causal forces behind those instances" (page 225). Can automata serve as such structures? An important test is the notion of a causal model, as described in Pearl 2009. How ST_{fa} might relate to causal models is explored in the next section,[1] which puts into context the development of ST_{fa} sketched in the subsequent sections.

2 Causal Models

A *causal model* $M = \langle \mathcal{U}, \mathcal{V}, \{F_X\}_{X \in \mathcal{V}} \rangle$ consists of a set $\mathcal{U}$ of *background variables*, a disjoint set $\mathcal{V}$ of *endogenous variables*, and for each endogenous variable $X \in \mathcal{V}$, a function F_X returning the value of X, given values for the other variables (in $\mathcal{U} \cup \mathcal{V} - \{X\}$). For $Y \in \mathcal{U} \cup \mathcal{V} - \{X\}$, we say X *depends on* Y if the value of Y may matter to the value of X in that there are two settings $\boldsymbol{v}$ and $\boldsymbol{v}'$ of values for variables in $\mathcal{U} \cup \mathcal{V} - \{X\}$ differing only on Y such that $F_X(\boldsymbol{v}) \neq F_X(\boldsymbol{v}')$. The *causal network* $\mathcal{G}(M)$ *of* M is the directed graph with the set $\mathcal{U} \cup \mathcal{V}$ of nodes and an edge from Y to X precisely if $X \in \mathcal{V}$ and X depends on Y. The intuition is that F_X describes a causal mechanism for X, and edges in $\mathcal{G}(M)$ represent (immediate) causal dependence. An example where the variables are Boolean (ranging over two values 1 and 0 for truth and falsity, respectively) is the shooting squad scenario from Schulz 2011, with four variables

> C for *the court orders the execution*, O for *the officer gives the signal*, R for *the rifleman shoots*, and P for *the prisoner dies* (page 248)

and $\mathcal{U} = \{C\}$, $\mathcal{V} = \{O, R, P\}$, causal network pictured by

$$\boxed{C} \rightarrow \boxed{O} \rightarrow \boxed{R} \rightarrow \boxed{P}$$

[1] I am indebted to the referee who set this task, with pointers that got me on my way.

and F_O returning the value of C, F_R returning the value of O, and F_P returning the value of R. But suppose

(∗) the court issues its execution orders *after* the prisoner dies

or (for the non-veridical case of "the prisoner dies before the order")

(†) the prisoner dies even though the court never orders his execution.

One option is to sharpen our reading of the variables C, O, R and P by associating times $\tau(C), \tau(O), \tau(R)$ and $\tau(P)$ with them such that (at the very least) whenever the causal network has an edge from Y to X, $\tau(X)$ is understood not to precede $\tau(Y)$. Exactly what times are (points or intervals?) or what precedence is, let us put aside for now, whilst heeding the point made in Halpern and Pearl 2005 (page 849) that

> it is always possible to timestamp events to impose an ordering on variables and thus construct a recursive model corresponding to a story

where a model M is said to be *recursive* if its causal network $\mathcal{G}(M)$ is acyclic. The idea for (∗) is that assuming the prisoner dies at $\tau(P)$ (making $P = 1$), the court's orders (afterwards) fall outside $\tau(C)$ (leaving $C = 0$), and rather than being in spurious compliance with the causal model, (∗) becomes an instance of an *external intervention modifying the causal model* (an eventuality fully accounted for in Pearl 2009). So too would (†), assuming the court orders clemency at $\tau(C)$ and the prisoner dies at $\tau(P)$. But what if the court never issues an order (one way or the other)? How are we to define $\tau(C)$? Were the court to order execution a moment after $\tau(C)$ and the prisoner then to die as a result, C would be worthless. Once the prisoner dies, however, $\tau(C)$ may as well be over. But what if the prisoner does not die? What if nothing happens, and the officer, rifleman, and prisoner are left waiting? Is $\tau(C)$ to go on indefinitely, along with $\tau(O)$, $\tau(R)$ and $\tau(P)$? Common sense suggests $\tau(X)$ is a bounded interval that may vary according to the value X takes, and the values of other variables Y and their times $\tau(Y)$.

It is noteworthy that causal models typically leave time out, exceptions being "dynamic" causal models where "variables are time indexed" (Halpern and Pearl 2005, page 863), with integer subscripts on variables distinguishing values pertaining to different times.[2] It is not obvious what the shooting squad example would gain by attaching subscripts on the variables C, O, R, P. And although

[2] This practice appears to run counter to Hiddleston 2005, footnote 5, which reads

> I take the variables to represent properties in the first instance. But I treat an event as an object's having a property at a time, so the variables can equally represent a particular case (events) or a repeatable set-up (properties). The variables must represent properties in the first instance if the graphs are capable of representing multiple situations: individual events (such as Janes being bitten by the snake at a given time and place) occur only once.

Hiddleston repeats this point in page 648 (and in another journal that same year).

the questions above about $\tau(X)$ are designed to clarify what a variable X represents, their answers come at a price: the loss of simplicity is accompanied by a loss of generality, making the causal model less interesting. Can we keep the shooting squad uncluttered by $\tau(X)$? Its causal network above neatly depicts a sequence of four events $C = 1$, $O = 1$, $R = 1$ and $P = 1$ of ordering, signalling, shooting and dying. But the denial $C = 0$ that the event $C = 1$ happens is, under a common view of aspect, a non-event: a state holding at *all* subintervals of an interval, rather than an event happening at *some* subinterval.

We can bring out the distinction here between events and states, forming a string $\alpha_1\alpha_2 \cdots \alpha_n$ of sets α_i of *fluents* (i.e., temporal propositions) to describe a time at which every fluent in α_1 holds, followed by one at which every fluent in α_2 holds, and so on through α_n. Using fluents $\dot{X}$ for the equation $X = 1$, the string $\boxed{\dot{C}}\boxed{\dot{O}}\boxed{\dot{R}}\boxed{\dot{P}}$ of length 4 records the sequence of events $C = 1$, $O = 1$, $R = 1$ and $P = 1$ at four successive times. And using fluents $\neg X$ for the equation $X = 0$, the string $\boxed{\neg C}\boxed{\neg O}\boxed{\neg R}\boxed{\neg P}$ also of length 4 describes a time at which $C = 0$ followed by a time at which $O = 0$, and then $R = 0$ and finally $P = 0$. Now, the difference between an event such as $C = 1$ and a state such as $C = 0$ is that whereas the fluent $\dot{C}$ expresses a *force* for change, the fluent $\neg C$ for the latter can be assumed to be *inertial*, where

(I) an inertial fluent persists forward and backward unless forced otherwise

(Fernando 2008). The upshot for $\boxed{\neg C}\boxed{\neg O}\boxed{\neg R}\boxed{\neg P}$ is that in the absence of external forces, (I) entails that the inertial fluents $\neg C$, $\neg O$, $\neg R$ and $\neg P$ flow forward and backward in $\boxed{\neg C}\boxed{\neg O}\boxed{\neg R}\boxed{\neg P}$ to yield the string

$$\hat{s} \;:=\; \boxed{\neg C, \neg O, \neg R, \neg P}\boxed{\neg C, \neg O, \neg R, \neg P}\boxed{\neg C, \neg O, \neg R, \neg P}\boxed{\neg C, \neg O, \neg R, \neg P}\,.$$

Furthermore, drawing on a relative conception of time as change, the string $\hat{s}$ *block-reduces* to $\boxed{\neg C, \neg O, \neg R, \neg P}$, the one symbol in $\hat{s}$. More on time-as-change and block reduction in the next section. For now, suffice it to say that in case $C = O = R = P = 0$, we are left with a block reduced string of length one, in contrast to the string $\boxed{\dot{C}}\boxed{\dot{O}}\boxed{\dot{R}}\boxed{\dot{P}}$ of length 4 for $C = O = R = P = 1$.[3] Other combinations of values for the variables are possible, but these arise from interventions to the causal model referred to by the "unless"-clause in (I) above. (I) arose in Fernando 2008 to account for the so-called *Imperfective Paradox* (e.g. Dowty 1979), the point being that interruptions to the progressive are, along with interventions to the causal model and *finks*, *masks* or *antidotes* in dispositions (e.g. Bird *ta*), external forces that may interfere with a causal picture (conceived in isolation from other pictures). Rather than take for granted some notion of

[3] A block-reduced string of length 4 can also be associated with $C = O = R = P = 0$ if, for instance, we add four non-inertial fluents marking the times at which the variables are determined separately to have value 0. As should become clear below, the conception time-as-change is relative to the fluents allowed to appear in boxes.

possible world, and interpret an event $X = x$ as a set of possible worlds (i.e., a proposition, as in footnote 6 of Halpern and Pearl 2005, and footnote 5 of Schulz 2011), the approach pursued under ST_{fa} below is to build up a temporal realm of occurrences from causal mechanisms that execute alongside other (possibly interfering) causal mechanisms. In ST_{fa}, a causal mechanism is a finite automaton that operates in discrete steps to induce a discrete notion of time. That notion of time is subject to refinement insofar as other automata may run before, during or after it (or indeed instead of parts of it).

To keep matters simple first, however, let us focus on a single causal model M that makes no mention of time. If we can restrict the causal model's set of variables to a finite set, and the values these variables take to a finite set (as we can with the shooting squad), then the set of strings that comply (atemporally) with M is accepted by a finite automaton (i.e. a regular language). To make all this precise, more notation is necessary. Given a causal model M in which each variable X takes values from a set $\mathsf{Val}(X)$, let us let us collect the possible variable-value pairs under M in

$$\mathsf{vv}(M) \ := \ \{\langle X, x\rangle \mid X \in \mathcal{U} \cup \mathcal{V} \text{ and } x \in \mathsf{Val}(X)\} \ .$$

(For example, the shooting squad fluents $\dot{C}$ and $\neg C$ amount to $\langle C, 1\rangle$ and $\langle C, 0\rangle$.) Next we look at strings of subsets of $\mathsf{vv}(M)$. Given a string $s \in Pow(\mathsf{vv}(M))^*$, let $\mathsf{vv}(s)$ be the subset of $\mathsf{vv}(M)$ that appears in s — that is,

$$\mathsf{vv}(\alpha_1 \cdots \alpha_n) \ := \ \bigcup_{i=1}^{n} \alpha_i \ .$$

We say s is *M-compliant* if $\mathsf{vv}(s)$ is a function $f : (\mathcal{U} \cup \mathcal{V}) \to \bigcup_{X \in \mathcal{U}\cup\mathcal{V}} \mathsf{Val}(X)$ such that for all $X \in \mathcal{V}$,

$$f(X) \ = \ F_X(f_{-X})$$

where f_{-X} is f minus the variable-value pair $\langle X, f(X)\rangle$. Finally, let us call M *finite* if there are only finitely many variables, and every set $\mathsf{Val}(X)$ of values is finite (i.e., the union $\mathcal{U} \cup \mathcal{V} \cup \bigcup_{X \in \mathcal{U}\cup\mathcal{V}} \mathsf{Val}(X)$ is finite). Since $\mathsf{vv}(M)$ is finite for finite M, it follows that

Fact. *Given a finite causal model M, the set of M-compliant strings is a regular language.*

Note that for the shooting squad M, a string s is M-compliant iff $\mathsf{vv}(s)$ is either $\{\langle C, 0\rangle, \langle O, 0\rangle, \langle R, 0\rangle, \langle P, 0\rangle\}$ or $\{\langle C, 1\rangle, \langle O, 1\rangle, \langle R, 1\rangle, \langle P, 1\rangle\}$. The next sections provide a finite-state system that brings temporal order, paring the language of M-compliant strings down to

$$\boxed{\dot{C}}\boxed{\dot{O}}\boxed{\dot{R}}\boxed{\dot{P}} \ + \ \boxed{\neg C, \neg O, \neg R, \neg P} \, .$$

3 Finite-State Temporality

The relative conception of time as change and the associated notion of block reduction mentioned in the previous section can be understood through the example of a calendar year, represented by the string

$$s_{mo} \ := \ \fbox{Jan}\fbox{Feb}\cdots\fbox{Dec}$$

of length 12 (with a month in each box), or (adding one of 31 days d1, d2,...d31) the string

$$s_{mo,dy} \ := \ \fbox{Jan,d1}\fbox{Jan,d2}\cdots\fbox{Jan,d31}\fbox{Feb,d1}\cdots\fbox{Dec,d31}$$

of length 365 (a box per day in a non-leap year). Unlike the points in the real line $\mathbb{R}$, a box can split if we enlarge the set B of (*b*oxable) symbols we can put in it, as the change from $\fbox{Jan}$ in s_{mo} to $\fbox{Jan,d1}\fbox{Jan,d2}\cdots\fbox{Jan,d31}$ in $s_{mo,dy}$ illustrates. Or, reversing direction, from $s_{mo,dy}$ to s_{mo}, let us define two functions ρ_{mo} and bc that respectively, restricts B to the months $mo = \{\text{Jan,Feb,...Dec}\}$

$$\rho_{mo}(s_{mo,dy}) \ = \ \fbox{Jan}^{31}\fbox{Feb}^{28}\cdots\fbox{Dec}^{31}$$

and compresses a block α^n to α

$$bc(\fbox{Jan}^{31}\fbox{Feb}^{28}\cdots\fbox{Dec}^{31}) \ = \ \fbox{Jan}\fbox{Feb}\cdots\fbox{Dec} \ = \ s_{mo}$$

so that $bc(\rho_{mo}(s_{mo,dy})) = s_{mo}$. More precisely, for $A \subseteq B$, ρ_A sees only the elements of A (discarding non-A's)

$$\rho_A(\alpha_1\alpha_2\cdots\alpha_n) \ := \ (\alpha_1 \cap A)(\alpha_2 \cap A)\cdots(\alpha_n \cap A)$$

whereas *block compression* bc sees only change (discarding repetitions/stuttering)

$$bc(s) \ := \ \begin{cases} bc(\alpha s') & \text{if } s = \alpha\alpha s' \\ \alpha bc(\alpha' s') & \text{if } s = \alpha\alpha' s' \text{ with } \alpha \neq \alpha' \\ s & \text{otherwise.} \end{cases}$$

Let bc_A be the composition $\rho_A; bc$ mapping s to

$$bc_A(s) \ := \ bc(\rho_A(s)),$$

so that

$$bc_{\{\text{Jan}\}}(s_{mo,dy}) = bc_{\{\text{Jan}\}}(s_{mo}) = \fbox{Jan}\fbox{} \qquad bc_{\{\text{d1}\}}(s_{mo,dy}) = (\fbox{d1}\fbox{})^{12}$$

$$bc_{\{\text{Feb}\}}(s_{mo,dy}) = bc_{\{\text{Feb}\}}(s_{mo}) = \fbox{}\fbox{Feb}\fbox{} \qquad bc_{\{\text{d2}\}}(s_{mo,dy}) = (\fbox{}\fbox{d2})^{12}\fbox{}\,.$$

We can delete any initial or final empty boxes by a function *unpad*, which we apply after bc_A to form π_A

$$\pi_A(s) \ := \ unpad(bc_A(s)).$$

A symbol $e \in B$ is then defined to be an *s-interval* if $\pi_{\{e\}}(s)$ is $\boxed{e}$

$$s \models \text{interval}(e) \quad :\Leftrightarrow \quad \pi_{\{e\}}(s) = \boxed{e}\,.$$

For relations between intervals, we apply $\pi_{\{e,e'\}}$ to the set of strings s where e and e' are s-intervals to form

$$\mathcal{L}_\pi(\{e,e'\}) \quad := \quad \{\pi_{\{e,e'\}}(s) \mid \pi_{\{e\}}(s) = \boxed{e} \text{ and } \pi_{\{e'\}}(s) = \boxed{e'}\}.$$

There are 13 strings in $\mathcal{L}_\pi(\{e,e'\})$, one per interval relation in Allen 1983, refining the relations $\prec$ of *(complete) precedence* and $\bigcirc$ of *overlap* used in the Russell-Wiener construction of time from events; see Table 1. We have

Table 1. From Russell-Wiener to Allen

RW	**Allen**	$Pow(\{e,e'\})^*$	**Allen**	$Pow(\{e,e'\})^*$	**Allen**	$Pow(\{e,e'\})^*$
$e \bigcirc e'$	`e = e'`	$\boxed{e,e'}$	`e fi e'`	$\boxed{e}\boxed{e,e'}$	`e f e'`	$\boxed{e'}\boxed{e,e'}$
	`e si e'`	$\boxed{e,e'}\boxed{e}$	`e di e'`	$\boxed{e}\boxed{e,e'}\boxed{e}$	`e oi e'`	$\boxed{e'}\boxed{e,e'}\boxed{e}$
	`e s e'`	$\boxed{e,e'}\boxed{e'}$	`e o e'`	$\boxed{e}\boxed{e,e'}\boxed{e'}$	`e d e'`	$\boxed{e'}\boxed{e,e'}\boxed{e'}$
$e \prec e'$	`e m e'`	$\boxed{e}\boxed{e'}$	$e < e'$	$\boxed{e}\boxed{}\boxed{e'}$		
$e' \prec e$	`e mi e'`	$\boxed{e'}\boxed{e}$	$e > e'$	$\boxed{e'}\boxed{}\boxed{e}$		

$$\mathcal{L}_\pi(\{e,e'\}) \quad = \quad \text{Allen}(e \bigcirc e') + \text{Allen}(e \prec e') + \text{Allen}(e' \prec e)$$

where $\text{Allen}(e \bigcirc e')$ consists of the 9 strings in which e overlaps e'

$$\text{Allen}(e \bigcirc e') \quad := \quad (\boxed{e} + \boxed{e'} + \epsilon)\boxed{e,e'}(\boxed{e} + \boxed{e'} + \epsilon)$$

(with empty string ϵ), and $\text{Allen}(e \prec e')$ consists of the 2 strings in which e precedes e'

$$\text{Allen}(e \prec e') \quad := \quad \boxed{e}\boxed{e'} + \boxed{e}\boxed{}\boxed{e'}$$

and similarly for $\text{Allen}(e' \prec e)$. If e and e' are s-intervals, we have

$$s \models e\ R\ e' \ \Leftrightarrow\ \pi_{\{e,e'\}}(s) \in \text{Allen}(e\ R\ e') \qquad \text{for } R \in \{\bigcirc, \prec, \succ\}$$

or, for example, in the case of the Allen relation `f` (for "*f*inish"),

$$s \models e\ \mathtt{f} e' \quad \Leftrightarrow \quad \pi_{\{e,e'\}}(s) = \boxed{e'}\boxed{e,e'}.$$

Returning to the previous section, we can strengthen the notion of an M-compliant string s by requiring that for all $\langle X,x\rangle, \langle Y,y\rangle \in \mathsf{vv}(s)$ such that X depends on Y, $\pi_{\{\langle X,x\rangle,\langle Y,y\rangle\}}(s)$ is one of the strings corresponding, under Table 1, to the desired relation of precedence. (That relation may say, be weakened from $\prec$ to include the Allen subrelations `=,si,s,fi,di,o` of Russell-Wiener overlap $\bigcirc$.)

To check if a string satisfies a formula φ, the approach taken above is to choose a finite set A of symbols such that the function π_A picks out a suitable granularity at which to analyze φ. The computational pressure is to make A as small as possible, but other formulas may require finer granularities, necessitating an enlargement of A. For this reason, it may be useful to consider an infinite set E of symbols and its family $Fin(E)$ of finite subsets, forming $Fin(E)$-indexed strings $(s_A)_{A \in Fin(E)}$ in which s_A can be calculated as $\pi_A(s_B)$ for any $B \supseteq A$, and collecting these in the *inverse limit of* the system $(\pi_A)_{A \in Fin(E)}$ of functions

$$\varprojlim (\pi_A)_{A \in Fin(E)} := \{(s_A)_{A \in Fin(E)} \in \prod_{A \in Fin(E)} Pow(A)^* \mid s_A = \pi_A(s_B) \text{ whenever } A \subseteq B \in Fin(E)\} .$$

As a function between strings over the alphabet $Pow(B)$, π_A is computable by a finite-state transducer (for finite B). Hence, if the π_A-based approach just outlined works for a formula φ, the set of strings satisfying φ is regular — making φ equivalent to a formula of Monadic Second-Order logic (e.g. Thomas 1997). While finite-state methods need not proceed via π_A, what the inverse limit above offers is a system of bounded but refinable temporal granularities in which intervals are conceptually prior to points (as a box splits into a string of boxes, upon closer examination). More in Fernando 2011a.

4 Inertia and Bounded Entailments

An assumption implicit in the previous section, for instance Table 1, is that a string such as $\boxed{e}\boxed{}\boxed{e'}$ can be read as $\boxed{e, \neg e'}\boxed{\neg e, \neg e'}\boxed{\neg e, e'}$, and more generally, $\alpha_1 \cdots \alpha_n$ is equivalent to $cl_\neg(\alpha_1 \cdots \alpha_n) := \alpha'_1 \cdots \alpha'_n$ where

$$\alpha'_i := \alpha_i \cup \{\neg e \mid e \in \bigcup_{j=1}^{n} \alpha_j - \alpha_i\} \quad \text{for } 1 \leq i \leq n.$$

As information tends to get conveyed partially, it is useful to accommodate underspecification in a string by retracting this assumption (replacing strings s in Table 1 by $cl_\neg(s)$). Rather than forming $cl_\neg(s)$ freely, we insert $\neg e$ whenever appropriate (as we do when replacing strings s in Table 1 by $cl_\neg(s)$).[4] Furthermore, we can treat inertial fluents e as follows, in accordance with condition (I) from section 2. We suppose there are forces acting for and against e, and fluents fe decreeing "there is a force for e" — at least, that is, in strings such that whenever fe occurs, e holds at the next moment unless some force opposes it. More precisely, let $\mathcal{F}_e$ be the set of strings $\alpha_1 \cdots \alpha_n$ such that

$$\mathsf{f}e \in \alpha_i \quad \text{implies} \quad e \in \alpha_{i+1} \text{ or } \mathsf{f}\neg e \in \alpha_i \qquad \text{for } 1 \leq i < n$$

[4] There is a distinction here between a string qua index versus a string qua denotation drawn at length in Fernando 2011. In the former case, it is natural to equate s with $cl_\neg(s)$; not so in the latter. Note $cl_\neg$ is computable by a finite-state transducer.

(where f$\neg e$ effectively says "there is a force opposed to e"). $\mathcal{F}_e$ can be expressed succinctly as

$$\boxed{\text{fe}}\boxed{} \;\Rightarrow\; \boxed{}\boxed{e} + \boxed{\text{f}\neg e}\boxed{}$$

using a binary operation $\Rightarrow$ that maps a pair of regular languages to a regular language (Fernando 2008, 2011). Closely related to $\mathcal{F}_e$ is the regular language

$$\boxed{e}\boxed{} \;\Rightarrow\; \boxed{}\boxed{e} + \boxed{\text{f}\neg e}\boxed{}$$

which intersected with

$$\boxed{}\boxed{e} \;\Rightarrow\; \boxed{e}\boxed{} + \boxed{\text{fe}}\boxed{}$$

yields a regular language $\mathcal{I}_e$ in which e persists forward and backward in the absence of forces on it. It is natural to view the languages $\mathcal{F}_e$ and $\mathcal{I}_e$ as constraints (satisfied by the strings belonging to them), associated with states (as opposed to events) represented by e.

An example of a state that an inertial fluent may represent is given by the habitual (3a), repeated below, followed by the episodic (3b).

(3) a. Tess eats dal.
    b. Tess is eating dal.

We must be careful not to confuse a string s representing an instance of Tess eating dal with the fluent e representing the habit of Tess eating dal. We may, however, expect the string s to belong to a language $\mathcal{L}_e$ associated with e. That is, $\mathcal{L}_e$ sits alongside the sets $\mathcal{F}_e$ and $\mathcal{I}_e$ above as languages associated (in different ways) with e. The obvious question is what is the status of such languages in our semantic theory?

I close with the suggestion that such languages are specifications of causal structures required by the *rules-and-regulations* view of Carlson 1995 to ground the truth of generic sentences. (These languages are underspecifications insofar as the causal structures are automata, but at present, I see no reason for that additional specificity.) Beyond $\mathcal{F}_e, \mathcal{I}_e$ and $\mathcal{L}_e$, there are *episodes* in the sense of Moens and Steedman 1988 consisting of "sequences of causally or otherwise contingently related sequences of events" that (lest we confuse these with episodic instances at the world) are "more related to the notion of a plan of action or an explanation of an event's occurrence than to anything to do with time itself" (page 26). The intuition is that these languages are

> resources for constructing local languages for *use* in particular situations

to quote Cooper and Ranta 2008 slightly out of context. A shift is made here from monolithic truth to use (or action) that is very much in line with the focus in ST_{fa} on finite-state transducers (rather than say, the real line $\mathbb{R}$ or possible worlds). More questions are certainly left open than answered, an important issue being inference. For entailments between formulas φ and ψ based on satisfaction $\models$, we can use a language L to relativize the inclusion

$$\varphi \vdash_L \psi \;:\Leftrightarrow\; (\forall s \in L)\ s \models \varphi \text{ implies } s \models \psi$$

which is decidable provided φ and ψ are in Monadic Second Order Logic and L is regular. In fact, the inclusion remains decidable for L context-free, as observed by Makoto Kanazawa. Regular or not, L may consist of strings that represent episodic instances beyond those of any single world. It is one thing to say a causal force for L exists at a world and time, another to spell out the consequences for the episodic instances at a world and time, and do justice to competing "inference tickets" (Ryle 1949).

References

Allen, J.F.: Maintaining knowledge about temporal intervals. Communications of the ACM 26(11), 832–843 (1983)

Bird, A.: Dispositional Expressions. In: The Routledge Companion to Philosophy of Language. Routledge (2012)

Carlson, G.N.: Truth conditions of generic sentences: two contrasting views. In: The Generic Book, pp. 224–237. University of Chicago Press (1995)

Cooper, R., Ranta, A.: Natural languages as collections of resources. In: Language in Flux: Dialogue Coordination, Language Variation, Change and Evolution. College Publications, London (2008)

Dowty, D.: Word Meaning and Montague Grammar. Reidel (1979)

Fernando, T.: A finite-state approach to events in natural language semantics. J. Logic and Computation 14(1), 79–92 (2004)

Fernando, T.: Branching from inertia worlds. J. Semantics 25(3), 321–344 (2008)

Fernando, T.: Regular relations for temporal propositions. Natural Language Engineering 17(2), 163–184 (2011)

Fernando, T.: Finite-state representations embodying temporal relations. In: 9th International Workshop on Finite State Methods and Natural Language Processing, Blois, pp. 12–20 (2011a), A revised, extended version is in the author's webpage

Halpern, J.Y., Pearl, J.: Causes and explanations: a structural-model approach, Part 1: Causes. Brit. J. Phil. Sci. 56, 843–887 (2005)

Hiddleston, E.: A causal theory of counterfactuals. Noûs 39(4), 632–657 (2005)

Lück, U.: Continu'ous Time Goes by Russell. Notre Dame J. Formal Logic 47(3), 397–434 (2006)

Moens, M., Steedman, M.: Temporal ontology and temporal reference. Computational Linguistics 14(2), 15–28 (1988)

Pearl, J.: Causality: Models, Reasoning and Inference, 2nd edn. Cambridge University Press (2009)

Ryle, G.: The Concept of Mind. University of Chicago Press (1949)

Schulz, K.: If you'd wiggled A, then B would've changed Causality and counterfactual conditionals. Synthese 179(2), 239–251 (2011)

Steedman, M.: The Productions of Time. Draft tutorial notes about temporal semantics, Draft 5.0 (May 2005), `http://homepages.inf.ed.ac.uk/steedman/papers.html`

Thomas, W.: Languages, automata and logic. In: Handbook of Formal Languages: Beyond Words, vol. 3, pp. 389–455. Springer (1997)

On Scales, Salience and Referential Language Use*

Michael Franke

Institute for Logic, Language & Computation
Universiteit van Amsterdam

Abstract. Kennedy (2007) explains differences in the contextual variability of gradable adjectives in terms of *salience* of minimal or maximal degree values on the scales that these terms are associated with in formal semantics. In contrast, this paper suggests that the attested contextual variability is a consequence of a more general tendency to use gradable terms to preferentially pick out *extreme-valued properties*. This tendency, in turn, can be explained by demonstrating that it is pragmatically beneficial to use those gradable properties in referential descriptions that are *perceptually salient* in a given context.

Keywords: gradable adjectives, scale topology, salience, game theory.

1 Scale Types, and "Kennedy's Observation"

A prominent line of current research in formal semantics links the meaning of gradable adjectives to *degrees on scales* (cf. Rotstein and Winter, 2004; Kennedy and McNally, 2005). In simplified terms, the denotation of a gradable adjective A is taken to be a function g_A : Dom(A) $\rightarrow D$ that maps any applicable argument of A to a degree $d \in D$, where $\langle D, \preceq \rangle$ is a suitably ordered *scale of degrees*. Different adjectives may be associated with different kinds of scales. Usually one-dimensional scales are assumed and a distinction is made as to whether these are: (i) totally open (*tall*, *short*), (ii) totally closed (*closed*, *open*), or (iii) half-open (*bent*, *pure*). Scale types explain a number of observations, such as which adjectives can combine with which modifiers. E.g., the expression *completely A* is felicitous only if A has a totally or upper-closed scale with a maximal element: compare the felicitous *completely closed* with the awkward ?*completely tall*.

Scale types also influence the licensing conditions of utterances involving gradable adjectives in positive form. Generally speaking, a simple positive sentence like "object x has property A" is considered true whenever the contextually supplied minimal degree of A-ness, $c(A)$, is no higher than $g_A(x)$. However, the contextual standard of applicability $c(A)$ is also affected by the scale type (c.f.

* This paper reconsiders and complements an earlier paper of mine on the same subject (Franke, 2012). I would like to thank the remaining *G-Team* (Gerhard Jäger, Roland Mühlenbernd, Jason Quinley), as well as Oliver Bott, Peter van Emde-Boas, Joey Frazee, Irene Heim, Graham Katz, Manfred Krifka, Louise McNally, Rick Nouwen and the anonymous reviewers for comments and discussion.

M. Aloni et al. (Eds.): Amsterdam Colloquium 2011, LNCS 7218, pp. 311–320, 2012.
© Springer-Verlag Berlin Heidelberg 2012

Kennedy, 2007): if there is a $\preceq$-maximal or -minimal degree contained in $\langle D, \preceq \rangle$, then $c(A)$ is bound to this; otherwise it is to be retrieved more flexibly from the context of utterance. In more tangible terms, "*Kennedy's observation*" (1) says that closed-scale adjectives compare rather inflexibly to endpoints of the associated scale (modulo the usual pragmatic slack where imprecision is conversationally harmless), while open-scale terms show more contextual variability.

(1) **"Kennedy's Observation":**

scale type		contextual standard of applicability
open	$\longleftrightarrow$	variable
closed	$\longleftrightarrow$	rigid, fixed to endpoints

For example, the contextual standard for the applicability of open-scale *tall* can vary considerably from one context (talking about jockeys) to another (talking about basketbal players), whereas that of closed-scale *closed* seems glued to the denotation of a minimal (zero) degree of openness.

2 Explaining "Kennedy's Observation"

Salience of Endpoints. Kennedy (2007) tries to explain the influence of scale topology on contextual usage conditions in terms of the *salience* of endpoints (2a) and a pragmatic principle called *Interpretive Economy* (2b) which demands pragmatic interpretation to make maximal use of the available semantic resources.

(2) **"Kennedy's Explanation":**

a. **Salience Assumption:**
End-points of closed scales are salient elements provided by the conventional semantic structure.

b. **Interpretive Economy:** (Kennedy, 2007, (66), p.36)
Maximize the contribution of the conventional meanings of the elements of a sentence to the computation of its truth conditions.

The idea seems quite natural: the evaluation of expression "x is A" requires us to fix a contextual standard $c(A)$; if A is associated with a closed scale, then by (2a) the semantic structure supplies some outstanding element, which by (2b) ought to be used to set $c(A)$; if A is associated with an open scale, the semantic structure carries no such salient points and $c(A)$ can be set more variably.

We should be fully satisfied with neither (2a) nor (2b). Firstly, as for (2a), it is not clear *a priori* whether endpoints on closed scales not only appear salient to us because they are the preferred denotation of the corresponding natural language expressions. In that case, the attempted explanation would be circular. The crucial problem is that it is very hard to determine, conceptually or empirically, when an element *of an abstract semantic structure* is salient. Phrased more constructively, if salience is to play a role in an explanation of (1) it should better be an empirically informed notion of *perceptual salience*, i.e., of salience of objects of perception measuring how much an object stands out or attracts attention relative to others. Secondly, we should not stop at the formulation of a pragmatic principle like (2b) even if it seems plausible and yields the desired results, but continue to ask for a *functional motivation*: what is the added pragmatic value of the principle in question that enabled its evolution and sustenance?

Evolution of Pragmatic Standards. Potts (2008) addresses the latter issue. While adopting (2a), he seeks to explain (1), not via (2b), but instead by an evolutionary argument why speakers and hearers conventionally coordinate on endpoints as the contextual standard for the use of closed-scale adjectives. Towards this end, Potts considers a strategic game in which speaker and hearer simultaneously choose a standard of application for a closed-scale adjective. Payoffs are proportional to how close the players' choices are to each other, so that the maximal payoff ensues when players choose the same standard of application. Potts then shows that if a population initially has a slight bias towards choosing the endpoints (his way of implementing focality of endpoints), then the replicator dynamics (Taylor and Jonker, 1978) will eventually lead to all of the speakers and hearers of the whole population choosing endpoints as standards of application.

Potts' account has some shortcomings, unfortunately. For one, it fails to make clear what the particular pragmatic benefit of endpoint use is: it is just a consequence of the assumption that the to-be-explained outcome is already predominant in the population initially. What is more, Potts' account is either silent about or makes wrong predictions for adjectives with open scales. If we assume that the only thing that differentiates open-scale and closed-scale adjectives is the presence or absence of endpoints, then, looking at open-scale adjectives in the same way, we would simply drop the assumption that there is an initial bias in the population for a particular standard of comparison. But in that case the replicator dynamics will still eventually gravitate towards some *arbitrary fixed* standard of application. But that does not seem quite right, as discussed next.

Extreme-Value Principle. Adjectives with open scales, though more contextually variable, are not entirely unconstrained. Take the open-scale adjective *tall*, say, and the question when an individual x is called *tall* when compared with a group of individuals Y. Although the precise rule of application is a question of current empirical research (e.g. Schmidt et al., 2009), it seems fair to say that x is more likely or more readily counted as tall, the more x's tallness falls within the *extreme* values of tallness within group Y. Usually it is not enough for x to be just slightly taller than the average tallness in Y. Rather, the further x's tallness is from the average or expected value of Y, the more it counts as *tall*.

These considerations suggest a different explanation for (1). If there is a tendency for gradable expressions to be used preferably to describe *extreme* values, then closed-scale terms will usually be used for values close to the endpoints, while open-scale terms could be used for a wider range of values simply due to the open-endedness of the scale. In other words, I suggest to explain (3), not (1).

(3) **Extreme-Value Principle:**
Gradable terms are preferably/usually used to describe extreme values, i.e., values far away from the median/mean of a given distribution.

The remainder of this paper is therefore concerned with two things: (i) a proof of concept that (3) indeed leads to a general association along the lines of (1), and (ii) an attempt of explaining (3) as a concomitant of pragmatic language use. The pragmatic rationalization for (3) that I will offer is superficially similar to Kennedy's explanation of (1) in (2), but conceptually different. My explanation of (3) involves a notion of salience of stimuli in context (4a), paired with an account of why the use of salience is actually beneficial in conversation (4b).

(4) a. **Salience of the Extreme:**
Salience of a stimulus in a given context is proportional to its (apparent/subjectively felt) *extremeness* or *outlieriness*, i.e., to the extent that the stimulus appears unexpected or surprising against the background of the other stimuli in the context.

b. **Benefit of the Extreme:**
Describing those properties of objects that are salient is pragmatically advantageous for coordinating reference.

The main intuition that inspires (4) is this: terms are associated with extreme values because we use them, among other things, to identify referents, and for doing so the use of extreme values is a very natural and easy, yet surprisingly effective solution.[1] In order to test this intuition, I propose a simple model of referential language use, to be introduced next.

3 Referential Games

A *referential game* is a game between a sender and a receiver, both of whom observe a context $c = \langle o_1, \ldots, o_n \rangle$ that consists of $n > 0$ objects. One of these objects is the *designated object* o_d that the sender wants to refer to. The receiver does not know which object that is. The goal is to describe the designated object by naming a property of o_d. For simplicity we assume that senders can choose only one property to describe o_d with, but can indicate whether o_d has a high or low degree of that property. If, after hearing the description, the receiver guesses the right referent, the game is a success for both players; if not, it's a failure.

More formally, let us assume that objects are represented as points in an m-dimensional *feature space* $\mathcal{F} \subseteq \mathbb{R}^m$, $m > 0$. Each dimension of $\mathcal{F}$ corresponds to some gradable property: the value of dimension j is the degree to which the object in question has property j. A context is thus a set of n points in $\mathcal{F}$, which can easily be represented as an $n \times m$-matrix c. For example, the context in (5) contains three "objects", namely Hans, Piet and Paul, which are represented as a triple of features, namely their degrees of tallness, weight, and intelligence.

[1] Elsewhere I tried to show that the use of extreme values would actually be detrimental if language was exclusively used to *describe* the actual degree of a given object as closely as possible (Franke, 2012). I focus here on the model of referential language use that was also discussed in that earlier work.

(5)

	tallness	weight	intelligence
Hans	0.2	-0.1	1.3
Piet	-0.1	0.0	0.3
Paul	0.3	-0.2	0.5

To make a distinction between open and closed scales, it is reasonable to assume that open-scale features take values in $\mathbb{R}$, while closed-scale features take values on some closed interval of reals. But open and closed scales should also plausibly differ with respect to the probability that a particular degree is observed. To keep matters simple, assume that a *random context* is obtained by sampling independently n random objects, and that a random object is obtained by sampling independently m random degree values for the relevant properties. Finally, let us assume, rather naïvely, that degrees are sampled from the distributions in (6) (see also Figure 2).

(6)

scale type	distribution type
open	normal ($\mu = 0$, $\sigma = 1/3$)
totally closed	uniform on $[0; 1]$
half-open	truncated normal on $\mathbb{R}^{\geq 0}$ ($\mu = 0.1$, $\sigma = 1/3$)

Together this yields a unique probability density $\Pr(c)$ for each context c (the exact nature of which will, however, not be of any relevance here). For each round of playing a referential game, a context is sampled with $\Pr(c)$ and from that context an object is selected uniformly at random as the designated one.

Finally, let the set of messages from which the sender can choose contain exactly one pair of antonymous terms for each property of the feature space. So, the set of messages is $M = \{1, \ldots, m\} \times \{\text{low}, \text{high}\}$, where, e.g., $m = \langle j, l \rangle \in M$ has a conventional meaning saying that property j is low. For example, if Hans is the designated object in context (5) above, the sender could describe him as being short or tall, skinny or fat, stupid or smart.

4 Solving Referential Games

Intuitively, I would describe Hans as *the smart guy* in the example above. (What about you?) This is because of his comparatively high value along that dimension, and his median values for the respective others. As usual in game theory, the subsequent question of interest is: does this intuition follow from an assessment of what is an *optimal* way of playing a referential game? – Unfortunately, as we will see presently, although there are infinitely many optimal ways of playing referential games, optimality comes at the price of plausibility. Therefore we should rather ask whether there is a *natural* solution to referential games that corroborates our intuitions. This section looks at one arguably natural solution to referential games, namely a strategy that exploits *perceptual salience.* I show that under this strategy extreme-value use, as in (3), emerges, and that communicative success is often not much worse than the theoretical optimum.

Optimal Solutions. Player behavior is captured in the notion of a *(pure) strategy*, as usual. A sender strategy is a function $\sigma : C \times \{1, \ldots, n\} \rightarrow M$ mapping a context and a designated object onto a message. A receiver strategy is a function: $\rho : C \times M \rightarrow \{1, \ldots, n\}$, mapping each context and each message onto an object. Given a context c with designated object o_d, the *utility* of playing with a sender strategy σ and receiver strategy ρ is simply:

$$\mathrm{U}(\sigma, \rho, c, o_d) = \begin{cases} 1 & \text{if } \rho(c, \sigma(c, o_d)) = o_d \\ 0 & \text{otherwise.} \end{cases}$$

The *expected utility* of σ and ρ is then just the averaged utility over all contexts and designated objects, weighted by the probability of occurrence:

$$\mathrm{EU}(\sigma, \rho) = \int \Pr(c) \times \mathrm{EU}(\sigma, \rho, c)\, \mathrm{d}c\,, \text{ where}$$
$$\mathrm{EU}(\sigma, \rho, c) = \sum_{i=1}^{n} \frac{1}{n} \times \mathrm{U}(\sigma, \rho, c, i)\,.$$

As usual, we say that $\langle \sigma, \rho \rangle$ is a *Nash equilibrium* iff (i) there is no σ' such that $\mathrm{EU}(\sigma, \rho) < \mathrm{EU}(\sigma', \rho)$ and (ii) there is no ρ' such that $\mathrm{EU}(\sigma, \rho) < \mathrm{EU}(\sigma, \rho')$.

Referential games can be considered an infinite collection of games G_c one for each context c that are standard Lewisean signaling games (Lewis, 1969): for fixed c, G_c has a set of states (here: objects) that are drawn from a uniform distribution; it also has a set of messages; finally, the receiver tries to guess the actual state that only the sender knows. The maximum possible payoff attainable in each G_c is $\min(1, \frac{2m}{n})$. This is because there are $2m$ messages to encode n states. If $n \leq 2m$, perfect communication is possible; otherwise only $2m$ of the n states can be named successfully. Consequently, call $\langle \sigma, \rho \rangle$ an *optimal solution* iff it scores perfectly in all games, i.e., iff $\mathrm{EU}(\sigma, \rho, c) = \min(1, \frac{2m}{n})$ for all contexts c. It is straightforward to show that each referential game has infinitely many optimal solutions each of which is a (Pareto-optimal) Nash equilibrium.

Optimal solutions are the theoretically conceivable maximum, they exist and even abound. Unfortunately, optimal solutions might be quite implausible. To see this, consider the context in (7) with Hans as designated object.

(7)

	tallness	weight
Hans	0.1	-0.1
Piet	-1.0	-1.0
Paul	1.0	1.0

Intuitively, there is no description that plausibly refers uniquely to Hans. Consider the alternatives: *the short guy* and *the skinny guy* most plausibly refer to Piet, and *the tall guy* and *the big guy* most plausibly refer to Paul. The problem is that poor average Hans doesn't stand out at all. However, by convention, telepathy or magic, an optimal solution would, for instance, yield perfect referential success by describing Hans as *tall* and using only *big* for Paul. This is not exclusive to this specific example and it is not the only reason why optimal solutions might be quite implausible. (Notice that referential games do not require agents to use messages "reasonably" in line with their "semantic meaning". Whence that optimal solutions need not stick with "reasonable semantic meaning" either.) The lesson to learn from this is that, when it comes to referential games, it is not necessarily helpful to look at optimal communication. Even if we know that there exist optimal solutions (and even if there are infinitely many), we should rather be interested in more *natural* strategies, but use optimal strategies as a yardstick to measure communicative success.

A Natural Solution: Salience. Much could be speculated about what a natural strategy is for referential games. But instead I want to look at just one strategy that strikes me as plausible and appealing because (i) it presupposes hardly any rationality on the side of the agents, as it merely exploits the agents' cognitive make-up, but still (ii) it is remarkably successful. The strategy in focus is one in which players simply choose whatever is most *salient* from their own perspective: the sender chooses the most salient property of the designated object; the receiver chooses the most salient object given that property. Neither player thus reasons strategically about what the other player does. Players merely exploit a shared cognitive bias of perception. Still, statistically this rather myopic choice rule does fairly well and also leads to the selection of extreme values. Both of these claims will be backed up below by numerical simulations.

But let me first elaborate on the notion of salience that I would like to use, which is a notion of contextualized perceptual salience inspired by recent research on visual salience in terms of *informativity* or *surprise* (e.g. Rosenholtz, 1999; Itti and Koch, 2001; Bruce and Tsotsos, 2009; Itti and Baldi, 2009). The general idea is that, when presented with a scene, those things stand out that are *unexpected.* This may be due to sophisticated world-knowledge, but may also be due to much less sophisticated expectations raised by the immediate contextual environment. In the spirit of the latter, I suggest that how salient object i's having property j to degree c_{ij} is, is a measure of how *unexpected*, c_{ij} appears against the background of the set c_j^{-1} of degrees for property j that occur in c. For example, given a context c, the set of degrees for property j are a vector of numbers c_j^{-1}, one for each object. Such a vector could be a tuple like $\langle 2, 3, 1, 1, 2, 1, 37 \rangle$ that could, for instance, represent abstractly the tallness of my 7 sons. Most values here lie around 1 or 2, so that the one value of 37 looks suspiciously like an *outlier.* I suggest here to explore the idea that the more a degree looks like an outlier in context, the more it is perceptually salient (4a).

Indeed much work in statistics has been devoted to the issue of outlier detection (c.f. Ben-Gal, 2005, for overview). For simplicity, I explore here only one very manageable approach to outlier detection in terms of the (linear) distance between points in the feature space (c.f. Knorr and Ng, 1998; Ramaswamy et al., 2000). So define the salience of object i having degree c_{ij} for property j as:

$$\text{Sal}_{\text{lin}}(c_{ij}, c) = \sum_{i'} |c_{i'j} - c_{ij}| \, .$$

We are then interested in the *salience matrix* s for context c, given by $s_{ij} = \text{Sal}_{\text{lin}}(c_{ij}, c)$. For example, the salience matrix for example context (5) is:

(8)

	tallness	weight	intelligence
Hans	0.4	0.2	1.8
Piet	0.7	0.3	1.2
Paul	0.5	0.5	1.0

The *salience-based choice rules* for sender and receiver are then defined as follows. The sender picks the most salient property j^* for the designated object o and chooses the corresponding message m^* indicating whether the property j^* of o_d is high or low (in the given context set $c_{j^*}^{-1}$). For example, if Hans is the designated object in (5), the sender looks at the first row of (8) and selects the column (i.e., property) with the highest number, leading to expression choice *the smart guy.* On hearing m^*, the receiver picks the object with the highest salience value for j^* from all those objects whose value for j^* is as indicated (high or low). In our example, when the receiver hears *the smart guy*, he consults the column for *intelligence* in (5) and (8), drops all rows whose values in (5) are below average and selects the row (i.e., referent) with the maximal salience value among the remaining, which is indeed the correct referent in this example. Whenever objects are exactly equally salient, players randomize.[2]

This procedure leads to communicative failure for the context (7) when the designated object is Hans, but is rather successful in general (see Figure 1), despite the fact that players blindly maximize salience from their own perspective, without taking each other's strategy into account. As shown in Figure 1, salience-based choices easily reach 90% of the theoretically possible amount of referential success unless the number of objects in context far exceeds the number of properties which to describe objects with.

Moreover, salience-driven choice of referential expressions also corroborate (3). The values c_{oj^*} selected by the sender's salience-based choice rule are indeed extreme (Figure 2). These results give a proof-of-concept that it is possible to think of (3) as a concomitant of pragmatically efficient language use. In other words, if we look at the actual objects that are described as *tall*, *smart* etc., we see that their respective degrees of tallness, intelligence etc. are indeed at the outer ends, so to speak, of their respective scales. The results in Figure 2 thus also make plausible that salience-driven referential communication leads to a difference between closed- and open-scale terms, in line with Kennedy's observation (1): closed-scale terms are used for a rather narrow range of values close to the end point of the scale; open scale terms are used more variably to describe a broader, in fact, open-ended range of values.

[2] To be precise, salience-based choices were implemented as follows. If o is the designated object, the sender selects property j^* uniformly at random from $\arg\max_j s_{oj}$. If $c_{oj^*} \geq \text{median}(c_{j^*}^{-1})$, she sends message $\langle j^*, h \rangle$, otherwise $\langle j^*, l \rangle$. If $\langle j^*, h \rangle$ is the observed message, the receiver selects uniformly at random from $\arg\max_i \{s_{ij^*} \mid c_{ij^*} \geq \text{median}(c_{j^*}^{-1})\}$. If $\langle j^*, h \rangle$ is received, the same applies, except with $<$ in the previous set restriction.

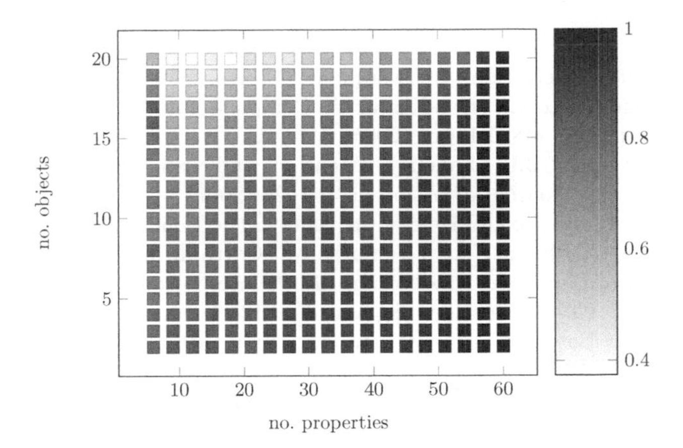

Fig. 1. Success of the salience-based choice rules relative to optimal solutions. Each square corresponds to a pair $\langle n, m \rangle$ of context size n and number of properties m (with $m/3$ properties for open, closed, and half-open scales each). For each pair, the salience-based choice rule was applied to 500 random $n \times m$-sized contexts. The color encodes the proportion of observed success rate divided by theoretical optimum for n and m.

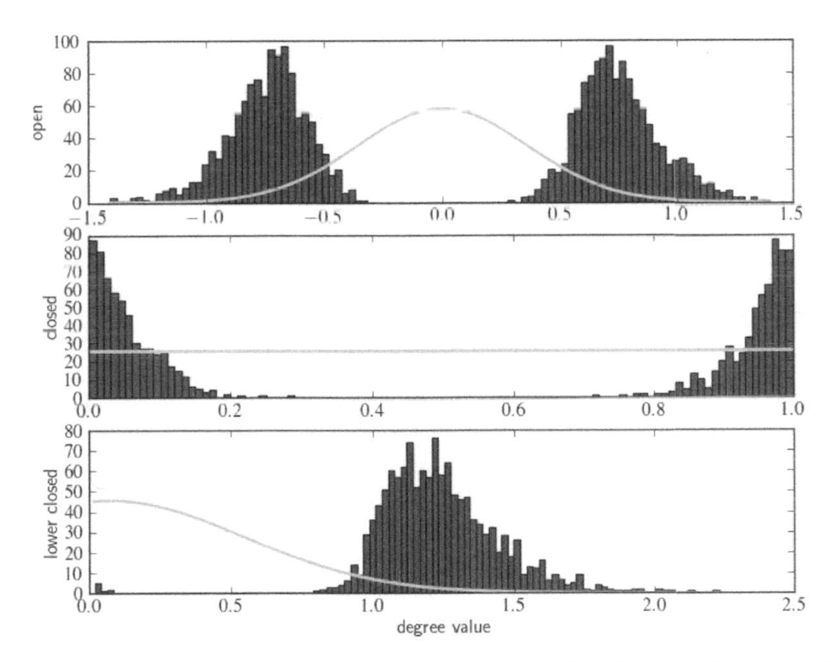

Fig. 2. Frequency with which degrees c_{oj^*} were choosen by the sender's salience-based choice rule in 5000 randomly sampled contexts with $n = 30$ and $m = 24$ (8 properties each for open, closed, and half-open scales with prior distributions indicated by the light gray lines).

References

Ben-Gal, I.: Outlier detection. In: Maimon, O., Rockach, L. (eds.) Data Mining and Knowledge Discovery Handbook: A Complete Guide for Practitioners and Researchers, pp. 131–148. Kluwer (2005)

Bruce, N.D.B., Tsotsos, J.K.: Saliency, attention, and visual search: An information theoretic approach. Journal of Vision 9(3), 1–24 (2009)

Franke, M.: Scales, Salience and Referential Safety: The Benefit of Communicating the Extreme. In: Scott-Phillips, T.C., Tamariz, M., Cartmill, E.A., Hurford, J.R. (eds.) The Evolution of Language: Proceedings of 9th International Conference (EvoLang 9), pp. 118–124. World Scientific, Singapore (2012)

Itti, L., Baldi, P.: Bayesian surprise attracts human attention. Vision Research 49 (2009)

Itti, L., Koch, C.: Computational modelling of visual attention. Nature Reviews Neuroscience 1, 194–203 (2001)

Kennedy, C.: Vagueness and grammar: The semantics of relative and absolute gradable adjectives. Linguistics and Philosophy 30, 1–45 (2007)

Kennedy, C., McNally, L.: Scale structure, degree modification, and the semantics of gradable predicates. Language 81(2), 345–381 (2005)

Knorr, E.M., Ng, R.T.: Algorithms for mining distance-based outliers in large datasets. In: Proceedings of the 24th VLDB Conference, pp. 392–403 (1998)

Lewis, D.: Convention. A Philosophical Study. Harvard University Press (1969)

Potts, C.: Interpretive Economy, Schelling Points, and evolutionary stability. UMass Amherst (2008) (manuscript)

Ramaswamy, S., Rastogi, R., Shim, K.: Efficient algorithms for mining outliers from large data sets. In: Proceedings of the ACM SIGMOD International Conference on Management of Data (2000)

Rosenholtz, R.: A simple saliency model predicts a number of motion popout phenomena. Vision Research 39, 3157–3163 (1999)

Rotstein, C., Winter, Y.: Total adjectives vs. partial adjectives: Scale structure and higher-order modifiers. Natural Language Semantics 12(3), 259–288 (2004)

Schmidt, L.A., Goodman, N.D., Barner, D., Tenenbaum, J.B.: How tall is *Tall*? compositionality, statistics, and gradable adjectives. In: Proceedings of the Thirty-First Annual Conference of the Cognitive Science Society (2009)

Taylor, P.D., Jonker, L.B.: Evolutionary stable strategies and game dynamics. Mathematical Bioscience 40(1-2), 145–156 (1978)

On the Semantics and Pragmatics of Dysfluency

Jonathan Ginzburg[1], Raquel Fernández[2], and David Schlangen[3]

[1] Univ. Paris Diderot, Sorbonne Paris Cité
CLILLAC-ARP (EA 3967), 75004 Paris, France
[2] Institute for Logic, Language & Computation
University of Amsterdam
P.O. Box 94242, 1090 GE Amsterdam, The Netherlands
[3] Faculty of Linguistics and Literary Studies
Bielefeld University
P.O. Box 10 01 31, 33615 Bielefeld, Germany

Abstract. Although dysfluent speech is pervasive in spoken conversation, dysfluencies have received little attention within formal theories of dialogue. The majority of work on dysfluent language has come from psycholinguistic models of speech production and comprehension (e.g. [10,3,1]) and from structural approaches designed to improve performance in speech applications (e.g. [14,8]). In this paper, we present a detailed formal account which: (a) unifies dysfluencies (self-repair) with Clarification Requests (CRs), without conflating them, (b) offers a precise explication of the roles of all key components of a dysfluency, including editing phrases and filled pauses, (c) accounts for the possibility of self-addressed questions in a dysfluency.

Keywords: repair, dysfluencies, dialogue semantics.

1 Introduction

Although dysfluent speech is pervasive in spoken conversation, dysfluencies have received little attention within formal theories of dialogue. The majority of work on dysfluent language has come from psycholinguistic models of speech production and comprehension (e.g. [10,3,1]) and from structural approaches designed to improve performance in speech applications (e.g. [14,8]).

Recent psycholinguistic studies have shown that both the simple fact that a dysfluency is occuring and its content can have immediate discourse effects—listeners interpret disfluent speech immediately and make use of the information it provides. E.g., [1] found that "filled pauses may inform the resolution of whatever ambiguity is most salient in a given situation", and [2] found that in a situation with two possible referents (yellow v. purple square), the fact that a description was self-corrected (e.g. 'yel- uh purple square') enabled listeners to draw the conclusion that the respective other referent ('purple square') was the correct one, before the correction was fully executed. Moreover, dysfluencies yield information: (1-a) entails (1-b) and defeasibly (1-c), which in certain settings (e.g. legal), given sufficient data, can be useful.

M. Aloni et al. (Eds.): Amsterdam Colloquium 2011, LNCS 7218, pp. 321–330, 2012.
© Springer-Verlag Berlin Heidelberg 2012

until you're |	*at the le-* ‖	*I mean* |	*at the right-hand* |	*edge*
start	reparandum ↑	editing term	alteration	continuation
	moment of interruption			

Fig. 1. General pattern of self-repair

(1) a. Andy: Peter was, well he was fired.
b. Andy was unsure about what he should say, after uttering 'was'.
c. Andy was unsure about how to describe what happened to Peter.

In this paper, we present a detailed formal account within the framework KoS [7,6, for example] which: (a) unifies dysfluencies (*self*-repair) with Clarification Requests (CRs), without conflating them, (b) offers a precise explication of the roles of all key components of a dysfluency, including editing phrases and filled pauses, (c) accounts for the possibility of self-addressed questions in a dysfluency.

We start with background on dysfluencies and on KoS. We then describe in turn our accounts of the two main types of dysfluencies, and end with brief conclusions.

2 Background

2.1 Dysfluencies: Structure and Taxonomy

As has often been noted (see e.g. [10], and references therein for earlier work), speech dysfluencies follow a fairly regular pattern. The elements of this pattern are shown in Figure 1, annotated with the labels introduced by [14] (who was building on [10]). Of these elements, all but the moment of interruption and the continuation are optional.

We partition the dysfluencies in two: (i) *backward-looking* dysfluencies (BLDs), as in (2-a,b)—the moment of interruption, which need not be followed by an explicit editing phrase, is followed by an alteration that refers back to an already uttered reparandum; (ii) *forward-looking* dysfluencies (FLDs), as in (2-c), where the moment of interruption is followed by a completion of the utterance which is delayed by a filled or unfilled pause (hesitation) or a repetition of a previously uttered part of the utterance (repetition).

(2) a. Flights to Boston I mean to Denver. (Shriberg 1994)
b. Have you seen Mark's erm earphones? Headphones. (British National Corpus, file KP0, l. 369-370)
c. Show flights arriving in uh Boston. (Shriberg 1994)

2.2 Dialogue GameBoards

We start by providing background on the dialogue framework we use here, namely KoS (see e.g. [7,6]). On the approach developed in KoS, there is actually no single context—instead of a single context, analysis is formulated at a level of information states, one per conversational participant. The dialogue gameboard represents information that arises from publicized interactions. Its structure is given in ((3))—the *spkr,addr* fields allow one to track turn ownership, *Facts* represents conversationally shared assumptions, *Pending* and *Moves* represent respectively moves that are in the process of/have been grounded, *QUD* tracks the questions currently under discussion, though not simply questions qua semantic objects, but pairs of entities which we call *InfoStrucs*: a question and an antecedent sub-utterance.[1] This latter entity provides a partial specification of the focal (sub)utterance, and hence it is dubbed the *focus establishing constituent* (FEC) (cf. *parallel element* in higher order unification–based approaches to ellipsis resolution e.g. [5].)[2]

(3) $$\text{DGBType} =_{def} \begin{bmatrix} \text{spkr: Ind} \\ \text{addr: Ind} \\ \text{utt-time : Time} \\ \text{c-utt : addressing(spkr,addr,utt-time)} \\ \text{Facts : Set(Proposition)} \\ \text{Pending : list(locutionary Proposition)} \\ \text{Moves : list(locutionary Proposition)} \\ \text{QUD : poset(Infostruc)} \end{bmatrix}$$

The basic units of change are mappings between dialogue gameboards that specify how one gameboard configuration can be modified into another on the basis of dialogue moves. We call a mapping between DGB types a *conversational rule*. The types specifying its domain and its range we dub, respectively, the *preconditions* and the *effects*, both of which are supertypes of DGBType.

Examples of such rules, needed to analyze querying and assertion interaction are given in (4). Rule (4-a) says that given a question q and ASK(A,B,q) being the LatestMove, one can update QUD with q as QUD–maximal. QSPEC is what characterizes the contextual background of reactive queries and assertions. (4-b) says that if q is QUD–maximal, then subsequent to this either conversational participant may make a move constrained to be q–specific (i.e. either About or Influencing q).[3]

[1] Extensive motivation for this can be found in [4,6], based primarily on semantic and syntactic paralleism in non-sentential utterances such as short answers, sluicing, and various other fragments.

[2] Thus, the FEC in the QUD associated with a wh-query will be the wh-phrase utterance, the FEC in the QUD emerging from a quantificational utterance will be the QNP utterance, whereas the FEC in a QUD accommodated in a clarification context will be the sub-utterance under clarification.

[3] We notate the underspecification of the turn holder as'TurnUnderspec', an abbreviation for the following specification which gets unified together with the rest of the rule: $$\begin{bmatrix} \text{PrevAud} = \left\{\text{pre.spkr,pre.addr}\right\} & : & \text{Set(Ind)} \\ \text{spkr} & : & \text{Ind} \\ \text{c1} & : & \text{member(spkr, PrevAud)} \\ \text{addr} & : & \text{Ind} \\ \text{c2} & : & \text{member(addr, PrevAud)} \\ & & \wedge \text{ addr} \neq \text{spkr} \end{bmatrix}$$

(4) a. Ask QUD–incrementation

$$\begin{bmatrix} \text{pre} & : & \begin{bmatrix} \text{I} & : & \text{InfoStruc} \\ \text{LatestMove} = \text{Ask(spkr,addr,I.q)} & : & \text{IllocProp} \end{bmatrix} \\ \text{effects} & : & \begin{bmatrix} \text{qud} = \left\langle \text{I.q,pre.qud} \right\rangle & : & \text{poset(InfoStruc)} \end{bmatrix} \end{bmatrix}$$

b. QSpec

$$\begin{bmatrix} \text{pre} & : \begin{bmatrix} \text{qud} = \left\langle \text{i, I} \right\rangle : \text{poset(InfoStruc)} \end{bmatrix} \\ \text{effects} & : \text{TurnUnderspec} \wedge_{merge} \begin{bmatrix} \text{r : AbSemObj} \\ \text{R: IllocRel} \\ \text{LatestMove} = \text{R(spkr,addr,r) : IllocProp} \\ \text{c1 : Qspecific(r,i.q)} \end{bmatrix} \end{bmatrix}$$

2.3 Grounding and Clarification

Given a setup with DGBs as just described and associated update rules, distributed among the conversationalists, it is relatively straightforward to provide a unified explication of grounding conditions and the potential for Clarification Requests (or CRification). In the immediate aftermath of a speech event u, **Pending** gets updated with a record of the form of (5) of type *locutionary proposition* (LocProp). Here T_u is a grammatical type for classifying u that emerges during the process of parsing u. The relationship between u and T_u—describable in terms of the proposition p_u given in (5)—can be utilized in providing an analysis of grounding/CRification conditions:

(5) $$p_u = \begin{bmatrix} \text{sit} = \text{u} \\ \text{sit-type} = \text{T}_u \end{bmatrix}$$

(6) a. Grounding: p_u is true: the utterance type fully classifies the utterance token.

b. CRification: p_u is false, either because T_u is weak (e.g. incomplete word recognition) or because u is incompletely specified (e.g. incomplete contextual resolution—problems with reference resolution or sense disambiguation).

We concentrate here on explicating the coherence of possible CRs. In the aftermath of an utterance u a variety of questions concerning u and definable from u and its grammatical type become available to the addressee of the utterance. These questions regulate the subject matter and ellipsis potential of CRs concerning u and generally have a short lifespan in context. To take one example, the non-sentential CRs in (7-a) and (7-b) are interpretable as in the parenthesized readings. This provides justification for the assumption that the context that emerges in clarification interaction involves the accommodation of an issue—one that for A's utterance in (7), assuming the sub-utterance 'Bo' is at issue, could be paraphrased as (7-c). The accommodation of this issue into QUD could be taken to license any utterances that are co-propositional with this issue, where

CoPropositionality is the relation between utterances defined in (8). Two questions are co-propositional if, modulo their domain, the questions involve similar answers. For instance 'Whether Bo left', 'Who left', and 'Which student left' (assuming Bo is a student) are all co-propositional. Hence the available follow up licensed in this way are CRs those that differ from MaxQud at most in terms of its domain, or corrections—propositions that instantiate MaxQud.

(7) A: Is Bo leaving?
  a. B: Bo? (= Who do you mean 'Bo'?)
  b. B: Who? (= Who do you mean 'Bo'?)
  c. *Who do you mean 'Bo'?*
  d. B: You mean Mo.

(8) *CoPropositionality*
  a. Two utterances u_0 and u_1 are *co-propositional* iff the questions q_0 and q_1 they contribute to QUD are co-propositional.
    (i) qud-contrib(m0.cont) is m0.cont if m0.cont : Question
    (ii) qud-contrib(m0.cont) is ?m0.cont if m0.cont : Prop
    (iii) q_0 and q_1 are co-propositional if there exists a record r such that $q_0(r) = q_1(r)$.

Repetition and meaning–oriented CRs can be specified by means of a uniform class of conversational rules, dubbed *Clarification Context Update Rules (CCURs)* in ([6]). Each CCUR specifies an accommodated MaxQUD built up from a sub-utterance u1 of the target utterance, the maximal element of Pending (*MaxPending*). Common to all CCURs is a license to follow up *MaxPending* with an utterance which is *co-propositional* with MaxQud. (9) is a simplified formulation of one CCUR, Parameter identification, which allows B to raise the issue about A's sub-utterance *u0*: *what did A mean by u0?* (9) underpins CRs such as those in (7).

(9) Parameter identification:

$$\left[\begin{array}{ll} \text{pre} & : \left[\begin{array}{l} \text{Spkr : Ind} \\ \text{MaxPending : LocProp} \\ \text{u0} \in \text{MaxPending.sit.constits} \end{array}\right] \\ \text{effects} & : \left[\begin{array}{l} \text{MaxQUD} = \left[\begin{array}{l} \text{q} = \lambda x\text{Mean(A,u0,x)} \\ \text{fec} = \text{u0} \end{array}\right] : \text{InfoStruc} \\ \text{LatestMove : LocProp} \\ \text{c1: CoProp(LatestMove.cont,MaxQUD.q)} \end{array}\right] \end{array}\right]$$

3 From CRs to Dysfluency: Informal Sketch

We argue that dysfluencies can and should be subsumed within a similar account, a point that goes back to [13]: in both cases (i) material is presented publicly, (ii) a problem with some of the material is detected and signalled (= there is a 'moment of interruption'); (iii) the problem is addressed and repaired, leaving (iv) the incriminated material with a special status, but within the discourse context. Concretely for dysfluencies—as the utterance unfolds incrementally questions can be pushed on to QUD about what has happened so far (e.g. *what did the speaker mean with sub-utterance u1?*) or what is still to come (e.g. *what word does the speaker mean to utter after sub-utterance u2?*).

By making this assumption we obtain a number of pleasing consequences. We can:

- **explain similarities to other-corrections:** the same mechanism is at work, differentiated only by the QUDs that get accommodated.
- **explain internal coherence of dysfluencies:** '#I was a little bit + swimming' is an odd dysfluency, it can never mean 'I was swimming' in the way that 'I was a little bit + actually, quite a bit shocked by that' means 'I was quite a bit shocked by that'. Why coherence? Because 'swimming' is not a good answer to 'What did I mean to say when I said 'a little bit'?'.
- **appropriateness changes implicate that original use unreasonable:** examples like (10) involve quantity implicatures. These can be explicated based on reasoning such as the following: *I could have said (reperandum), but on reflection I said (alteration), which differs only in filtering away the requisite entailment.*

(10) it's basically (the f- + a front) leg [implicature: no unique front leg]

4 Dysfluency Rules

As we have seen, there are various benefits that arrive by integrating CRs and dysfluencies within one explanatory framework. In order to do this we need to extend PENDING to incorporate utterances that are *in progress*, and hence, incompletely specified semantically and phonologically. Conceptually this is a natural step to make. Formally and methodologically this is a rather big step, as it presupposes the use of a grammar which can associate types word by word (or minimally constituent by constituent), as e.g. in Categorial Grammar [15] and Dynamic Syntax [9]. It raises a variety of issues with which we cannot deal in the current paper: monotonicity, nature of incremental denotations, etc.

For our current purposes, the decisions we need to make can be stated independently of the specific grammatical formalism used. The main assumptions we are forced to make concern where PENDING instantiation and contextual instantiation occur, and more generally, the testing of the fit between the speech events and the types assigned to them. We assume that this takes place incrementally. For concreteness we will assume further that this takes place word by word, though examples like (11), which demonstrate the existence of word-internal monitoring, show that this is occasionally an overly strong assumption.

(11) From [11] *We can go straight on to the ye-, to the orange node.*

BLDs are handled by the update rule in ((12)). This indicates that if u0 is a sub-utterance of the maximally–pending utterance, QUD may be updated so that the issue is 'what did A mean by u0', whereas the FEC is $u0$, and the follow up utterance needs to be be co-propositional with MaxQud:

(12) Backwards looking appropriateness repair:

$$\left[\begin{array}{ll} \text{pre} & : \left[\begin{array}{l} \text{spkr : Ind} \\ \text{addr : Ind} \\ \text{pending} = \left\langle \text{p0,rest} \right\rangle\text{: list(LocProp)} \\ \text{u0 : LocProp} \\ \text{c1: member(u0, p0.sit.constits)} \end{array}\right] \\ \text{effects} & : \left[\begin{array}{l} \text{MaxQUD} = \\ \left[\begin{array}{l} \text{q} = \lambda x \text{ Mean(pre.spkr,pre.u0,x)} \\ \text{fec} = \text{u0} \end{array}\right]: \\ \text{InfoStruc} \\ \text{LatestMove : LocProp} \\ \text{c2: Copropositional(LatestMove}^{content}, \\ \text{MaxQUD)} \end{array}\right] \end{array}\right]$$

Given ((12)), (2a,b) can be analyzed as follows: in (2-a) the alteration 'I mean to Denver' provides a direct answer to the issue *what did A mean with the utterance 'to Boston'*; in (2-b) we analyze 'headphones' as a bare fragment ('short answer') which gets the reading 'I mean headphones' given the QUD-maximality of the issue *what did A mean with the utterance 'earphones'*.

Consider now (13). This differs from (2-a) in one significant way–a different editing phrase is used, namely 'no', which has distinct properties from 'I mean'.

(13) From [11]: From yellow down to brown - no - that's red.

Whereas 'I mean' is naturally viewed as a syntactic constituent of the alteration, 'no' cannot be so analyzed. Arguably the most parsimonious analysis[4] involves assimilating this use to uses such as:

(14) a. [A opens freezer to discover smashed beer bottle] A: No! ('I do not want *this* (the beer bottle smashing) to happen')
b. [Little Billie approaches socket holding nail] Parent: No Billie ('I do not want *this* (Billie putting the nail in the socket) to happen')

This use of 'no' involves the expression of a negative attitude towards an event and would, in particular, allow 'no' to be used to express a negative attitude towards an unintended utterance event. We could analyze (13) as involving the

[4] An extended version of this paper considers and rejects resolution based on a contextually available polar question or proposition.

unintended utterance 'brown'. Following this, the rule (12) is triggered with the specification QUD.q = what did A mean by FEC? and the FEC = 'brown.' The analysis then proceeds like the earlier cases.

We specify FLDs with the update rule in (15)—given a context where the LatestMove is a forward looking editing phrase by A, the next speaker—underspecified between the current one and the addressee—may address the issue of what A intended to say next by providing a co-propositional utterance:[5]

(15) Forward Looking Utterance rule:

$$\left[\begin{array}{l}\text{preconds} : \left[\begin{array}{l}\text{spkr} : \text{Ind}\\ \text{addr} : \text{Ind}\\ \text{pending} = \left\langle \text{p0,rest} \right\rangle : \text{list(LocProp)}\\ \text{u0} : \text{LocProp}\\ \text{c1: member(u0, p0.sit.constits)}\\ \text{LatestMove}^{content} = \text{FLDEdit(spkr,u0)} : \text{IllocProp}\end{array}\right]\\ \text{effects} : \text{TurnUnderspec} \wedge_{merge}\\ \left[\begin{array}{l}\text{MaxQUD} =\\ \left[\begin{array}{l}\text{q} = \lambda x\ \text{MeanNextUtt(pre.spkr,pre.u0,x)}\\ \text{fec} = \text{u0}\end{array}\right] : \text{InfoStruc}\\ \text{LatestMove} : \text{LocProp}\\ \text{c2: Copropositional(LatestMove}^{content}\text{,MaxQUD)}\end{array}\right]\end{array}\right]$$

(15) differs from its BLD analogue, then, in two ways: first, in leaving the turn underspecified and second, by the fact that the preconditions involves the LatestMove having as its content what we describe as an *FLDEdit* move, which we elucidate somewhat shortly. Words like 'uh', 'thee' will be assumed to have such a force, hence the utterance of such a word is a prerequisite for an FLD. To make this explicit, we assume that 'uh' could be analyzed by means of the lexical entry in (16):

(16)

$$\left[\begin{array}{l}\text{phon} : \texttt{uh}\\ \text{cat} = \textit{interjection} : \text{syncat}\\ \text{dgb-params} : \left[\begin{array}{l}\text{spkr} : \text{IND}\\ \text{addr} : \text{IND}\\ \text{MaxPending} : \text{LocProp}\\ \text{u0} : \text{LocProp}\\ \text{c1: member(u0, MaxPending.sit.constits)}\\ \text{rest} : \text{address(spkr,addr,MaxPending)}\end{array}\right]\\ \text{cont} = \left[\text{c1} : \text{FLDEdit(spkr,addr,MaxPending)}\right] : \text{Prop}\end{array}\right]$$

[5] This rule is inspired in part by Purver's rule for *fillers*, (91), p. 92, ([12]).

We demonstrate how to analyze (17):

(17) From [14]: A: Show flights arriving in uh Boston.

After A utters u0= 'in', she interjects 'uh', thereby expressing FLDEdit(A,B,'in'). This triggers the Forward Looking Utterance rule with MaxQUD.q = λx MeanNextUtt(A,'in',x) and FEC = 'in'. 'Boston' can then be interpreted as answering this question, with resolution based on the short answer rule.

Similar analyses can be provided for (18). Here instead of 'uh' we have a lengthened version of 'a', which expresses an FLDEdit moves:

(18) From [11]: A vertical line to a- to a black disk.

Let us return to consider what the predicate 'FLDEdit' amounts to from a semantic point of view. Intuitively, (19) should be understood as 'A wants to say something to B *after* u0, but is having difficulty (so this will take a bit of time)':

(19) FLDEdit(A,B,u0)

This means we could unpack (19) in a number of ways, most obviously by making explicit the utterance-to-be-produced $u1$, representing this roughly as in (20):

(20) $\exists u1$[After(u1,u0) $\wedge$ Want(A,Utter(A,B,u1))]

This opens the way for a more 'pragmatic' account of FLDs, which we will sketch here, one in which (15) could be *derived* rather than stipulated. Once a word is uttered that introduces FLDEdit(A,B,u0) into the context, in other words has an import like (20), this leads to a context akin to ones like (21), that license *inter alia* elliptical constructions like sluicing and anaphora:

(21) a. A: A woman phoned. introduces issue: 'who is the woman that phoned'.
 b. A: Max drank some wine. introduces issue: 'what wine did Max drink' .

Indeed a nice consequence of (15), whether we view it as basic or derived, is that it offers the potential to explain cases like (22) where in the aftermath of a filled pause an issue along the lines of the one we have posited as the *effect* of the conversational rule ((15)) actually gets uttered:

(22) a. Carol 133 Well it's (pause) it's (pause) er (pause) what's his name? Bernard Matthews' turkey roast. (BNC, KBJ)
 b. Here we are in this place, what's its name? Australia.
 c. They're pretty ... um, how can I describe the Finns? They're quite an unusual crowd actually. `http://www.guardian.co.uk/sport/2010/sep/10/small-talk-steve-backley-interview`

On our account such utterances are licensed because these questions are co-propositional with the issue 'what did A mean to say after u0'. Such examples also highlight another feature of KoS's dialogue semantics: the fact that a speaker can straightforwardly answer their own question, indeed in these cases the speaker is the "addressee" of the query. Such cases get handled easily in KoS because turn taking is abstracted away from querying: the conversational rule QSpec, introduced earlier as (4-b), allows either conversationalist to take the turn given the QUD-maximality of q.

Concluding Comment. Finally, the account we provide has a strong methodological import: editing phrases like 'no' and 'I mean' select *inter alia* for speech events that include the discompetent products of performance. This means that the latter are also integrated within the realm of semantic competence.

Acknowledgements. Raquel Fernández acknowledges support from NWO (MEERVOUD grant 632.002.001). David Schlangen acknowledges support from DFG (Emmy Noether Programme) Some portions of this paper were presented at Constraints in Discourse 2011 in Agay. We thank the audience there as well as the reviewers for the Amsterdam Colloquium for their comments.

References

1. Bailey, K.G.D., Ferreira, F.: The processing of filled pause disfluencies in the visual world. In: van Gompel, R.P.G., Fischer, M.H., Murray, W.S., Hill, R.L. (eds.) Eye Movements: A Window on Mind and Brain, pp. 485–500. Elsevier (2007)
2. Brennan, S.E., Schober, M.F.: How listeners compensate for disfluencies in spontaneous speech. Journal of Memory and Language 44, 274–296 (2001)
3. Clark, H., FoxTree, J.: Using uh and um in spontaneous speech. Cognition 84, 73–111 (2002)
4. Fernández, R.: Non-Sentential Utterances in Dialogue: Classification, Resolution and Use. Ph.D. thesis, King's College, London (2006)
5. Gardent, C., Kohlhase, M.: Computing parallelism in discourse. In: IJCAI, pp. 1016–1021 (1997)
6. Ginzburg, J.: The Interactive Stance: Meaning for Conversation. Oxford University Press, Oxford (2012)
7. Ginzburg, J., Fernández, R.: Computational models of dialogue. In: Clark, A., Fox, C., Lappin, S. (eds.) Handbook of Computational Linguistics and Natural Language. Blackwell, Oxford (2010)
8. Heeman, P.A., Allen, J.F.: Speech repairs, intonational phrases and discourse markers: Modeling speakers' utterances in spoken dialogue. Computational Linguistics 25(4), 527–571 (1999)
9. Kempson, R., Meyer-Viol, W., Gabbay, D.: Dynamic Syntax: The Flow of Language Understanding. Blackwell, Oxford (2000)
10. Levelt, W.J.: Monitoring and self-repair in speech. Cognition 14, 41–104 (1983)
11. Levelt, W.J.: Speaking: From intention to articulation. The MIT Press (1989)
12. Purver, M.: The Theory and Use of Clarification in Dialogue. Ph.D. thesis, King's College, London (2004)
13. Schegloff, E., Jefferson, G., Sacks, H.: The preference for self-correction in the organization of repair in conversation. Language 53, 361–382 (1977)
14. Shriberg, E.E.: Preliminaries to a theory of speech disfluencies. Ph.D. thesis, University of California at Berkeley, Berkeley, USA (1994)
15. Steedman, M.: The Syntactic Process. Linguistic Inquiry Monographs. MIT Press, Cambridge (1999)

Pragmatic Constraints on Gesture Use: The Effect of Downward and Non Entailing Contexts on Gesture Processing*

Gianluca Giorgolo and Stephanie Needham

Institute of Cognitive Science, Carleton University, Ottawa, Canada
gianluca_giorgolo@carelton.ca, sneedham@connect.carleton.ca

Abstract. We report on ongoing research on the semantic and pragmatic factors that influence the interpretation of co-speech spontaneous gestures. We extend the semantic theory introduced by Giorgolo in [1] by proposing a pragmatic principle that controls the felicitousness of spontaneous gestures in different linguistic contexts. The principle is based on the idea of rationality in communication and will result in an extension of the gricean Maxim of Quantity [2]. We also present an experiment that we designed to test the predictions of the combined semantico-pragmatic principles.

Keywords: gesture, multimodality, semantics, pragmatics.

1 Introduction

This paper is concerned with the identification of some of the principles that govern the interaction between language and gesture at the semantic level. The main contribution of this paper is the extension of the model for gesture semantics proposed by [1] to take into account pragmatic factors. Our extension is quite conservative as it is based on the assumption of rational behaviour in communication, a fairly standard assumption in current pragmatic theory [3]. These extensions allows us to make interesting predictions about the possibility of observing gestures in certain linguistic contexts. The paper presents the first results of an ongoing experiment designed to test these predictions.

In this paper, we will concentrate on the spontaneous manual gestures that typically accompany verbal language. More specifically we will focus on the class of gestures usually referred to as *iconic gestures* [4]. This class of gestures includes those hand movements that are used spontaneously by speakers and that visualize physical properties of the entities or the events referred to in

* This research is supported by an Early Researcher Award from the Ontario Ministry of Research and Innovation and NSERC Discovery Grant #371969. The authors thank the two anoynmous reviewers for their insightful comments and the audience at the 18th Amsterdam Colloquium. The authors would also like thank Sebastien Plante for acting in the experiment stimuli and Raj Singh and Deidre Kelly for useful feedback on the experiments and the analysis.

M. Aloni et al. (Eds.): Amsterdam Colloquium 2011, LNCS 7218, pp. 331–340, 2012.
© Springer-Verlag Berlin Heidelberg 2012

the utterance. This type of gesture lacks a codified form of execution (in this sense they are spontaneous and free form), even though they tend to be used consistently in the same stretch of discourse [5].

In Section 2 we present in more detail the framework proposed by [1] and discuss its limitations. Section 3 introduces the simple extension we propose to the framework and discusses some of the implications of this extension. The current state of the experimental work is discussed in Section 4. We conclude with Section 5 by discussing the significance of our findings and what they mean for a theory of gesture meaning.

2 The Interpretation of Spontaneous Co-speech Gestures

Our starting point is the *semantics* for gesture proposed by Giorgolo [1]. This analysis assumes that gestures contribute to the interpretation of an utterance by providing additional information expressed in terms of an iconic representation. According to this analysis, the iconic representation is a process that identifies the salient spatial features of the referent of a gesture (be it an entity or an event) and that encodes them as visible actions. The representation identifies an equivalence class of spatial configurations that are *indistinguishable* from the virtual space created by the hand movements at a context dependent level of description (determined by the salient features picked for the gestural representation). In this semantic model the two modalities are interfaced by an operation that *intersects* the semantic content of the gesture with the content of its verbal anchor point (roughly the semantic constituent connected both temporally and semantically with the gesture).

The process can be visualized as the diagram in Figure 1. The speech component σ of a multimodal utterance is interpreted in the usual way. Each verbal constituent is assigned an abstract object taken from an ontology F of entities, events and truth values. A family of (partial) mappings *Loc* connects the ontology F to a spatial ontology S by linking each entity, event, property and relation to its spatial extension (a spatial region, a spatio-temporal region, a set of regions or a set of tuples of regions). On the other side, the gesture γ is translated from a collection of motoric configurations into a virtual spatial object. This virtual space is then used to create the equivalence class of the spaces that are sufficiently similar to the represented one. The characteristic function of this class is taken to be the core meaning of the gesture. Finally the meaning contribution of the verbal component and the one provided by gesture are combined via a generalized meet operation.

This model makes already a number of strong predictions partially confirmed by experiments reported in [1]. For instance the model restricts the distribution of gestures by constraining their co-occurrence with verbal expressions. According to the model gestures can co-occur only with those linguistic constituents whose interpretation (under one of the *Loc* mappings) is of a type that can be intersected with the meaning of a gesture (basically the model restricts the verbal correlate of gestures to expressions that denote properties or relations). At the

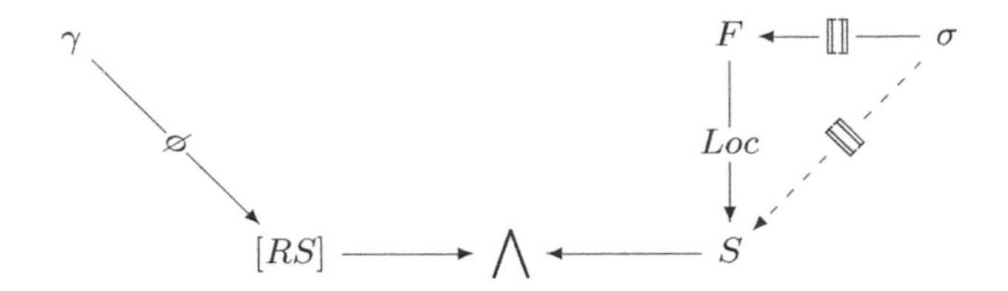

Fig. 1. Interpretation process for a multimodal utterance

same time the model precludes the possibility of introducing discourse referents with gestures (see [6] for a different approach in which gestures have the possibility to introduce discourse referents and are in general treated as discourse segments).

However, in its current form, the model cannot predict the distribution of information between the two modalities. The extension we present in the next section allows us to make simple predictions of this kind. A full model of the distribution of information between modalities would require the introduction of more theoretical constructs, such as a way of quantifying the cost of expressing a piece of information in a given modality. In this paper we present just a first attempt in this direction that does not require the introduction of any additional theoretical assumptions besides the notion of rationality in communication, already independently motivated by the study of verbal language.

3 Pragmatic Constraints on Gesture Interpretation

Gesture is not the primary mode of communication in most cases. This is particularly true for the spontaneous gestures we are dealing with in this paper. Although we pay attention to the hands of our interlocutor, our attention is focused primarily on the facial area with quick glimpses at what the hands are doing [7]. We expect that communication has evolved so that both listeners and speakers take into consideration this fact when they engage in a conversation.

We can couple this observation with the general rules governing communication that have been proposed when considering language in isolation. In particular the various incarnations of the Maxim of Quantity seem to be particularly related to the situation we are dealing with. In its most general form the maxim assigns a lower and upper bound to the amount of information that a contribution to the conversation should convey. The general idea is that the contribution should be such that it presents our interlocutor with enough information to move the conversation towards the desired goal (some ideal informational state) without including irrelevant or excessive details. By observing naturally occurring gestural data, one has pretty soon the impression that gestures often play the role of providing some more details about the situation under discussion without crossing the upper bound limit imposed by the Maxim of Quantity. In a sense gestures allow us to smuggle some more information into the conversation without making it too heavy. The freedom of adding more information, however, is combined with the lower saliency that information expressed gesturally shows. A speaker encoding a piece of information in gesture runs the risk of seeing that information disregarded by her interlocutor. We expect this fact to be built in the rules of communication and to determine how people use gestures.

We can express this rule in the following terms: *encode a piece of information as a gesture only if it provides additional information moving you closer to your goal and if it is not necessary to achieve your goal.* The idea is therefore that gestures can only monotonically increase the amount of information available and that they can do so only if their contribution is not vital for the success of the conversation. This rule allows us to preserve the validity of the Maxim of Quantity. With gesture we are apparently allowed to break it by introducing some information that often (if paraphrased in verbal terms) would make our contribution too heavy. At the same time we are not breaking the rule by marking this additional information as not particularly salient and by allowing our interlocutor to disregard it safely.[1]

Assuming the modifier-like semantics for gesture we sketched above, we can find linguistic contexts in which the use of a gesture would contravene the rule just stated. The cases we focus on in this paper is the one of downward monotone and non-monotone contexts. Given the non propositional content of gestures we focus on the contexts induced by determiners and quantifiers. In those contexts deciding to use a gesture to convey additional information may result in a communicative failure. In fact in the case of a downward monotone context failing to integrate the gestural information may result on the part of the interlocutor in an interpretation that is too strict and that may even not include the state of information that is the goal of the speaker. Similarly in the case of a non monotone context the adjoined information may be crucial to determine the correct interpretation. This is never the case in an upward monotone context, as any interpretation that includes the additional contribution of the gesture is in a sense a subset of the laxer interpretation without gesture.

An alternative but very similar way of looking at our predictions is in terms of the distribution of information between modalities, and from the perspective of the generation of the communicative message. As already mentioned, the gestural modality can only be used to convey information that is not considered crucial *at the moment of the contribution.* The information conveyed by gesture may have been crucial at a previous point in time, may be crucial at a future moment or may never be relevant for the communicative goals of the speaker. Combining this constraint with the intersective nature of gestural meaning we arrive at the same conclusions outlined before. In terms of information distribution, in fact, a speaker can choose to add in an upward monotone context a piece of information

[1] There are linguistic expressions that signal the necessity of shifting attention from language and towards other modalities such as gesture. For example the use of deictics like "this" and "so" and similar phrases like "shaped like this" or "this big" create the effect of marking the gestural information as necessary for the communicative goals. We expect that in those cases this rule and the predictions depending on it would not apply.

that is not at the moment relevant but that may become relevant at a later point in time. In contrast in a downward monotone context the intersection of information is not compatible with the addition of information (what a speaker may do is to repeat information that is already part of the common ground, for instance using a gesture together with a definite description that is linked to a specific entity already presented as having the properties encoded by the gesture).

According to both interpretations the fact that gestures cannot co-occur with downward and non entailing contexts is a derived property, depending on the assumptions that gestures are interfaced at the level of semantics via an intersection operation and that their function is to communicate peripheral information.

To show that that downward and non monotone contexts are bad candidates for the use of gestures, we designed an experiment that takes advantage of the fact that listeners are attuned to the dispreference of speakers for using gestures in those contexts. We expect listeners to lack a strategy to interpret gestures in those contexts or to have at least a preference for not integrating them.

4 Experiment

The experiment is designed to measure whether subjects integrate or not the information conveyed by gesture in their mental representation, depending on the monotonicity of the linguistic contexts in which the gesture appears. The prediction is that preference for integration is dependent on the context and that the upward monotone contexts will show a strong preference for integrating gesture (around 75% according to previous measurements of [1]). On the other hand, downward and non monotone contexts will show a strong dispreference for an integrated interpretation.

4.1 Experimental Setup

We designed a set of stimuli to test our hypotheses. The experiment consists of 12 stimuli grouped in three conditions corresponding to the three possible monotonic behaviours of determiners and quantifiers (upward entailing, downward entailing and non-entailing). 9 stimuli presented gestures accompanying the restrictor of the determiner, and 3 gestures accompanying the predicate of the determiner/quantifier. By measuring the responses in both the nominal (restrictor position) and the verbal (predicate position) domains we aim at testing whether indeed a uniform intersective interface is a plausible model for the integration of gesture and speech, and provide therefore additional evidence for the model of [1].

The experiment consists in a decision task. Each stimulus is a short audio and video recording of a confederate engaged in a seemingly natural conversation. The part of the conversation shown to the subject is a short monologue introducing a context ended with a quantified statement about the main topic of the monologue. The monologues are all constructed according to a single template. A class of entities or (possible) events is introduced together with a number of relevant subcategories. Each subcategory is identified by some physical property and an iconic gesture is associated with that property. The final statement contains a reference to the general category accompanied by a gesture associated with one of the subcategories. The clip is cut in such a way that the statement gives the impression of being the first part of a longer utterance. Subjects were asked to pick the most likely continuation of the monologue from a list of sentences. To clarify the setup we present below one of the stimuli used, the "spiders" stimulus. This stimulus exemplifies the case of a gesture used in a downward monotone context. The speech segments accompanied by a gesture are annotated with square brackets ([]) and with a tag identifying the gesture. The gestures are shown in Figure 2.

> *"Spiders" transcript*: Last week I booked a flight down to South America 'cause I'm going there for vacation this summer and I was talking to some friends about —you know— what it's like there and one of my friends was telling me about the spiders there especially in the jungle and he was saying that there's these [LITTLE : little spiders] that are everywhere all the time and these [BIG : big fat spiders] that pretty much only come out at night and fortunately none of the [BIG : spiders] are actually poisonous...

The possible continuations for this stimulus are listed below:

1. but the small ones are deadly poisonous
2. in fact some people get bitten on purpose by the small ones because they think it helps prevent heart diseases
3. so I definitely need to buy some serum in case I get bitten
4. the spiders living there are among the deadliest on the planet

(a) LITTLE (b) BIG

Fig. 2. Still frames of the gestures used in the "spiders" stimulus

Each stimulus included from 3 to 6 possible continuations, all corresponding to three possible types of continuation associated with the condition and explained in details below. This means that in all stimuli created with more than 3 conditions each group was represented by more than one continuation. This was done to avoid having the subject rely on the underlying assumptions of the experiment. In this way we hoped to obtain more natural responses based simply on the plausibility of the continuation.

The continuations are constructed in a way that allows us to use them to infer the interpretation that subjects assign to the stimuli. We assume that there are always two available interpretations: one that includes the gestural information (what we call the *integrated interpretation*) and one constructed without it (the *non integrated interpretation*). To detect which interpretation is associated with each stimulus the continuations are designed to be compatible with only one of the two available interpretations and in contradiction with the other. In other words, given two interpretations Γ and Δ each continuation p is constructed in such a way that only one of $\Gamma \wedge p$ and $\Delta \wedge p$ is satisfiable.

However notice that, given the semantics of gesture, it is possible to construct this type of continuation only in the case of non-entailing contexts. In the case of upward and downward entailing contexts this is not possible. The reason in both cases is the fact that one interpretation entails the other. In the case of an upward entailing context $\Gamma[\cdot]^{\uparrow}$ we have that the common ground including the gestural information is entailed by the one that does not include it, in symbols $\Gamma[G(A)]^{\uparrow} \leq_t \Gamma[A]^{\uparrow}$, where G is the denotation of a gesture and A the denotation of the linguistic constituent that the gesture modifies. This follows from the definition of upward entailing function and from the intesective semantics of gestures (we have in general that $G(A) \leq A$, where A is an object of some boolean type and G a gesture modifying it). This means that whenever the addition of a proposition p to a common ground $\Gamma[G(A)]^{\uparrow}$ is satisfiable so it is its addition to the weaker common ground $\Gamma[A]^{\uparrow}$ (this is a consequence of the fact that conjunction with a fixed proposition is an upward monotone operation). It is however possible to construct a proposition whose addition to $\Gamma[A]^{\uparrow}$ is satisfiable, while its addition to $\Gamma[G(A)]^{\uparrow}$ leads to a contradiction. In the case of downward entailing contexts we have the dual situation: given that $\Gamma[A]^{\downarrow} \leq_t \Gamma[G(A)]^{\downarrow}$ we can only find a proposition that is compatible with $\Gamma[G(A)]^{\downarrow}$ without being compatible with $\Gamma[A]^{\downarrow}$, as any proposition compatible with $\Gamma[A]^{\downarrow}$ will be compatible also with $\Gamma[G(A)]^{\downarrow}$ (where with compatible we mean that its conjunction with a proposition is satisfiable).

In total we have 4 possible types of possible continuations, depending on whether they are compatible with the two available interpretations, the integrated and the non integrated one. The four possible types of continuations are summarized below. The types are codified as pairs that define whether the continuation is compatible (expressed by +) or not (expressed by -) with the interpretation that includes the information conveyed by gesture (first component of the pair) and the interpretation that does not include the gestural contribution (second component of the pair).

1. $\langle +, + \rangle$, a continuation compatible with both the integrated and the non integrated interpretation,
2. $\langle +, - \rangle$, a continuation compatible only with the integrated interoperation,

3. $\langle -, + \rangle$, a continuation compatible only with the non integrated interpretation,
4. $\langle -, - \rangle$, a continuation in contradiction with both the integrated and the non integrated interpretation.

Table 1 groups the types of continuations according to their use in the three conditions. The first column lists the two possible interpretations, while the other three columns show which type of continuation was used as a witness of the interpretation in the different conditions (continuations of type $\langle -, - \rangle$ are not present in the table because they do not correspond to any possible interpretation).

Table 1. Types of continuations used in the three experimental conditions

Gesture interpretation	Conditions		
	Upward monotone	Downward monotone	Non monotone
Integrated	$\langle +, + \rangle$	$\langle +, - \rangle$	$\langle +, - \rangle$
Non integrated	$\langle -, + \rangle$	$\langle +, + \rangle$	$\langle -, + \rangle$

Returning to the "spiders" example, continuation 1 and 3 can be added to the common ground only if the gesture has been integrated in the interpretation (an effect similar to the one obtained by substituting the last sentence of the monologue with 'none of the big spiders are poisonous'). Continuation 2 is the only continuation compatible with the non-integration of the gesture, but at the same time it is also compatible with an integrated interpretation. The fact that none of the big spiders are poisonous does not entail that the small ones are, even though pragmatic factors may favor this interpretation (a fact that would make our result stronger).

To each list of continuations we added a distractor in the form of a continuation of type $\langle -, - \rangle$ (see continuation 4 in the "spiders" example).[2]

The experiment was run online with subjects recruited mainly among the undergraduate population of Carleton University. The instructions made no mention of gestures to avoid an unnatural focus of subjects on the manual modality. The sequence of stimuli was randomized as was the order in which the proposed continuations were presented. Subjects were allowed to watch each clip more than one time.

4.2 Results

We report on the results of a first experiment involving 20 subjects. For each subject we measured the response on all 12 stimuli. Three stimuli (two testing

[2] The distractor was added mainly to control for low quality entries in the results, given we could not control for particularly poor experimental conditions.

the non monotone condition and one testing the upward monotone condition) have been removed from the data analysis due to problems in their construction.

Figure 3 shows the distribution of the preferences for the integrated interpretation in the three conditions. The preferences are computed as a single score by summing the number of times the integrated reading was chosen in each condition for each subject, and normalized between 0 and 1 (where 0 represents complete dispreference for the integrated interpretation and 1 represents complete preference). It is clear that subjects strongly preferred the integrated interpretation in the upward monotone condition (score of 0.802), while in the case of the downward and non monotone contexts subjects dispreferred the integrated interpretation (scores respectively of 0.296 and 0.351). Performing a Tukey's HSD test shows the mean preference for the integrated interpretation in upward monotone contexts is significantly different from the one of downward and non monotone contexts (p-values respectively of 0.00013 and 0.00008), while the distribution of preference for integrated interpretations in downward and non-monotone contexts is to all purposes the same ($p = 0.991$), as predicted by the model.

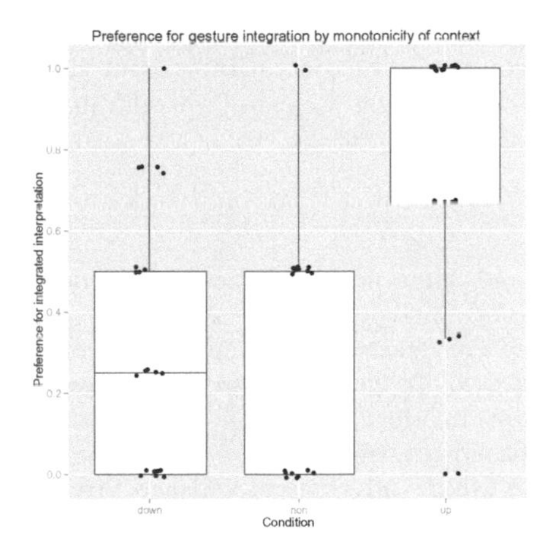

Fig. 3. Distribution of the preference for the integrated interpretation by condition

Other factors do not influence the preference for the integrated reading. In particular the fact that the gesture modifies the restrictor or the predicate of the determiner does not influence the reading preferred (one-way ANOVA, $F(1,52) = 0.7416, p = 0.3931$).

The results are consistent with the model we propose. We are at the moment performing additional experiments to confirm the results of this pilot study. In particular we are measuring the responses to a baseline condition that does not include gesture. In this way we hope to find whether our results could be explained in terms of general common sense reasoning or if indeed the gestural dimension is necessary to obtain the results observed.

5 Conclusion

We extended the semantics presented in [1] by adding a rule that deals with the pragmatic constraints of gesture use. The extension is based on the notion of rational behaviour in communication and more particularly on the Maxim of Quantity. This new framework makes strong and unexpected predictions about the possibility of encoding information as gesture in certain linguistic contexts. To test the validity of these predictions we designed an experiment that confirmed our expectations. The results are possibly compatible with other theories of gesture meaning. For instance it is possible to adapt the semantics proposed by Lascarides and Stone in [6] to deal with the data we collected. However such an adaptation would require very strong assumptions as it would be based on a theory of anaphoric-like connections between gesture and language. Our theory does not require any such stipulation as it is based on the simplest possible semantics we can associate with gesture and a general principle about human communication.

We are in the process of extending our empirical explorations to further confirm our hypotheses. In particular we want to use a different experimental setting that does not force our subjects to focus too much on logical operations. At the same time we want to confirm our expectations about the possibility of breaking the pragmatic rule we proposed by means of specific linguistic expressions.

References

1. Giorgolo, G.: Space and Time in Our Hands. PhD thesis, Utrecht Institute for Linguistics OTS, Utrecht: LOT publications 262 (2010)
2. Grice, H.P.: Logic and conversation. In: Cole, P., Morgan, J. (eds.) Syntax and Semantics. Speech Acts, vol. 3, pp. 41–58. Academic Press, New York (1975)
3. Horn, L.R.: Implicature. In: Horn, L.R., Ward, G. (eds.) The Handbook of Pragmatics. Blackwell Publishing, Oxford (2004)
4. McNeill, D.: Hand and Mind. University of Chicago Press, Chicago (1992)
5. McNeill, D.: Gesture and Thought. University of Chicago Press, Chicago (2005)
6. Lascarides, A., Stone, M.: A Formal Semantics Analysis of Gesture. Journal of Semantics (2009)
7. Mancas, M., Pirri, F., Pizzoli, M.: Human-motion Saliency in Multi-motion Scenes and in Close Interaction. In: Proceedings of Gesture Workshop, Athens (2011)

Sameness, Ellipsis and Anaphora

Daniel Hardt[1], Line Mikkelsen[2], and Bjarne Ørsnes[3,*]

[1] Copenhagen Business School
dh.itm@cbs.dk
[2] University of California at Berkeley
mikkelsen@berkeley.edu
[3] Freie Universität Berlin/Copenhagen Business School
Bjarne.Oersnes@fu-berlin.de

Abstract. We compare explicit assertions of sameness with analogous elliptical and anaphoric expressions, and find striking differences in their interpretation. We account for those differences with a two part proposal: first, we propose that *same* is additive, similar to *too*. Second, *same* must take scope over a containing event-denoting expression. We give evidence that the scope-taking of *same* is subject to standard island constraints, and we also show that *same* always compares two event-denoting clauses that differ in a relevant property.

Keywords: Ellipsis, anaphora, *same*.

1 Introduction

It is a truism that elliptical and anaphoric expressions are in some sense interpreted the same as the antecedent. For example, one widely cited work on ellipsis and anaphora poses as its central question: "Under what circumstances can bits of a syntactic structure be said to be the same as or different from other bits of a syntactic structure?" ([6][p xi]). In this paper we examine expressions involving explicit assertions of sameness (*do the same* and *the same* N), and we find that they differ in surprising ways from analogous elliptical and anaphoric expressions (VP ellipsis, *do it/that/so* and pronouns).

We observe that *same* always compares a described event with an antecedent event. The two events must be distinct, and the antecedent event description must be true in context. Our analysis is that *same* is additive and furthermore must take scope over a containing event-denoting expression. We show that this scope-taking is subject to standard island constraints on syntactic movement. We then examine differences between *same* and the additive particle *too*, and suggest that these differences have to do with the fact that *same* must compare events, and thus is subject to constraints on how events are individuated. We end with a brief discussion of related work on so-called "internal" readings of *same*.

* Thanks to Maribel Romero, Irene Heim, Maziar Toosarvandani and an anonymous reviewer for useful feedback and discussion. Bjarne Ørsnes was supported by the *Deutsche Forschungsgemeinschaft* under the grant nr. DFG (MU 2822/2-1).

M. Aloni et al. (Eds.): Amsterdam Colloquium 2011, LNCS 7218, pp. 341–350, 2012.
© Springer-Verlag Berlin Heidelberg 2012

2 Same vs. Ellipsis and Anaphora

We begin with example (1);

(1) I feel it's important to vote in favor (although I don't have a vote). I appeal to my colleagues to do so/it/that/*the same, for the good of European citizens.

Here, *do the same* is infelicitous, while the related forms (*do so/it/that*) are felicitous. Intuitively, *do the same* is ruled out because the antecedent voting event did not actually occur. If the example is modified so that the antecedent event did occur, *do the same* is acceptable.

(2) I voted in favor. I appeal to my colleagues to do the same, for the good of European citizens.

Example (3) illustrates another case where *do the same* is ruled out where related forms are all acceptable.

(3) John caught a big fish, and he did VPE/so/it/that/*the same without any fishing equipment[1].

Here, it seems that what rules out *do the same* is the very fact that it is referring to the same event; if the example is modified so that there are two different events, *do the same* is fine.

(4) John caught a big fish last week, and he did the same yesterday without any fishing equipment.

So, far we have seen that *do the same* describes an event that must differ from the antecedent event description, and second, the antecedent event description must be true in context.

We follow [4] in claiming that *same* requires an event-based account. It is perhaps not surprising that *the same* is naturally described in terms of events when appearing as the object of *do*, since it must describe an event (in this case, an action) in such cases. However in our view *same* always requires reference to events. To show this, we turn now to cases where *same* appears in ordinary NP's like *the same book*, as in the following examples (Irene Heim, p.c.):

(5) I read *War and Peace* on my last vacation, and I hope that you will read {it/the same book} (next year).
(6) I never got around to reading *War and Peace*, but I hope that you will read {it/*the same book} (next year).

---

[1] In this example VPE (VP ellipsis) is acceptable, in addition to the other VP anaphora forms. VPE is not acceptable for example (1), presumably because of specific constraints on VPE when introduced by *to* in embedded clauses. See [8,10] for discussion.

This contrast is very similar to the one observed between (1) and (2) – unless the antecedent event, *reading War and Peace*, actually happened, *the same book* is infelicitous, while the pronoun is acceptable. Example (7) affirms the contrast observed in (3): unlike a pronoun, *same* does not allow the antecedent event to be the same as the current event.

(7) I read *War and Peace* on my last vacation, and I read {it/*the same book} in a single sitting.

Here, *the same* book is infelicitous, except perhaps on the reading where one has read *War and Peace* on two separate occasions.

These last two contrasts show that, even when *same* occurs in an ordinary, individual-denoting NP, it involves comparison of events.[2] We conclude that *same* places two constraints on the event being described: first, the described event must be distinct from the antecedent event, and second, the antecedent event description must be true in context.

3 Same Is Additive

In our view, the above observations suggest that *same* is additive, similar to *too*. This is supported by the following variants of (1) and (2) involving *too*:

(8) I feel it's important to vote in favor (although I don't have a vote).
I appeal to my colleagues to do so/it/that *too, for the good of European citizens.

(9) I voted in favor. I appeal to my colleagues to do so/it/that too, for the good of European citizens.

Too patterns with *same* here, generating infelicity unless the antecedent event description is true in context. It is well known that *too* is "additive", in that it adds to the current context a clause that is closely related to an antecedent clause, but differing in some way. A recent characterization can be found in [14][p 5] (see also [13,9]):

(10) Let ϕ *too*$_i$ be an LF, with *too* co-indexed with LF ψ_i. Then $c + \phi\ too_i$ is defined iff:

$$\text{(a) } [[\psi]] \neq [[\phi]], \text{ (b) } [[\psi]] \in [[\phi]]^F, \text{ (c) } c + \psi = c \text{ (i.e. } \psi \text{ is true in } c)$$

(a) ensures that the antecedent is distinct from the phrase occurring with *too*
(b) identifies an "appropriately contrasting antecedent"
(c) ensures that the antecedent is true in context

[2] Note that we follow ([11]) in understanding the term "event" to cover processes and states as well as events in a narrow sense.

We propose that *same* gives rise to the same conditions (a) - (c); furthermore, ϕ is always an event-denoting expression containing *same.* We state the conditions imposed by *same* as follows:

(11) Let $same_i$ ϕ be an LF, with *same* co-indexed with LF ψ_i. Then $c+$ $same_i$ ϕ is defined iff:

(a) $[[\psi]] \neq [[\phi]]$, (b) $[[\psi]] \in [[\phi]]^F$, (c) $c + \psi = c$ (i.e. ψ is true in c)

Returning to examples (1) and (2), we have

(12) [my colleagues do the same (= vote in favor)]$_{VP}$

We assume that *do the same* is interpreted as *vote in favor.* With *same* adjoined to the containing VP, we have:

(13) [same [my colleagues vote in favor]$_{VP}$]$_{VP}$

Assuming there is focus on "colleagues", for (2):
$[[\phi]]$ = my COLLEAGUES vote in favor; $[[\phi]]^F$ = x vote in favor; $[[\psi]]$ = I vote in favor

(a) $[[\psi]] \neq [[\phi]]$ (ok)
(b) $[[\psi]] \in [[\phi]]^F$ (ok)
(c) c + ψ = c (ψ true in current context) (ok)

For example (2), conditions (a) - (c) are all satisfied and the example is thus acceptable. For example (1), condition (c) is not satisfied – there is not an appropriately contrasting antecedent ψ that is true in context.

Example (3) is ruled out by condition (a):

(14) [john did the same (= caught big fish)]$_{VP}$
[same [john caught big fish]]

ϕ = JOHN caught big fish; $[[\phi]]^F$ = x caught big fish; ψ = john caught big fish

a) $[[\psi]] \neq [[\phi]]$ (false)
b) $[[\psi]] \in [[\phi]]^F$ (ok)
c) $c + \psi$ = c (ok)

We turn now to the contrast in (5) and (6), repeated here:

(15) I read *War and Peace* on my last vacation, and I hope that you will read {it/the same book} (next year).
(16) I never got around to reading *War and Peace*, but I hope that you will read {it/*the same book} (next year).

For both (5) and (16), we have ϕ = YOU read War and Peace; $[[\phi]]^F$ = x read War and Peace; ψ = I read War and Peace

a) $[[\psi]] \neq [[\phi]]$ (ok)
b) $[[\psi]] \in [[\phi]]^F$ (ok)

For (5) the antecedent *I read War and Peace* is true in context (condition c), but not for (6), accounting for the difference in acceptability. The account of (7) is completely parallel to that of (3). Note that this depends on *same* taking scope over an event-denoting constituent that contains the NP in which it appears. We turn now to this issue.

4 Scope

The most obvious difference between *same* and *too* stems from the fact that *too* is an adverb that can be overtly adjoined to a variety of constituents, including clauses. *Same* is an adjective that moves covertly to take scope over a clausal element. Here we show that this covert movement is subject to island constraints. The example in (17) is a case in point.

(17) John knows why Mary killed a fish.
Harry knows why she did so/it/that, too.
*Harry knows why she did the same thing.[3]

The *why* clause is an island for movement being a *wh*-clause, thus *same* cannot take scope over the matrix clause; instead it takes scope over [she did the same thing]. This leads to a violation of our condition (a), distinctness. This problem does not arise with *too*, since it is adjoined to the matrix clause. Here the contrast between *John* and *Harry* satisfies distinctness.

Compare this with the example in (18): where *same* appears in an untensed VP:

(18) John asked Mary to catch a fish.
Harry asked her to VPE/do so/do it/do that, too.
Harry asked her to do the same thing.

We find (18) completely acceptable with *same*. This parallels the contrast found with wh-movement:

(19) *What does John know why Mary caught?
What did John ask Mary to catch?
What did John say Mary caught?

The *wh*-clause is a barrier to movement, so movement out of the *wh*-clause is impossible. This explains why (17) is bad with *same*. However, (19) shows that movement is possible out of an untensed clause and out of a finite clause embedded under the bridge-verb *say*. This explains why (18) is acceptable with *same* and also why the example in (20) is acceptable with *same*.

[3] Some of our examples have *do the same* while others have *do the same thing*. We treat them as being interpretively equivalent, and have nothing to say about what bears on the choice between them.

(20) John said Mary caught a fish.
Harry said she did VPE/so/it/that, too.
Harry said she did the same thing.

Similar contrasts are found with other well-known island constraints:

Complex NP Constraint:

(21) Peter rejected the claim that Mary had caught a fish.
Harry rejected the claim that she did VPE/so/it/that too.
*Harry rejected the claim that she did the same.

Sentential Subject Constraint:

(22) That Mary caught a fish bothers Harry.
That she did VPE/so/it/that bothers Frank, too.
*That she did the same bothers Frank.

Adjunct Constraint:

(23) Mary laughs when she catches a fish
Harry laughs when she does VPE/so/it/that, too.
*Harry laughs when she does the same thing.

In all of these examples with island environments we find that *same* is impossible or highly degraded. As an adjective *same* moves covertly to take scope over *Mary had caught a fish* or *Mary catches a fish.* But *same* cannot move out of the island. In all cases we find a violation of our principle (a): distinctness. The two compared events are not distinct. *too*, being an adverb that can adjoin to any clause, is less restricted in its scope. Thus, with *too* we can have a contrast in the referents of the matrix subjects, ensuring that the two compared events are sufficiently distinct.

5 Individuation

We have seen that *same* sometimes differs from *too* for a structural reason: since *same* must move to take scope over the material to be compared, it is subject to island constraints. Here we observe other cases where *same* differs from *too*: but in these cases we argue that the difference is semantic – we argue that *same* must compare events that differ in a relevant property, while *too* is more flexible.

(24) Harry should vote, and
he will VPE/do so/do it/do that, too.
*he will do the same.

Here *too* is acceptable because it compares two propositions that differ in modality (*should* vs *will*). Modality, however, is not a property of events, so that it is not available as a basis for comparison with *same.* Tense on the other hand is a property of events, which explains the acceptability of *same* in the following example:

(25) Harry voted last year, and
he will next year, too.
he will do the same next year.

More generally, we expect the contrast requirement of *same* to track the Davidsonian conception of events in (26); from [11].

(26) Events are particular spatiotemporal entities with functionally integrated participants.

Only elements that individuate events (event type, participants, time and location) can satisfy the contrast requirement of *same*. For both *same* and *too* the contrast requirement is satisfied by the element expressing new or contrastive information, i.e. a focussed element. As a clause-level adjunct, *too* does not place any restrictions on what elements of that clause may be focussed. In contrast, *same* occurs as part of an anaphoric VP or NP, making certain constituents unavailable for focussing and contrast. This is illustrated by (27):

(27) I encourage all my colleagues to go to the meeting.
I encourage them to VOTE, too.
*I encourage them to do the same.

Here, *too* contrasts the VP's *go to the meeting* and *vote*. *Same* cannot do that here – it cannot contrast the VP that it appears within. Similarly, in (28), *too* contrasts the NP's *early meeting* and *late meeting*. But it is not possible for *same* to contrast the NP it appears within.

(28) I encourage all my colleagues to boycott the EARLY meeting.
I encourage them to boycott the LATE meeting, too.
*I encourage them to boycott the same LATE meeting.

In general, then, the contrasted element cannot appear in the same constituent as *same*. One interesting variation involves *orphans* ([5][p 285-6]), as illustrated by (29).

(29) I encourage all my colleagues to boycott the EARLY meeting.
I encourage them to do the same with the LATE meeting.

In [12], it is argued that the NP *the LATE meeting* functions semantically as an argument to the VP, although it appears syntactically as an adjunct to VP. Being outside the VP anaphor, it can be focussed and being adjoined to VP, *same* can scope over it, explaining the felicity of (29).

6 Related Work

Several authors have proposed that *same* can move to take scope over a clause. [7] briefly considers *same* and *different*, and argues that they can move to take clausal scope, much as she argues for comparative morphemes (see also [1] on comparative uses of *same*). [4] also argues that *same* and *different* can take clausal scope, and furthermore argues that *same* and *different* always compare events. Carlson distinguishes between "internal" and "external" readings. In the current paper, we have dealt exclusively with external readings, that is, readings where there are two separate clauses to be compared. An internal reading arises in a clause that "provides its own context", as Carlson puts it [p 532]. An example of this is (30), discussed by [3]:

(30) Anna and Bill read the same book.

The internal reading of (30) is described by Barker as follows: (30) is true "just in case there exists some book x – any book x – such that Anna read x and Bill read x" ([3][p 2]). The internal reading in (30) relies on the existence of the conjoined NP *Anna and Bill*, which can be interpreted distributively. More generally, Barker argues that *same* must raise in the derivation of internal readings, and that movement is "parasitic" on the presence of another quantificational element (in the case of (30), a conjoined NP).

Both Carlson and Barker focus their attention on internal readings. In the current paper, we have restricted our attention to "external readings", in which the interpretation of *same* relies on an antecedent in prior discourse, and we have focused on differences with corresponding anaphoric and elliptical forms. [2] performs a similar comparison, arguing that for pronouns, the anaphoric link is *stipulated*, while for *same*-NP's it is *asserted*. This is based on three differences observed: first, *same*-NP's can appear with *almost*, second, they can appear in existential *there*-sentences, and third, they often carry certain existential implications. Both the observations and the proposed account of them are quite different from those of the current paper, and we will have to leave to future work the task of integrating these two perspectives.

Barker considers external readings briefly, but does not attempt to extend his account to them – in fact, he argues against doing this ([3][p 6-9]). Similarly, in this paper we have not attempted to extend our account of external readings to apply to internal readings. However, in this section we make some observations about the relation between the two readings.

First, the internal readings are subject to the same island effects we have observed for external readings. This is observed by [4][p 534]: "the licensing NP must appear within the same 'scope domain' as the dependent expression". For example, in (17) above, *same* could not take scope over the matrix clause, making the external reading of *same* infelicitous. In (31), below, the same effect can be observed with an internal reading:

(31) *John and Harry know why Mary killed the same fish.

Similar effects can be seen with internal versions of all of the island effects we have observed for external readings.

We also observed for external readings that the two events must differ in a particular event-property. Thus example (24) shows that a contrast between modals is not felicitous, since modality is not a property of events. The same is true of internal readings, as shown by the following contrast:

(32) *Harry necessarily reads the same book.
Harry usually reads the same book.

While we won't attempt any specific analysis of such internal readings, we suggest that the contrast in (32) is similar to that observed between (24) and (25). *same* compares events in terms of a distinction in properties, thus quantification over times (*usually*) can license an internal reading, while a modal quantification (*necessarily*) cannot.

Finally we have this observation: while Barker correctly observes that the internal reading of *same* requires a quantificational element that can function as a "trigger" for that reading, his notion of "parasitic" scope suggests that the movement of *same* is only possible when such a trigger is present. Discussing an example involving the quantificational element *everyone*, Barker notes, "The reason I call this parasitic scope is that the scope target for same does not even exist until everyone has taken scope." ([3][p 21]) Our account of external readings casts doubt on this suggestion, since we have proposed that *same* moves in a similar way for external readings, in the absence of any such trigger.

7 Conclusions

Much of the literature on elliptical and anaphoric expressions rests on a notion of sameness: in some sense the elliptical or anaphoric expression is to be understood as the same as the antecedent. One might naturally expect, therefore, that explicit assertions of sameness would be interpretively indistinguishable from analogous elliptical or anaphoric expressions. In this paper we have seen that this is emphatically not the case; unlike their anaphoric and elliptical analogues, explicit sameness expressions require a comparison of two event-denoting expressions. The truth of the antecedent event-denoting clause is presupposed, and there is a requirement that the two events differ in a relevant property.

As far as we know, these facts have not been previously observed – however, we suspect that our account of these facts can be fruitfully related to a body of work accounting for so-called internal readings with *same* and related terms. While we have not attempted to extend our account to such internal readings, we have pointed out several points of commonality that we plan to explore in subsequent work.

References

1. Alrenga, P.: Dimensions in the semantics of comparatives. Ph.D. thesis, University of California Santa Cruz (2007)
2. Alrenga, P.: Stipulated vs. asserted anaphora. Paper Presented at the 83rd Annual LSA Meeting in San Francisco, January 11 (2009)

3. Barker, C.: Parasitic scope. Linguistics and Philosophy 30(4), 407–444 (2007)
4. Carlson, G.: Same and different: Consequences for syntax and semantics. Linguistics and Philosophy 10(4), 531–565 (1987)
5. Culicover, P.W., Jackendoff, R.: Simpler Syntax. Oxford University Press (2005)
6. Fiengo, R., May, R.: Indices and Identity. MIT Press, Cambridge (1994)
7. Heim, I.: Notes on comparatives and related matters, ms., University of Texas, Austin (1985)
8. Johnson, K.: What VP ellipsis can do, and what it can't, but not why. In: Baltin, M., Collins, C. (eds.) The Handbook of Contemporary Syntactic Theory, pp. 439–479. Blackwell, Oxford (2001)
9. Krifka, M.: Additive particles under stress. In: Proceedings of the Conference on Semantics and Linguistic Theory, pp. 111–128. Cornell University, Ithaca (1999)
10. Lobeck, A.: Ellipsis: Functional Heads, Licensing, and Identification. Oxford University Press, Oxford (1995)
11. Maienborn, C.: Event semantics. In: Maienborn, C., von Heusinger, K., Portner, P. (eds.) Semantics: An International Handbook of Natural Language Meaning. De Gruyter Mouton (2011)
12. Mikkelsen, L., Hardt, D., Ørsnes, B.: Orphans hosted by VP anaphora. In: Proceedings of 29th West Coast Conference on Formal Linguistics (2011)
13. Riester, A., Kamp, H.: Squiggly Issues: Alternative Sets, Complex DPs, and Intensionality. In: Aloni, M., Bastiaanse, H., de Jager, T., Schulz, K. (eds.) Logic, Language and Meaning. LNCS, vol. 6042, pp. 374–383. Springer, Heidelberg (2010)
14. Singh, R.: VP-deletion, obligatory *too*, and focus semantics, MIT (2008) (unpublished manuscript)

As Simple as It Seems

Vincent Homer

ENS-IJN
29, rue d'Ulm 75005 Paris, France

Abstract. Even when linearized after it, *seem* can take scope over the modal *can,* provided that a downward-entailing trigger is present locally. Downward-entailingness plays a key role in the scope reversal: *seem* takes scope not only over *can,* but also over the trigger, because it is a mobile positive polarity item, which has the ability to raise covertly above a potential anti-licenser. This movement permits aspectual configurations that are otherwise disallowed in the complement of *seem.*

Keywords: Polarity, Raising, Modal, Tense, Infinitive.

1 Introduction

It has been claimed (Homer (2010, 2012b), Iatridou & Zeijlstra (2010)) that a number of intensional verbs, e.g. deontic *must,* are PPIs which undergo covert raising in order to scope out of an anti-licensing environment; this claim helps explain an English phenomenon known under the name of 'can't seem to' construction, a case of syntax-semantics mismatch (Langendoen (1970), Jacobson (2006)):

(1) John can't seem to lose weight.
Paraphrasable as: It seems that John can't lose weight. SEEM≫NEG≫CAN

This phenomenon consists in the reversal of the relative scope of *seem* (and for some speakers, *appear*) and *can* (in the schema, E_{DE} is a downward-entailing expression):

(2)	Surface order (ignoring V-to-T movement):	E_{DE}	...	can	...	seem
		1		2		3
	Scopal relations:	SEEM	≫	E_{DE}	≫	CAN
		3		1		2

Some conditions are necessary (though not sufficient). 1. Only a few modals can take part in the scope reversal with *seem:* in fact, previous researchers claim that only *can,* and more specifically yet, only ability *can,* lends itself to the scope reversal (Property 1). Although it is certain that deontic and epistemic *can* are not involved, this description is too restrictive, in view of (3):

(3) There can't seem to be enough vampire movies. ✓SEEM≫NEG≫CAN

I also wish to point out that for a number of speakers, *will* enters the scope reversal too:

(4) He won't seem to give me a straight answer. ✓SEEM≫NEG≫WILL

M. Aloni et al. (Eds.): Amsterdam Colloquium 2011, LNCS 7218, pp. 351–360, 2012.
© Springer-Verlag Berlin Heidelberg 2012

Paraphrasable as: It seems that he cannot give me a straight answer.

2. The reversal only occurs in the presence of an expression (henceforth the 'reversal trigger') which denotes a DE function (Prop. 2):

(5) #John can seem to lose weight. *SEEM≫CAN
Not paraphrasable as: It seems that John can lose weight.

3. Furthermore, *seem* achieves wide scope over both the trigger and *can* (in that order) (2) (Prop. 3); 4. contrary to what normally happens in present-tense sentences (6), the predicate under *seem* need not be stative, witness *lose weight* in (1) (Prop. 4):

(6) *John seems/doesn't seem to lose weight.[1]

5. The two verbs have to be relatively 'close' to each other (Prop. 5). 6. Semantically, more than just scope reversal is involved: sentences that instantiate the phenomenon yield an actuality implication (Prop. 6; lack of space doesn't allow me to discuss it):

(7) This man can't seem to climb Mount Everest.
Context A: The speaker just looks at a man and only bases her judgment on his apparent health condition.
Context B: The speaker knows that the man tried to climb Mt. Everest and failed.

Sentence (7) cannot be uttered felicitously in context A, but it can in context B.

I argue that scope reversal is not illusive (*contra* Jacobson (2006)) and that a fully compositional analysis is possible. First, I discard two possible accounts: the phenomenon is not idiomatic (Sect. 2), and neg-raising is not involved (Sect. 3). Sect. 4 offers an analysis in terms of a covert movement motivated by positive polarity; this movement also explains why in the presence of *can* non-statives are exceptionally available in the complement of *seem* in present tense sentences (Sect. 5).

2 Not an Idiom

There are a number of reversal triggers besides negation: they form a substantial set, which is included in the familiar natural class of (Strawson) DE expressions:

(8) a. <u>Few</u> can seem to fathom how he could be so popular. [Jacobson 2006, ex. 9]
b. <u>At most five people</u> can seem to understand this.
c. John can <u>never</u> seem to speak in full sentences. [Jacobson 2006, ex. 7]
d. I just bought this lens, and I can <u>rarely</u> seem to get a clear picture.
e. <u>Only</u> John can seem to stomach watching reruns of the 6th game of the 1986 Series. [Jacobson 2006, ex. 10]

They are thus characterized by their variability and their predictability: this allows us to eliminate the hypothesis that *can, seem,* and the trigger jointly form an idiom. Still, a possible rejoinder would be that there really are two parts to consider, a negatively polarized idiom *can seem* (SEEM≫CAN) on the one hand, and some licenser (the

[1] The sentence is grammatical under a non episodic reading.

trigger) on the other. This is a non-starter as well. In effect, the trigger can (and in fact must) take intermediate scope between the other two elements: (8-d) for example is paraphrasable as (9-a), not as (9-b):

(9) a. It seems that I rarely can get a clear picture.
 b. It seems upon rare occasions that I can get a clear picture.

In (8-d), the adverb *rarely* binds the closest situation variable in its scope, which is an argument of *can,* and hence quantifies over situations of being able to get a clear picture, not over *seeming* eventualities. Even if *can seem* were an idiom in its own right, it would be expected to combine compositionally with the rest of the sentence. Then the trigger would have to scope over the semantic construct *can seem* (SEEM≫CAN)—since it is generated higher than *can* and *seem*—contrary to fact. A proponent of the idiom line would then be forced to say that the trigger makes a non-compositional contribution to meaning in combination with *can* and *seem.* In other words, the three elements would have to form an idiom. But we have already rejected this possibility: therefore the scope reversal is not idiomatic. Still one might have in mind yet another option, which is eliminated in the next section, namely that the reversal is due to neg-raising.

3 Not Neg-Raising

Seem can achieve wide scope over negation and negative quantifiers by the *semantic* route of neg-raising. *Think* is an example of a neg-raising predicate (NRP): a homogeneity inference, which is responsible for the effect, is attached to it (Gajewski (2005)):

(10) John doesn't think that I understand French.
 Paraphrasable as: John thinks that I don't understand French.
 a. *Assertion:* It is not the case that John thinks that I understand French.
 b. *Homogeneity inference:* John thinks that I understand French or John thinks that I don't understand French.
 ∴. John thinks that I don't understand French.

The neg-raised reading of a sentence containing a negated NRP is often favored, but it is not mandatory. Importantly, when it obtains, negation is syntactically higher than the NRP, it is not transported back into a low position. A distinctive property of NRPs is that only they pass the *cyclicity test,* as shown for *want* in (11). The test uses a negation and an NRP embedded under another NRP: it is passed when negation is interpreted below the lower NRP although it surfaces above the higher one (Horn 1978):

(11) I don't think that John wants to help me. ✓THINK≫WANT≫NEG

(12) I don't think that John seems to understand the situation. ✓THINK≫SEEM≫NEG

Seem passes the test too, and is thus an NRP. But it is covert raising (syntax), not neg-raising (semantics), which explains the scope reversal: the predicted neg-raised reading of (1) is (13) (due to the universal projection of the homogeneity inference associated with *seem,* see Gajewski (2005)), and it is not the desired reading:

(13) In all worlds w' compatible with J.'s abilities in w*, it seems in w' that J. isn't losing weight.
≠ It seems that there are no worlds w' compatible with J.'s abilities in w* such that J. is losing weight in w'.

4 Movement

We have seen that *seem* gives rise to neg-raised readings under DE expressions; these readings require *seem* to be syntactically *under* these expressions. But it is also possible for *seem* to be interpreted in a higher position than scope-bearing elements linearized before it, a fact that has not been documented yet, as far as I can tell.

We will need to distinguish two kinds of movements: (i.) *seem* can be interpreted in a higher position than certain elements, non-DE adverbs in particular (but the identity of the mover and the direction of the movement are unclear) (Sect. 4.1); (ii.) *seem, qua* mobile PPI (like deontic *must* and *supposed*), raises covertly out of the scope of a potential anti-licenser (Sect. 4.2 through 4.4).

4.1 Non-DE Adverbial Expressions

Because *seem* is a raising-to-subject verb, its wide scope over quantified subjects can always be due to A-reconstruction. But adverbs are not usually assumed to A-move, let alone reconstruct. Therefore the fact that *seem* can outscope a number of adverbs, e.g. *often, always, easily,* that precede it on the surface, might suggest that it raises covertly past them:

(14) a. *Context:* Just looking at the hospital's visitors register, a doctor says...
People often seem to visit the patient of room 32. ✓SEEM≫OFTEN
b. Some of you guys easily seem to forget that football is a team sport. ✓SEEM≫EASILY

But there is a fact that could point to A-movement and reconstruction of adverbs after all: remarkably, the only option is surface scope when *seem* takes a tensed complement:

(15) a. It often seems that people visit this patient.*SEEM≫OFTEN; OFTEN≫SEEM
b. It easily seems that some of you guys forget that football is a team sport. *SEEM≫EASILY; EASILY≫SEEM

Now, whatever the mechanism of wide scope over non-DE elements may be, it doesn't suffice to account for the scope reversal with DE expressions: Prop. 2 (a DE expression is needed) and Prop. 3 (*seem* outscopes both the trigger and the modal) can only be understood in light of the fact that certain intensional verbs are mobile PPIs.

4.2 Mobile PPIs

The deontic modal *must* is a mobile PPI (as claimed in Israel (1996), Iatridou & Zeijlstra (2010)): this fact is established in Homer (2010, 2012b). The property of *must* that bears directly on the present discussion is its ability to raise covertly past an potential anti-licenser, e.g. negation (*must* can be shown to be generated under negation; it raises covertly from this position; V-to-T is semantically idle, see Chomsky (2000)):

(16) a. John must_{deon}n't leave. MUST≫NEG;*NEG≫MUST
b. LF: [John$_1$ must_{deon} not ___ [t_1 leave]]

Evidence for the movement of $must_{deon}$, which I label *escape* in Homer (2010, 2012b), comes from occurrences of a quantified subject scopally sandwiched between raised $must_{deon}$ and a clausemate negation. I show in Homer (2012b) that the particular scopal configuration instantiated in (17) cannot obtain through a purely semantic route and has to be syntactic (in fact, *must* is not an NRP); therefore the intermediate scope of quantified subjects is a test for movement:

(17) *Context:* The rules of this bowling game state that exactly one pin must remain standing, no matter which one...
Exactly one pin must_{deon}n't be knocked down. ✓MUST≫EXACTLY_ONE≫NEG

PPIs can be 'shielded' from an anti-licenser by interveners, e.g. *every, always,* conjunction, because-clauses, etc. Remarkably, when $must_{deon}$ is shielded, it cannot raise (compare with *a single person,* which is not an intervener):

(18) a. Not everyone must_{deon} leave. *MUST≫NEG; NEG≫MUST
b. Not a single person must_{deon} leave. MUST≫NEG; *NEG≫MUST

Escape is a last resort, i.e. it is blocked when unnecessary (due to shielding). Importantly, it is also clause-bound:

(19) You don't think John must_{deon} be friendly. *MUST≫NEG≫THINK

The syntactic mechanism whereby PPIs acquire wide scope over potential anti-licensers is not to be confused with the semantic route to wide scope, *viz.* neg-raising. The two processes are distinct but not incompatible: nothing in principle precludes the conjunction of the two properties in a given predicate. Such is indeed the case of *should* and—for some speakers—of *supposed* (Homer 2012b). Such is also the case of *seem,* which has already been shown to be an NRP: I now set out to show that it is also a mobile PPI.

4.3 PPIs Interpretable under a Clausemate Negation

First, it bears saying that it is generally assumed, and wrongly so, that in order to be a PPI, a given expression must be unable to be interpreted under a clausemate negation. In Homer (2012a) I propose a theory of polarity item licensing which predicts that there can exist PPIs which are licit in such a position. This theory of licensing has three main ingredients: (i.) polarity items are sensitive to the monotonicity of their syntactic environment (rather than to c-command by a DE expression); (ii.) to be licensed, a PI needs to find itself in at least one constituent that has the appropriate monotonicity w.r.t. its position (upward for a PPI, downward for an NPI); (iii.) only certain constituents are eligible for the evaluation of the acceptability of PIs, but a PI can be licensed in any of the eligible constituents in which it is acceptable. Regarding the eligibility of constituents, the presence of the Pol head, which hosts the polarity operator of the sentence (negation for negative sentences, and a positive operator for positive ones) is required in the constituents in which the acceptability of *some,* a well-known PPI, is checked. Specifically, for each CP γ that contains *some,* only the constituents that contain the Pol head of γ are eligible. This condition on eligibility seems to be lexically determined because it is PI-specific. The contrast between (20-a) and (20-b) falls out from this hypothesis:

(20) a. John didn't understand something. *NEG≫SOME
b. It is impossible that John understood something. ✓IMPOS.≫SOME

In (20-a), all eligible constituents are DE w.r.t. the position of *some* under negation, whereas in (20-b), *some* is acceptable in at least one eligible constituent, e.g. the embedded TP. Just like *must*$_{deon}$, French *devoir*$_{deon}$ 'must' is a mobile PPI (Homer 2010); but unlike *must,* it is interpretable under a clausemate negation (without shielding), which suggests that the smallest eligible constituents for its evaluation are smaller than PolP. Licensing is liberal: it can occur in any constituent chosen for evaluation. This means that depending on which constituent gets evaluated, i.e. a constituent that contains negation vs. a constituent that doesn't, *devoir* either has to raise or—since escape is a last resort—cannot raise; this alternative gives rise to an illusive air of optionality:

(21) Jean ne doit$_{deon}$ pas faire de jogging. NEG≫DEVOIR; DEVOIR≫NEG
Jean NEG must NEG do of jogging
'Jean need not jog/must not jog.'

4.4 Escape of *Seem*

Like *devoir, seem* can take scope above or below a clausemate negation. To show that *seem* can raise, let us consider a sentence where the presupposition trigger *longer* is merged between it and negation. I assume *longer* to be a fixed point; the make-up of the presupposition that gets computed is an index of the position where *seem* is licensed and interpreted, i.e. if *seem* is part of the presupposition, it is interpreted in the c-command domain of *longer* at LF; if not, it is not:

(22) a. *Context:* James Bond is wending his way across a deserted warehouse; he trips over an unidentifiable body and says:
Well, this man no longer seems to be alive. SEEM≫NEG≫LONGER
Presupposition: This man was alive.
b. Uttered in 2006: In view of the latest recording, Bin Laden no longer seems to be dead. LONGER≫SEEM
Presupposition: Bin Laden used to seem to be dead.

In (22-a), *seem* is interpreted in a position above *longer,* since it is not used in the computation of the presupposition; it is also interpreted above negation, which suggests that it raises past negation.[2] In (22-b), *seem* gets interpreted lower than *longer* and by transitivity, since the NPI *longer* needs licensing, lower than negation; neg-raising remains possible and explains the availability of a wide scope reading of *seem* over negation concomitant with a narrow scope reading under *longer*.

The 'pin' test (17) confirms the movement of *seem* past negation:

(23) *Context:* There are as many guests as there are seats in an auditorium and the speaker knows it. She takes a quick look: exactly one seat is not taken...
Exactly one guest doesn't seem to have arrived. ✓SEEM≫EXACTLY_ONE≫NEG

The PPI hypothesis explains why *seem* only outscopes *can* when both *can* and *seem* are in the scope of a DE expression (Prop. 2). Compare (24) and (25):

(24) a. #John can seem to lose weight. *SEEM≫CAN
b. *Context:* John just rose from under the water. Incredible though it may sound...
#John can not seem to breathe for 2 minutes. *SEEM≫CAN≫NEG

(25) a. John can't seem to lose weight. ✓SEEM≫NEG≫CAN
b. He can no longer seem to live without her. ✓SEEM≫NEG≫LONGER≫CAN

We verify that *seem* not only can but has to outscope the reversal trigger (Prop. 3):

(26) *Context:* John had been bragging that someday he would levitate; and one day he rose above ground at a party, to his friends' amazement. But Peter later demonstrated to everyone that John used a mechanical trick at that party...
#John can no longer seem to levitate. SEEM≫NEG≫LONGER≫CAN
Paraphrasable as (only reading): John used to be able to levitate and he seems to have lost the ability.
Not as: It no longer seems that John can levitate.

That *seem* cannot be sandwiched between *longer* and *can,* but has to scope over negation when it outscopes *can,* leads us to conclude that the scope reversal is due to *polarity sensitivity.* The shielding effect of *every* lends further support to the PPI hypothesis:

[2] If, as I assume, *no longer* comprises sentential negation and the element *longer* attached to the clausal spine, the widest scope of *seem* can be envisaged in an alternative way: first the modal raises covertly past *longer* and stays there, below negation, an option open to it since it can raise past scope-bearing elements, and it is the kind of PPI that is interpretable under negation; second, neg-raising kicks in. Under this view, *seem* doesn't end up higher than negation. This doesn't seem to be a possible derivation, though. The reason is that neg-raising is optional (Homer 2012b)—true, it is sometimes required to license a strict NPI, e.g. *until,* in the scope of a neg-raising predicate (*'I don't (*particularly) think that John left until 5'*); but the NPI *longer,* whatever its strength may be, can be 'licensed through' a non neg-raising predicate (*'I am not sure I any longer agree with this conclusion, maybe one could turn that question around'*). So neg-raising would be expected to be optional when *seem* outscopes *longer.* However a sentence like *'This man no longer seems to be alive'* lacks the expected non neg-raised reading, i.e. a reading paraphrasable as: this man was alive, and it is not true that he seems to be alive. This suggests that *seem* can raise past negation, and that when *longer* intervenes, it raises past negation in one fell swoop.

(27) a. #Not everyone can seem to lose weight. *SEEM≫NEG≫CAN
b. Not a single person can seem to lose weight. SEEM≫NEG≫CAN

Escape, the covert raising of *must* (*should* and *supposed*) is constrained: it appears to be clause-bound; likewise, *seem* cannot raise past a superordinate negation (28-a) (even with a neg-raised reading of negated *think*), and doesn't acquire maximal scope in (28-b):[3]

(28) a. #You don't think John can seem to lose weight. *SEEM≫THINK≫NEG≫CAN
*THINK≫NEG≫SEEM≫CAN
b. You think John can't seem to lose weight. *SEEM≫THINK≫NEG≫CAN

While the reversal can occur in weakly negative environments, such as the one created by the strictly DE *at most five* (8-b), it is not triggered by a superordinate negation (28-a). Due to neg-raising, the environment in the embedded clause in (28-a) is one in which NPIs of all strengths can be licensed, therefore it appears to be negative enough to precipitate the reversal (provided that the acceptability of *seem* is evaluated in a constituent which contains this superordinate negation). The lack of scope reversal must therefore be due to the presence of a clause boundary.

Another fact militates in favor of the movement approach: scope reversal is possible even when *can* is itself part of a well attested idiom, e.g. *can help:*

(29) John can't seem to help falling asleep. ✓SEEM≫NEG≫CAN

If *seem* didn't raise covertly to give rise to scope reversal, as claimed in this article, it would intervene between *can* and *help,* in contravention of the definition of idioms as connected sequences of strings (Sportiche 2005): *can help* conforms to this definition, but at LF, after movement of *seem* has taken place.

In sum, *seem* appears to be able to raise past anti-licensing elements. But observe that surface scope is forced with a tensed complement:[4]

(30) *Same context as (22-a)...*
#Well, it no longer seems that this man is alive. *SEEM≫LONGER
Presupposition: It used to seem that this man is alive.

This suggests that when it embeds a tensed complement, *seem* loses the ability to raise, or that it ceases to be a PPI. If the former is the case, then, by parity of reasoning, it is probably also via a covert movement—sensitive to the presence of a tensed complement (15)—that it takes scope over certain non-DE adverbs (14), rather than via reconstruction of the adverbs. Settling the issue is beyond the scope of this article, but if we assume that there is a movement of *seem* past non-DE adverbs (e.g. *often* in (14-a)), it differs from escape (for a reason that is still mysterious at this stage):

(31) a. #John can often seem to open the door. *SEEM≫OFTEN≫CAN

[3] At this stage, I am not in a position to state in what syntactic position *seem* lands.

[4] Lacking covert movement, it is neg-raising which gives rise to wide scope in (i):

(i) It doesn't seem that this man is alive. ✓SEEM≫NEG

b. John can rarely seem to open the door. ✓SEEM≫RARELY≫CAN

Seem cannot take wide scope over *often* if *can* intervenes, while it can take scope (via escape) both over the DE adverb *rarely* and over *can*.

This section has established that *seem* can raise covertly out of an anti-licensing environment, and that this movement is the source of the scope reversal. This hypothesis also explains why the complement of *seem* at merge can contain non-stative predicates only in the 'can't seem to' constellation.

5 Temporal and Aspectual Properties of the Infinitive

The infinitival complement of *seem* normally has two remarkable properties: (i.) its tense is identical with the matrix tense ((32-d) follows this rule, for it has a *present* perfect), and (ii.) when the matrix tense on *seem* is present, the main embedded predicate must be stative (32-b), or else receive a non episodic reading, just like predicates in present tense clauses (32-e), hence the oddness of certain sentences where reversal fails ((5), (27-a), etc.); if the matrix tense is past, an embedded eventive is acceptable (32-f):

(32) a. John seems to be happy (*yesterday/*tomorrow).
b. *John seems to fall.
c. *John seems to sleep.
d. John seems to have fallen.
e. John is happy/*falls/*sleeps.
f. John seemed to fall.

The 'can't seem to' constellation provides the only exception to constraint (ii.): we have seen examples of non-statives in the present, e.g. *lose weight* (1). In the sentences that exhibit scope reversal, the infinitive is undoubtedly c-selected by *seem,* not *can,* because of the presence of *to*. Irrespective of the particular hypothesis developed here (movement for polarity reasons), it seems inevitable that the source of the aspectual restriction is not linked to the subcategorization of *seem* (against the received view, Stowell (1982), Wurmbrand (2011) among many others) but is semantic. This constraint remains to be explored, but one thing is already clear: in our perspective, once *seem* moves out, it no longer composes with the complement, but *can* does (Prop. 4). This is how the semantic constraint is respected when the complement contains an otherwise illicit non-stative. The facts discussed here call for a revision of the received view on the temporal-aspectual properties of the infinitival complement of *seem,* and of infinitives in general. This received view has an important shortcoming anyway: it focuses exclusively on the properties of the *infinitival* complement of *seem,* but ignores that the same puzzling contrast (32-b)-(32-f) also holds in tensed complements:

(33) a. *It seems that John falls. b. It seemed that John fell.

The eventive predicate in (33-b) is interpretable as denoting an eventuality simultaneous with the *seeming* eventuality; (nearly) everywhere else, past tense complements of past tense verbs cannot receive this interpretation. There has to be 'past-shifting': e.g. the time of the leaving is ordered before the time of the saying in (34); exceptions can be found in the complements of perception verbs, see examples of 'belief based on perception' by B. Partee, cited e.g. in von Stechow (2009)):

(34) Peter said that John left the building.

6 Conclusion

The hypothesis that *seem* can move covertly out of an anti-licensing environment provides a simple explanation to the scope reversal of *can* and *seem.* One interesting aspect of the phenomenon is that it may help further our understanding of the temporal properties of infinitives. In closing, the movement hypothesis, conjoined with the clause-boundedness of escape, leads me to propose that there is no clause boundary between *can* and *seem* (Prop. 5). A novel distinction among English root modals is thus needed: *can* (and *will*) creates a monoclausal structure, but only under certain construals (ability *can* is one of them). With other sorts of *can* and other root modals, a biclausal structure is created (Prop. 1). Therefore the syntax of *can* varies depending on its modal base and ordering source. *Able* does not enter into the scope reversal:

(35) #John isn't able to seem to lose weight. *SEEM$\gg$NEG$\gg$ABLE

It differs from *can* in that it takes a complement headed by *to;* it might also differ from it along the control/raising distinction; but it is too early to venture an explanation. Epistemic modals have been argued to create monoclausal structures (Homer 2010); yet, they do not allow scope reversal. It is possible that this reversal is blocked by Epistemic Containment, the constraint whereby an epistemic modal needs to outscope other clausemate scope-bearing elements (von Fintel & Iatridou 2003).

References

Chomsky, N.: Minimalist inquiries: The framework. In: Martin, R., Michaels, D., Uriagereka, J. (eds.) Step by Step: Essays on Minimalist Syntax in Honor of Howard Lasnik, pp. 89–155. MIT Press, Cambridge (2000)

von Fintel, K., Iatridou, S.: Epistemic containment. Linguistic Inquiry 34, 173–198 (2003)

Gajewski, J.: Neg-raising: Polarity and presupposition. Ph.D. thesis. MIT, Cambridge (2005)

Homer, V.: Epistemic modals: High ma non troppo. In: Proceedings of NELS 40 (2010)

Homer, V.: Domains of polarity items. Journal of Semantics, under revision (2012a)

Homer, V.: Neg-raising and positive polarity: The view from modals. Ms., ENS (2012b)

Horn, L.: Remarks on neg-raising. Syntax and Semantics 9, 129–220 (1978)

Iatridou, S., Zeijlstra, H.: Modals, negation and polarity. Talk given at Glow Asia 8 (2010)

Israel, M.: Polarity sensitivity as lexical semantics. Linguistics and Philosophy 19, 619–666 (1996)

Jacobson, P.: I can't seem to figure this out. In: Birner, B.J., Ward, G. (eds.) Drawing the Boundaries of Meaning: Neo-Gricean Studies in Pragmatics and Semantics in Honor of Laurence R. Horn. Studies in Language Companion Series, vol. 80. John Benjamins Publishing Company (2006)

Langendoen, D.T.: The 'can't seem to' construction. Linguistic Inquiry 1(1), 25–35 (1970)

Sportiche, D.: Division of labor between Merge and Move. In: Proceedings of the Workshop Division of Linguistic Labor, the La Bretesche Workshop, pp. 206–285 (2005)

von Stechow, A.: The (non)-interpretation of subordinate tense. Talk given at Georg-August-Universität Göttingen (2009)

Stowell, T.: The tense of infinitives. Linguistic Inquiry 13, 561–570 (1982)

Wurmbrand, S.: Tense and aspect in English infinitives. Ms., University of Connecticut. Storrs (2011)

On the Non-licensing of NPIs in the *Only*-Focus

I-Ta Chris Hsieh

University of Connecticut, Storrs, USA
i-ta.hsieh@uconn.edu

Abstract. This paper focuses on the ungrammaticality of NPIs in the focus associated with *only*. I first show that a naïve combination of the Strawson Downward Entailing (SDE) condition of NPI licensing (von Fintel 1999) and a Horn-style semantics of *only* (Horn 1969; a.o.) fails to predict that NPIs are ungrammatical in the focus associated *only*. To solve this problem, I suggest an analysis that appeals to a semantics of *only* implemented with the notion of innocent exclusion (Fox 2007) and a revision of the SDE condition that refers to assignment functions. The proposed analysis suggests that NPI licensing should be independent of the information provided by the discourse context.

Keywords: *only*, innocent exclusion, negative polarity item, NPI licensing.

1 Introduction

Having a limited distribution, negative polarity items (NPIs) such as *any* and *ever*, as shown in (1), are grammatical in the scope of *only*. These items, as shown in (2) however, are ungrammatical in the focus[1].

(1) a. Only $[\text{John}]_f$ ate <u>any</u> vegetables. b. Only $[\text{Mary}]_f$ has <u>ever</u> been to Paris.

(2) a. *Only [<u>any</u> student]$_f$ ate vegetables.

b. *Only [that Mary has <u>ever</u> been to Paris]$_f$ is new to John.

While the grammaticality of NPIs in the scope of *only* has received great attention (Horn 1996; von Fintel 1999; Wagner 2006; a.o.), their ungrammaticality inside the focus is not much discussed (Beaver 2004; Wagner 2006; a.o.).

The focus of this paper is the ungrammaticality of NPIs in the *only*-focus. Building on a Downward-Entailing-based (DE-based) approach of NPI licensing (Fauconnier 1975, 1979; Ladusaw 1979; von Fintel 1999; a.o.), I first show that a naïve combination of the Strawson Downward Entailing (SDE) condition (von Fintel 1999; a.o.) and a Horn-style semantics of *only* (Horn 1969; a.o.) fails to predict the ungrammaticality of NPIs inside the *only*-focus. To solve this problem along with the SDE approach, this paper suggests a solution that appeals to two crucial ingredients: one is a semantics of *only* implemented with the notion of innocent exclusion (Fox 2007), and the

[1] Some seemingly counterexamples to this generalization are presented by Linebarger (1987) and Geurt and van der Sandt (2004). As pointed out by Horn (1996), Beaver (2004) and others, however, those examples should be analyzed as involving other licensors than *only*.

M. Aloni et al. (Eds.): Amsterdam Colloquium 2011, LNCS 7218, pp. 361–370, 2012.
© Springer-Verlag Berlin Heidelberg 2012

other is a revision of the SDE condition that refers to assignment functions and is context independent.

To be more specific, by using the term ‘context’ in the introduction above, I refer to the contextual variable *C*, which is claimed in the current theories (von Fintel 1994; Rooth 1992; a.o.) to be syntactically present in quantificational sentences and sentences with focus. For instance, in (3) the contextual variable *C* limits the quantificational domain to the salient individuals in the discourse context; in (4) the contextual variable *C* introduces the alternative set.

(3) a. Every student had a good time. b. LF: [[*every-C-student*][*had a good time*]]
(4) a. Even [John]$_f$ passed the exam. b. LF:[*even-C* [[*John*]$_f$ *passed the exam*]]]
C:={*John passed the exam, Mary passed the exam, Bill passed the exam, …*}

The claim I would like to make in this paper is that NPI licensing should be independent of the specification of *C*; an NPI is grammatical in a linguistic environment only if its SDE-ness is guaranteed solely by the lexically specified information, namely the presuppositions (definedness conditions) and the truth conditions. In an environment where an SDE inference is supported only with the implementation of information from the discourse context, NPIs cannot be licensed.

The rest of this paper is structured as the following. In section 2, I show that while a Horn-style semantics of *only* together with the SDE condition of NPI licensing captures the grammaticality of NPIs in the scope of *only*, such a combination fails to predict the ungrammaticality of these items in the focus. The proposed solution to this problem is given in section 3. Section 4 is the conclusion.

2 NPIs and *Only*

2.1 NPIs in the Scope of *Only* and the SDE Condition

Fauconnier (1975, 1979) and Ladusaw (1979) suggest that a downward entailing inference (i.e. an inference from a set to its subset) is crucial for an environment to license NPIs. For instance, a downward entailing inference is supported in the restrictor of the universal quantifier *every* but not that of the existential quantifier *some* (see (5)); NPIs hence are grammatical in the restrictor of *every* but not in that of *some* (see (6)). Here I adopt the notion of cross-categorial entailment (=>) given in (7).

(5) books on NPIs$\subseteq$books
a. Every student who read a book passed the exam. =>
Every student who read a book on NPIs passed the exam.

b. Some student who read a book passed the exam.=/=>
Some student who read a book on NPIs passed the exam.

(6) Every/*some student who read <u>any</u> books on NPIs passed the exam.

(7) Cross-categorial entailment (=>)
a. for any p, q of type t, p=>q iff p=0 or q=1
b. for any f, h of type <σ, τ>, f =>h iff for all x of type σ, f(x)=>h(x)

A DE inference, however, is not intuitively supported in the scope of *only* despite the fact that NPIs are grammatical in this environment (see (1)). Take (8) for instance; it could be the case that nobody other than John ate vegetables but what John ate was not kale; in this case, the inference in (8) is not truth-preserving.

(8) kale$\subseteq$vegetables
Only [John]$_f$ ate vegetables. =/=> Only [John]$_f$ ate kale.

To account for the licensing of NPIs in the scope of *only*, von Fintel (1999) suggests that NPI licensing is subject to a weaker notion of entailment, which he dubbed as Strawson entailment; the premise Strawson-entails the conclusion iff the premise together with the presuppositions (i.e. definedness conditions) of the conclusion entail the conclusion. A licensing condition of NPIs based on Strawson Entailment is formalized as in (9).

(9) a. The SDE condition of NPI licensing
An NPI is only grammatical if it is in the scope of α such that $[\![\alpha]\!]$is SDE.
b. Strawson Downward Entailingness (von Fintel (1999))::
A function f of type $<\sigma, \tau>$ is Strawson downward entailing (SDE) iff for all x and y of type σ such that x=>y and f(x) is defined: f(y)=>f(x)

The SDE condition in (9) together with a Horn-style semantics of *only* like (10) (Horn 1969) provide a straightforward account for the licensing of NPIs in the scope of *only*. Here I assume that *only* takes as its first argument a contextually provided alternative set *C* whose members are of the same type as that of the focalized constituent (Rooth 1985, 1992); the focalized constituent then moves at LF to adjoin to *only*-*C* and serves as the second argument of *only*[2]; a predicate that is created by λ-abstraction over the trace of the focalized constituent serves as the third argument (see (10a))[3]. Here I term the proposition that *only* quantifies in 'the prejacent'; for instance, the prejacent of *Only* [*John*]$_f$ *ate vegetables* is *John ate vegetables*. Based on (10b), an *only*-sentence is defined only if its prejacent is true; if defined, it is true iff there is no alternative proposition that is true and not entailed by the prejacent. (11) is an example that demonstrates the semantics in (10). I further assume that *C* is a free variable bearing an index and receives its value from a contextually provided assignment function g, which maps an integer to a set of alternatives to the focus.

(10) a.

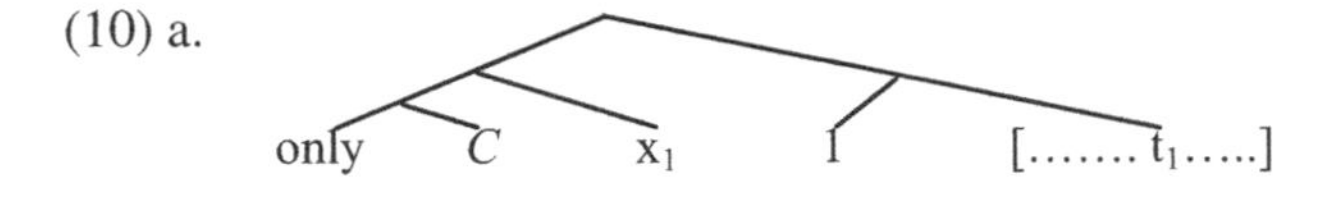

[2] In this paper I adopt a focus-movement approach for the syntax of an *only*-sentence (Wagner 2006; a.o.). Adopting a propositional approach (Rooth 1992; a.o.) instead, however, would not lead to different predictions regarding the issue discussed.

[3] I assume Intensional Functional Application (IFA; see (i); see Heim and Kratzer (1998)).

(i) Intensional Functional Application (IFA): If α is a branching node and $\{\beta, \gamma\}$ the set of its daughters, then, for any world w and assignment g: if $[\![\beta]\!]^{w,g}$ is a function whose domain contains $[\lambda w'. [\![\gamma]\!]^{w',g}]$, then $[\![\alpha]\!]^{w,g} = [\![\beta]\!]^{w,g}(\lambda w'. [\![\gamma]\!]^{w',g})$

b. for any $C_{<\sigma, t>}$, x_σ, $P_{<s, <\sigma\, t>>}$, $[\![\text{only}]\!]^w(C)(x)(P)$ is defined only if $P(w)(x)=1$; if defined, $[\![\text{only}]\!]^w(C)(x)(P)=1$ iff $\forall y_\sigma[y \in C$ & $P(w)(y)]$:
$\{w': P(w')(x)=1\} \subseteq \{w': P(w')(y)=1\}$

(11) a. Only [John]$_f$ ate vegetables.

b. LF: [[[*only*-C_8] *John*$_3$][3 [t_3 ate vegetables]]]

c. $[\![$ (11a)$]\!]^{w,g}$ is defined only if $[\![$*ate vegetables*$]\!]^w$(John)=1; if defined, $[\![$ (11a)$]\!]^{w,g}$=1 iff $\forall y_e[y \in g(8)$ & $P(w)(y)]$:
$\{w': [\![$*ate vegetables*$]\!]^{w'}$(John)$\} \subseteq \{w': [\![$*ate vegetables*$]\!]^{w'}(y)\}$

The semantics in (10) renders the scope of *only* (i.e. the third argument of *only*) SDE (in other words, $[\![\text{only}]\!]^w(C)(x)$ is an SDE function). Take (8) again for instance; while the premise *Only John ate vegetables* alone does not entail the conclusion *only John ate kale*, the premise together with the presupposition of the conclusion, namely *John ate kale*, entail the conclusion. The licensing of NPIs in the scope of *only* hence follows from the SDE condition (see (9)), which states that NPIs are only grammatical in SDE environments.

2.2 NPIs in the Focus Associated with *Only*

As shown in (2), unlike in its scope, NPIs are ungrammatical in the focus associated with *only*. While the combination of a Horn-style semantics of *only* like (10b) and the SDE condition in (9) correctly predicts the grammaticality of NPIs in the scope of *only*, they fail to capture the ungrammaticality of these items in the focus.

In order to show this, there are a couple of assumptions that need to be laid out. I assume that focus evokes a presupposition on the alternative set *C*: *C* must contain the focalized element x and at least one other member than x (Rooth 1992). Furthermore, when checking the entailment relation between any two distinct propositions, *C* should be kept constant in the premise and the conclusion; otherwise, there would be no constant context within which to access the downward or upward inference. Therefore, when checking the entailment relation between (12a, b), the alternative set *C* should be kept constant in the premise and the conclusion and contain as its members both *some student* and *some linguistics student*.

Based on (10), the focus in an *only*-sentence is an SDE environment as well. As shown in (12), the premise (12a) contradicts the presupposition of the conclusion (12b); while (12a) asserts that no alternative proposition not entailed by its prejacent *some student cried* is true, (12b) presupposes that its prejacent *some linguistics student cried*, one of the alternative propositions stronger than the prejacent of (12a), is true. Given that the premise (12a) together with the presupposition of the conclusion (12b) is a contradiction and a contradiction entails any propositions (see (7)), an SDE inference from (12a) to (12b) is valid (in (12a, b), $[\![\Sigma P]\!]^w = [\lambda f_{<e, <e, t>>}. f([\![\textit{cried}]\!]^w)]$).

(12) a. Only [some student]$_f$ cried.
LF: [[*only*-C_8 [*some student*]$_3$] [$_{\Sigma P}$ 3 [t_3 *cried*]]
Presupposition: There is a student x such that x cried in w.

Assertion: $\forall f_{<e,<e,t>>}[f \in g(8)$ and $[\![\Sigma P]\!]^w(f)=1]$:
$\{w'$: there is a student x such that x cried in $w'\} \subseteq \{w': [\![\Sigma P]\!]^{w'}(f)=1\}$

b. Only [some linguistics student]$_f$ cried.
LF: [[*only*-C_8 [*some linguistics student*]$_3$] [$_{\Sigma P}$ 3 [t_3 *cried*]]
Presupposition: There is a linguistics student x such that x cried in w.
Assertion: $\forall f_{<e,<e,t>>}[f \in g(8)$ and $[\![\Sigma P]\!]^w(f)=1]$:$\{w'$: there is a linguistics student x such that x cried in $w'\} \subseteq \{w': [\![\Sigma P]\!]^{w'}(f)=1\}$

c. g(8):= {*some student, some linguistics student, ...*}
(12a) + the presupposition of (12b) => (12b)
Contradiction

That the focus associated with *only*, based on a Horn-style semantics like (10), is SDE further leads to the prediction that NPIs are grammatical in this environment. As already shown in (2) however, this prediction is incorrect.

Given that the SDE-ness in the focus associated with *only* results from a contradiction between the premise and the presupposition of the conclusion and hence is trivial, one might try to save a Horn-style semantics like (10) from this wrong prediction by claiming that while NPI licensing is subject to SDE-ness, an environment that is trivially SDE cannot license NPIs. Such an analysis, however, cannot be an adequate solution. In the literature (e.g., Horn 1996; Ippolito 2008; a.o.), the claim that an *only*-sentence presupposes its prejacent has been challenged; many researchers have suggested that the presupposition triggered by *only* should be something weaker than the prejacent (see (13) for some of such proposals). Once the presupposition in (10b) is replaced by a weaker one like (13a, b), the focus associated with *only* would be non-trivially SDE, which again leads to the wrong prediction that NPIs are grammatical in this environment.

(13) a. Horn 1996: *Existential Presupposition*
$[\![\text{only}]\!]^w(C)(x)(P)$ is defined only if $\exists y[y \in C$ and $P(w)(y)=1]$

b. Ippolito 2008: *Conditional Presupposition*
$[\![\text{only}]\!]^w(C)(x)(P)$ is defined only if $\exists y[y \in C$ and $P(w)(y)=1] \rightarrow P(w)(x)$

Due to the controversy on the presupposition triggered by *only*, I suggest that the key to a fundamental solution should lie in its exclusive component. In the following, I suggest that a semantics of *only* implemented with the notion of innocent exclusion (Fox 2007) together with a revision of the SDE condition that refers to assignment functions provide a solution to the problem pointed above.

3 The Proposal

3.1 *Only* and Innocent Exclusion

Fox (2007) suggests that in an *only*-sentence, the alternatives that are excluded must be innocently excludable. In a nutshell, the set of innocently excludable alternatives (henceforth, I.E.) is the intersection of the maximal sets the conjunction of negation of whose members is consistent with the prejacent. Take (14) for instance; assuming that the alternative set *C* introduced in (14a) is (14b), the only innocently excludable alternative is *John and Mary*. The other two alternatives, namely *John* and *Mary*, cannot be in the same maximal set of excludable alternatives, for excluding both these two alternatives together would lead to contradiction to the prejacent (i.e. the conjunction of *John did not cry* and *Mary did not cry* contradicts to the prejacent *John or Mary cried*).

(14) a. Only [John or_f Mary] cried.

b. The alternative set *C*: {*John, Mary, John or Mary, John and Mary*}
Max. sets of excludable alternatives: {*John, J. and M.*}, {*Mary, J. and M.*}
I.E.:= {*John, John and Mary*}∩{*Mary, John and Mary*}={*John and Mary*}

Building on Fox's (2007) proposal, I suggest the semantics of *only* in (15), which is implemented with a cross-categorial notion of innocent exclusion[4]. Based on (15), the *only*-sentence (12a) translates as (16a′).

(15) a. for any $C_{<\sigma, t>}$, x_σ, $P_{<s, <\sigma t>>}$, $[\![\text{only}]\!]^w(C)(x)(P)$ is defined only if P(w)(x)=1; if defined, $[\![\text{only}]\!]^w(C)(x)(P)=1$ iff $\neg\exists y_\sigma \in$ I.E.(x, *C*, P): P(w)(y)=1

b. for any x_σ and its alternative set $C_{<\sigma, t>}$ and any $P_{<s, <\sigma, t>>}$, the set of innocent excludable alternatives to x w.r.t *C* and P (henceforth, I.E.(x, *C*, P)) is defined as the following:

I.E.(x, *C*, P):= $\cap\{C' \subseteq C$: C' is a maximal set in *C* such that $\cap(\{[\lambda w. \neg P(w)(y)]: y \in C'\} \cup \{[\lambda w: P(w)(x)]\}) \neq \varnothing\}$

(16) a. Only [some student]$_f$ cried.
LF: [[*only*-C_8 [*some student*]$_3$] [$_{\Sigma P}$ 3 [t_3 *cried*]]

a′. $[\![(16a)]\!]^{w,g}$ is defined only if there is a student x such that x cried in w; if defined, $[\![(16a)]\!]^{w,g}=1$ iff $\neg\exists f_{<e, <e, t>>} \in$ I.E.(*some-stu*, g(8), $[\lambda w'. [\![\Sigma P]\!]^{w'}]$): $[\![\Sigma P]\!]^w(f)=1$

The I.E. in (16a′) is determined by the content of g(8) and may vary from one discourse context to another. This is evidenced in the examples (17) and (18). Consider the discourse context in (17) first. In (17), the discourse context provides an alternative set g(8) as in (17a′); by asserting (16a) in this context, one excludes all the alternatives in g(8) other than the focus *some student*. In this case, the set of innocently excludable alternatives include all the members in g(8) other than *some student*[5].

---

[4] The lexical entry for *only* and the original definition of innocent exclusion suggested in Fox (2007) are propositional (see (i)).

(i) $[\![\text{only}]\!](A_{<<s, t>, t>})(p_{<s, t>})= [\lambda w: p(w)=1. \forall q \in \text{I.E.}(p, A) \rightarrow q(w)=0]$
I.E.(p, A)=$\cap\{A' \subseteq A$: A' is a max. set in A s.t. $A'^{\neg} \cup \{p\}$is consistent$\}$; $A'^{\neg}=\{\neg p: p \in A\}$

[5] How an alternative set of a generalized quantifier in focus is still controversy (see Beaver and Clark (2008) and others). Stepping aside from this issue, In the examples given in (17) and (18), the alternatives only include those that are salient in the discourse context.

(17) Discourse context: The college was holding a campus-wide reception for the new president. Faculty, staff, and students were all invited. You know the faculty, the staff, and you also know the students from linguistics. Nonetheless, you didn't know the rest of the students by name. You saw that the faculty, staff and the linguistics students had no interest in the vegetables that were served, but one of the other students in the reception was eating them voraciously. The next day someone asks you who at the reception ate vegetables....

You: Only [some student]$_f$ ate vegetables.

a. g(8):= {*some student, some linguistic students, some faculty, some staff*}
max. set of excludable alt.: {*some ling. student, some faculty, some staff*}
I.E.:= {*some linguistics student, some faculty, some staff*}

Now consider (18); this time the discourse context provides an alternative set g(8) as in (18a′). In this context, the set of students is exhaustified by the set of linguistics students and the set of philosophy students. By asserting (16a) in this context, one merely excludes the alternatives *some faculty* and *some staff* but not the other two (namely *some linguistics student* and *some philosophy student*). In (18a′), given that the conjunction of negation *of some linguistics student ate vegetables* and that of *some philosophy student ate vegetables* contradicts to the prejacent *some student ate vegetables*, these two alternatives cannot be in the same maximal set of excludable alternatives. The set of innocently excludable alternatives hence merely contains the alternatives *some faculty* and *some staff*.

(18) Discourse context: The college was holding a reception to honor the students in linguistics and philosophy only. In addition to those students, the faculty and staff of the entire college were invited. While at this reception you saw that the faculty and staff showed no interest in the vegetables, someone you identified as being a student ate vegetables voraciously. The next day someone asks you who at the reception ate vegetables...

You: Only [some student]$_f$ ate vegetables.

a′. g(8):={*some stu, some phil. stu, some ling. stu, some faculty, some staff*}
(the set of students contains only linguistics and philosophy students)
max. excludable alt. set:= {*some phil. stu, some faculty, some staff*},
{*some ling. stu, some faculty, some staff*}
I.E.:={*some faculty, some staff*}

Incorporating the idea of innocent exclusion in the semantics of *only*, as in (15), the focus associated with *only* (i.e. the second argument of *only*) is no longer always SDE. With the discourse context and the value of g(8) given in (17), the focus of *only* is an SDE context: (16a) together with the presupposition of (19a) entails (19a). On the other hand, in the discourse context and the value of g(8) given in (18), an SUE rather than SDE inference is supported in the *only*-focus: (19a) together with the presupposition of (16a) entail (16a).

(19) a. Only [some linguistics student]$_f$ cried.
LF: [[*only-C*$_8$ [*some linguistics student*]$_3$] [$_{\Sigma P}$ 3 [t_3 *cried*]]

a′. $[\![(19b)]\!]^{w,g}$ is defined only if there is some linguistics student x such that x cried in w; if defined, $[\![(19b)]\!]^{w,g}=1$ iff

$$\neg\exists f_{<e,<e,t>>} \in \text{I.E.}(\textit{some-ling-stu}, g(8), [\lambda w'. [\![\Sigma P]\!]^{w'}]): [\![\Sigma P]\!]^{w}(f)=1$$

b. with the alternative set g(8) in (17a):
I.E.(*some-ling-stu*, g(8), [λw′. $[\![\Sigma P]\!]^{w'}$]):={*some faculty*, *some staff*}

c. with the alternative set g(8) in (18a):
I.E.(*some-ling-stu*, g(8), [λw′. $[\![\Sigma P]\!]^{w'}$]):={*some phil. stu.*, *some faculty*, *some staff*}

3.2 NPI Licensing, the SDE Condition, and Context (In)Dependency

The second tool that is needed to account for the ungrammaticality of NPIs inside the *only*-focus is a licensing condition that is context-independent. As discussed above, an SDE inference may be valid in the focus associated with *only* with certain specification for *C* (i.e. the first argument of *only*). Based on the SDE condition in (10), we are led to the prediction that NPIs are grammatical in the focus associated with *only* in the discourse context that renders *only-C* an SDE function (e.g., (17)). This prediction is not borne out; NPIs are ungrammatical in the focus of an *only*-sentence regardless the discourse context. A revision of the SDE condition in (10) hence is required.

Here I suggest a revision of the SDE condition as in (20). Based on (20), a syntactic object is qualified as an NPI licensor iff it denotes an SDE function with respect to any assignment function g. Given that an assignment function g is given by the discourse context, (20) suggests that an NPI licensor should denote an SDE function across discourse contexts.

(20) *The Revised SDE Condition*: An NPI is only grammatical in the scope of α such that, for any assignment function g, $[\![\alpha]\!]^g$ is SDE.

Based on the semantics of *only* in (15), *only-C* denotes an SDE function only with certain specification for *C*. Given that the focus associated with *only* is not always SDE with respect to a contextually provided assignment function g, NPIs are ungrammatical in this environment.

The analysis suggested above not only predicts the ungrammaticality of NPIs inside the focus associated with *only* but also that of these items in a DP-constituent containing the focus. As observed by Wagner (2006) and others, NPIs are ungrammatical in a DP-constituent containing the *only*-focus (see (21a)). Wagner (2006) and others suggest that at LF the whole DP-constituent that contains the focus is pie-pied and serves as the second argument of *only* (see (21b)); with this assumption, the ungrammaticality of NPIs in a DP-constituent containing the focus associated with *only* follows from the semantics in (15) and the SDE condition in (20): given that based on the semantics in (15), the second argument of *only* is not always an SDE environment with respect to an assignment function g, NPIs are ungrammatical in a DP-constituent containing the focus associated with *only*.

(21) a. *Only <u>any</u> inhabitant of [Twin Earth]$_f$ met Particle Man.
b. LF: [[*only-C* [*<u>any</u> inhabitant of Twin Earth*]$_3$ [3 [t$_3$ *met Particle Man*]]]

4 Concluding Remarks

In this paper, I discuss the ungrammaticality of NPIs in the focus associated with *only*. I have shown that a naïve combination of a Horn-style semantics of *only* and the SDE condition of NPI licensing fails to predict the ungrammaticality of these items in the *only*-focus. I further suggest that this problem can be solved with a semantics of *only* with the implementation of the notion of innocent exclusion (Fox 2007) and a revision of the SDE condition that refers to assignment functions.

The proposal of this paper reaches the conclusion that NPI licensing is independent of the discourse context; for a linguistic expression to license an NPI, it must denote an SDE function without the implementation of the information provided by the discourse context. This, however, is not an uncontroversial claim; one challenge that this claims faces is the contrast between definite and indefinite superlatives with respect to NPI licensing. Herdan and Sharvit (2006) observe that although the distribution is limited, indefinite superlatives do exist (see (22b)); nevertheless, NPIs are only licensed in definite but not in indefinite superlatives (see (23)).

(22) a. This class has the best student. b. This class has a best student.

(23) This class has the/*a best student with any knowledge of French

To provide a unified semantics for *-est* in both definite and indefinite superlatives (see (22)), Herdan and Sharvit suggest the semantics in (24).

(24) $[\![\text{-est}]\!] = \lambda S_{<et,\,t>}.\ \lambda R_{<d,\,et>}.\ \lambda P_{et}.\ \lambda x_e$: (i) there is an $X \in S$ s.t. $x \in X$; and (ii) $x \in P$. For some $X \in S$ such that $x \in X$, there is a degree d such that $\{z\text{: } R(d)(z)=1 \text{ and } P(z)=1\}=\{x\}$

Based on this semantics, the superlative morpheme $[\![\text{-est}]\!](S)(R)$ is not an SDE function. As Herdan and Sharvit (2006) point out however, an SDE inference, based on (24), is supported in the restrictor P of *-est* when S is a singleton set. Given that this is only guaranteed in a definite superlative due to the uniqueness requirement of *the*, NPIs are grammatical only in definite superlatives.

The contrast in (22) poses a challenge to the conclusion drawn from the discussion above, for based on Herdan and Sharvit's analysis, NPIs in definite superlatives are licensed because of a particular specification for the contextual variable S, which can be assigned though a contextually provided assignment function. To account for the case of superlatives, a further look at the semantics of superlatives is required and I hence leave this for future research.

References

1. Beaver, D., Clark, B.: Sense and Sensitivity: How Focus Determines Meaning. Blackwell, Oxford (2008)
2. Beaver, D.: Five only pieces. Theoretical Linguistics 30, 45–64 (2004)
3. Fauconnier, G.: Polarity and the Scale Principle. Chicago Linguistics Society 1 (1975)
4. Fauconnier, G.: Implication Reversal in a Natural Language. In: Guenthner, F., Schmidt, S.J. (eds.) Formal Semantics and Pragmatics for Natural Languages, Dodrecht, pp. 289–302 (1979)
5. Fox, D.: Free Choice Disjunction and the Theory of Scalar Implicature. In: Sauerland, U., Stateva, P. (eds.) Presupposition and Implicature in Compositional Semantics. Palgrave Macmillan, New York (2007)
6. Gajewski, J.: Neg-raising and polarity. Linguistics and Philosophy 30, 289–328 (2007)
7. Geurt, B., van der Sandt, R.: Interpreting focus. Theoretical Linguistics 30, 1–44 (2004)
8. Heim, I., Kratzer, A.: Semantics in Generative Grammar. Blackwell, Oxford (1998)
9. Heim, I.: A Note on Negative Polarity Items and DE-ness. NELS 14 (1984)
10. Herdan, S., Sharvit, Y.: Definite and nondefinite superlatives and NPI licensing. Syntax 9, 1–31 (2006)
11. Horn, L.R.: A presuppositional analysis of only and even. Chicago Linguistics Society 5 (1969)
12. Horn, L.R.: Exclusive company: Only and the dynamics of vertical inference. Journal of Semantics 13, 10–40 (1996)
13. Ippolito, M.: On the meaning of only. Journal of Semantics 25, 45–91 (2008)
14. Ladusaw, W.A.: Polarity Sensitivity as Inherent Scope Relations. Ph.D Dissertation. The University of Texas at Austin (1979)
15. Linebarger, M.C.: Negative polarity and grammatical representation. Linguistic and Philosophy 10, 325–387 (1987)
16. Rooth, M.: Association with Focus. Ph.D. Dissertation, Umass, Amherst (1985)
17. Rooth, M.: A theory of focus interpretation. Natural Language Semantics 1, 75–116 (1992)
18. von Fintel, K.: Restrictions on Quantifier Domains. Ph.D. Dissertation, Umass Amherst (1994)
19. von Fintel, K.: NPI-licensing, Strawson entailment, and context dependency. Journal of Semantics 16, 97–148 (1999)
20. Wagner, M.: Association by movement: Evidence from NPI-licensing. Natural Language Semantics 14, 297–324 (2006)
21. Zwarts, F.: Facets of Negation. In: van der Does, J., van Eijck, J. (eds.) Quantifiers, Logic and Language, pp. 385–421. CSLI Publication, Standford (1996)

Now: A Discourse-Based Theory

Julie Hunter

l'Université de Pau et des Pays de l'Adour and
l'EHESS/l'Institut Jean-Nicod, Paris
juliehunter@gmail.com

Abstract. This paper offers a semantic theory of uses of *now* in which *now* refers to a time introduced in discourse. I argue that the interpretation of an anaphoric use of *now* is determined by the rhetorical structure of the discourse in which the token of *now* figures.

Keywords: anaphora, *now*, rhetorical structure, discourse structure.

In general, *now* is interpreted as the utterance time and cannot refer to a time made salient in the discourse in the way that a third person pronoun can refer to an individual made salient in the discourse:

(1) I like to think about my grandmother. I always had a great time with **her**.
(2) I like to think back on the summer of '97. I was so happy ***now**.

Yet there are exceptions (cf. Banfield 1982, Hunter 2010, Kamp & Reyle 1993, Lee & Choi 2009, Predelli 1998, Recanati 2004, Schlenker 2004). In the following examples, *now* denotes a time that lies in the past of the utterance time and is introduced at some prior point in the discourse:

(3) Five months later, I sat with her as she lay in bed, breathing thin slivers of breath and moaning... I was alone in her bleak room. Alone, because there was none of her in it, just a body that **now** held no essence of my mum.[1]
(4) The letter is marked "personal and private" and is addressed to President Franklin D. Roosevelt's secretary, Grace Tully, who was with the ailing chief executive in Warm Springs, Ga., that Thursday in 1945. The writer was Lucy Mercer Rutherfurd, who decades before had been FDR's mistress and who **now** was making arrangements for what would be their last fateful meeting at the president's rural retreat.[2]

(3) is taken from an article in which the author describes her mother's struggles with Alzheimer's. Throughout the article, it is clear that the author is recounting past events. Her use of *now* does not denote the utterance time in any sense; it

[1] 'Her misery was now so deep, her existence so shallow – Fiona Phillips on dealing with Alzheimer's', from *Daily Mail*, 28.08.2010.
[2] 'What was for FDR's eyes only is now for yours', *The Washington Post*, 29.07.2010.

M. Aloni et al. (Eds.): Amsterdam Colloquium 2011, LNCS 7218, pp. 371–380, 2012.
© Springer-Verlag Berlin Heidelberg 2012

rather denotes a time in the past at which she visited her ailing mother. The two sentences in (4) are about a letter to FDR that was acquired by the National Archives. The author of the article describes the writing of the letter as an event in the past and clearly distances that event from the time of the acquisition. Still, he can use *now* to denote the time of the past letter writing event.

This paper offers a semantic theory of anaphoric uses of *now*; that is, uses of *now* in which *now* refers to a time introduced in discourse. Contrary to existing theories of *now*, I argue that the interpretation of an anaphoric use of *now* is determined by the rhetorical structure of the discourse in which the token of *now* figures. The details of my theory are presented in *Segmented Discourse Representation Theory* (Asher & Lascarides 2003).

1 Previous Discourse Accounts

Kamp & Reyle (1993) recognize that *now* can be used anaphorically to refer to a time introduced in discourse. However, *now* can only be so used, they claim, to modify past tense clauses that describe states. The idea underlying this distinction is that clauses describing states exploit the incoming reference time while clauses describing events introduce reference times of their own. While I cannot elaborate on Kamp & Reyle's view here, the important point is that because state-denoting clauses exploit incoming reference points, they are able to shift the *temporal perspective point* (TPpt) of a discourse, where the TPpt is the time relative to which events are described as unfolding. Event denoting sentences, because they do not exploit incoming reference times, do not shift the TPpt. As *now* depends on the TPpt of a discourse, it can refer to a past time only when the TPpt has been shifted to the past by a state-denoting sentence.

Kamp & Reyle's account does not do justice to the data on *now*, however. First, *now* can be used to modify event denoting clauses.

(5) Before being dipped into the liquid air, it had not caught fire; but **now** it exploded, it was consumed so rapidly.[3]

A second problem stems from a claim that clauses that describe events shift the time relative to the input context while clauses that describe states inherit the time of the incoming context. A third problem is that Kamp & Reyle use a discourse theory that updates the TPpt for a discourse sentence by sentence, or perhaps clause by clause, without taking into account the relations between these sentences or clauses.[4] What we see when we look at data on *now* is that (a) a past tense clause modified by *now* may fail to stand in a temporal relation to the clause that has come before it, regardless of its aspect, and (b) even if there is a relation between the two clauses, this relation may not be enough to

[3] Variation on example in 'Liquid Air Experiments,' *The New York Times*, 13.05.1899.

[4] Lee & Choi (2009) also suffers from the second and third problems for Kamp & Reyle's view. Their treatment of aspect is more nuanced than Kamp & Reyle's, but their account retains the division between state and event denoting clauses as well as a discourse theory that simply updates the reference point clause by clause.

determine the interpretation of *now* because sometimes the time denoted by *now* is the time introduced by a clause much earlier in the discourse. Such long distance relationships can hold regardless of the aspect of the *now*-modified clause.

When t_β Is Independent of t_α: Kamp & Reyle hold that the time denoted by a clause β depends on the tense and aspect of β together with a reference point introduced by the previous sentence. If α is the clause preceding β, β should stand in a temporal relation with α. Yet sometimes, a clause immediately preceding β can denote a time that is completely irrelevant to the interpretation of β.

(6) *Asked in 2012*
   a. [Why was the left so much more accepting of the 2011 budget than of the 2010 budget?]$_\chi$
   b. [First, many on the left took a cue from conservatives,]$_\gamma$ [who had assailed the 2011 budget as falling short of the cutting that was needed.]$_\alpha$
   c. [Second, Mr. Obama was **now** in better standing with liberals than he had been in 2010]$_\beta$ [having recently repealed "don't ask, don't tell".]$_\eta$

Kamp & Reyle predict that because *now* modifies a stative clause in β, it should refer to the temporal perspective point determined by the discourse through clause α. Because α is in the past perfect, it in turn exploits the time introduced by the simple past clause γ, thereby making t_γ (the time denoted by γ) the temporal perspective point. t_γ, of course, must have started before t_χ—the left first took the cue and then accepted the budget—but given that it comes after the conservative reaction to the president's budget, we know that t_γ did not begin too long before the beginning of t_χ. (I assume the question presupposes that the left was more accepting of the 2011 budget.) Intuitively, t_β is independent of t_α and, therefore, t_γ; *now* refers to the time at which the left accepted the budget not the time at which they took a cue from conservatives. The tense and aspect of the sentences in (6) do not alone determine a temporal relation between γ (or α) and β, though the discourse does enforce a temporal relation between β and χ: t_β overlaps t_χ and because β is presented as an answer, or in this case an explanation, of χ, t_β must begin before the beginning of t_χ. If we view (6) as a question with multiple independent answers, as opposed to a mere sequence of sentences, the temporal relation between β and χ on the one hand coupled with the temporal independence of γ/α and β on the other is no surprise.

(7) demonstrates the same point using a different rhetorical structure.

(7) a. [When Mr. Kaine agreed to run the DNC in 2009]$_\chi$ —[even while finishing his last year as governor]$_\gamma$ —[his closest advisers were stunned]$_\eta$ [and they counseled him to renege.]$_\alpha$
   b. [**Now** Mr. Kaine was facing an unwanted repeat of the same, uncomfortable situation.]$_\beta$[5]

[5] Variation on example from 'Will Obama Ask Kaine to Seek Virginia Senate Seat?', *The New York Times*, 10.02.2011.

Again, because β describes a state using the past perfect, it should set the time of α as the temporal perspective point and then inherit that time. Yet intuitively, *now* in β refers to a time introduced in the discourse prior to the introduction of clauses χ - α, which describe eventualities holding well in the past of t_β.

When t_β Is the Time Denoted by a Clause Preceding α: Even when a clause β is temporally related to the preceding clause α, Kamp & Reyle can still fail to make the right predictions. They predict, for example, that event denoting clauses in the simple past, like β in (8) below, will denote a time other than that denoted by the preceding clause. While this is the case in (8), Kamp & Reyle miss the stronger claim that *now* refers to t_χ.

(8) a. [The scientist dipped the felt into liquid air]$_\gamma$ [and the result was astonishing]$_\chi$.
  b. [Before being dipped into the liquid air, it had not caught fire;]$_\alpha$ [but **now** it exploded, it was consumed so rapidly.]$_\beta$

While Kamp & Reyle are right that a past tense use of *now* must refer to a time already available in the discourse and that this time will not in general be made available by the preceding clause if the *now* modified clause denotes an event, (8) shows that *now* need not find its referent in the preceding clause. Kamp & Reyle's prediction that *now* cannot modify past tense clauses describing past events—a prediction discredited by examples like (8)—is explained in part by the fact that they only considered the temporal relation between a clause β and the temporal perspective point used by the previous clause α.

2 Rhetorical Contexts and *now*

The temporal relations in a text—which determine the time to which a past use of *now* will refer—are not determined, at least not entirely, by the tense and aspect of individual sentences together with the order in which they appear in the text. To make predictions about the interpretation of *now* in past tense clauses, we need a theory that allows a clause β to stand in a temporal relation to a clause χ even if there is a clause α introduced in the discourse between χ and β such that (a) α stands in no temporal (or rhetorical) relation to β or (b) the temporal or rhetorical relation between α and β is not sufficient for determining the time of β. We need the structure offered by a theory of rhetorical contexts and relations between clauses in a discourse. I will show that in particular, the temporal relations offered by *Segmented Discourse Representation Theory* (Asher & Lascarides 2003), or SDRT, can be used to make more accurate predictions about the interpretation of *now* than can theories of tense and aspect alone.

To capture the semantics of *now*, we need a theory of discourse content that uses structured contexts. I begin with Kamp & Reyle's *Discourse Representation Theory* and add structure to discourse contexts in two ways. First, to each DRS, I add a level, call it K_0, that represents information about utterance events. K_0 is the most global level of a given DRS K; the content of utterances, i.e., the content that is normally treated by discourse theories like DRT, is added to sub-contexts of this 'extra-linguistic' level. The notion of K_0 is introduced in Hunter (2010) in order to handle indexicals, among other expressions, and we need it to handle examples in which *now* picks out the time of utterance rather than a time introduced in discourse.[6]

Second, I expand on Hunter (2010)'s contexts by adding rhetorical structure as developed by Asher & Lascarides (2003) and use Asher & Lascarides' semantics for discourse relations. We start by dividing a discourse into *elementary discourse units* or EDUs, where an EDU is a minimal unit in a discourse that can stand in a rhetorical relation with another unit—EDUs are, in a sense, the 'words' of a discourse. Next, each EDU is represented as a DRS. Finally, each DRS for each EDU is related to another EDU, or chunk of EDUs, via a rhetorical relation. Both the content of the *segmented* DRSs and the relations between them are recorded in our contexts below level K_0.

SDRT represents the rhetorical structure of a discourse in a two-dimensional graph space that distinguishes between subordinating and coordinating relations. In a subordinating relation, $\text{SUB}(\alpha, \beta)$, β does not move the discourse forward but simply provides more information about α, in the form of, e.g., background, explanation or elaboration. In a coordinating relation, $\text{COOR}(\alpha, \beta)$, β does not provide more information about α but rather moves the discourse forward relative to a topic shared by α and β.

With our structured discourse contexts in place, I propose that we treat *now* as an anaphoric, presuppositional expression along the lines of van der Sandt (1992). While I will not provide a complete motivation for this proposal here—see Hunter (2010), Hunter & Asher (2005), Maier (2009), Roberts (2002), and Zeevat (1999) for arguments—the general idea is that *now* triggers a presupposition that must be bound to, or otherwise satisfied by, an antecedent time. *Now*, like other indexicals, does not bring along its own interpretation, but depends on the incoming context—generally the K_0 level of the incoming context—to provide one. The fact that it is up to the incoming context to provide an antecedent for standard indexicals is seen most clearly with *you* and *that*, which can fail to refer if the context does not provide an antecedent.

Now can find its antecedent time either from the extra-linguistic context, K_0, or from the discourse context, K_1-K_n. *Now* exhibits a strong preference for resolution to the utterance time, but this preference can be over-ridden when resolution at K_0 is blocked. In the examples considered here, the past tense blocks resolution at the extra-linguistic level. To capture *now*'s preference for resolution at K_0, I use the operator $\uparrow$ introduced in Hunter & Asher (2005), which requires material in its scope to be resolved at the highest context possible. Given DRT's treatment of existential formulas as introducers of discourse referents, the presupposition of *now* will look like this:

(9) $\uparrow \exists t(t = ?)$

'?' signals that t is anaphoric and must be identified with a discourse referent for a time already available in the context in which *now*'s presupposition is triggered.

[6] See Hunter (2010) and Maier (2009) for a motivation of such structured contexts.

Now, like other indexicals and third person pronouns, is incapable of local accommodation. It requires that there be a super-ordinate time that it can treat as the 'current' time. Even when it refers to the utterance time, its presupposition is triggered in a sub-context of K_0 and then bound in K_0. When *now* cannot be resolved to the utterance time, I claim that it is resolved to the time of its immediately super-ordinate antecedent clause. That is, when β in a subordinating relation SUB(α, β) is modified by *now*, the time denoted by *now* will be determined by the time of the eventuality described in α such that the relation between the two times will be as close to identity as possible given the semantics of the rhetorical relation relating α and β. As all subordinating relations in SDRT would allow for temporal overlap between α and β, temporal overlap will be required when β is modified by *now*. For ELABORATION and BACKGROUND, which require that one of their arguments be temporally included in the other, the addition of *now* will require identity between the times denoted by the arguments. For EXPLANATION, identity is ruled out: a cause must begin before its effect. The addition of *now* will, however, entail that the cause started right before its effect. These features of *now*'s semantics have the following consequences:

***Now* Restricts the Temporal Relations Predicted by SDRT:** *Now* restricts the temporal relations that a theory of rhetorical structure like SDRT predicts will hold between a subordinate clause and its super-ordinate antecedent. For example, if a clause β explains a clause α, SDRT allows that the time of β (t_β) might start well before t_α and it might even end before t_α begins.

(10) [I hit him today]$_\alpha$ [because he hit me last week.]$_\beta$

In (10), the event described in β ended before that described in α began. If β is modified by *now*, however, t_β must overlap t_α and must start just before t_α.

(11) [I hit him]$_\alpha$ [because he (***now**) hit me.]$_\beta$

(12) a. [This became apparent in Darwin's reaction to Jenkin's critique]$_\gamma$...
b. [Darwin gave up his original assumption that evolution occurred best in small, isolated populations]$_\chi$,
c. [because he **now** feared that small populations would not throw up enough individual variants for selection to be effective.]$_\beta$

In (11), the cause ended before the effect began, so *now* is not licensed. In (12), the cause (β) immediately brought about its effect (χ) and t_χ overlaps t_β. χ and β together elaborate on Darwin's reaction to Jenkin's critique, introduced in γ. The semantics of ELABORATION require that t_χ and t_β together be included in t_γ.

COOR(α, *now*-β), then SUB(χ, (α, *now*-β)): If a past-tense, *now*-modified clause β is related to a preceding clause α via a coordinating relation, the whole unit (α+β) will be subordinate to another clause χ whose time will serve as *now*'s antecedent.[7] In (4), α and β provide information about the same individual and

[7] *Now* can be used as a modifier of past tense sentences in narratives without an explicit super-ordinate antecedent.

so would be related via CONTINUATION, a coordinating relation in SDRT, while the unit $\alpha+\beta$ would be related to χ by BACKGROUND, a subordinating relation.

(4) a. [The letter is marked "personal and private" and is addressed to...]$_\eta$
b. [The writer was Lucy Mercer Rutherfurd,]$_\chi$
c. [who decades before had been FDR's mistress]$_\alpha$ [and who **now** was making arrangements for what would be their last fateful meeting at the president's rural retreat.]$_\beta$

The semantics of BACKGROUND in SDRT allow that if a clause β is subordinate to a clause χ via BACKGROUND, then t_β can start well before t_χ. Unlike EXPLANATION, however, t_χ must always be included in t_β. When we add *now* to β, temporal overlap is taken care of by the semantics of BACKGROUND, but t_β must start when t_χ starts. Because χ elaborates on or provides background for η by providing information about *who* wrote the letter, χ in turn inherits η's time. I assume that since χ and η are both about the writing of the letter under discussion, the time that they both denote is the time of the letter writing event. *Now* is thus interpreted as the time of the letter writing event, as desired.

***Now* Can Be Used to Enforce a Temporal Break:** Suppose a *now*-modified clause β is in a coordinating relation with a clause α where $\alpha+\beta$ is subordinate to another clause χ as described above. If the time of α is different from the time of χ and the rhetorical relation between α and β does not enforce a temporal break between α and β then *now* will be licensed to enforce a break and a return to the time of χ. The felicity of (4), for example, is greatly aided by *now* because CONTINUATION does not impose a temporal relation between its arguments. With *now*, it is clear that while on the one hand, the *now*-modified clause is still providing background, it's providing information about what was going on at the time of the letter writing event (indirectly) introduced in χ, not at the time of α. Similar remarks can be made for (8):

(8) a. [The scientist dipped the felt into liquid air]$_\gamma$ [and the result was astonishing]$_\chi$.
b. [Before being dipped into the liquid air, it had not caught fire;]$_\alpha$ [but **now** it exploded, it was consumed so rapidly.]$_\beta$

The example is much clearer with *now* because *now* helps to separate the time of β from the time of α and to enforce a tie between t_β and t_χ.

---

(13) But Rokiroki gripped the strangers wrists so that he could not draw his hatchet. And **now** he called again to his little daughter...[8]

However, SDRT posits topics for narrations, so *now* in (13) will have a super-ordinate antecedent determined by the discourse topic. *Now*, like *next* and *then*, is easily used to modify a sentence related to another sentence via NARRATION because all of the sentences that figure in a narration elaborate on the topic event. Because they all elaborate on the topic, they must all share in the topic time, but none can be identical to the topic time because the semantics of NARRATION ensure that there is no temporal overlap between two clauses related by NARRATION.

***Now* Can Be Omitted in Certain Subordinating Structures:** When a clause β elaborates on a clause α, for example, it is ensured by the semantics of ELABORATION that t_β is included in t_α. Sometimes, it is also clear that t_α is included in t_β whether or not β is modified by *now*. This is the case in (3):

(3) [Five months later,]$_\eta$ [I sat with her as she lay in bed]$_\gamma$... [I was alone in her bleak room.]$_\chi$ [Alone, because there was none of her in it,]$_\alpha$ [just a body that (**now**) held no essence of my mum.]$_\beta$

β does not figure in a complex subordinate unit, i.e. β is not related via a coordinating relation to any other discourse units, and there are no markers such as *for example* to suggest that t_β is only properly included in t_α. Regardless of whether *now* is used in (3), it is understood that t_β is t_α.[9] In this case, the requirements of *now* are already satisfied by β, so *now* can be omitted without affecting the truth conditions. This observation can be generalized to other subordinating relations: if a clause β is subordinate to a clause α via BACKGROUND or EXPLANATION and t_α and t_β are as close to identical as allowed by the semantics of these relations without *now*, then the requirements of *now* are satisfied and *now* can be omitted without changing the truth conditions for the discourse.

3 A Note about Contrast and Change of State

If we remove *now* from (3), the truth conditions of the example do not change, but something is nonetheless lost. *Now* suggests that the state described in β began recently; the change from the author's mother's body having an essence to its not holding an essence is important for the story the author is recounting and the use of *now* reinforces this theme. Similarly, in (12), *now* makes it clear that Darwin did not always have the fear described in the *now*-modified clause. *Now* suggests a change in Darwin's thinking and so aids the tie between the *now* modified clause and its antecedent, which mentions the catalyst for the change in Darwin's thinking. As a final illustration, *now* in (8) emphasizes that fact that the felt's exploding is a new event and so reinforces the contrast that holds between the *now* modified clause and the preceding clause.

As Hunter (2010), Lee & Choi (2009), and Recanati (2004) have observed, *now*, at least when it modifies past tense clauses, often signals a recent change or a contrast of some sort. In opposition to Hunter (2010) and Recanati (2004), however, I maintain that this effect does not arise from a semantic requirement that the eventuality described by the *now*-modified clause be contrasted, either explicitly or implicitly, with some other eventuality. Rather, the contrastive feel of so many *now* examples is a pragmatic effect that arises naturally from the

[9] The chances are high that the author's mother's room contained an essence-less body long before the time at which the author paid the visit under discussion in α. Nevertheless, the discourse only demands that t_α be the same as t_β. The discourse could be true in a scenario in which the mother's body lost its essence at exactly the moment that the author walked through the door to pay the visit mentioned in α. While this scenario is implausible, it is allowed by the discourse structure.

semantics that I have laid out so far together with certain features of a discourse. There are multiple reasons to resist the claim that *now* requires a contrast or a change of state. First, even in examples that have a contrastive feel, it is often difficult to say in what sense these examples contain a contrast. It is certainly not the case that the *now*-modified clause must be related to another clause via the CONTRAST relation defined in SDRT, for example. Amongst the English examples that I have discussed in this article, only in (8) would SDRT say that the *now*-modified clause is an argument for CONTRAST.

Second, there are many examples in which *now* modifies a past tense clause that do not give rise to a contrastive effect.

(14) [2011 was a great year for computer science.] In attacking the problem of the ambiguity of human language, computer science was **now** closing in on what researchers refer to as the "Paris Hilton problem".[10]

(15) But Rokiroki, exerting all his strength, gripped the strangers wrists so that he could not draw his hatchet. And **now** he called again to his little daughter, who stood trembling on the bank above.

In (14), it is obvious from the context that computer science was not closing in on the Paris Hilton problem before whatever time serves as *now*'s antecedent. In this sense, there is a kind of opposition implicit between the time at which computer science is said to have been closing in on the Paris Hilton problem and the times before. Nevertheless, (14) does not have a contrastive feel because the change of state is not at issue in the discourse. *Now* simply serves to emphasize the period under discussion in the discourse and other times are not relevant. Note that we could replace *now* in this example with *at this time* and the implicit opposition between the time of the eventuality described by the *now*-modified clause and previous times would still be there. Yet we would not for this reason want to build a requirement of contrast into the semantics of *at this time*. In (15), *now* again signals a change from one eventuality to another, but again, this is not a reason to argue that *now* requires a contrast between two times. For one thing, *now* could not felicitously be replaced by *but*, a marker for contrast in discourse theories like SDRT. For another, *and now* could be replaced by *next* or *and then* and the discourse would have the same effect of signalling a shift from one eventuality to another. But as with *at this time*, we would not want to argue that *next* or *then* requires a contrast.

The contrastive feel of examples involving past tense uses of *now* is better explained as a natural consequence of the semantics of *now* combined with certain features of the discourse in which the *now* modified clause figures. *Now* inherits the time of its super-ordinate antecedent. Sometimes this feature of *now*'s semantics allows it to play an indispensable structuring role in a discourse. Other times, *now* is not needed to structure a discourse, but simply serves to emphasize the temporal relation between the clause it modifies and its antecedent. *Now*'s semantics stop here and will not give rise to a contrastive effect on their own.

[10] Variation on example in 'A Fight to Win the Future: Computers vs. Humans,' *The New York Times*, 14.02.2011.

If, however, within the discourse, the *now*-modified clause and its antecedent clause fall on one side of a larger contrastive structure, then the fact that *now* emphasizes the temporal relation between the clause that it modifies and its antecedent will naturally give rise to an emphasis on the temporal nature of the contrasted eventualities. In (5) and (8), the *now*-modified clause enters into a local contrast relation with the previous clause in the discourse. Because the use of *now* emphasizes the temporal nature of one side of the contrast, the temporal nature of the other side is brought to light. In (3) and (12), the discourse is about a change of a body state and a change in a set of beliefs, respectively. In both cases, the discourse sets up a much higher-level contrastive structure. Again, *now* serves to emphasize the temporal relations on one side of this structure, which naturally gives rise to a 'then' vs. 'now' reading of the contrastive structure.[11]

References

Afantenos, S., et al.: An empirical resource for discovering cognitive principles of discourse organization: the ANNODIS corpus. In: Proceedings of LREC 2012 (2012)

Asher, N., Lascarides, A.: The Logics of Conversation. Cambridge University Press (2003)

Banfield, A.: Unspeakable Sentences: Narration and Representation in the Language of Fiction. Routledge & Kegan Paul, London (1982)

Hunter, J.: Presuppositional Indexicals. Ph.D. thesis, The University of Texas (2010)

Hunter, J., Asher, N.: A presuppositional account of indexicals. In: Dekker, P., Franke, M. (eds.) The Proceedings of the 15th Amsterdam Colloquium, pp. 119–124 (2005)

Kamp, H.: Formal properties of now. Theoria 37, 227–273 (1971)

Kamp, H., Reyle, U.: From Discourse to Logic. Kluwer Academic Publishers (1993)

Kaplan, D.: Demonstratives. In: Almog, J., Perry, J., Wettstein, H. (eds.) Themes from Kaplan. Oxford University Press, USA (1989)

Lee, E., Choi, J.: Two nows in korean. Journal of Semantics 26 (2009)

Maier, E.: Proper names and indexicals trigger rigid presuppositions. Journal of Semantics 23, 253–315 (2009)

Predelli, S.: Utterance, interpretation and the logic of indexicals. Mind & Language 13(3), 400–414 (1998)

Recanati, F.: Indexicality and context shift. Harvard University (2004) (manuscript)

Roberts, C.: Demonstratives as definites. In: van Deemter, K., Kibble, R. (eds.) Information Sharing. CSLI Press (2002)

Rooth, M.: A theory of focus interpretation. Natural Language Semantics 1, 75–116 (1992)

van der Sandt, R.: Presupposition projection as anaphora resolution. Journal of Semantics 9, 333–377 (1992)

Schlenker, P.: Context of thought and context of utterance. Mind & Language 19(3), 279–304 (2004)

Zeevat, H.: Demonstratives in discourse. Journal of Semantics 16, 279–313 (1999)

---

[11] Another factor that might encourage a contrastive reading of *now* is focus. Following the work of Rooth (1992), a focused element in a sentence gives rise to a set of alternatives, which in turn gives rise to a contrast. How exactly focus affects the interpretation of *now* would be an interesting topic for further study, though I doubt that focus affects the fundamentals of the theory that I am presenting here.

Obligatory Implicatures and Grammaticality*

Natalia Ivlieva

MIT
ivlieva@mit.edu

Abstract. The paper explores some puzzling data on number agreement with disjunctive noun phrases in Russian. Specifically, it is shown that plural agreement can be blocked as a result of scalar implicature calculation. More generally, I propose that a sentence can be judged ungrammatical when it has scalar implicatures that contradict each other and that cannot be disregarded due to relevance or to the requirements of certain scalar items.

Keywords: Scalar Implicature Calculation, Formal Alternatives, Dependent Plurals, Number Agreement.

1 The Puzzle

The core observation in this paper comes from Russian number agreement with disjunctive noun phrases[1]. Typically, when the subject is a disjunction and both disjuncts are singular, plural agreement on the verb is ungrammatical, as shown in (1):

(1) Bill ili Fred prišel-Ø /*-i.
Bill or Fred came-SG/*-PL
'Bill or Fred came.'

However, as first noted in [3], plural agreement with singular disjuncts is fine in certain environments. For instance, it becomes possible under modals and frequency adverbials, as shown in (2):

* I am very grateful to Irene Heim, Danny Fox, Gennaro Chierchia for invaluable comments and all the discussions we had. Special thanks to Sam Alxatib, Anton Ivanov and Yasu Sudo for their friendly help.

[1] It is well-known that disjunctive noun phrases number agreement with disjunctive NPs shows a great amount of variability (see [4], [7], [13], [14] for English). In Russian, the agreement facts also become quite complicated the moment we take into consideration cases of agreement with disjuncts conflicting in number/gender, which probably has to do with the fact that in such cases other factors (like the possibility of First Conjunct Agreement) come into play. For the purposes of this paper, I am abstracting away from those complications and I am only discussing cases of agreement that cannot be the result of FCA. As far as those cases are concerned, it seems that the judgements are quite robust. However, they do not seem to hold for English. I am leaving the question of why languages like English behave differently in this respect for future research.

M. Aloni et al. (Eds.): Amsterdam Colloquium 2011, LNCS 7218, pp. 381–390, 2012.
© Springer-Verlag Berlin Heidelberg 2012

(2) Každyj vtornik Bill ili Fred prixodil-Ø/-i k Saše.
every Tuesday Bill or Fred came-SG/-PL to Sasha
'Every day Bill or Fred came to Sasha.'

Number agreement makes a semantic difference in (2). When the verb shows singular morphology, the disjunction can have either wide or narrow scope with respect to the quantificational adverbial, but when the agreement is plural, the disjunction can only take narrow scope. So, in a scenario where every day the same person came to Sasha, but the speaker is unsure which one, only singular agreement is possible, as demonstrated by (3):

(3) Každyj vtornik Bill ili Fred prixodil$_{SG}$/*-i$_{PL}$ k S. — ja ne pomnju kto imenno.
'Every Tuesday Bill or Fred came to S., but I don't remember who exactly.'

In the following sections, I am going to show that the ungrammaticality of plural agreement in (1) and (3) follows from independently motivated principles of scalar implicature calculation. To do so, I will first show that plural agreement with disjunction is a special case of *dependent plurality*, which has been analyzed as a scalar phenomenon in [19].

2 Zweig's Dependent Plurality

It has been observed since [2] that sentences like (4a), in which a bare plural NP appears in the scope of another plural element (either a NP or an adverbial), can have a so-called dependent plural reading. The sentence in (4a) doesn't necessarily mean that any of my friends attends more than one school. It only requires that each of my friends attend one school, but at the same time it is required that more than one school be referred to overall. If all of my friends attend the same school, (4b) must be used:

(4) a. My friends attend good schools.
b. My friends attend a good school. (examples from [19])

Zweig in [19] gives an account of dependent plurality based on the following two assumptions. First, plurals do not have the 'more than one' component as part of their meaning, i.e. they are number neutral predicates truth-conditionally (cf. [11], [16], [18]). Second, the 'more than one' component arises as a scalar implicature (*multiplicity implicature*), based on the scalemate relationship between the bare plural and its singular alternative[2] (cf. [18]).

An independent piece of evidence in favor of viewing 'more than one' as a scalar implicature comes from the fact that precisely this component disappears in environments where known scalar implicatures do too, for example, under negation:

[2] Note that Zweig takes singular NPs to denote atomic individuals.

(5) John doesn't own dogs. (see [16], [18] etc.)

According to [19], implicature calculation takes place at the level of an event predicate, namely before event closure is applied. The event predicate with a plural variable is weaker than its singular counterpart, giving rise to a scalar implicature. This assumption is crucial for accounting for the multiplicity implicature, since after event closure applies, plural and singular alternatives become equivalent, so the multiplicity implicature cannot be generated any more. Zweig, however, does not give any explanation of why the multiplicity implicature is obligatory, unlike other scalar implicatures. I should acknowledge that this is the weakest part of the proposal but I have to leave it as a stipulation for now.

There is a striking similarity between the reading of (2) with plural agreement and the dependent plural reading of (4a). Specifically, the sentence (2) with plural agreement requires that each Tuesday at least one of the guys come to Sasha, while at the same time it is inappropriate if each Tuesday, it is the same person who is coming (in that case singular agreement must be used). One could say that plural agreement makes the dependent plural reading obligatory.

3 The Proposal

As we just saw, Russian disjunctions behave similarly to English indefinite NPs (either singular or plural). To capture this correlation, I will make the following assumptions:

- Disjunction is a GQ consisting of a covert existential quantifier and a predicate [a or b]. The semantics of the predicate is given in (6)[3]:

(6) $[\![\text{a or b}]\!] = \lambda x.\ x{=}a \vee x{=}b$

- The predicate [a or b] can be singular or plural, triggering singular or plural agreement on the verb respectively.
- The plural feature denotes the closure of the predicate under sum formation. The predicate [a or b] with a plural feature will have the following denotation:

(7) $[\![(\text{a or b})\text{-}PL]\!] = {}^{*}(\lambda x.\ x{=}a \vee x{=}b) = \lambda x.\ x{=}a \vee x{=}b \vee x{=}a{\oplus}b$

It seems reasonable to postulate that sentences like (1) or (2) have two implicatures: a multiplicity implicature (MI) generated by the plural feature (following Zweig's logic) and an *exclusivity implicature* (EI) generated by a scalar item *or*.

The MI and EI for (1), repeated here as (8), are shown in (9):

(8) *[Bill or Fred]$_{PL}$ came.

[3] In (6) and (7), as well as in (14) type e expressions *a* and *b* on the left-hand side must be understood as shifted to the type <e,t> by means of Partee's IDENT.

(9) a. Bill or Fred came and *it's not true that only one of them came* =
= Bill *and* Fred came (MI)

b. Bill or Fred came and *it's not true that both Bill and Fred came* (EI)

Obviously, these two implicatures contradict each other, and this is precisely why the plural agreement (which reflects the plural feature on disjunction) is ungrammatical.

In cases like (2)/(10) the situation is different:

(10) Every Tuesday [Bill or Fred]$_{PL}$ came.

The two implicatures we get in this case are given below:

(11) a. Every Tuesday Bill or Fred came and *it's not true that every Tuesday Bill came and it's not true that every Tuesday Fred came* (MI)

b. Every Tuesday Bill or Fred came and *it's not true that every Tuesday Bill and Fred came.* (EI)

These two implicatures are consistent with each other, giving rise exactly to the dependent plural reading: both boys have to come overall, but on no Tuesday, both boys have to come.

In order to formalize this intuition, I will explicitly lay out my assumptions on the implicature calculation process. First, I assume that scalar implicatures are brought about by a covert exhaustivity operator EXH akin to 'only' (see [10] a.o.):

(12) $[[\text{EXH}_{ALT}]] = \lambda P_{<s,t>}.\lambda e. P(e) \;\&\; \forall Q \in \text{ALT} \;\&\; Q \in \text{NW}(P)$[4]$: [\neg Q(e)]$

Second, I assume, following [15], that the set of alternatives for a sentence with two occurrences of a scalar item $\varphi(X, Y)$, where X is an element of the scale Q_X and Y an element of the scale Q_Y, is defined as follows:

(13) $\text{Alt}(\varphi(X, Y)) = \{\varphi(X', Y') | X' \text{ an element of } Q_X, Y' \text{ an element of } Q_Y\}$

Third, I follow [19] in assuming that implicatures are calculated before event closure.

Also, I adopt from [19] the idea that plural and singular are scalar alternatives, with singular being the strongest element of the scale.

To these I add a new assumption, namely that the predicative OR ([a or b]) has the non-Boolean conjunction [a⊕b], defined in (14), as its non-weaker alternative:

---

[4] NW(P) denotes the set of non-weaker alternatives, i.e., those that are not entailed by the prejacent. There is a debate in the literature regarding whether the exhaustivity operator negates only stronger alternatives or non-weaker ones. I will stick to the second option, following [5].

(14) $[\![a\oplus b]\!] = \lambda x.\ x=a\oplus b$ (cf. [9])

Below I give some examples from [9] which serve as evidence for the existence of non-Boolean conjunction as it applies to <e,t> predicates (for the details I refer the reader to [9]):

(15) John and Mary are husband and wife.

(16) The flag is green and white.

Now let us see how these assumptions, taken together, explain the data in (1)/(8) and (2)/(10). First, let's examine the "non-quantificational" case repeated below:

(17) *[Bill or Fred]$_{PL}$ came.

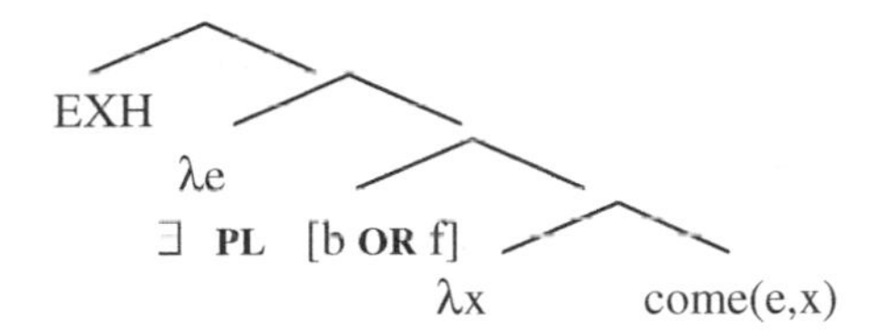

Before existential event closure applies, the sentence denotes the following:

(18) $\lambda e.[{}^{*}came(e, b) \lor {}^{*}came(e, f) \lor {}^{*}came(e, b\oplus f)]$ [PL, OR]

As shown in the LF, we are dealing with two scalar items: predicative OR and PL associated with two scales <OR, AND> and <PL, SG> respectively.

Based on the assumption in (13), the set of scalar alternatives for (18) consists of (19) and (20) (the [SG, AND] alternative is contradictory):

(19) $\lambda e.[{}^{*}came(e, b) \lor {}^{*}came(e,f)]$ [SG, OR]

(20) $\lambda e.\ {}^{*}came\ (e, b\oplus f)$ [PL, AND]

The result of exhaustification of (18) with respect to the alternatives given above is shown in (21):

(21) $\lambda e.[{}^{*}came(e, b) \lor {}^{*}came(e, f) \lor {}^{*}came(e, b\oplus f)]$ &
$\&\neg[{}^{*}came(e, b) \lor {}^{*}came(e,f)]$ &
$\&\neg[{}^{*}came\ (e, b\oplus f)]$

(21) is contradictory, which I take to be the reason of the ungrammaticality of (17).

Now let's turn to the "quantificational" case repeated below:

(22) Every Tuesday [Bill or Fred]$_{PL}$ come.

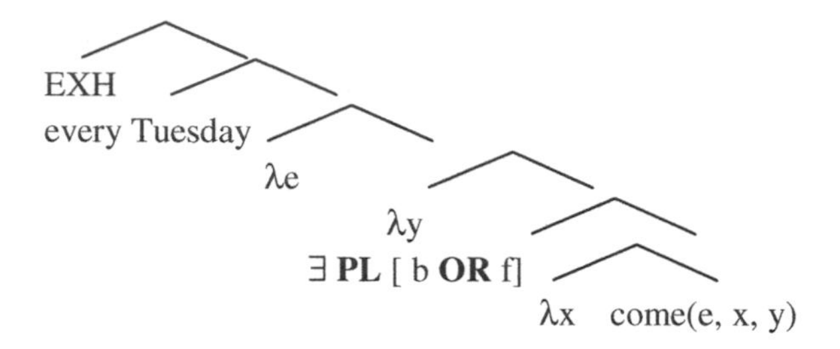

In order to make the analysis work, we need to adopt the following denotation for the adverbial universal quantifier (based on [8]) [5]:

(23) $[\![\text{every Tuesday}]\!] = \lambda P.\lambda e. [{}^*P(e)(\sigma^*Tu) \& \forall y\, [Tu(y) \rightarrow \exists e': e' \subset e \,\& P(e')(y)]]$

The denotation in (23) consists of two parts – besides universal quantification over events, it also introduces a plural event which is the sum of smaller events we quantify over. This will be crucial in the analysis of (22).

Kratzer in [8], building on [17], argues that a denotation along the lines of (23) is needed in order to account for cumulative readings of sentences containing universal quantifiers. For example, to account for the cumulative reading of (24), which can be informally stated as "between them, three copy editors found all the mistakes in the manuscript", we need to make reference to the bigger event in order to be able to state that this event is not just an event in which every mistake was caught but, crucially, an event of catching mistakes. This will guarantee that nothing irrelevant gets included in that event.

(24) Three copy editors found every mistake in the manuscript.

Let's come back to (22). Before event closure the sentence denotes the following:

(25) $\lambda e. [{}^*come(e, b, \sigma^*Tu) \vee {}^*come(e, f, \sigma^*Tu) \vee {}^*come(e, b \oplus f, \sigma^*Tu)] \&$
$\& \forall y[Tu(y) \rightarrow \exists e'[e' \subset e \& [{}^*come(e', b, y) \vee {}^*come(e', f, y) \vee {}^*come(e', b \oplus f, y)]]]$

The alternatives of (25) are given below:

(26) a. *SG*-alternative:
$\lambda e. [{}^*come(e, b, \sigma^*Tu) \vee {}^*come(e, f, \sigma^*Tu)] \&$
$\& \forall y\, [Tu(y) \rightarrow \exists e'[e' \subset e \,\& [{}^*come(e', b, y) \vee {}^*come(e', f, y)]]]$

b. *AND*-alternative:
$\lambda e.\, {}^*come(e, b \oplus f, \sigma^*Tu) \,\&\, \forall y\, [Tu.(y) \rightarrow \exists e'[e' \subset e \,\& {}^*come\, (e', b \oplus f, y)]]$

[5] The formula in (23) makes use of Link's σ operator.

The exhaustification of (25) with respect to these alternatives is in (27):

(27)

a. $\lambda e.\ [[{*}come(e, b, \sigma{*}Tu) \vee {*}come(e, f, \sigma{*}Tu) \vee {*}come(e, b\oplus f, \sigma{*}Tu)]\ \&$
$\& \forall y[Tu.(y) \rightarrow \exists e'[e' \subset e \& [{*}come(e', b, y) \vee {*}come\ (e', f, y) \vee {*}come\ (e', b\oplus f, y)]]]$
&

b. $\neg[\mathbf{{*}come(e, b, \sigma{*}\mathit{Tu}) \vee {*}come(e, f, \sigma{*}\mathit{Tu})}] \vee$
$\vee \neg\forall y\ [Tu(y) \rightarrow \exists e'[e' \subset e\ \& [{*}come(e', b, y) \vee {*}come\ (e', f, y)]]]$
&

c. $\neg{*}come(e, b\oplus f, \sigma{*}Tu) \vee \mathbf{\neg\forall y\ [\mathit{Tu}(y) \rightarrow \exists e'[e' \subset e\ \& {*}come\ (e', b\oplus f, y)]]]}$

What we get in (27) is equivalent to the conjunction of the assertion with the bold-faced disjuncts in (b) and (c). The second disjunct in (b) says that there is a Tuesday on which neither Bill came nor Fred came, which contradicts the assertion and must therefore be false, and the first disjunct in (b) says that neither Bill nor Fred is the complete agent of the big coming event. Thus, the conjunction of the assertion and the first disjunct in (b) guarantees that the agent of the big event is the sum of Bill and Fred, which accounts for the multiplicity implicature, presented informally in (11a).

The first disjunct in (c) says that Bill and Fred are not the agent of the big coming event, thus it is incompatible with the conjunction of (a) and (b) and hence should be false, making the second disjunct, which says that 'There is a Tuesday for which it's not the case that both Bill and Fred came', true. This is exactly the exclusivity implicature we presented informally in (11b).

4 Implications for the Theory of Scalar Implicatures

The analysis we proposed in the previous section raises a question: can the scalar implicatures of a sentence be a reason for its ungrammaticality?

It is well known that in general scalar implicatures are not obligatory, i.e. they can be cancelled, as illustrated below:

(28) John ate five cookies. In fact, he ate seven.

If the implicature generated by the numeral "five" ('It's not the case that John ate 6 cookies') was obligatory, we would get a contradictory sequence of sentences, but the common intuition is that these sentences are compatible.

One possible way of thinking about implicature cancellation is the following. Following [12], I will assume that the exhaustification operator is obligatory. But certain alternatives are subject to relevance. If they are not relevant, we are allowed to prune them. So, the availability of an implicature depends on the relevance of the alternative that gives rise to it.

This assumption allows to account for the contrast between the two dialogues in (29) and (30):

(29) A: Did John eat 5 cookies?
B: Yes, he ate 5 cookies. In fact, he ate 7.

(30) A: How many of the cookies did John eat?
B: # He ate 5 cookies. In fact, he ate 7.

In (29) the alternative is not relevant, thus no implicature is generated and no contradiction arises. But when we force the alternative to be relevant, as in (30), the implicature associated with it will no longer be cancellable.

Given the assumptions that we committed to in Section 3, we also need a way of incorporating the fact that an implicature associated with plural is obligatory, cf. oddness of the sequence in (31) (cf. [1]):

(31) I saw boys. # In fact, I saw only one boy.

This suggests that certain alternatives are not subject to relevance and they cannot be pruned. In our case, we will have to say that alternatives to plural cannot be pruned, thus obligatorily leading to a multiplicity implicature.

Now let's come back to the explanation of the ungrammaticality of (32):

(32) *[Bill or Fred]$_{PL}$ came.

In the previous section, I argued that the ungrammaticality of (32) is due to a conflict between implicatures generated by two scalar items: PL and OR.

(33) $\lambda e.\ [{}^*came(e, b) \vee {}^*came(e, f)]$ SG-alternative

(34) $\lambda e.{}^*came\ (e, b \oplus f)$ AND-alternative

But as I said above, the status of these implicatures must be different: the one generated by the plural (33) cannot be pruned, whereas the one generated by *or* (34) is in fact prunable. Why do we get the conflict then?

Note that if we were able to prune the *AND*-alternative, we wouldn't get a contradiction and the sentence would be grammatical, meaning 'Bill and Fred came', which is equivalent to the alternative we wanted to prune. Intuitively, it seems that such a situation should be blocked. In order to account for that, I will use the constraint on pruning proposed in [6], stated in (35):

(35) If we have two symmetric[6] alternatives before contextual pruning applies, we can not prune one without also pruning the other.

[6] S_1 and S_2 are symmetric alternatives of S if S is equivalent to the disjunction of S_1 and S_2 and S_1 and S_2 contradict each other.

As Fox and Katzir argue, the principle can be derived from the fact that contextual pruning can only eliminate alternatives which are not relevant from a formally defined set of alternatives and from certain assumptions about relevance, namely that the prejacent of the exhaustivity operator is always relevant and that relevance is closed under negation and conjunction.

Imagine we have A and two symmetric alternatives to A: B and A&¬B.

(36) A < B, A&¬B

Given our assumptions about relevance, it is clear that if B is relevant, then its symmetric alternative A&¬B must be relevant as well. Conversely, if A&¬B is relevant, B is relevant too (using assumptions about relevance and the fact that B=A&¬(A&¬B)). Hence, they must either both be relevant or they must both be irrelevant. So, when contextual pruning applies, it either eliminates both of them or neither of them. This leads to the principle in (35).

Returning to our case, note that our alternatives, presented schematically in [37], are symmetric as well: they contradict each other and their disjunction is equal to the prejacent.

(37) $[AvB]_{PL}$ < $[AvB]_{SG}$ (*singular alternative*), [A⊕B] (*AND-alternative*)

Hence, if we were to prune A⊕B, we would also have to prune $[AvB]_{SG}$, but we are not allowed to prune the singular alternative, thus we cannot prune A⊕B either. From this we derive the conflict between the implicatures, and the ungrammaticalty of (32).

The following generalization describes how grammaticality is related to implicature calculation:

(38) Ungrammaticality arises in those cases when the implicatures of a sentence lead to a contradiction and there is no possibility of obviating the contradiction by pruning the alternatives which give rise to those implicatures (either because of the requirements of certain scalar items, or because of the constraint on pruning in (35)).

The difference between cases of implicature cancellation like those in (29) and cases like (32) follows from (38): in cases like (29), the contradiction can be obviated by pruning "malicious" alternatives, as nothing blocks that operation. In cases like (32), the contradiction created by implicatures cannot be saved by pruning of alternatives because of the requirements of the plural (alternatives to plural cannot be pruned) and because of the constraint on pruning (it is impossible to prune the and-alternative without also pruning the singular alternative).

5 Conclusion

In this paper, I raised the question of whether scalar implicatures can ever lead to ungrammaticality. Based on puzzling data on agreement with disjunctions, I argued that ungrammaticality results in cases when the implicatures of a sentence lead to a contradiction that cannot be obviated by pruning the alternatives that give rise to those implicatures.

References

1. Chierchia, G., Fox, D., Spector, B.: The Grammatical View of Scalar Implicatures. In: Pornter, P., Maienborn, C., von Heusinger, K. (eds.) Semantics: An International Handbook of Natural Language Meaning. Mouton de Gruyter (to appear)
2. Chomsky, N.: Questions of form and interpretation. In: Austerlitz, R. (ed.) Scope of American Linguistics, pp. 159–196. The Peter De Ridder Press (1975)
3. Crockett, D.B.: Agreement in Contemprorary Standard Russian. Slavica Publishers (1976)
4. Eggert, R.: Disconcordance: The syntax, semantics and pragmatics of or-agreement. University of Chicago dissertation (2002)
5. Fox, D.: Free Choice and the Theory of Scalar Implicatures. In: Sauerland, U., Stateva, P. (eds.) Presupposition and Implicature in Compositional Semantics, pp. 71–120. Palgrave (2007)
6. Fox, D., Katzir, R.: On the Characterization of Alternatives. Natural Language Semantics 19, 87–107 (2011)
7. Jennings, R.: The Genealogy of Disjunction. Oxford University Press (1994)
8. Kratzer, A.: The Event Argument (in prep.)
9. Krifka, M.: Boolean and Non-Boolean And. In: Kálman, L., Polos, L. (eds.) Papers from the Second Symposium on Logic and Language. Akadémiai Kiadó, Budapest (1990)
10. Krifka, M.: The Semantics and Pragmatics of Polarity Items. Linguistic Analysis 25, 209–257 (1995)
11. Krifka, M.: Bare NPs: kind-referring, indefinites, both, or neither? In: Young, R., Zhou, Y. (eds.) Proceedings of SALT 13, pp. 180–203. CLC Publications, Cornell University (2004)
12. Magri, G.: A Theory of Individual-Level Predicates based on Blind Mandatory Implicatures. MIT dissertation (2009)
13. Morgan, J.: Some problems of determination in English number agreement. In: Alvarez, B., McCoy, T. (eds.) Proceeding of the First Eastern States Conference on Linguistics, pp. 69–78. Ohio State University (1985)
14. Peterson, P.: Establishing verb agreement with disjunctively conjoined subjects: Strategies vs. Principles. Australian Journal of Linguistics 6, 231–249 (1986)
15. Sauerland, U.: Scalar Implicatures in Complex Sentences. Linguistics and Philosophy 27, 367–391 (2004)
16. Sauerland, U., Andersen, J., Yatsushiro, K.: The plural involves comparison. In: Kesper, S., Reis, M. (eds.) Linguistic Evidence. Mouton de Gruyter (2005)
17. Schein, B.: Plurals and Events. MIT Press (1993)
18. Spector, B.: Aspects of the pragmatics of plural morphology: on higher-order implicatures. In: Sauerland, U., Stateva, P. (eds.) Presupposition and Implicature in Compositional Semantics, pp. 243–281. Palgrave (2007)
19. Zweig, E.: Number-Neutral Bare Plurals and the Multiplicity Implicature. Linguistics and Philosophy 32, 353–407 (2009)

Only *Only*? An Experimental Window on Exclusiveness*

Jacques Jayez[1] and Bob van Tiel[2]

[1] ENS de Lyon and L2C2, CNRS, Lyon, France
[2] Radboud University, Nijmegen, The Netherlands

Abstract. Beaver and Clark (2008) recently argued that *only* φ does not presuppose the proposition in its scope, contrary to the 'standard' theory articulated by Horn (1969). Their rejection of the standard theory is partially based on results of a survey test. We present new experimental evidence challenging Beaver and Clark's interpretation of this survey test and suggesting that dropping the standard theory altogether might be too radical a move.

1 Introduction

According to Horn's (1969) famous analysis, a sentence of the form *only* φ presupposes its prejacent φ and asserts that all alternatives to φ are false. For example, (1a) presupposes ($\rightsquigarrow$) that Paul smokes and asserts ($\Rightarrow$) that nobody else does. In support of this view, the prejacent appears to project from under sentential negation, as (2) suggests.

(1) Only Paul smokes.
$\rightsquigarrow$ Paul smokes.
$\Rightarrow$ Nobody else smokes.

(2) It is not the case that only Paul smokes.
$\rightsquigarrow$ Paul smokes.
$\not\Rightarrow$ Nobody else smokes.

Beaver and Clark (2008) (B&C) raise a number of objections against the standard theory and argue that *only* presupposes that the prejacent is the weakest proposition on a scale and asserts that it is the strongest proposition on the same scale. 'Weak' and 'strong' are not defined in terms of logical entailment, but in a more liberal way which is compatible with scales based on degrees of importance or cardinality.

In this paper, we reconsider the empirical and especially the experimental evidence laid out by B&C to back up their claim. The gist of B&C's experimental data is that the prejacent of *only* behaves differently from other presuppositions.

* We gratefully acknowledge the financial support of the European Science Foundation, ESF travel grant Euro-Xprag 4273.

M. Aloni et al. (Eds.): Amsterdam Colloquium 2011, LNCS 7218, pp. 391–400, 2012.
© Springer-Verlag Berlin Heidelberg 2012

We present results of a more extensive experiment that challenge this conclusion on two counts. First, other presupposition triggers with an entirely different semantics behave in the same way as *only*. Second, a semantically very similar trigger (*seulement* in French) behaves quite differently. This runs counter to B&C's claim that *only* has a somehow special status compared to other presupposition triggers.

In sections 2.1, 2.2 and 2.3, we present B&C's objections against the standard theory and describe their alternative proposal. In section 2.4, we show that their empirical arguments are inconclusive. In section 3, we present the results of an extensive survey, which will be discussed in section 4. These results undermine B&C's critique of the standard theory.

2 Beaver and Clark's Approach

2.1 Linguistic Observations

According to the standard theory, *only* φ presupposes its prejacent. This means that the prejacent is predicted to project. Against this prediction, B&C observe that there are cases in which the prejacent is not preserved under negation. Consider (3). It is clear that the speaker does not imply that the person in question is a blond bimbo with no brains.

(3) She's one of the first that really represents the country and isn't only some blond bimbo with no brains.

Another piece of evidence in the same direction is provided by examples like (4), which does not entail that Mary invited Susan and Paul, since she invited their six cousins instead.

(4) Last year, Mary invited Susan and Paul. This year, she did not invite only Susan and Paul, but preferred to invite their six cousins.

These examples seem to endanger the standard theory. If *only* φ presupposes its prejacent, why does it not project in the examples above?

2.2 The Tequila Test

To further underpin the view that the prejacent of *only* is more 'fragile' than presuppositions of other triggers, B&C devised an experiment based on the interpretation of a little story:

One year there were 90 students in Arroyo.
30 drank Tequila and nothing else.
30 drank non-alcoholic beverages and nothing else.
30 drank everything, no matter what.

Subjects had to answer the following two questions: *How many students didn't only drink Tequila* (VP-*only*) and *How many students didn't drink only Tequila* (NP-*only*). They could choose between the following answers: '30', '60' and 'Don't know'.

The standard theory predicts most participants to opt for the '30' answer. After all, only the 30 students who drank everything satisfy both the prejacent (i.e., they drank Tequila) and the asserted content (i.e., they drank something else). If, on the contrary, the prejacent is not genuinely presupposed, participants might also include the 30 students who drank non-alcoholic beverages and nothing else in the denotation of the question. While these students do not satisfy the prejacent, they do fulfill the asserted content. In that case, participants should choose the '60' answer. Assuming that the first set is ruled out in any case, it might seem that the 'Tequila test' is an adequate instrument to decide whether the prejacent is presupposed.

B&C report the following results (absolute numbers between brackets):

	'30'	'60'	Don't know
NP-*only*	22% (9)	76% (31)	2.4% (1)
VP-*only*	41% (17)	56% (23)	2.4% (1)

There is a substantial difference between the results in the NP-*only* and VP-*only* condition. Because the subjects were not divided into two independent or paired samples, it is difficult to interpret these results in a reliable way. It is possible to run a McNemar's test on the results, under the assumption that the subjects are 'coherent'; that is, that the subjects who chose '30' for NP-*only* are a subset of those who chose '30' for VP-*only* and that the subjects who chose '60' for VP-*only* still chose '60' for NP-*only*. In that case, the difference between the two positions for *only* is significant at the 0.05 threshold (p-value ≈ 0.012).[1] That is, the answer '60' was chosen significantly more often in the NP-*only* condition than in the VP-*only* condition.

Regardless of this difference, there was overall a high number of '60' answers, particularly in the NP-*only* condition. These results appear to jeopardise the view that the prejacent is presupposed. But for this argument to go through, it still has to be shown that other presupposition triggers lead to significantly different results. To show this, B&C used a comparable testing procedure for four other presupposition triggers: *stop*, *realize*, *regret* and *their*. The set-up was basically the same. Participants read a short story followed by a question in which the trigger was embedded under negation. The story involved 90 students, 30 of which satisfied the presupposition and the asserted content, 30 of which satisfied only the asserted content, and 30 of which satisfied neither the presupposition nor the asserted content (or in some cases the presupposition but not the asserted

[1] B&C report a non-significant result for a chi-square test. The problem with using this test is twofold: if the subjects are coherent, in the sense considered here, the chi-square is not a good indicator. If they are not coherent, to a degree that falsifies our assumption, the question is more complex because the shift in perception that this incoherence suggests has to explained.

content).[2] The critical question was whether participants include the students who falsified the presupposition but not the asserted content in the denotation of the question. The target item for *stop* is provided below:

There are 90 students:
  30 used to drink but gave up.
  30 never drank Tequila.
  30 currently drink Tequila.
How many students didn't stop drinking Tequila?

If the question presupposes that the students used to drink Tequila, the correct answer is '30'. If this presupposition is as 'fragile' as the prejacent for *only*, participants might opt for '60', thus including the students that never drank Tequila. The results for these presupposition triggers clearly differ from the results that were found for *only*. For *stop*, *realize* and *their*, 9, 12 and 10 participants out of a total of 13, chose the answer which is compatible with the projection of the presupposition, namely '30'. These findings seem to indicate that the prejacent has a somewhat different status than ordinary presuppositions. This compromises the standard theory of *only*, but fits in neatly with B&C's counterproposal, to which we now turn.

2.3 The Proposal

B&C propose to amend the standard theory by exploiting the scalar character of *only*. To put it concisely, *only* φ presupposes that the prejacent is at most as strong and asserts that it is at least as strong as any true alternative. More precisely, we have:

> *Only* p presupposes (asserts) that for every proposition q in an appropriate set of alternatives to p, $ALT(p)$, if q is true then p is at most (at least) as strong as q. In symbols: *only* p presupposes the propositions defined by $\lambda w \forall q \in ALT_\sigma(p)(w \models q \Rightarrow q \geq_\sigma p)$, and asserts the propositions defined by $\lambda w \forall q \in ALT_\sigma(p)(w \models q \Rightarrow q \leq_\sigma p)$, where σ is the belief state of the speaker.

Let us apply this definition to (1), assuming that the set of alternatives is calculated on the basis of alternatives of the form 'x smokes', which are ordered by entailment. Here, x ranges over a set of possible persons or groups. The presupposition eliminates worlds in which 'Paul smokes' is stronger than some alternative

[2] The reason for this variation is that it is very difficult for some triggers to describe a character that falsifies both the asserted content and the presupposition. For *Who didn't stop drinking?*, for example, this would amount to a character who at the very moment of uttering the question started drinking. Such a character falsifies both the presupposition (i.e., she didn't drink before) and the asserted content (i.e., she doesn't drink now). But it is hard to exclude the possibility that this character actually did drink at least some time before the moment of utterance, thus verifying the presupposition. To avoid this ambiguity, B&C construed a character that verifies the presupposition in these cases.

which is true at the same world. The common ground is then updated with the main content. This move eliminates worlds in which there is a proposition of the form 'x smokes' which is stronger than 'Paul smokes'. For instance it eliminates worlds in which Paul and Mary, or Paul and John smoke. The net result is a set of worlds where, for each true proposition $q \in ALT_\sigma(p)$, $q =_\sigma p$. If we apply a negation to *Only Paul smokes*, the presupposition is (normally) preserved but the main content is negated. So, the negated sentence asserts that $\lambda w \exists q \in ALT_\sigma(p)(w \models q \,\&\, q >_\sigma p)$, in other words, that Paul *and* someone else smoke.

When alternatives are not ordered by entailment, a different result can obtain. For instance, if a cardinality-based ordering is used, the presupposition is that the prejacent concerns at most as many individuals as any true alternative. This delivers the required reading for (4). The negated main content entails that Mary invited more persons than just two. However, it does *not* entail that the guests include Susan and Paul, since the alternatives are compared on the basis of cardinality and not of entailment.

Summarising, B&C (i) replace the prejacent with a lower bound on the relative strength of the true alternatives, and (ii) assume that the main content sets an upper bound on the strength of the true alternatives. The derivation of the prejacent is thus an effect of the interaction between these two constraints, not an intrinsic semantic property of *only*.

2.4 Preliminary Discussion

One might wonder why it is necessary at all to gather experimental data in order to decide between different theories of *only*. In examples like (4), the prejacent manifestly does not project. Doesn't this itself conclusively disprove the standard theory? In fact not. Examples like (4) can be construed for any presupposition trigger. For example, B's answer in (5) clearly does not imply that John has been smoking recently. Such examples can be explained as local accommodation or metalinguistic negation (e.g., Geurts 1999).

(5) [Context: it is common belief that John never smoked. B is trying to quit.]

A – John seems to be much more relaxed than you are.
B – He didn't stop smoking a week ago!

On the whole, B&C's empirical observations are not as conclusive as they may seem at a first glance because they are not restricted to *only*, but concern rather the pragmatic conditions on the felicity of presupposing. So, it is not clear that a specific theory should be constructed for *only* on the basis of examples such as (3) or (4).[3]

This leaves us with the results of the Tequila test. Although B&C's empirical observations do not support the view that the prejacent is different from an

[3] A similar remark applies to Ippolito's (2008, 50-52) discussion about *it's possible that only* φ. As shown by Herburger (2000, 95) the observations that would tend to show that the prejacent is suspended are not specific to *only*. B&C mention Herburger's work and add further examples which suggest that *only* is not the main factor in those cases (Beaver and Clark 2008, 245-246).

ordinary presupposition, the results of their Tequila test clearly do. Hence the importance of carefully evaluating the validity of B&C's experimental results.

3 An Experimental Approach

The Tequila test, as carried out by B&C, faces several more or less severe problems. First, one of their surveys included a very limited number of 13 participants. Second, no fillers were included in any of the surveys. This was especially pertinent for the experiment that involved *only*. The entire experiment consisted of two nearly identical questions, differing only in the position of *only* in the sentence (before the VP or just before the NP). It is quite possible that this juxtaposition led participants to explicitly contrast the two questions. As a final worry, there were some presentational and interpretational differences between the story involving *only* and the stories involving the other presupposition triggers.

A perhaps more interesting issue concerns the scope of B&C's Tequila test. First, the selection of presupposition triggers B&C employed to compare with *only* was rather limited. Second, it might be hypothesised that B&C's results are somehow peculiar to English *only*. This idea is fueled by our intuitions about the Dutch and French equivalents of *only*. These considerations led us to form the following hypotheses: (i) For English, there are presupposition triggers that behave just like *only*; (ii) The prejacent of the Dutch and French equivalents of *only* are not as 'fragile' as the prejacent of English *only*. The truth of either hypothesis would disprove B&C's conclusion that the prejacent of *only* has a special status compared with other presuppositions.

3.1 The Basic Protocol

In order to remedy some of the methodological issues of B&C's survey, we made some changes to the design of the Tequila test. Instead of using numbers, we used characters. Just like in B&C's version, one character satisfied neither the asserted content nor the presupposition of the question (or in some cases the presupposition but not the asserted content). One character verified the asserted content but falsified the presupposition. One character verified both the asserted content and the presupposition. These characters correspond to respectively A, B and C in the examples below. Again, the critical issue was whether participants include the B character in the denotation of the question.

We ran the experiment in three languages: Dutch, English and French. The target stimuli were interspersed with filler stimuli in the same vein but using various quantifiers such as *at most three* or *often*. The stimuli and the attribution of the actions or situations to A, B and C were pseudo-randomised. We had 16 presupposition triggers and 16 fillers for Dutch, the same numbers for English, and 15 triggers and 23 fillers for French. The triggers included focus particles like *only* or *also*, factives like *know* or *regret*, implicatives like *manage* or *succeed*, aspectuals like *stop* or *start* and definites like *the* or *all*. For English, participants were recruited through Amazon MTurk. For French and Dutch, university students were asked to fill out the experiment. After we got the results,

Only	Other triggers
Three people were in the cafeteria A drank orange juice and nothing else B drank coffee and nothing else C drank orange juice and coffee	Three people are riding a bus A had a job at the bank but quit B never had a job in her life C has a job at the bank and still works there
Who didn't drink only orange juice? □ C □ C and B □ I don't know	Who didn't resign from the bank? □ C □ C and B □ I don't know

Fig. 1. Two target stimuli

we decided to eliminate the *démissionner* ('resign') case from the French data, because the little story associated with it was problematic.

3.2 The Image-Based Protocol

When it turned out that the results for English *only* were markedly distinct from the results for its Dutch and French equivalents, as will be explained in the next section, we decided to run an additional experiment for English speakers, based on the expectation that the linguistic presentation of possible answers may have influenced participants. In this experiment, participants were presented with series of three images and had to answer a question that was completely analogous to the question asked in the Tequila paradigm. Participants could tick any number of boxes they liked. The critical stimulus is shown in figure 2.

Who does not have only an apple?

Fig. 2. An image-based target stimulus

25 participants were drafted through Amazon MTurk. The experiment consisted of 6 stimuli, including 1 target stimulus featuring *only* and 5 fillers involving quantifiers. The stimuli were pseudo-randomised.

3.3 Results

The comparison between different kinds of English triggers is summarised graphically in figure 3. The left (black) column represents the percentage of 'C' answers, the middle (white) column the percentage of 'B and C' answers and the right one (grey) the percentage of 'I don't know' answers. Because there are very few 'I don't know' answers, it is possible to binarize the results by dividing the answers into 'C' versus other answers ('B and C' and 'I don't know'). The dependent variable is the proportion of 'C' answers. The independent variables are language and type of stimulus, e.g. implicatives, factives, etc.

We analysed these data by means of a logistic regression analysis. Using the 'lme4' package in R, we fitted a simple model of mixed logistic regression, having the subjects as random effect and adding a post hoc comparison based on the 'multcomp' package. The results are summarised in figure 3 for English *only*. The number of 'C' responses for this item is compared to the number of 'C' responses for other kinds of triggers. The difference is significant for factive and focus elements and non-significant for other categories.

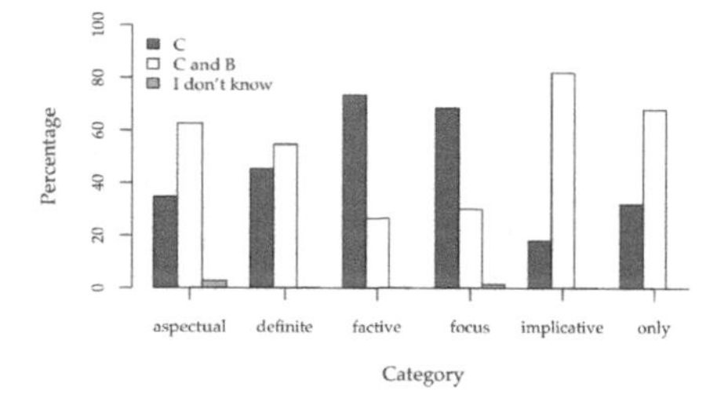

	z value	Pr(>|z|)
vs. aspectual	-0.261	0.99983
vs. definite	-1.249	0.80606
vs. factive	-3.892	0.00139**
vs. focus	-3.289	0.01233*
vs. implicative	1.419	0.70749

Fig. 3. English triggers

The pictorial task for English illustrated in figure 2 gave totally consonant results. For the critical item involving *only*, 72% of speakers chose the 'B and C' answer. So it seems unlikely that the linguistic nature of the test was an issue.

The Dutch and French counterparts of *only*, *alleen* and *seulement*, do not behave like *only* in English. The two relevant histograms and the post hoc contrasts are shown in figure 4. The post hoc contrasts on a simple logistic regression with the response binary variable restricted to the *only* case show that Dutch and French are not significantly different whereas they are both different from English ($\Pr(>|z|) = 0.93, 0.0004, 0.0008$).

4 Discussion and Perspectives

The language-based and image-based results for English are consonant with B&C's observations for *only* and the other presupposition triggers they investigated. This shows that our results were not affected by the methodological changes we made to the Tequila paradigm.

Our overall results call into question B&C's conclusion that the prejacent of *only* behaves differently from ordinary presupposition triggers. Figure 3 shows no difference between *only* and aspectuals, implicatives, or definites. This is unexpected if the relation of *only* to its prejacent is specific. If we assume, with B&C, that the preponderance of 'C and B' answers for *only* suggests that the prejacent

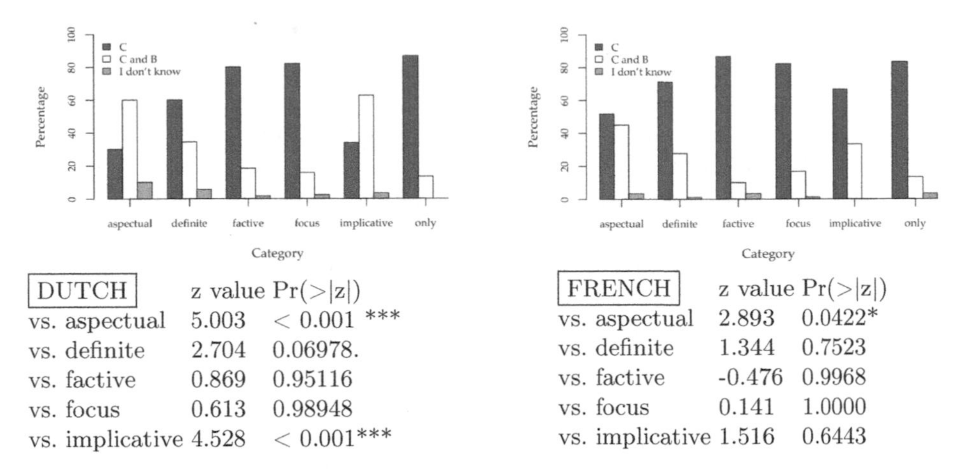

DUTCH	z value	Pr(>|z|)
vs. aspectual	5.003	< 0.001 ***
vs. definite	2.704	0.06978.
vs. factive	0.869	0.95116
vs. focus	0.613	0.98948
vs. implicative	4.528	< 0.001***

FRENCH	z value	Pr(>|z|)
vs. aspectual	2.893	0.0422*
vs. definite	1.344	0.7523
vs. factive	-0.476	0.9968
vs. focus	0.141	1.0000
vs. implicative	1.516	0.6443

Fig. 4. Dutch and French triggers

is not genuinely presupposed, we would arrive at the counterintuitive conclusion that the presuppositions ordinarily associated with aspectuals, implicatives, and definites are not genuine presuppositions either. Whatever conclusions one might draw from the behavior of *only*, it is clear that these do not hold for *alleen* and *seulement.* For these triggers, participants almost unanimously opted for the 'C' answer. Adopting B&C's interpretation again leads to the implausible conclusion that the prejacent of *alleen* φ and *seulement* φ is genuinely presupposed but the prejacent of *only* φ is not.

One might wonder whether the observed profiles coincide with the 'weak' versus 'strong' trigger distinction made by Abusch (2010). Abusch contrasts examples like those in (6). *Win*, which presupposes a participation in the competition, allows for the suspension of its presupposition and is, in this respect, 'weak', in contrast to *again*, which is a 'strong' trigger.

(6) a. I don't know whether John finally participated in the race, but if he <u>won</u> it he may be very proud!

b. ?? I don't know whether John won this race before, but if he won <u>again</u>, he may be very proud!

At a first glance, it might seem that strong triggers evoke mostly 'C' answers whereas weak triggers evoke mostly 'C and B' answers. This indeed holds for the strong triggers *again* and *also*. Unfortunately, the parallelism breaks down when it comes to aspectuals, which are presumably weak. We found perhaps surprisingly that *start* leads to mostly 'C and B' answers whereas *stop* evokes mostly 'C' answers. Overall, the data do not correspond to a systematic weak/strong distinction.

An important issue in the semantics of exclusives is their scalar character. It is well-known that *only* is scalar in that it can be interpreted as entailing that the degrees above or below a certain threshold, as expressed by the prejacent, are not reached. French *seulement* has the same property (e.g., Beyssade 2010), whereas *alleen* is not scalar, see (7). It is remarkable, then, that *seulement* patterns with *alleen* and not with *only*, with which it shares its scalar nature.

(7) a. Paul is only a first-year student.
 b. Paul est seulement un étudiant de première année.
 c. *Paul is alleen een eerstejaars student.

The reported observations raise a more general question. It is often (partly) implicitly assumed that the behaviour of presupposition triggers should be a reflection of their formal semantics, because assuming the contrary would lead us to renounce any explanation. In our opinion, this dilemma lacks serious foundations. The high cross-linguistic variability of certain triggers, which sound otherwise quite comparable, comes as a surprise under this view, but remains compatible with an approach that is not (entirely) representational, in which triggers *in addition* to a descriptive content (main content + presupposed content) have a statistical profile with respect to, say, suspension under negation or other environments. It remains to see what factors cause these differences in statistical profile.

In future work, we intend to tighten the experimental conditions, by having a homogeneous pool of subjects in the three languages and controlling the stimuli and the choice of answers even more carefully. We are also planning two new experiments, one using reaction times, in order to determine whether there is any correlation between the 'B and C' answer and the choice duration. We also intend to test whether the observed difference might be connected with the 'loneliness' flavour associated with *seulement* and *alleen*, both of which provide adjectives meaning 'alone', in contrast to English (**Paul is only*). To this aim, we will turn to languages similar to English in this respect like Chinese.

References

Abusch, D.: Presupposition triggering from alternatives. Journal of Semantics 27, 37–80 (2010)

Beaver, D., Clark, B.Z.: Sense and Sensitivity. How Focus Determines Meaning. Wiley-Blackwell, Chichester (2008)

Beyssade, C.: Seulement et ses usages scalaires. Langue Français 165, 103–124 (2010)

Geurts, B.: Presuppositions and Pronouns. Elsevier, Amsterdam (1999)

Herburger, E.: What Counts: Focus and Quantification. MIT Press, Cambridge (2000)

Horn, L.: A Presuppositional Analysis of only and even. Papers from the Fifth Regional Meeting of the Chicago Linguistics Society, pp. 98–107 (1969)

Ippolito, M.: On the Meaning of Only. Journal of Semantics 25, 45–91 (2008)

Roberts, C.: Only, presupposition and implicature. MS., Ohio State University (2006), `http://ling.osu.edu/~croberts/only.pdf`

On the Information Status of Appositive Relative Clauses

Todor Koev*

Department of Linguistics, Rutgers University
tkoev@eden.rutgers.edu

Abstract. Existing semantic theories of appositive relative clauses (ARCs) assume that ARCs contribute asserted but not at-issue content (Böer & Lycan [4], Bach [3], Chierchia & McConnell-Ginet [5], Potts [13], AnderBois et al. [2], Murray [12]). In this paper I demonstrate that the information status of ARCs depends on their linear position in the clause: clause-medial ARCs are not at-issue whereas clause-final ARCs can behave like regular at-issue content. I propose a uniform one-dimensional semantics under which ARCs are conjuncts that can acquire at-issue status if the issue raised by the main clause has been terminated. The idea is formally implemented in Dynamic Predicate Logic (Groenendijk & Stokhof [9]) enriched with propositional variables (AnderBois et al. [2]).

Keywords: appositive relative clauses, at-issue/not-at-issue content.

1 Introduction

Existing semantic theories of appositive relative clauses (ARCs) assume that ARCs express asserted but NOT-AT-ISSUE content, often referred to as "deemphasized", "backgrounded", or "secondary" content (Böer & Lycan [4], Bach [3], Chierchia & McConnell-Ginet [5], Potts [13], AnderBois et al. [2], Murray [12]). Intuitively, not-at-issue content is not part of the main point of the utterance. The main point of the utterance is typically expressed by the main clause, which is AT-ISSUE.

In this paper I demonstrate that the information status of ARCs depends on their linear position in the clause. I show that CLAUSE-MEDIAL ARCs, as in (1), are not at-issue whereas CLAUSE-FINAL ARCs, as in (2), can behave like regular at-issue content.[1] (To improve readability, ARCs in linguistic examples have been underlined.)

* I would like to thankfully acknowledge the contributions of Maria Bittner, Veneeta Dayal, Carlos Fasola, Jane Grimshaw, Roger Schwarzschild, and Kristen Syrett to the current work. I also thank two anonymous reviewers for their suggestions and criticism.

[1] This has previously been noticed but not explained in Cornilescu [6] and AnderBois et al. [2].

M. Aloni et al. (Eds.): Amsterdam Colloquium 2011, LNCS 7218, pp. 401–410, 2012.
© Springer-Verlag Berlin Heidelberg 2012

(1) Maradona, who scored a goal with his hand, won the Golden Boot.

(2) Sally admires Maradona, who scored a goal with his hand.

Previous researchers (AnderBois et al. [2], Murray [12]; both building on Stalnaker [19]) have modeled the contrast between at-issue and not-at-issue assertions by the way those two types of content enter the context set. Main clauses put forward an update proposal (a set of worlds) which can be accepted or rejected by the addressee; hence, they are at-issue. If accepted, the context set is reduced to the worlds in which the proposal proposition holds. In contrast, appositives directly restrict the context set to the worlds in which their content holds, hence they are not at-issue. I argue for an alternative account according to which both main clauses and ARCs introduce proposals. Proposals are introduced/accepted at the left/right clause boundary, respectively. Since the right boundary of clause-medial ARCs falls inside the main clause, proposals associated with such ARCs are never at-issue. The right boundary of clause-final ARCs can be construed as either inside or outside the main clause. When the latter is the case, proposals introduced by such ARCs become at-issue. Those ideas are formally implemented in Dynamic Predicate Logic (Groenendijk & Stokhof [9]) enriched with propositional variables, following AnderBois et al. [2].

The paper is structured as follows. Section 2 discusses the empirical facts. Section 3 evaluates previous accounts and presents my own account. Section 4 concludes and addresses two broader issues.

2 Two Tests for At-issue/Not-at-issue Content

In this section I demonstrate that clause-medial ARCs are invariably not at-issue while clause-final ARCs can be at-issue. This is argued on a basis of two empirical tests, both of which underscore the inability of not-at-issue content to address the main topic of the conversation, or, in Roberts' [14] terms, the *question under discussion.*[2]

2.1 The Direct Reply Test

Since not-at-issue content is not part of the primary assertion of a sentence, one would expect that such content cannot be a direct target for subsequent discourse. I call this the Direct Reply Test.

(3) *Direct Reply Test*
Only at-issue content can be directly targeted by the addressee, e.g. by replies like "Yes", "No", "That's not true", etc.

[2] See Tonhauser [20] for more discussion on those tests for at-issue/not-at-issue content.

Note that it *is* possible to target not-at-issue content but only indirectly. Indirect targeting has a more severe conversational effect: it disrupts the natural flow of discourse. It also involves different grammatical tools, including hedges like "Actually, ...", "Well, ...", etc.

Clause-medial ARCs fail the Direct Reply Test and thus are not at-issue.

(4) (cf. Amaral et al. [1])
A: Edna, who is a fearless leader, started the descent.
B: #No, she isn't. She is a coward.

In contrast, clause-final ARCs pass the Direct Reply Test. In this way, they behave like regular at-issue content.

(5) A: Jack invited Edna, who is a fearless leader.
B: No, she isn't. She is a coward.

As an aside, the very same contrast crops up in cases in which the main clause and the ARC are of a different sentence type. Consider cases in which one of the clauses is a declarative and the other is an interrogative.[3]

(6) A: Has John, who was talking to Mary a minute ago, gone home?
B: No, he hasn't. He is still at the party.
B′: #No, he wasn't. He was talking to Stacy.

(7) A: Marcia, who you wanted to meet, didn't you?, has just arrived.
B: No, she hasn't. She is still in San Francisco.
B′: ??No, I didn't. I wanted to meet Sarah.

(8) A: Did you see John, who was talking to Mary a minute ago?
B: No, I didn't. I had no idea he was at the party.
B′: ??No, he wasn't. He was talking to Stacy.

(9) A: Jack invited Marcia, who you wanted to meet, didn't you?
B: ??No, he didn't. He invited Sarah.
B′: No, I didn't. I wanted to meet Sarah.

One can make intuitive sense of this paradigm as follows. Clause-medial ARCs, whether declarative or interrogative, are not at-issue and cannot be directly addressed by the hearer in (6), (7). However, clause-final ARCs can be at-issue. Given the reasonable assumption that answering a question has higher discourse urgency than rejecting an assertion, one can explain why reacting to the asserted content in (8) and (9) is significantly degraded. In short, the Direct Reply Test delivers similar results here too.

2.2 The Answerability Test

Since not-at-issue content cannot address the main topic of the conversation, it is to be expected that it cannot answer questions. I call this test the ANSWERABILITY TEST.

[3] The example in (6A) is from McCawley [11].

(10) *Answerability Test*
Only at-issue content can be employed by interlocutors to answer questions.

Both types of ARCs fail the Answerability Test when a singleton question is asked.[4]

(11) A: What disease did Tammy's husband have?
B: ??Tammy's husband, who had prostate cancer, was treated at the Dominican Hospital.

(12) A: What disease did Tammy's husband have?
B: ??The doctors of the Dominican Hospital treated Tammy's husband, who had prostate cancer.

The general problem with such answers seems to be that the information contributed by the main clause is not relevant to the question being asked. A way around this problem is to frame the dialogue in such a way that it involves two questions, one addressed by the main clause and the other addressed by the appositive.[5] In such contexts the contrast between clause-medial and clause-final ARCs reemerges. Clause-medial ARCs fail the Answerability Test and hence are not at-issue.

(13) A: What did Paula bring and when did she leave the party?
B: ??Paula, who brought cookies, left after midnight.

Clause-final ARCs pass the Answerability Test and once again behave like at-issue content.

(14) A: Who did you meet at the party and what did they bring?
B: I met Paula, who brought cookies.

2.3 Intermediate Conclusions

We see that clause-medial ARCs are heavily restricted in their uses since they cannot address the main topic of the conversation, i.e. express not-at-issue content. In contrast, clause-final ARCs seem to be able to serve any pragmatic purpose that main clauses can serve. In that respect they behave like regular at-issue content.

3 The Formal Account

In this section I first review existing accounts of apposition and then introduce my own account.

[4] The examples below are based on AnderBois et al. [2].
[5] I owe this observation to Veneeta Dayal.

3.1 Previous Accounts of ARCs

There are two main semantic approaches to ARCs. Sometimes ARCs are analyzed as conjuncts that are part of the at-issue proposition contributed by the entire sentence (Frege [8], Böer & Lycan [4], Rodman [15], Sells [17], Kempson [10]). According to this approach, sentences with appositives have a single truth value. Alternatively, ARCs are viewed as contributing secondary propositions which are independent of the at-issue proposition expressed by the main clause (Bach [3], Dever [7], Potts [13]). According to this approach, sentences with appositives have multiple truth values (one contributed by the main clause and the rest contributed by the appositives).

One can immediately see that neither approach would straightforwardly account for the data discussed above. Conjunctive approaches will be at pains to explain why clause-medial ARCs are not at-issue and multi-dimensional approaches will be hard-pressed to tell us why clause-final ARCs behave like at-issue content. An extra pragmatic mechanism would be needed to explain the deviant examples in each case, although it is not clear what this mechanism would be. One could also try to combine the two approaches and simply claim that clause-medial ARCs project a secondary proposition while clause-final ARCs are conjunctions. Yet this would amount to giving up the idea that clause-medial and clause-final ARCs share a unified semantics.

There are also mixed approaches which treat ARCs as conjuncts that are interpreted in a special way (AnderBois et al. [2], Murray [12]; see also Schlenker [16]). Those are one-dimensional dynamic accounts which seek to explain the not-at-issue status of (clause-medial) ARCs in terms of the direct way appositives restrict the context set. Main clauses put forward an update proposal (a set of worlds) which can be accepted or rejected by the addressee; hence, they are at-issue. If accepted, the context set is reduced to the worlds in which the proposal proposition holds. In contrast, appositives directly restrict the context set to the worlds in which their content holds; hence, they are not at-issue. The main problem with those accounts is similar to the main problem of two-dimensional accounts: since ARCs express restrictions on the context set, it is incorrectly predicted that ARCs are invariably not at-issue. Below, I offer an account which captures the different information status of clause-medial and clause-final ARCs.

3.2 The Formal System

In Dynamic Predicate Logic (Groenendijk & Stokhof [9]) formulas are interpreted as binary relations between assignment functions. This allows values of variables to be passed on to subsequent conjuncts. AnderBois et al. [2] enrich the set of variables with propositional variables, interpreted as sets of worlds. This makes it possible that lexical relations posit conditions on the worlds in some proposition and not on the worlds in the entire context set. I state the semantic rules that are necessary to interpret the examples below.[6]

(15) a. $g[R_p(t_1, ..., t_n)]h$ iff $g = h$ & $\forall w \in h(p) : \langle [\![t_1]\!]^h, ..., [\![t_n]\!]^h \rangle \in [\![R]\!]^w$
b. $g[\neg R_p(t_1, ..., t_n)]h$ iff $g = h$ & $\forall w \in h(p) : \langle [\![t_1]\!]^h, ..., [\![t_n]\!]^h \rangle \notin [\![R]\!]^w$
c. $g[t_1 = t_2]h$ iff $g = h$ & $[\![t_1]\!]^h = [\![t_2]\!]^h$
d. $g[p_1 = p_2]h$ iff $g = h$ & $h(p_1) = h(p_2)$
e. $g[p_1 \subseteq p_2]h$ iff $g = h$ & $h(p_1) \subseteq h(p_2)$
f. $g[\phi \wedge \psi]h$ iff $\exists k : g[\phi]k$ & $k[\psi]h$
g. $g[\exists u]h$ iff $\forall v : v \neq u \Rightarrow g(v) = h(v)$

Lexical relations are relativized to propositions and require that the relation holds in all the worlds in that proposition (15a). A negation of a lexical relation relativized to some proposition requires that the relation does not hold in any of the worlds in that proposition (15b).[7] The rest of the semantic rules are as expected. Atomic formulas which do not represent lexical relations are tests on the input/output assignment (15c)-(15e).[8] Conjunction is interpreted as relational composition (15f) and random assignment (here singled out as a separate formula) non-deterministically updates the value of its variable while keeping all other variable values intact (15g).

Below, I make use of this logic but present a very different account from that of AnderBois et al. [2].

3.3 The Account

Adding propositional variables to the logic allows us to model the intuition that main clauses place restrictions on the proposals they make and not directly on the context set. A logical translation for a simple sentence is illustrated in (16), where p is the proposal proposition and p_{cs} is a distinguished variable representing the current context set.

(16) A: Edna is a fearless leader. $\exists p \wedge \exists x \wedge x = edna \wedge fearless.leader_p(x)$
B: (OK.) $p \subseteq p_{cs} \wedge \exists p_{cs} \wedge p_{cs} = p$

The speaker puts forward a new proposal, formally represented as updating the value of the propositional variable p to a non-empty set of worlds.[9] The last

[6] I use (with or without subscripts) $p, q, ...$ for propositional variables; t for terms (i.e. individual variables or constants); $u, v, ...$ for individual or propositional variables. Reference to models is suppressed.

[7] AnderBois et al. [2] do not provide a rule for negated formulas. The rule proposed here (due to Robert Henderson, p.c.) is limited to atomic formulas expressing lexical relations. Formulating a general rule for negation turns out not to be a trivial matter and I will not propose one here.

[8] I use the same relational symbols in the object language and the metalanguage.

[9] I require that assignment functions always assign non-empty sets of worlds to propositional variables. Given the rules in (15a)-(15b), an empty proposal would always pass the test expressed by lexical relations and thus has to be blocked.

conjunct restricts the values assigned to p by requiring that in all of p's worlds, Edna is a fearless leader. The hearer accepts the proposal, overtly or silently. This is modeled as checking whether p is a subset of the current context set and assigning the value of p to the new context set. I will refer to the three conjuncts in (16B) as the ACCEPTANCE UPDATE. The acceptance update updates the context set to some non-empty subset of it, such that in all of the worlds in this new context set the output value of x, Edna, is a fearless leader. This new context set is passed on (i.e., is anaphorically available) to subsequent discourse.[10]

The hearer need not, however, accept the proposal that has been introduced by the speaker. In (17) B rejects the original proposal introduced by A and makes a counterproposal. Since A agrees with the counterproposal, the updated context set, a subset of the old context set, will only consist of worlds in which the output value of x, Edna, is *not* a fearless leader.

(17)	A: Edna is a fearless leader.	$\exists p \wedge \exists x \wedge x = edna \wedge fearless.leader_p(x)$
	B: No, she isn't.	$\exists q \wedge \neg fearless.leader_q(x)$
	A: (OK.)	$q \subseteq p_{cs} \wedge \exists p_{cs} \wedge p_{cs} = q$

Let us now consider sentences with ARCs. I make two important assumptions. First, I assume that *both* main clauses and ARCs introduce new proposals. Also, I make the principled assumption that proposals are introduced/accepted at the left/right *clause boundary*, respectively. Since the right boundary of clause-medial ARCs is included in the main clause, proposals associated with such ARCs will always be terminated. The right boundary of clause-final ARCs can, however, be construed as either inside or outside the main clause.[11] If the latter, the proposal introduced by the ARC will be up for negotiation. Those two assumptions derive the fact that clause-medial ARCs are invariably not at-issue whereas clause-final ARCs can be at-issue. I first discuss in detail the examples in (4)-(5), which involve the Direct Reply Test, and then sketch a similar solution for the data in (13)-(14), which involve the Answerability Test.

Sentences with clause-medial ARCs, as in (18), make two proposals. Given the assumption that acceptance updates are introduced at clause boundaries, A's utterance updates the context set with the appositive content, i.e. the information that Edna is a fearless leader is no longer at-issue. Hence, a counterproposal by B that Edna is not a fearless leader will have no chance of being accepted and thus is correctly ruled out. B could, however, introduce the counterproposal that

[10] Since values to variables are assigned non-deterministically, the discourse in (16) will produce not a single p_{cs} value but rather a class of such values, which will consist of the maximal set and all of its non-empty subsets. The fact that all non-maximal sets survive is crucial for choosing the correct new context set. This does not lead to loss of information because the maximal set will always be present.

[11] I will tentatively assume that the "floating" right boundary of main clauses in sentences with clause-final ARCs is due to the availability of two different syntactic structures. In one, the appositive is attached to its anchor and is inside the main clause. In another, the appositive is extraposed to the end of the sentence and is outside the main clause.

Edna did not start the descent, as in (19). This is because this piece of information is not part of the current context set and thus is still at-issue. If A accepts the counterproposal, the updated context set will record the information that Edna did not start the descent, thus closing the discussion about whether Edna did or did not start the descent. (Below, main clause boundaries are marked with square brackets.)

(18) A: [Edna, <u>who is a fearless leader</u>, started the descent].

$\exists p \wedge \exists x \wedge x = edna \wedge \exists q \wedge fearless.leader_q(x) \wedge$
$q \subseteq p_{cs} \wedge \exists p_{cs} \wedge p_{cs} = q \wedge start.descent_p(x)$

B: #No, she isn't. $\exists r \wedge \neg fearless.leader_r(x)$

(19) A: [Edna, <u>who is a fearless leader</u>, started the descent].

$\exists p \wedge \exists x \wedge x = edna \wedge \exists q \wedge fearless.leader_q(x) \wedge$
$q \subseteq p_{cs} \wedge \exists p_{cs} \wedge p_{cs} = q \wedge start.descent_p(x)$

B: No, she didn't. $\exists r \wedge \neg start.descent_r(x)$

A: (OK.) $r \subseteq p_{cs} \wedge \exists p_{cs} \wedge p_{cs} = r$

Sentences with clause-final ARCs can have two different logical representations. If the ARC is construed as inside the main clause, as in (20), the logical representation of the original sentence will contain the acceptance update for the ARC, making only the main clause at-issue. If, however, the ARC is construed as being outside the main clause, as in (21), the main clause proposal will be terminated and only the ARC will be at-issue.[12]

(20) A: [Jack invited Edna, <u>who is a fearless leader</u>].

$\exists p \wedge \exists x \wedge x = jack \wedge \exists y \wedge y = edna \wedge invite_p(x, y) \wedge$
$\exists q \wedge fearless.leader_q(y) \wedge q \subseteq p_{cs} \wedge \exists p_{cs} \wedge p_{cs} = q$

B: No, he didn't. $\exists r \wedge \neg invite_r(x, y)$

A: (OK.) $r \subseteq p_{cs} \wedge \exists p_{cs} \wedge p_{cs} = r$

(21) A: [Jack invited Edna], <u>who is a fearless leader</u>.

$\exists p \wedge \exists x \wedge x = jack \wedge \exists y \wedge y = edna \wedge invite_p(x, y) \wedge$
$p \subseteq p_{cs} \wedge \exists p_{cs} \wedge p_{cs} = p \wedge \exists q \wedge fearless.leader_q(y)$

B: No, she isn't. $\exists r \wedge \neg fearless.leader_r(y)$

A: (OK.) $r \subseteq p_{cs} \wedge \exists p_{cs} \wedge p_{cs} = r$

The current account predicts that in sentences with clause-final ARCs, either the main clause or the appositive is at-issue, but never both. It follows that clause-final ARCs can be, but need not be at-issue.

I would like to suggest that the divergence between clause-medial and clause-final ARCs with respect to the Answerability Test can receive a parallel explanation. When speakers ask questions, they expect to be able to accept or reject the

[12] Note that acceptance updates do double duty in my analysis: they are part of the logical translations of sentences or they represent the hearer's agreement with a proposal. Either way, acceptance updates signal not-at-issue content, i.e. state that the discussion about an issue has been terminated. (I thank Maria Bittner for bringing up this point.)

answers the hearer provides. Given the analysis presented above, clause-medial ARCs make proposals that cannot be directly rejected and thus are not good linguistic means for addressing issues. In contrast, clause-final ARCs can introduce proposals that are open to discussion, hence such ARCs can felicitously be used by interlocuters to answer questions.

4 Conclusion and Beyond

I have argued for a conjunctive one-dimensional account of ARCs according to which both main clauses and ARCs introduce proposals. Which proposal will acquire at-issue status is purely a matter of linear order. Clause-medial ARCs restrict the possible values for the context set as part of the semantics of the sentence and are invariably not at-issue. Clause-final ARCs need not restrict the possible values for the context set as part of the semantics of the sentence and thus can acquire at-issue status. Importantly, the contrast between clause-medial and clause-final ARCs is explained by a uniform semantic mechanism. In the remaining part of this conclusion I address two far-reaching issues.

One consequence of the current analysis is that it suggests a fundamentally different semantics for nominal appositives and ARCs. This is because nominal appositives seem to remain not at-issue independently of their linear position. E.g., "Mike Stewart, a comedian, was at the party last night" and "Last night I talked to Mike Stewart, a comedian" fail both tests for at-issue content discussed above. A natural suggestion would be that, due to their non-clausal status, nominal appositives do not introduce proposal propositions but acquire their not at-issue status by some related mechanism, e.g. by directly restricting the possible values for the context set.

Second, the notion of not-at-issueness is often used interchangeably with PROJECTION, the ability of certain implications to arise even when their triggers occur in non-assertive contexts. Some authors even try to explain projection in terms of not-at-issueness (Simons et al. [18]). Even though there is a significant overlap between the two categories—presupposition being the prime example that comfortably fits both—those two categories differ. First, meanings can be not at-issue without projecting. E.g., the embedded clause of "Jack imagined that it was raining outside" behaves like not-at-issue content with respect to both tests for at-issue content: it cannot be directly challenged by "That's not true" and it cannot answer the question "Is it raining outside?". Also, as Simons et al. [18] themselves observe, there seem to be meanings that project but are at-issue, e.g. clause-final ARCs. Thus, the explanation proposed above need not also address the projective properties of ARCs.

References

1. Amaral, P., Roberts, C., Smith, E.A.: Review of *The Logic of Conventional Implicatures* by Chris Potts. Linguistics and Philosophy 30, 707–749 (2007)

2. AnderBois, S., Brasoveanu, A., Henderson, R.: Crossing the Appositive/At-issue meaning Boundary. In: Proceedings of SALT 20, elanguage, pp. 328–346 (2010)
3. Bach, K.: The Myth of Conventional Implicature. Linguistics and Philosophy 22, 327–366 (1999)
4. Böer, S., Lycan, W.: The Myth of Semantic Presupposition. Indiana University Linguistics Club, Bloomington (1976)
5. Chierchia, G., McConnell-Ginet, S.: Meaning and Grammar: An Introduction to Semantics. MIT Press, Cambridge (1990)
6. Cornilescu, A.: Non-Restrictive Relative Clauses, An Essay in Semantic Description. Revue Roumaine de Linguistique 26, 41–67 (1981)
7. Dever, J.: Complex Demonstratives. Linguistics and Philosophy 24, 271–330 (2001)
8. Frege, G.: Über Sinn und Bedeutung. Zeitschrift für Philosophie und philosophische Kritik 100, 25–50 (1892)
9. Groenendijk, J., Stokhof, M.: Dynamic Predicate Logic. Linguistics and Philosophy 14, 39–100 (1991)
10. Kempson, R.: Nonrestrictive Relatives and Growth of Logical Form. In: WCCFL 22 Proceedings, pp. 301–314. Cascadilla Press, Somerville (2003)
11. McCawley, J.D.: The Syntactic Phenomena of English, vol. 2. The University of Chicago Press, Chicago (1988)
12. Murray, S.: Evidentiality and the Structure of Speech Acts. Unpublished Dissertation, Rutgers University (2010)
13. Potts, C.: The Logic of Conventional Implicatures. OUP, New York (2005)
14. Roberts, C.: Information Structure: Towards an Integrated Formal Theory of Pragmatics. In: Yoon, J.H., Kathol, A. (eds.) OSU Working Papers in Linguistics, vol. 49, pp. 91–136. The Ohio State University, Department of Linguistics (1996)
15. Rodman, R.: Scope Phenomenon, Movement Transformations and Relative Clauses. In: Partee, B. (ed.) Montague Grammar, pp. 165–176. Academic Press, New York (1976)
16. Schlenker, P.: Supplements within a Unidimensional Semantics I: Scope. In: Aloni, M., Bastiaanse, H., de Jager, T., Schulz, K. (eds.) Amsterdam Colloquium 2009. LNCS, vol. 6042, pp. 74–83. Springer, Heidelberg (2010)
17. Sells, P.: Restrictive and Non-Restrictive Modification. CSLI-85-28 (1985)
18. Simons, M., Tonhauser, J., Beaver, D., Roberts, C.: What Projects and Why. In: Proceedings of SALT 20, elanguage, pp. 309–327 (2010)
19. Stalnaker, R.: Assertion. In: Cole, P. (ed.) Syntax and Semantics 9: Pragmatics. LNCS, pp. 315–332. Academic Press, New York (1978)
20. Tonhauser, J.: Diagnosing (Not-)at-issue Content. In: Proceedings of SULA 6 (2011)

Definiteness in Superlatives

Sveta Krasikova*

University of Konstanz
svetlana.krasikova@uni-konstanz.de

Abstract. A new analysis of superlatives is presented, which assigns two functions to the definite article. It reflects the definiteness of the DP in absolute superlatives, and the definiteness of the DegP in comparative superlatives. The analysis accounts for the distribution of *the* in attributive, amount, predicative and adverbial superlatives in English.

Keywords: superlatives, definiteness, absolute-comparative ambiguity.

1 Introduction

Most semantic theories of superlatives are concerned with resolving the comparative-absolute ambiguity superlative sentences are notorious for. Comparative superlatives are considered a thorny case because their semantic indefiniteness seems to clash with the presence of the definite article. The standard solution is to replace *the* by an abstract A in such cases.

However, the empirical landscape is much more intricate. The definite article is optional in some superlatives, e.g. with superlative adverbs. In amount superlatives, it disambiguates between comparative and proportional readings. For example, the bare superlative in (1) is interpreted as (1-a), and the insertion of *the* leads to the interpretation in (1-b) .

(1) John climbed (the) most mountains.
  a. John climbed more than half of the mountains.
  b. John climbed more mountains than anyone else.

In this work, we aim at a more accurate account of the contribution of the definite article in superlatives, which can explain the observed distribution and establish a correlation between the superlative meaning and definiteness.

The paper is structured as follows. In Section 2 and 3 we review the existing analyses of superlatives and motivate a new approach, which is developed in Section 4 and compared to other theories in Section 5. Section 6 discusses the predictions of the new analysis on the distribution of the definite article.

* I am grateful to Arnim von Stechow, Doris Penka, Irene Heim, Maribel Romero, Brian Leahy for very insightful comments and the discussion of various aspects of this work. The errors are all mine. Support for this research was provided by the Deutsche Forschungsgemeinschaft (DFG).

M. Aloni et al. (Eds.): Amsterdam Colloquium 2011, LNCS 7218, pp. 411–420, 2012.
© Springer-Verlag Berlin Heidelberg 2012

2 Absolute and Comparative Superlatives

Adjectival superlatives are known to be ambiguous between an absolute and a comparative reading.[1] For example, sentence (2-a) can be used to convey that John climbed a mountain that is higher than any other mountain, cf. (2-b). On the other hand, (2-a) can be used to compare John's achievement to the other people's achievements, cf. (2-c).

(2) a. John climbed the highest mountain.
b. John climbed Mount Everest. *absolute*
c. John climbed a higher mountain than anyone else. *comparative*

Comparative readings are not so easily distinguishable from absolute ones. In the example at hand, we are ultimately comparing the heights of mountains in both cases. The comparative reading may therefore be thought of as a case of the absolute one in which the mountains compared are the ones climbed by the salient people. There are two strands in the literature, depending on whether comparative readings are subsumed under absolute ones or not. I will refer to them as pragmatic and structural approaches (PA and SA, henceforth).

Before we look at the difference between PA and SA, let us identify their common features. Be it pragmatic or structural, any existing analysis assumes that the superlative meaning is contributed by the superlative morpheme which is restricted by a contextual variable corresponding to the comparison class. According to the lexical entry in (3) from [4], *-est* applies to a comparison class C and a gradable predicate A to return a predicate of individuals that is true of those entities whose A-ness degrees exceed A-ness degrees of everyone else in C. Thus, the role of the superlative is to existentially bind the degree variable projected by the gradable predicate and restrict it by an appropriate type of comparison. C is also presupposed to consist of entities to which A applies.[2]

(3) a. $[\![\text{-est}]\!] = \lambda C \in D_{et} \lambda A \in D_{d(et)} \lambda x \in D_e \exists d[A(d)(x) \wedge \forall y[y \in C \wedge y \neq x \rightarrow \neg A(d)(y)]]$
b. $[\![\text{-est}]\!](C)(A)(x)$ is only defined iff $\forall y \in C[\exists d[A(d)(y)]]$

PA ([4], [2], [11], [12]) maintain the view that to resolve the ambiguity, it suffices to pragmatically fix the value of C. The corresponding analysis of (2-a) is outlined in (4), where (4-a) is the assumed LF, (4-b) is the meaning of the NP, (4-c) are the resulting truth conditions, and (5-a) and (5-b) are the value assignments of C under the absolute and the comparative reading, respectively.

(4) a. John climbed [the [-est C [high mountain]]]
b. $[\![\text{high}]\!]([\![\text{mountain}]\!]) = \lambda d \lambda x\ x$ is a mountain $\wedge\ height(x) \geq d$
c. John climbed the unique x, s.t. x is a mountain $\wedge\ \exists d[height(x) \geq d$ $\wedge\ \forall y[y \in C \wedge y \neq x \rightarrow \neg[y$ is a mountain $\wedge\ height(y) \geq d]]]$

[1] This observation goes back to [10].
[2] I disregard other presuppositions usually assumed to be triggered by the superlative as they are orthogonal to the present discussion.

(5) a. $C_{ABS} = \{x : x \text{ is a salient mountain}\}$
b. $C_{COMP} = \{x : x \text{ is a mountain climbed by a salient individual}\}$

SA ([13], [4]) agree that ambiguity is related to the value of the comparison class, but argue that the latter depends on the structural position of *-est* at LF. Thus, to derive the comparative reading of (2-a), *-est* is interpreted DP-externally. Its movement creates the right kind of property for the derivation of the comparative reading, namely the relation holding between an individual x and a degree d if x climbed a d-high mountain. Correspondingly, it is not mountains but mountain climbers that are compared, and the resulting value of C_{COMP} in (6-c) is different from the one in (5-b).[3] The superlative DP is assumed to be indefinite, which accords with the mobility of *-est* .[4] Accordingly, an abstract indefinite determiner A heads the DP. Note that the uniqueness requirement contributed by *-est* renders the definite article superfluous, so the replacement of *the* by A does not have a considerable semantic effect.

(6) a. John [-est C] [$\lambda d \lambda x$ x climbed [A d high mountain]]
b. $\exists x[x \text{ is a mountain} \wedge \text{John climbed } x \wedge \exists d[height(x) \geq d \wedge \forall y[y \in C \wedge y \neq \text{John} \rightarrow \neg \exists z[z \text{ is a mountain } \wedge y \text{ climbed } z \wedge height(z) \geq d]]]]$
c. $C_{COMP} = \{x : \exists d[x \text{ climbed a } d\text{-high mountain}\}$

The main argument in favour of SA, and contra PA, is the availability of the so called upstairs *de dicto* reading in intensional superlatives. Consider sentence (8) uttered in the context in (7).

(7) – How high a mountain do you need to climb?
John: – I need to climb a 6000 m high mountain.
Bill: – I need to climb a 4000 m high mountain.
Mary: – I need to climb a 3000 m high mountain.

(8) John needs to climb the highest mountain.

Two things go without question: the superlative DP receives a *de dicto* interpretation, and the resulting reading is a comparative one. However, the task of providing a value for C under PA has proved a challenge in such cases. No matter how we constrain the mountains in C and whether we replace *the* by an abstract indefinite article, the uniqueness requirement on *-est* leads to a definite interpretation, which is incompatible with the given scenario, see [4].[5] SA do not have this problem. The relevant interpretation can be derived by assigning *-est*

[3] Both PA and SA concur that focus plays a role in promoting this or that reading. That is, if *John* is focussed in (2-a), the contextual variable is assigned C_{COMP} as its value. We will return to the issue of focus in section 4.

[4] Movement out of a definite DP is subject to syntactic island constraints.

[5] However, the problem of upstairs *de dicto* readings has been addressed within PA. [11] develop an analysis based on a higher order meaning of *the*. The fact that the type-shifted definite article and the additional machinery coming with it is only required for upstairs *de dicto* readings makes their approach stipulatory.

wide scope relative to *need*, cf. (9-a). This move guarantees that C corresponds to the set in (9-b), which does not create a problem in combination with the uniqueness requirement.

(9) a. John [-est C] [$\lambda d \lambda x$ x needs to climb [A d high mountain]]
b. C = {x: $\exists d[x$ needs to climb a d-high mountain]}

We can draw two lessons from this kind of sentence. First, unless we make sure that we technically compare different kinds of entities under an absolute and a comparative reading, the uniqueness requirement clashes with the indefiniteness of the DP in comparative superlatives. Second, the definite article, though merely superfluous in other kinds of superlatives, is at odds with an indefinite interpretation of the upstairs *de dicto* superlatives.

3 Distribution of the Definite Determiner

The previous section revealed a weakness of existing approaches. They don't assign any role to *the*. Moreover, *the* appears incompatible with an apparently indefinite interpretation of some comparative superlatives. Anna Szabolsci was one of the first to observe that only comparative superlatives are indefinite, cf. [13].[6] Along with some syntactic tests, she considers how superlatives behave in "definiteness effect contexts". It turns out that comparative superlatives may occur in such contexts, while absolute superlatives are unacceptable. For example, putting focus on *John* in (10-c) promotes the comparative reading and creates a contrast to (10-b), which is as bad as the variant with a non-superlative definite in (10-a).

(10) a. *John has the sister.
b. *John has the smartest sister.
c. JOHN has the smartest sister.

However, the occurrence of *the* in superlatives does not seem to correlate with their definiteness. Three classes of English superlatives may be distinguished relative to how *the* is distributed.

First, attributive superlatives require the presence of the article regardless of the interpretation available, cf. (2-a).[7] Adopting Szabolsci's generalisation that absolute superlatives are definite and comparative ones are indefinite, it looks as if *the* were interpreted in the former and were spurious in the latter case.

Second, in amount superlatives, which are indefinite under comparative and proportional readings, *the* has been observed to play a disambiguating role in English, cf. [3]. (11) and (12) are examples of a comparative and a proportional

[6] See [11] for an alternative view.

[7] [5] also discuss indefinite attributive superlatives, e.g. "This class has a best student." Such examples are only acceptable in contexts with a salient partition over the extension of the noun; in this example students are partitioned into classes.

reading of amount superlatives, the paraphrases given in (11-c) and (12-c) follow the analysis in Hackl [3].

(11) a. John climbed the most mountains.
 b. John climbed more mountains than anyone else. *comparative*
 c. There is a plurality of mountains John climbed whose cardinality exceeds the cardinalities of pluralities of mountains others climbed.

(12) a. John climbed most mountains.
 b. John climbed more than half of the mountains. *proportional*
 c. There is plurality of mountains that John climbed that exceeds the cardinalities of all non-overlapping pluralities of mountains.

Finally, adverbial and predicative superlatives do not require the definite article, cf. (13). The optionality of *the* is not disambiguating in this case. In fact, neither of the constructions can exhibit the relevant kind of ambiguity. Adverbial superlatives can be ambiguous, but comparison classes associated with different readings all involve different kinds of eventualities, modified by the adverb. In predicative superlatives, there is no room for variation either: alternatives in comparison classes are always shaped after the subject.

(13) a. John ran (the) fastest.
 b. John is (the) tallest.

There is a fair amount of variation in how superlatives are formed across languages. However, the fact that the presence of the definite determiner does not reflect the definiteness of the superlative DP holds across languages. For example, in German the definite article can never be optional, regardless of the reading and the kind of superlative, cf. (14). Note that (14-a) and (14-b) have both an absolute/proportional and a comparative reading.

(14) a. Hans hat *(den) höchsten Berg bestiegen.
 Hans has the highest mountain climbed
 b. Hans hat (die) meisten Berge bestiegen
 Hans has the most mountains climbed
 c. Hans ist *(am) schnellsten gelaufen.
 Hans is at.the fastest run
 d. Hans ist *(am) schnellsten.
 Hans is at.the fastest

To give an intermediate summary, most of the existing theories are successful in accounting for the differences in absolute and comparative superlatives; however, the presence of the definite article remains a mystery under any approach. The distribution of *the* does not seem incidental, considering its role in amount superlatives, but it does not reflect the definiteness of the superlative DP in all cases. Ideally, we would like to understand why comparative superlatives are marked for definiteness despite their semantic indefiniteness, and why some languages make the definite article optional in some contexts.

4 Definite DP and Definite DegP

We would like to suggest that the distribution of the definite article is complicated by the fact that it plays a double role in superlatives. It either heads a definite determiner phrase, or provides a definite standard of comparison, i.e. heads a definite degree phrase. In absolute superlatives, which behave as definites, *the* requires the uniqueness of the referent of the superlative DP. In comparative superlatives, it requires the uniqueness of the degree to which the gradable predicate relates its individual argument.

4.1 DP-Internal Superlatives

We assume that absolute and proportional superlatives involve an abstract superlative degree modifier whose presence is licensed by the uninterpretable superlative morphology on the adjective. Let us call this modifier SUP. The following definition is based on the entry for the superlative morpheme in (3).

(15) $[\![\text{SUP}]\!] = \lambda A \in D_{d(et)} \lambda P \in D_{et} \lambda x \in D_e P(x) \wedge \exists d[A(d)(x) \wedge \forall y[P(y) \wedge y \neq x \rightarrow \neg A(d)(y)]]$

According to this definition, SUP applies to a gradable predicate A and a predicate P, expressed by the head noun, to return a predicate holding of x iff P is true of x, and has A to a degree to which no other entity in the extension of P does. Line (16) illustrates the application of SUP to *highest mountain* in the derivation of the absolute reading of (2-a).

(16) $[\![\text{SUP}]\!]([\![\text{highest}]\!])([\![\text{mountain}]\!]) = \lambda x\ x \text{ is a mountain} \wedge \exists d[height(x) \geq d \wedge \forall y[y \text{ is a mountain} \wedge x \neq y \rightarrow height(y) < d]]$

Due to the uniqueness requirement of SUP, the resulting predicate in (16) can only be true of one mountain. Consequently, a principle in the spirit of "maximise presuppositions" requires that the DP be realised as definite. The definite article therefore emphasises that the referent of the superlative DP is unique. In (17) we sketch the complete LF and the truth conditions for the absolute reading of (2-a).

(17) a. John climbed [the [SUP highest mountain]]
b. John climbed the unique x, such that x is a mountain $\wedge$ $\exists d[height(x) \geq d \wedge \forall y[y \text{ is a mountain } \wedge x \neq y \rightarrow height(y) < d]]$

This derivation is almost identical to the PA derivation given in (4). The difference is that (4-a) is unspecified, corresponding to the absolute or the comparative reading depending on the value of C. (17-a) is not ambiguous in the same sense and is meant to represent the absolute reading only.[8]

[8] An anonymous reviewer observes that there ought to be a way for the context to constrain the comparison class. We agree that SUP, as defined in (15), is too context-insensitive and may need to additionally depend on a contextual variable, which, together with the head noun, determines the comparison class. This may be useful for the treatment of indefinite superlatives, see [5] and fn.7, where the contextual restriction would be "in the same partition as x."

In proportional amount superlatives, the uniqueness condition on SUP does not have the same effect. Since we are dealing with plurals, there may be several pluralities falling under the superlative description. The use of the definite article would therefore be unmotivated and the DP is realised as a bare plural, see (18).

(18) a. John climbed [$\exists$ [SUP most mountains]]
b. $\exists X[X$ are mountains $\wedge$ John climbed $X \wedge \exists d[card(X) \geq d \wedge \forall Y[Y$ are mountains $\wedge X \neq Y \rightarrow card(Y) < d]]]$

Given that two pluralities are distinct if they do not overlap, as we assume with [3], the truth conditions in (18-b) are met iff John climbed a set of mountains that has more elements than any complement set of mountains. This can only be true if John climbed more than half of the mountains.

4.2 Superlatives as Plural Definites

We propose that comparative superlatives involve a different kind of degree binding. While the degree argument of the gradable predicate is existentially bound by SUP in absolute superlatives, it is saturated by a definite degree description in comparative superlatives. We assume the structure in (19) for the comparative reading of (2-a). The definite article restricted by a maximalised contextual degree property C fills the degree argument slot of *highest*, whose morphology is again not interpreted but rather indicates the presence of the maximality operator. The resulting predicate can be intersectively combined with the head noun. The entire DP is realised as definite due to the definiteness of the DegP.

(19) $[_{NP}[_{AP}[_{DegP}$ the max C] highest] mountain]

Following Szabolsci's insight, which guided most of the existing theories of superlatives, we propose that the value of C is determined by the focus structure of the sentence. For concreteness, assume that focus is evaluated by Rooth's squiggle operator ([9]), which comes with a variable corresponding to the focus semantic value of the constituent in its scope. The value of the free variable C on the definite determiner is restricted to be a subset of the value of the variable introduced by the squiggle. Suppose, the focus falls on the subject in (2-a), then C ends up a subset of the set of degree sets defined in (20-b).

(20) a. [λd JOHN climbed [$\exists$ d highest mountain] $\sim C$]
b. C = $\{D{:}\exists x[D = \lambda w \lambda d\ x$ climbed a d-high mountain in $w]\}$[9]

Turning to the contribution of *the*, we assume that it applies to a set of degree properties C, modified by the maximality operator, defined in (21-a), and returns its unique element, if there is one. Thus, we suggest that the superlative

[9] We intensionalise the focus value to avoid the situation in which alternatives corresponding to different individuals are indistinguishable, e.g. if the highest mountains different individuals happen to have climbed are equally high; see [8], who handles the same problem with *only* in a similar way.

morphology is responsible for maximalising the restriction set of the definite article, and *the* performs a "uniqueness test" on the maximalised set, see (21-b). In the given example, if John happens to have made the best achievement, *max* turns the set of degree properties corresponding to the achievements of different climbers to the singleton containing the property of degrees ranging up to the maximal height reached by John, and *the* returns that property.

(21) a. $max(Q) = \lambda D_{dt}[Q(D) \wedge \forall D'[Q(D') \rightarrow D'(w) \subseteq D(w)]]$
b. $[\![\text{the}]\!] = \lambda Q : \exists! D[Q(D)]. \iota D[Q(D)]$

Following a proposal by Sigrid Beck ([1]), we treat the extension of the resulting property as a plurality of degrees. Beck assumes that a set of degrees may saturate the degree argument of some degree predicate by acting as a plurality of degrees interpreted distributively. To derive distributive readings, she introduces Link's star operator, to the effect that the plurality of degrees receives sentential scope. We adopt this proposal here. The definite article together with the restricting property C moves out of the DegP of the adjective and combines with the degree property that has been modified by *. Given the standard definition of *, the resulting truth conditions boil down to (22-b).

(22) a. [the *max* C] *[λd JOHN climbed [$\exists$ d highest mountain]$\sim$ C]
b. $\exists x[x \text{ is a mountain in } w \wedge \text{ John climbed } x \text{ in } w \wedge$
$\forall d[d \in [\![\text{the } max \text{ C}]\!](w) \rightarrow height(x) \geq d]]$

5 Comparison of Approaches

The current approach shares with SA the ability to derive the right truth conditions for upstairs *de dicto* readings. By moving the definite term over the modal, we create the comparison class C that contains properties holding of d in w iff some salient individual climbs a d-high mountain in every world accessible from w, cf. (23-b). If John happens to have a requirement exceeding that of Bill or Mary, the sentence is correctly predicted true, cf. (23-c). Thus, the analysis does not face the problem of PA with the uniqueness requirement in intensional contexts.

(23) a. [the *max* C] *[λd need JOHN climbed [$\exists$ d highest mountains]$\sim$ C]
b. $\{D\text{: } \exists x[D = \lambda w \lambda d \forall w' \in Acc_w[x \text{ climbed a } d\text{-high mountain in } w']]\}$
c. $\forall d[d \in [\![\text{the } max \text{ C}]\!](w) \rightarrow \forall w' \in Acc_w[\exists x[x \text{ is a mountain in } w'$
$\wedge \text{ John climbed } x \text{ in } w' \wedge height_{w'}(x) \geq d]]]$

Under the present approach, the definite article is not superfluous, but makes a meaningful contribution to the analysis of a superlative construction. In accordance with Szabolsci's generalisation, absolute superlative DPs are analysed as definite descriptions, *the* acting as a DP determiner in the presence of SUP. In a comparative superlative, *the* reflects the definiteness of the DegP, so the DP need not be definite.

Another feature of the present analysis worth mentioning is the degree-based nature of comparison classes in comparative superlatives. Note that the analyses discussed in section 2 treat C_{COMP} as a set of individuals, whereas the present analysis assumes that C_{COMP} contains properties of degrees.[10] One of the advantages of degree-based comparison classes, discussed by Maribel Romero in [7], is in the treatment of the so called modal superlatives. For example, the reading of (24-a), paraphrased in (24-b), expresses a comparison to the maximally possible height from the relevant set. The paraphrase suggests that here, like in a clausal comparative construction, the standard of comparison is degree-based rather than individual-based, see [7] for discussion.[11]

(24) a. John climbed the highest mountain possible.
b. John climbed as high a mountain as was possible to climb.

6 Optionality

One of the two remaining issues to be addressed is the optionality of *the* in adverbial and predicative superlatives. Under the present analysis, both an absolute and a comparative construal are possible in such cases and derive the same truth conditions, cf. (25) and (26). For the absolute case, we can assume a covert contextual variable that plays the role of the head noun, see (26-c).

(25) a. [the *max* C] *[λd JOHN ran [A d fastest] $\sim$ C]
b. $\exists e[e \text{ is a running by John} \wedge \forall d[d \in [\![\text{the C}]\!](w) \rightarrow speed(e) \geq d]]$
c. $\text{C} = \{D : \exists x[D = \lambda w \lambda d \exists e[e \text{ is a running by } x \wedge speed(e) \geq d]]\}$

(26) a. John ran [SUP fastest C]
b. $\text{C} = \{e : \exists x[e \text{ is a running by } x]\}$
c. $\exists e[e \text{ is a running by John} \wedge \exists d[speed(e) \geq d \wedge$
$\forall c'[e' \in C \wedge e' \neq e \rightarrow speed(e') < d]]]$

Predicative superlatives are always absolute. The optionality of *the* may be related to different types of interpretations available for a predicative construction, as well as the possibility of a plural definite construal, cf. (27-c). We do not explore this issue further here.

(27) a. John is [the [SUP tallest C]]
b. John is [SUP tallest C]
c. [the *max* C] *[λd JOHN is d tallest $\sim$ C]

Finally, the present approach predicts that the definite article is optional in amount superlatives, which are indefinite regardless of the superlative reading available. Though right for English, this prediction is not borne out for German

[10] This makes our treatment of comparative cases parallel to Heim's analysis of the superlative as a focus sensitive operator restricted by a set of degree sets, cf. [4].

[11] Another construction convincingly argued by [6] to require a degree-based comparison are superlatives with relative clauses containing NPIs.

and other languages where the definite article is never omitted. For reasons of space, we refrain ourselves to a couple of informal remarks concerning German. The most straightforward explanation for the distribution of the definite article in German is that the superlative DegP is always a plural definite in that language. This kind of approach faces two tasks. It first needs to provide an appropriate type of restriction for the definite determiner in non-comparative cases, see [7] for a derivation of absolute reading under a degree-based approach to comparison classes; and, most importantly, it has to explain why English cannot receive the same account as German.

7 Conclusion

We argued that the definite article in superlatives may reflect definiteness at the DP as well as DegP level. According to the present approach, only absolute attributive superlatives are genuinely definite; in comparative superlatives, *the* marks the definiteness of the degree argument, and does not necessarily reflect the definiteness of the host DP. This analysis is not only successful in deriving all possible readings of superlatives, it also accounts for the distribution of *the* in different kinds of superlatives in English.

References

1. Beck, S.: Lucinda Driving Too Fast Again – the Scalar Properties of Ambiguous Than-Clauses. Journal of Semantics (2012)
2. Farkas, D., Kiss, K.E.: On the Comparative and Absolute Readings of Superlatives. Natural Language and Linguistic Theory 18(3), 417–455 (2000)
3. Hackl, M.: On the Grammar and Processing of Proportional Quantifiers: 'Most' Versus 'More Than Half'. Natural Language Semantics 17, 63–98 (2009)
4. Heim, I.: Notes on Superlatives. MIT (1999) (unpublished manuscript)
5. Herdan, S., Sharvit, Y.: Definite and Nondefinite Superlatives and NPI Licensing. Syntax 9, 1–31 (2006)
6. Howard, E.: The most alternative analysis of superlative NPIs ever. Handout of a talk delivered at Workshop on degree semantics and its interfaces, Utrecht (2011)
7. Romero, M.: Modal Superlatives and 2-place vs. 3-place '-est' . University of Konstanz (2011) (unpublished manuscript)
8. Rooth, M.: Association With Focus. Ph.D. Dissertation. University of Massachusetts at Amherst (1985)
9. Rooth, M.: A Theory of Focus Interpretation. Natural Language Semantics 1(1), 75–116 (1992)
10. Ross, J.R.: A Partial Grammar of English Superlatives. MA thesis. University of Pennsilvania (1964)
11. Sharvit, Y., Stateva, P.: Superlative Expressions, Context and Focus. Linguistics and Philosophy 25, 453–505 (2002)
12. Stateva, P.: How Different Are Different Degree Constructions. PhD Dissertation. University of Connecticut (2002)
13. Szabolsci, A.: Comparative Superlatives. In: Fukui, N., Rapoport, T., Sagey, E. (eds.) MIT WPL, vol. 8, pp. 245–266 (1986)

The Accommodation Potential of Implicative Verbs

Noor van Leusen

Radboud University, Nijmegen, The Netherlands
n.vanleusen@phil.ru.nl

Abstract. We present an analysis of *implicative verbs*,[1] complement-taking verbs which induce entailment-like inferences, but which are also claimed to trigger presuppositions. What is presupposed, however, is much more variable than with e.g. factive verbs. Sketching a formal treatment in Logical Description Grammar we consider the role of pragmatic reasoning and accommodation in deriving these presuppositions.

Keywords: Implicative verbs, presupposition, accommodation, LDG.

1 Introduction

Since Karttunen's work on *implicatives* in English [K71a] [K71b], verbs such as 'to manage to', 'to forget to', 'to happen to', or 'to force to', etc., have become known for their characteristic inference pattern. On the one hand, they are claimed to *entail or implicate* the truth or falsity of their complement sentence, depending on the polarity of the embedding construction. For instance, we have

(1) a. Jim managed to button his coat.
   $\rightarrow$ Jim buttoned his coat.
  b. Jim did not manage to button his coat.
   Jim did not happen to manage to button his coat.
   $\rightarrow$ Jim did not button his coat.

Various subclasses can be distinguished. Following [NCK06], 'to manage to' might be called a *two-way* implicative, because it triggers an entailment both in its unnegated and its negated form; moreover, it may be called *affirmative* because it implies the truth of the complement sentence in its bare, unnegated form.

On the other, implicative verbs are often claimed to evoke *presuppositions*. Verbs of success or failure are taken to presuppose that there is or was an attempt on the subject's part to achieve the state or event described by the complement sentence. In addition, 'to manage to' is often said to presuppose that there is some difficulty to be overcome in order to achieve the complement state or event. Thus, both (1a) and (1b) would presuppose[2], variously,

[1] This research was supported by a grant from the Netherlands Organisation for Scientific Research (NWO), which is gratefully acknowledged.

[2] Some papers, notably [KP79], classify these inferences as conventional implicatures.

M. Aloni et al. (Eds.): Amsterdam Colloquium 2011, LNCS 7218, pp. 421–430, 2012.
© Springer-Verlag Berlin Heidelberg 2012

(2) $\rightsquigarrow$ Jim made an attempt to button his coat.
$\rightsquigarrow$ It was difficult for Jim to button his coat.

Verbs of forgetting or remembering are often taken to presuppose that there is or was an obligation on the subject's part to achieve the state or event described by the complement sentence. More tentatively, it is suggested that these verbs evoke the supposition that the subject intends to achieve the complement state or event, or that, less specifically, he is expected to do so, or ought to do so.

Moreover, most implicatives evoke what may be called characteristic causal or explanatory presuppositions: reasons why the state or event denoted by the complement sentence is or is not achieved in the case at hand. This comes out perhaps most clearly in cases such as (3) where the implicative is negated,

(3) Ed didn't manage / remember / bother / dare / happen to open the door.

all of which imply that Ed did not open the door but presuppose a different *reason* for this fact. Ed did not make a sufficient effort or was not sufficiently skilled to open the door, he did not keep in mind his plan or obligation to close it, he did not care enough or did not take the trouble, he did not have sufficient courage, or somehow the right circumstances did not apply.

While most of the early descriptive literature [K71a] [Giv73] and some more recent papers [Luz99] [NCK06] concentrate on the entailments or implications of sentences with implicative verbs, the focus of this paper is on their presuppositional side. In particular, we are concerned with the *variability* of the presuppositions — if indeed that is what they are. We will investigate how these inferences may come about, and consider what that means for the interaction between general pragmatic reasoning and the satisfaction of lexical presuppositional conditions in a dynamic semantic perspective on sentence meaning.

Existing treatments of implicatives within the dynamic semantic paradigm, e.g. [Bea01], tend to treat them exclusively as presupposition triggers and often only provide an analysis of the bench-mark case 'to manage to'. We aim to improve on this in the following way. Firstly, if we want to account for the presuppositions of implicatives, we cannot pass by their implications or entailments. Only on the basis of their full inferential signature may the core semantic content of implicative sentences be teased apart from any requirements on context they induce. What is presupposed then follows from the interaction with general pragmatic reasoning. Secondly, it is fruitful to look at implicatives other than 'to manage to', which happens to be one that induces virtually uncancellable implications both in its bare and negated form. Not all implicative verbs are like that, and we will zoom in on a slightly weaker instance, namely 'to remember to'.

Finally, the status of the inferences under discussion as presuppositions can be called into question. While some of them can be viewed as scalar and conversational implicatures, we propose that most of them are 'pragmatic presuppositions' (Stalnaker), or 'secondary inferences' that result from the need to satisfy and explain a basic appropriateness condition of the implicative verb. They constitute accommodated material in the wide sense of [Tho90].

Our analysis is couched in Logical Description Grammar for discourse, a model of text interpretation which combines underspecification with discourse theory, and uses compositional DRT as semantic representation language. Unlike most dynamic semantic formalisms, it supports a liberal notion of accommodation, where what is accommodated can be more than what is minimally required to satisfy conditions on contexts, in terms of logical strength. What comes out as a most preferred context specification follows from the interaction with the interpreting agent's common knowledge and pragmatic reasoning. Thus, on the basis of a sufficiently general requirement on a context, a variety of more specific suppositions can be abduced.

We start off with a sketch of the LDG formalism, highlighting what is relevant for the topic at hand. Our analysis of implicative verbs is laid down in subsequent sections. For reasons of space, we concentrate on the core proposal and economize on discussion of data.

2 Discourse Meaning and Accommodation in LDG

The LDG *framework of discourse interpretation*, laid down in [LM03] [Leu07], consists of a description grammar, which embodies a language user's linguistic knowledge, a representation of his world knowledge and beliefs, and a preference system, which models his capacity to assign preferences over different potential interpretations of a discourse and to draw default inferences. In processing a discourse, the language user incrementally constructs a *discourse description* from input sentence descriptions. It describes the specific syntactic, semantic, and pragmatic properties of the discourse.

Descriptions are sets of statements in a logical language (classical type logic, in our case). The objects described by sentence and discourse descriptions are fully specified linguistic tree structures. Interpretation is a *reasoning* task, in which the hearer infers what tree structures fit or verify the discourse description given his grammar and nonlinguistic knowledge. As descriptions can be partial, there can be more than one verifying tree structure. Each structure comes with a potential reading of the sentence or discourse, and one or more of them may come out as most preferred.

See [Leu07] for an explication of the incrementality of the formalism and the specifics of the parsing process. A central feature of the model is that it makes available *local contexts* as a parameter in the compositional semantics. Each node k in a sentence or discourse tree comes not just with the usual syntactic categories and semantic values (σ_k), but also carries a local context (Γ_k). Local contexts are constructed from semantic values going top-down and from left to right through the discourse tree. The local context of the root of a discourse tree is identified with B, the *implicit background* to the discourse. B stands for whatever is taken for granted or pragmatically presupposed by the discourse participants in the view of the interpreting agent.

Crucially, local contexts are underspecified objects; *accommodation* is the (partial) specification of local contexts as it results from the satisfaction of constraints stated on them in the grammar and the discourse description. The implicit background is a largely underspecified body of information which is described bit by bit in the course of the interpretation process. We call this *context specification*.[3]

The discourse grammar specifies a few general conditions on local contexts, and elements the lexicon may introduce specific ones. Typically, context-sensitive elements such as anaphors and presupposition triggers contribute the latter. *Discourse meaning* is defined as $B \oplus \sigma_r$, i.e., what was 'taken for granted' merged or updated with 'what was said', where all the syntactic, semantic and pragmatic conditions collected in the discourse description must be satisfied, including constraints on local contexts. To illustrate, the discourse meaning of the out-of-the-blue sentence in (4a) is shown in (4b), where, among others, the requirements on B in (4c) and (4d) must be satisfied.

(4) a. Ed closed the door.
   b. $B \oplus [\ | \ w_r\!: o_0 \ \mathit{closed} \ \sigma_3^\pi \ ]$
   c. triggered by the name: $[o_0 \mid w_r\!: \mathit{Ed} \ o_0] \sqsubseteq B$
   d. triggered by the definite description: $[\sigma_3^\pi \mid] \sqsubseteq B, \quad B \models [\mid w_r\!: \mathit{door} \ \sigma_3^\pi]$

Semantic values are in a variant of compositional DRT put forward in [LM03]. σ_k^π is an underspecified discourse marker. [..|...] is a DRS, to the left of the | sign is the universe, to the right the conditions. We use $o_k, u_k, w_r, \ldots$ for different types of discourse markers; u_k represent *new* referents (generated in the discourse); o_k represent *old* ones; w_r stores the actual world. 'w_r: *door* u_3' should be read as 'the occupant of u_3 is a door in the world occupying w_r'. The operator $\oplus$ *merges* DRSs, $\sqsubseteq$ denotes *inclusion* and $\models$ denotes *entailment*.

The hearer may satisfy (4c) and (4d) by accepting $[o_0 \mid w_r\!: \mathit{Ed} \ o_0] \sqsubseteq B$ and $[\ u_7 \mid w_r\!: \mathit{door} \ u_7] \sqsubseteq B$ (where u_7 is an arbitrary fresh discourse marker). By further reasoning the hearer may obtain (5) as final discourse meaning:

(5) $B \oplus [w_r, o_0, u_7 \mid w_r\!: \mathit{Ed} \ o_0, \ w_r\!: o_0 \ \mathit{closed} \ u_7, \ w_r\!: \mathit{door} \ u_7]$

Note that this interpretation builds on *accommodation*: the hearer adjusts his representation of the implicit background in order to satisfy linguistically generated requirements.

A hearer may freely abduce background information, selecting a context scenario which is most likely to explain what is presupposed, and consistent with what he considers to be common knowledge and the speaker's beliefs and intentions in making his assertion. What is accommodated in any given case is not just a matter of satisfying presuppositional conditions, it depends on the total of the hearer's beliefs and preferences. Pragmatic reasoning is indispensible for generating preferences over accommodation options. We assume the preference system of the model implements a form of defeasible reasoning about what the speaker is committing to given his beliefs and intentions and what he said so far. The pragmatic rules are not formally specified, but we consider their impact in the interpretation process.

[3] [Leu07] provides a treatment of local and intermediate accommodation as well.

3 Reasoning towards Culmination

Right from the birth of 'implicative verbs' as a distinct class in [K71a], there has been discussion about the strength or status of the derived complements. In [NCK06], Karttunen and co-authors point out that "it can be difficult to distinguish **entailments** that is, what the author is actually committed to, and **conversational implicatures**, that is, what a reader/hearer may feel entitled to infer." Indeed, judgements may vary. For instance, 'to remember to' is viewed as a two-way implicative in [NCK06], but we prefer to categorise it as a one-way implicative, in line with [Horn72]. Consider

(6) a. Martha remembered to turn out the lights.
  b. Martha turned out the lights.
  c. Martha didn't remember to turn out the lights.
  d. Martha didn't turn out the lights.
  e. (i) ... so I had to remind her.
   (ii) ... but luckily she brushed against the switch.

While (6a) implies (6b) in a strong sense, (6c) only 'invites the inference' in (6d). As can be seen from the continuations in (6e), the inference is defeasible. Horn categorises it as a subspecies of implicature.

The strong inferences can be recognised by their resistance to cancellation:

(7) a. # Though Jim managed to button his coat, he did not button it.
  b. ?? Though Martha remembered to call the dean, she didn't call him.

It can be hard to determine whether the inference arises and is cancelled or not, however. For one, some verbal constructions, e.g. 'to be able to', are ambiguous between implicative and nonimplicative senses. In other cases, strong contrastive marking may override the implicative inference, or select a nonimplicative sense of the construction. Moreover, the inferences are temporally specific [Giv73]; they inherit, among more, the temporal setting of the implicative predicate. Thus, (8a) constitutes no evidence of cancellation, the continuation in (8a) is simply consistent with the inference (8b):[4]

(8) a. Martha didn't remember to turn out the lights (t_1), but she turned them out later on ($t_2, t_1 < t_2$).
  b. Martha didn't turn out the lights (at t_1)

Finally, the inferences depend on the tense and aspect of the implicative predicate. The clearcut 'actuality entailments' evoked with simple past predicates do not surface with a generic or progressive use of the same verb. Arguably, these uses evoke generic and progressive implicative inferences.

[4] Perhaps a two-way analysis of 'to remember to' can be defended along these lines.

Analysis As [Luz99] observes, implicatives always occur as part of a sequence of verbs expressing a single process. We may add that in general, that process is resultative. The implicative verb refers to an attitude or state of the subject or an activity he is involved in, which in the view of the speaker is a necessary prestate, or causes, or culminates in the state or event referred to by the complement sentence. In line with the intuitive analysis of [K71a], we suggest that in the context of interpretation, the implicative state or activity figures as a sufficient and/or necessary condition for the achievement or culmination (or lack thereof) of the complement state or event. This explains the implicative inferences.

We'll come to a formal implementation of this shortly, but like to emphasize that, ultimately, pragmatics is what drives these inferences, and what explains why implicative verb senses come into existence in a language. There is a general pragmatic drift to expect that a process culminates or a purpose is achieved or an intention made true, when a speaker asserts that a precondition for the result or goal is satisfied or that an intentional act took place. And when he conveys that a precondition is not fulfilled, we expect that the result or the goal is not achieved. In the course of time, for some verbs the implicativeness can become conventionalised and part of their lexical description. Specialisation may take place, resulting in groups of semantically closely related verbs which differ only in the optionality or obligatoriness of the implication they evoke. Verbs which express the same core meaning across languages can be expected to differ in the strength of the implicative inference.

4 The Implicative Condition

We propose that implicative verbs lexically introduce an *implicative condition*, a requirement on their local context which defines the resultative, causal or conditional relation between the state or action the implicative verb refers to, and the state or event described by the complement sentence. In combination with the semantic content of the sentence containing the verb, it accounts for the implicative inferences in context. The one-way implicative 'to remember to' presupposes, we propose, a sufficient condition: in the relevant local context, remembering to close the door results in a door-closing. Suppose a hearer computes the discourse meaning of out-of-the-blue sentence (9a). At some point he may have inferred (9b), while the implicative condition (9c) must still be satisfied.

(9) a. Ed remembered to close the door.

b. $B \oplus [w_r\, o_0\, t_1 \,|\, w_r\!: Ed\ o_0,$
$w_r : remember.at\,(o_0, \lambda w, t[\,|\, w\!:\!o_0\ close.door.at\ t], t_1),\ t_1 < n]$

c. Implicative condition, contributed by the inflected verb:
$B \models\!\!\!<$
$[\,|\,[t_1 \,|\, w_r : remember.at\,(o_0, \lambda w, t[\,|\, w\!:\!o_0\ close.door.at\ t], t_1),\ t_1 < n] \Rightarrow$
$[t_2 \,|\, w_r : o_0\ close.door.at\ t_2,\ t_1 \bigcirc t_2,\ t_2 < n]]$

Various accommodation options arise. The one shown in (10a) can be excluded because it sins against a grammatical constraint: B must be a *proper DRS* but would contain free discourse marker t_1. The other two are viable options.

(10) a. $[t_2 \,|\, w_r : o_0\; close.door.at\, t_2,\; t_1 \bigcirc t_2,\; t_2 < n] \sqsubseteq B$
b. $[\,|\,[t_1 \,|\, w_r : remember.at\,(o_0, \lambda w, t[\,|\, w : o_0\; close.door.at\, t], t_1),\; t_1 < n] \Rightarrow$
$[t_2 \,|\, w_r : o_0\; close.door.at\, t_2,\; t_1 \bigcirc t_2,\; t_2 < n]] \sqsubseteq B$
c. $[\,|\,[t_1 \,|\, w_r : remember.at\,(o_0, \lambda w, t[\,|\, w : o_0\; close.door.at\, t], t_1),\; t_1 < n] \Leftrightarrow$
$[t_2 \,|\, w_r : o_0\; close.door.at\, t_2,\; t_1 \bigcirc t_2,\; t_2 < n]] \sqsubseteq B$

Considering what would explain the speaker's use of an implicative in the given context, the hearer might assume that he simply intends to describe a state of affairs and takes for granted that Ed remembering to close the door in the situation at hand is sufficient to make him close it. Thus the hearer would accept option (10b). However, he may go on to ask what makes the speaker mention Ed's remembering at all, if he could have said right away that Ed opened the door. Quite possibly because in the situation under discussion, the speaker considers it a *necessary* requirement as well: without remembering to close the door Ed would not have closed it. If, in addition, Ed's remembering is relevant in the sense that it is an open issue in the discourse whether he did or not, the speaker has good reasons to mention it. Now the hearer accepts (10c).

The indefeasibility of the implicative inference with the bare positive use of the verb is now accounted for, because there is no grammatical interpretation in which the implicative condition is not satisfied. With the negated form, however, only the interpretation corresponding to option (10c) predicts the implicative inference. We suggest that this is the preferred option in principle.[5] The implicative inference is then predicted to arise with the negated form in out-of-the-blue use and in linguistic contexts such as (6e.i) which reinforce the inference. But when it is disconfirmed, as in (6e.ii), the hearer selects accommodation option (10b). Typically, the continuation in (6e.ii) conveys that there is another action beside remembering that would result in Martha turning out the lights on the relevant occasion. So the speaker does not consider Martha's remembering a necessary requirement. The availability of accommodation option (10b) accounts for the defeasibility of the implicative inference with the negated form.

Because the implicative condition is 'presuppositional', i.e. a requirement on local context, the resultative or conditional link constitutes non-at issue information. It will commonly project to global context, unless forced to accommodate locally, e.g. through metalinguistic denial. As such it is employed by the hearer in the computation of discourse meaning. More complex examples in which one implicative construction embeds another will be interesting test-cases. If the various preconditions project to global context, the implicative inferences evoked by the sentence should follow from their interaction in the spirit of [K71b].

Furthermore, the implicative condition is stated in the domain of locality of the implicative verb and fires relative to its local context. This has some important consequences, which space restrictions allow us only to mention here. One is that entailments which hold locally do not necessarily surface globally. Thus, it can be explained that when triggered in the scope of e.g. modal and

[5] In terms of pragmatic rules, it provides a 'better explanation' of the speaker's use of an implicative.

generic operators the implicative inference is not evoked. Another is that it solves the notorious 'binding problem', detected in [KP79].

Every implicative verb comes with its own characteristic implicative condition. In the case of 'to manage to', we follow the general trend (cf. [Bea01]) and assume that its semantic content says that the complement state or event obtains. The entailments with both the positive and the negated use of the verb follow from this straight away, and are predicted to be indefeasible. Instead of the usual attempt/difficulty-presupposition, however, we'll have an implicative condition which says that the complement state or event holds only if the subject individual makes an effort (of at least degree d) to obtain that result:

(11) a. Someone managed to close the door.

b. $B \oplus [u_1 \,|\, w_r : human\, u_1,\ w_r : u_1\ close.door.at\, t_1,\ t_1 < n]$

c. $B \oplus [u_1 \,|\, w_r : human\, u_1\,] \models\!\!\!\!\sim [\,|\, [t_1 | \, w_r : u_1\ close.door.at\, t_1,\ t_1 < n\,] \Rightarrow$
$[\,t_2 | \, w_r : put.effort.d^{\leq}.in.at\, (u_1, \lambda w, t[\,|\, w : u_1\ close.door.at\, t], t_2),$
$t_2 \bigcirc t_1,\ t_2 < n\,]]$

This differs minimally from standard treatments in that the requirement that an effort be made is a *post*condition on the succes of the complement state or event, and is formulated in somewhat less specific terms than attempt or difficulty.

5 Linguistic Presuppositions or Pragmatic Strengthening?

While any specific implicative verb can carry additional presuppositions, we are interested to see how far the implicative condition may take us in accounting for the presuppositions claimed to be evoked by implicatives, given interaction with pragmatic reasoning. Remember the presuppositions are a diverse lot; focusing on 'to manage to' we will go through them one by one.

The implicative condition accounts directly for the characteristic *causal* or *explanatory* presuppositions mentioned in the introduction. Satisfying the implicative condition in (11c), the hearer may accommodate that for anybody to close the door on the particular occasion under discussion, he must make a certain effort, and doing so will cause the closing of the door. If we are told that a subject did not manage to close the door, then not having made that effort is a likely explanation of why he didn't close it. Thus, the satisfaction of the implicative condition induces the accommodation of explanatory suppositions.

Now for the *attempt suppositions*. The hearer may very well satisfy the implicative condition by accommodating that opening the door on this occasion requires an intentional, active attempt to do so on the agent's part. This is a plausible instance of putting in a certain effort. Arguably, with a positive use of the verb the attempt-inference is defeasible in examples like the following. The less specific inference signals that the implicative condition is still satisfied:

(12) Without intending it, Bill managed to insult the dean.
$\not\rightsquigarrow$ Bill attempted to insult the dean.
$\rightsquigarrow$ Bill was involved in some effort which resulted in insulting the dean.

With the negated use of the implicative (and a neutral stress pattern), as in

(13) Jack didn't manage to convince the dean.

the hearer is predicted to infer that the subject did not make the effort necessary to cause convincing the dean. In general, however, a more specific inference is evoked, namely that Jack did make an effort, though an insufficient one. This reading results from selecting a logically stronger context specification. Among the factors which can explain the preference for this reading, we suggest Gricean-style scalar reasoning is a central one. Given satisfaction of the implicative condition, in context 'Jack did not put a sufficient effort into convincing the dean' is a stronger alternative than 'Jack did not convince the dean.' Presumably, the speaker didn't communicate the stronger alternative because he doesn't believe that Jack did not put a sufficient effort into convincing the dean. If the speaker is informed about the matter, it follows that he believes that Jack tried to convince the dean. Thus analysing the inference as a scalar implicature, its defeasibility in constructions such as (14) is predicted.

(14) Jack didn't manage to convince the dean, in fact, he didn't even try.

The *difficulty suppositions* are naturally accounted for as an accommodation effect of satisfying the implicative condition. If it requires an effort on Jack's part to convince the dean, then probably there is some difficulty or challenge involved in this, which explains why the effort is needed. A most preferred context does not just make the implicative condition true but also plausible in the situation under discussion. Testing in projection and cancellation contexts reveals that the difficulty suppositions are generally persistent, often generic, but always adapted to the situation under discussion given common knowledge. The difficulty may be specific to the subject given his opportunities and skills, as in (2), it can be a difficulty in the action or goal to be achieved for anybody in general, or it can be completely contingent on the situation at hand. All of this is to be expected when the inference is not hard-wired as a linguistic presupposition, but rather the product of context specification. Finally, the implicative condition is sufficiently unspecific to account for cases like the following in which the effort made by the subject on the event referred to does not involve any concrete difficulty.

(15) John generally runs 10 miles per hour. Small wonder he managed to run 5 miles in 30 minutes yesterday.

$\not\rightsquigarrow$ It was difficult for John to run 5 miles in 30 minutes yesterday.
$\rightsquigarrow$ It is difficult for most people to run 5 miles in 30 minutes.
$\rightsquigarrow$ For John to run 5 miles in 30 minutes requires a certain effort.

A discussion of suppositions of *expectation* or *norm* and those of *obligation* attached to 'to remember to' shall have to wait for another occasion. We hope to have shown that an analysis in terms of implicative condition, pragmatic reasoning and accommodation effect is well-suited to account for the variability and defeasibility of the inferences under discussion.

6 Conclusion

We looked into the inferential behavour of implicative verbs, and argued that they contribute a presuppositional constraint that characterises the process they describe as resultative, culminating in the state or event described by the complement sentence. It was proposed that what are often claimed to be presuppositions can be accounted for as implicatures or secondary inferences resulting from the need to satisfy and explain the implicative condition of the verb. The analysis builds on a liberal notion of accommodation, where pragmatic reasoning rules serve to generate preferences over accommodation options.

Further research is needed, especially the interactions with tense, aspect and modality must be spelled out. More extensive testing of the projection behaviour of inferences and cancellation contexts is needed. Cross-linguistic research will be helpful to get a grasp on the diversity of implicitive word senses within the language system. Moreover, the explicit specification of pragmatic reasoning rules that justify the selection of accommodation options constitutes a theoretical challenge. The investigation of implicative verbs will be of interest for the theory of presupposition as well as the theory of pragmatics/accommodation. If implicative verbs are to find their place within the family of projective elements, a rich semantic-pragmatic interpretational system is called for. This might be LDG for discourse, or a model of comparable strength.

References

Bea01. Beaver, D.I.: Presupposition and Assertion in Dynamic Semantics. CSLI Publications, Stanford (2001)

Giv73. Givón, T.: The time-axis phenomenon. Language 49(4), 890–925 (1973)

Horn72. Horn, L.R.: On the Semantic Properties of the Logical Operators in English. Ph.D. thesis, University of California at Los Angeles (1972)

K71a. Karttunen, L.: Implicative verbs. Language 47(2), 340–358 (1971)

K71b. Karttunen, L.: The logic of English predicate complement constructions. Distributed by the Indiana University Linguistics Club (1971), `http://www2.parc.com/istl/members/karttune/publications//english_predicate.pdf`

KP79. Karttunen, L., Peters, S.: Conventional implicature. Syntax and Semantics 11, 1–56 (1979)

Luz99. Luzón Marco, M.J.: A semantic-syntactic study of implicative verbs based on corpus analysis. Estudios Inglese de la Universidad Complutense 7, 69–87 (1999)

NCK06. Nairn, R., Condoravdi, C., Karttunen, L.: Computing relative polarity for textual inference. In: Bos, J., Koller, A. (eds.) Proceedings of ICoS-5, Inference in Computational Semantics, Buxton, England, pp. 67–76 (2006)

Leu07. van Leusen, N.: Description Grammar for Discourse. Ph.D. thesis, Radboud University Nijmegen (2007)

LM03. van Leusen, N., Muskens, R.: Construction by Description in Discourse Representation. In: Peregrin, J. (ed.) Meaning, the Dynamic Turn, pp. 33–65. Elsevier (2003)

Tho90. Thomason, R.: Accommodation, meaning, and implicature: Interdisciplinary foundations for pragmatics. In: Cohen, P.R., Morgan, J., Pollock, M.E. (eds.) Intentions in Communication, pp. 325–363. MIT Press (1990)

Tropes, Intensional Relative Clauses, and the Notion of a Variable Object

Friederike Moltmann

IHPST (Paris 1/ENS/CNRS)
fmoltmann@univ-paris1.fr

Abstract. NPs with intensional relative clauses such as *the impact of the book John needs to write* pose a significant challenge for trope theory (the theory of particularized properties), since they seem to refer to tropes that lack an actual bearer. I will propose a novel semantic analysis of such NPs on the basis of the notion of a variable object. This analysis avoids a range of difficulties that an alternative analysis based on the notion of an individual concept would face.

Keywords: tropes, intensional verbs, individual concepts, relative clauses, situations.

1 Introduction

It is a common view, since Aristotle, that terms of the sort in (1) refer to tropes or particularized properties, that is, particular, non-sharable features of individuals (Williams 1953, Strawson 1959, Woltersdorff 1977, Campbell 1990, Lowe 2006, Mertz 1996):

(1) a. the wisdom of Socrates
b. the originality of the book
c. the simplicity of the dress

According to that view, (1a) refers to the particular manifestation of wisdom in Socrates, that is, a wisdom trope with Socrates as its bearer.

Given general diagnostics for trope reference, there are equally good reasons to take the terms below to be terms referring to tropes, namely quantitative tropes (Campbell 1990, Moltmann 2009, to appear a):

(2) a. the number of planets
b. the height of the building
c. the length of the vacation

According to that view (2a) refers to the instantiation of the property of being eight in the plurality of planets, a feature not shared by any equally numbered plurality.

M. Aloni et al. (Eds.): Amsterdam Colloquium 2011, LNCS 7218, pp. 431–440, 2012.
© Springer-Verlag Berlin Heidelberg 2012

There are closely related terms, however, that present a significant challenge to trope theory. These are NPs of the sort below with relative clauses containing an intensional verb:

(3) a. the impact of the book John needs to write
b. the simplicity of the dress Mary needs for the occasion
c. the wisdom of the director that the institutes should hire
(4) a. the number of people that fit into the car
b. the height of the desk John needs
c. the length of the time John might be away

I will call apparent trope-referring NPs with intensional relative clauses of this kind IR-NPs.

Tropes as discussed in philosophy are meant to be real entities, involving real objects as bearers. In fact, tropes generally are taken to depend for their existence and their identity on their bearer. But the tropes that the terms in (3) and (4) seem to refer to lack an actual bearer. In this paper, I will argue that nonetheless the terms in (3) and (4) refer to tropes, or rather, in most cases, what I will call variable tropes. Central on this account is the notion of a variable object, a particular case of the notion of a variable embodiment of Fine (1999). IR-NPs either refer to tropes with a variable object as bearer or else they themselves refer to variable tropes whose bearer is driven by the variability of the bearer. I will argue that making using variable objects avoids a range of serious difficulties for the more standard alternative account that would make use of individual concepts.

2 Trope Reference with NPs Containing Intensional Relative Clauses

The NPs in (3) and (4) share a range of diagnostics for trope reference with the NPs in (1) and (2). Tropes, unlike properties, generally are taken to be perceivable and causally efficacious (Williams 1953, Lowe 2006). In fact, both sorts of NPs allow for the application of perceptual predicates, as in (5), and predicates describing causal relations, as in (6):

(5) a. John observed Mary's politeness.
b. John noticed the number of screws that were missing.
(6) a. The weight of the lamp caused the table to break.
b. The great number of screws that were missing caused the table to fall apart.

Moreover both sorts of terms accept predicates of neutral evaluation such as *exceed, great, high*, or *negligible*, predicates which are not naturally applicable to the corresponding abstract objects (properties or numbers):

(7) a. The number of women exceeds the number of men.
b. The number of people that fit into the bus exceeds the number of people that fit into the car.
c. ?? Eight exceeds seven.

(8) a. The impact of John's book is great / negligible.
b. The impact of the book John needs to write has to be great / negligible.
c. ?? The property of having an impact is great / negligible.

Finally, trope reference is reflected in the application of the *be* of identity as opposed to the predicate is the same as. The observation is that whereas (8a) and (8c) can be true, (8b) and (8e) cannot:

(8) a. The number of men is the same as the number of women.
b. ?? The number of men is the number of women.
c. The number of people that fit into the bus is the same as the number of people that fit into the car.
d. ?? The number of people that fit into the bus is the number of people that fit into the car.

The same as in fact expresses close or exact similarity not numerical identity, which is expressed only by identity *be* (Moltmann 2009, to appear). Tropes with different bearers that instantiate the same property are similar but not identical. Tropes that instantiate the same 'natural' property (for example the same number property) are exactly similar, and thus 'the same'.

3 Approaches Based on Individual Concepts

Given standard semantics, an obvious approach to the terms in (3)-(4) would be to consider them terms referring to tropes with individual concepts as bearers, that is, (partial) functions from worlds and times ('circumstances' for short) to individuals (or collections of individuals) (Montague 1974). That individual concepts of some sort are the denotations of certain types of NPs with intensional relative clauses has in fact been argued by Moltmann (2008) for NPs as in (9a) and, for the closely related construction in (9b), by Grosu/Krifka (2007):

(9) a. The assistant John needs must speak French fluently.
b. The gifted mathematician that you claim to be could solve this problem in no time.

However, using reference to individual concepts and of individual concepts as bearers of tropes raises a range of problems, ontologically, conceptually, empirically, and regarding the compositional semantics of IR-NPs.

The ontological problem concerns the notion of a trope itself: tropes are entities in the world that are potentially causally efficacious and perceivable. This means that tropes have objects as bearers, not intensions or functions (unless of course the tropes are features of abstract objects, but this is not what is at stake).

The conceptual problem concerns substitution problems that reference to individual concepts in general give rise to: an abstract function has quite different properties (that is, is a bearer of quite different tropes) than 'the book that John needs to write'.

The empirical problem concerns the particular behavior of NPs as in (3)-(4) with respect to the requirement that the predicate contain a modal. Sometimes IR-NPs are subject to the requirement, as in (10a), sometimes they are not, as in (10b):

(10) a. The impact of the paper John needs to write ?? exceeds /ok must exceed the impact of the papers he has so far written.
b. The number of people that fit into the bus exceeds the number of people that fit into the car.

Finally, there are problems regarding the compositional semantics of IR-NPs on the basis of individual concepts. There are two options of analyzing *the book John needs to write* as standing for an individual concept. For reasons of space, I can discuss those only briefly and only in their roughest outline.

The first option would be an extension of Grosu/Krifka's (2007) analysis *of the gifted mathematician that John claims to be*. Their analysis involves several assumptions. First, it involves type-lifting of all predicates to predicates of individual concepts and all singular terms (including proper names) to terms for individual concepts. Second, it requires treating all intensional verbs as operators quantifying over possible worlds. Finally, it interprets the head noun *book* in the upper position, rather than reconstructing it into the lower position inside the relative clause. Greatly simplifying, this analysis would yield the following as the denotation of *the book John needs to write*:

(11) $\min(\{f \mid \text{book}(f)\} \cap \{f \mid \text{John need to write } (f)\})$

(The second set would be the set of functions mapping a world w compatible with the satisfaction of John's needs to an object John writes in w.) This analysis raises a range of problems. First of all, it involves an excessive use of individual concepts as well as the assumption that all intensional verbs be analyzable as operators quantifying over words, an assumption that a great number of philosophers will find problematic. Furthermore, it poses a problem of uniqueness (a problem that did not arise for the construction for which Grosu and Krifka's analysis was originally developed[1]). In a given word in which John's needs are satisfied, John may have written more than one book meeting his need. To account for uniqueness, not entire worlds should be taken into account in which John's needs are satisfied, but rather situations exactly satisfying the

[1] Grosu/Krifka (2007) have no problem of uniqueness because they analyse *the gifted mathematician John claims to be* as involving identity *be*, which for them takes two individual concepts as arguments.

need. A given world in which John's needs are satisfied may then contain several situations satisfying his need.

The second option of analysing *the book John needs* to write as standing for an individual concept would involve reconstructing the head noun into the lower position inside the relative clause yielding the analysis indicated below:

(12) The f [for any world w compatible with the satisfaction of John's needs, $write^w$(John, f(w)) & $book^w$(f(w)]

This analysis raises the very same problem of uniqueness. Also, just like the first analysis, it is forced to treat all intensional verbs as modal operators quantifying over worlds. Moreover, in its attempt of avoiding type-shifting the analysis cannot go very far. Even though it is plausible that the head noun reconstructs into the lower position, reconstruction of the functional trope noun into a position inside the relative clause is in general impossible. There is no place inside the relative clause for a noun like *impact* in (3a), repeated below:

(3) a. the impact of the book John needs to write

Impact will have to be interpreted in the upper position. But this means that it will have to denote a function applying to individual concepts.

4 The Variable-Objects Approach

The account I would like to propose is based both on the notion of a variable object and the notion of a variable trope. Variable objects are entities that fall under Fine's (1999) more general notion of a variable embodiment. The notion of a variable embodiment for Fine is a central notion in metaphysics and accounts for a great variety of 'ordinary' objects. But Fine himself (p.c.) also meant to apply the notion of a variable embodiment to the semantic values of functional NPs as in (13) as well as NPs with intensional relative clauses such as the book John needs to write:

(13) a. John changed his trainer.
b. The temperature is rising.
c. The number of students has increased.

According to the standard Montagovian view, functional NPs such as (13a)-(13c) are of a different type than singular terms: they are of type <e, t> rather than of type <e>. Functional NPs, that is, stand for individual concepts: functions from world-time pairs to objects. Some predicates such as *change, rise, increase* will apply to individual concepts directly. Other predicates will be type-shifted to predicates of individual concepts subject to the following meaning postulate:

(14) For any predicate of individuals P and any individual concept f,
$P'^{w,t}(f) = 1$ iff $P^{w,t}(f(w, t)) = 1$.

There are various reasons to consider NPs of the sort in (13) as standing for objects (variable objects) rather than being of a different type than singular terms. For example, object-related predicates can apply to such NPs just as they apply to individuals (such as the predicates *change, rise,* and *increase*). This also holds for NPs with intensional relative clauses. Most strikingly, the predicate *count* can apply with such NPs just as it applies with ordinary singular terms:

(15) a. John counted the books he needs to write.
b. John counted the screws that are missing.

Moreover, functional NPs can naturally provide the bearers of tropes:

(16) a. The impact of the increasing number of students is noticeable.
b. The rise of the temperature caused the drought.

The notion of a variable embodiment allows an account of functional NPs that avoids type-shifting of predicates and also avoids treating their referents as abstract functions.

A variable embodiment, according to Fine, is an entity that allows for the replacement of constituting material or parts, and more generally that may have different manifestations in different circumstances. Organisms and artifacts are variable embodiments, but also entities like 'the water in the river'. 'The water in the river' is a variable embodiment that has different manifestations as different quantities of water at different times. Variable embodiments differ from 'rigid embodiments', entities which do not allow for a replacement of their immediate parts. An example is a token of the word *be*, which has as its immediate parts a token of *b*and a token of *e*, neither of which allows replacement.

Fine's theory of variable embodiments as formulated in Fine (1999) applies to variable embodiments that may have different manifestations at different times. But the theory is also meant to apply to entities that have different manifestations in different worlds and in fact may lack a manifestation in the actual world. 'The book John needs to write' is such an entity. It is an entity that has different manifestations as different objects in various counterfactual worlds. My term of a variable object is meant to apply to entities that have different manifestations as different objects at different times or in different worlds.

Let us then adopt the following conditions from Fine (1999) for variable objects:

(17) a. *Existence*
A variable object e exists in a circumstance i iff e has a manifestation in i.
b. *Location*
If a variable object e exists in a circumstance i, then e's location in is that of its manifestation in i.
c. *Property inheritance 1*
A variable object e has a (world- or time-relative) property P in a circumstance i if e's manifestation in i has P.

In addition to local properties, which they obtain in the way of (17c), variable objects may have global properties, properties that they may have on the basis of several of their manifestations at different times (for example properties of *change, rise, or increase*). Variable objects moreover may have properties that are not time- or world-relative (though may be attributed at a time or in a world):

(17) d. *Property inheritance 2*
A variable object has a (world- and time-independent) property P if all its manifestations in any circumstances have P.

When the property in consideration is understood as a particularized property (a trope), these two conditions can be reformulated as follows:

(18) a. *Trope 'inheritance' 1*
A variable object e bears a trope t relative to a circumstance i if e's manifestation in i bears a trope in i that is exactly similar to t.
b. *Trope 'inheritance' 2*
A variable object bears a trope t if for any circumstance i, e's manifestation in i bears a trope exactly similar to t in i.

Using variable objects in this sense has a significant advantage over the individual-concept approach to the compositional semantics of functional NPs and NPs with intensional relative clauses by avoiding a type ambiguity among predicates entirely. Let us first apply the account to (19a):

(19) a. the impact of the number of students
b. the increase of the number of students

The functional trope noun in the upper position applies to a variable object and maps it onto a local trope based on a single circumstance. The two functional trope nouns in (19a) denote a function from variable objects to tropes, as below, where F is the function mapping a variable object e and a circumstance <w, t> to the manifestation of e in <w, t>:

(20) For a variable object e, $\text{impact}^{w,t}(e)$ = the trope that has e as its bearer and is exactly similar to $\text{impact}^{w,t}(F(e, (w, t)))$.

A different case is that of *the impact of the book John needs to write*, which refers not to a single trope but rather to a variable trope. Let us first focus on what the variable object is that *the book John needs to write* stands for. Assuming that the head noun *book* is interpreted in the lower position inside the relative clause, the lower variable will stand for a variable object, an object to which the relative clause attributes certain properties in particular circumstances. But this variable object cannot be the variable object each of whose manifestations is a paper John writes in a world in which John's needs are satisfied. A world in which John's needs are satisfied may contain several papers that John writes in that world. Moreover, some of those papers may not qualify as 'the paper John's needs to write': the complement of *need* gives only a partial characterization of the exact need. Rather to obtain uniqueness, use must be made of situations exactly satisfying John's needs. That is, 'the paper John needs to write' stands for the variable object each of whose manifestations is a paper John writes in a situation exactly satisfying John's needs. Uniqueness then holds relative to a situation of satisfaction of the need. The situation may impose various constraints on the paper John writes in it (constraints the speaker in fact need not know about). Given its dependence on satisfaction situations, the variable object that is 'the paper John needs to write' is an object that itself depends on a need.

But what is a need? A need is not a state of needing and thus not a Davidsonian event argument. The reason is that only a need, but not a state of needing, can be 'satisfied' by a situation. How then can a 'need' be obtained in the interpretation of a sentence so that the variable object in question could depend on it? Without going into a greater discussion, I would simply like appeal to a particular syntactic proposal concerning the verb *need* by Harves/Kayne (to appear). According to their view, the verb *need* is the result of incorporating the copula *have* and the noun *need*. Given this proposal, an entity that is a need would be made available as part of the compositional semantics of the complex predicate *have+need*, as below, where the variable object d_e is dependent on a need e:

(21) the book [John needs to write e] = the book [John has a need to write e] =
the e [John has a need to write [e book]] = ιd[$\exists$e (have(e, John) &
need(e, ^ write(John, d_e) & book(d_e))]

5 The Modal Compatibility Requirement

Let us turn to the question of when a modal is required in the main clause of a sentence with an IR-NP. Following Grosu/Krifka (2007), who noticed the requirement for a related construction (see also Moltmann 2008), I will call this the Modal Compatibility Requirement:

(22) *The Modal Compatibility Requirement (MCR)*
IR-NPs require an appropriate modal in the main clause to 'access' the entities in the counterfactual circumstances.

The MCR does not hold for all sentences with IR-NPs. It does not hold in (23a), though it does hold in (23b):

(23) a. The number of people that can fit into the bus exceeds the number of people that can fit into the car.
b. The impact of the book John needs to write ?? exceeds / ok must exceed / OK might exceed the impact of the book he has already written.

This might suggest that IR-NPs referring to quantitative tropes are not subject of the MCR. But this is not right. The MCR is in place below:

(24) ?? The number of people John might invite exceeds the number of people Mary might invite.

Yet the distinction between quantitative and qualitative trope does matter, as illustrated by the contrast between (25a) and (25b):

(25) a. The number of papers a student has to write during this program is too high.
b. The quality of the paper John must write ?? is very high / ok must be high.

I propose an explanation of the MCR and exceptions to it based on general conditions on when a variable object can bear a trope on the basis of its manifestations. The cases in which the MCR is in place are cases in which the head noun applies to a variable object and maps it onto a variable trope. A variable trope driven by the variability of its bearer e has as its manifestation in a circumstance i the trope t that has as its bearer the manifestation of e in i. A variable trope that has manifestations only in counterfactual circumstances requires a modal in the main clause in order to be attributed local properties in the first place. The noun *impact* in *the impact of the book John needs to write* thus denotes a function mapping a variable object onto a variable trope, as below:

(26) For a variable object e,
$impact^{w,t}(e)$ = the variable trope o such that for any circumstance s in which e has a manifestation F(e, s), $impact^{w,t}(F(e, s))$ = the manifestation of o in s.

Regarding the case of quantitative tropes not subject to the MCR, it is plausible to assume that the same number of people fit into the bus / the car in the various relevant circumstances. This means that the number tropes in the various circumstances are exactly similar and thus that the variable object itself can bear an exactly similar number trope. This is then not a case of a variable trope, but of an ordinary trope with a variable object as its bearer. Such cases are restricted to quantitative tropes because exact similarity among qualitative tropes is unlikely to obtain, given that natural language predicates do not express fully specific qualitative properties, but unspecific, determinable ones.

6 Conclusion

To summarize, the notion of a variable object allows an account of an otherwise very puzzling construction of apparent trope-referring terms. The notion of a variable object as such is not a peculiar notion, though, invoked only for the analysis of IR-NPs. Rather, it falls under the more general and ontologically central notion of a variable embodiment (in Fine's metaphysics). As subject, it is subject to the very same ontological conditions as drive variable embodiments in general.

References

1. Aristotle: The Categories
2. Campbell, K.: Abstract Particulars. Blackwell, Oxford (1990)
3. Fine, K.: Things and Their Parts. Midwest Studies of Philosophy 23, 61–74 (1999)
4. Grosu, A., Krifka, M.: The Gifted Mathematician that you Claim to be. Linguistics and Philosophy 30, 445–485 (2007)
5. Harves, S., Kayne, R.S.: Having *need* and needing *have*. Linguistic Inquiry (to appear)
6. Lowe, J.: The Four-Category Ontology. A Metaphysics Foundation for Natural Science. Oxford UP (2006)
7. Mertz, D.W.: Moderate Realism and Its Logic. Yale UP, New Haven (1996)
8. Moltmann, F.: Intensional Verbs and Quantifiers. Natural Language Semantics 5(1), 1–52 (1997)
9. Moltmann, F.: Intensional Verbs and Their Intentional Objects. Natural Language Semantics 16(3), 239–270 (2008)
10. Moltmann, F.: Degree Structure as Trope Structure: A Trope-Based Analysis of Comparative and Positive Adjectives. Linguistics and Philosophy 32, 51–94 (2009)
11. Moltmann, F.: Reference to Numbers in Natural Language. Philosophical Studies (to appear)
12. Montague, R.: The Proper Treatment of Quantification in Ordinary English. In: Hintikka, J., et al. (eds.) Approaches to Natural Language, Reidel, Dordrecht (1973)
13. Strawson, P.: Individuals. An Essay in Descriptive Metaphysics. Methuen, London (1959)
14. Williams, D.C.: On the Elements of Being. Review of Metaphysics 7, 3–18 (1953)
15. Woltersdorff, N.: On Universals. Chicago UP, Chicago (1970)

A Theory of Names and True Intensionality*

Reinhard Muskens

Tilburg Center for Logic and Philosophy of Science
r.a.muskens@uvt.nl
http://let.uvt.nl/general/people/rmuskens/

Abstract. Standard approaches to proper names, based on Kripke's views, hold that the semantic values of expressions are (set-theoretic) functions from possible worlds to extensions and that names are rigid designators, i.e. that their values are *constant* functions from worlds to entities. The difficulties with these approaches are well-known and in this paper we develop an alternative. Based on earlier work on a higher order logic that is *truly intensional* in the sense that it does not validate the axiom scheme of Extensionality, we develop a simple theory of names in which Kripke's intuitions concerning rigidity are accounted for, but the more unpalatable consequences of standard implementations of his theory are avoided. The logic uses Frege's distinction between sense and reference and while it accepts the rigidity of names it rejects the view that names have direct reference. Names have constant denotations across possible worlds, but the semantic value of a name is not determined by its denotation.

Keywords: names, axiom of extensionality, true intensionality, rigid designation.

1 Introduction

Standard approaches to proper names, based on Kripke (1971, 1972), make the following three assumptions.

(a) The semantic values of expressions are (possibly partial) functions from possible worlds to extensions.
(b) These functions are identified with their *graphs*, as in set theory.
(c) Names are rigid designators, i.e. their extensions are world-independent.

In particular, the semantic values of names are taken to be *constant* functions from worlds to entities, possibly undefined for some worlds.

The difficulties resulting from these assumptions are well-known. On the one hand, there are general 'logical omniscience' problems with the possible worlds approach resulting from (a) + (b). Since functions, in the set-theoretic conception, are extensional entities, with their identity criteria given by input-output

* I would like to thank the anonymous referees for their excellent advice.

M. Aloni et al. (Eds.): Amsterdam Colloquium 2011, LNCS 7218, pp. 441–449, 2012.
© Springer-Verlag Berlin Heidelberg 2012

behaviour, the semantic values of far too many expressions will be identified. Implications and their contrapositives, for example, will be lumped together. That is incorrect since one may very well believe $p \to q$ but fail to believe $\neg q \to \neg p$, so that there is at least one property the semantic values of these expressions do not have in common.

Adding (c) as a further restriction makes things worse, since if the semantic value of a name depends only on its bearer it is predicted that names with the same bearer can be substituted for one another in *any* context. This leads to philosophers claiming and dogmatically defending the position that the Ancients *did* know that Hesperus was Phosphorus before that identity was actually discovered, an armchair intuition that does not seem to be shared by many outside the profession. It also leads to the prediction that the following are equivalent.

(1) a. We do not know *a priori* that Hesperus is Phosphorus
  b. We do not know *a priori* that Phosphorus is Phosphorus

(1a) is asserted in Kripke (1972, page 308); (1b) is obviously false. Traditional theorists are therefore confronted with the challenge to come up with a logic in which the values of (1a) and (1b) can be distinguished. No precise system seems to have been developed thus far.

The substitutivity problems that follow from the adoption of (a)–(c) show that this combination cannot stand, but this does not mean, of course, that (c), the idea that names denote rigidly, has to go. In this paper I will sketch a theory that does not suffer from the many problems that are connected with identifying intensions with certain functions in extension, but in which it is still possible to consistently formalize the intuition that names denote rigidly.

2 A Truly Intensional Logic

We move to a (higher order) logic that is *truly intensional*. By this we mean that the following axiom (schema) of extensionality fails.[1]

(2) $\forall XY(\forall \vec{x}(X\vec{x} \leftrightarrow Y\vec{x}) \to \forall Z(ZX \leftrightarrow ZY))$

This axiom schema says that whenever two relations X and Y of arbitrary arity are co-extensional, $\forall \vec{x}(X\vec{x} \leftrightarrow Y\vec{x})$, they are in fact identical in the sense that they have the same properties, i.e. $\forall Z(ZX \leftrightarrow ZY)$.[2]

[1] Our notion of *true intensionality* is just the notion of *intensionality* defined in Whitehead and Russell (1913), but using that term without modification may lead to confusion nowadays, as the word is now widely used for Carnap's imperfect approximation of the original concept. I will mostly, but not always, use *true intensionality* for *intensionality* in this paper. *Hyperintensionality*, another word for the same idea, is less than felicitous, as it suggests a property stronger than intensionality, while it is only stronger than Carnap's approximation.

[2] The *Leibniz identity* $\forall Z(ZX \leftrightarrow ZY)$ can be abbreviated as $X = Y$. Note that $\forall Z(ZX \to ZY)$ is in fact equivalent.

There are now several approaches to type theory that manage to avoid making (2) valid. Fitting (2002) and Benzmüller et al. (2004) are two of them, but since both interpret the central machinery of type logic in some non-standard way,[3] the logic used here will be the ITL of Muskens (2007). In this logic all operators have standard interpretations and in fact the interpretation of the logic is a rather straightforward generalisation of that of Henkin (1950), making (2) invalid but retaining all classical rules for logical operators. The following somewhat impressionistic description mainly highlights ITL's minor differences with standard simple type theory. For precise definitions consult Muskens (2007).

Type System. ITL's type system is *relational*, rather than *functional*. Given some set of basic types, further types are formed by the rule that $\langle \alpha_1 \ldots \alpha_n \rangle$ is a type if $\alpha_1, \ldots, \alpha_n$ are. Objects of type $\langle \alpha_1 \ldots \alpha_n \rangle$ are n-ary relations in intension that take objects of type α_k in their k-th argument place. Readers familiar with functional type logics may identify $\langle \alpha_1 \ldots \alpha_n \rangle$ with $\alpha_1 \rightarrow \ldots \rightarrow \alpha_n \rightarrow t$, where t is the type of propositions and truth values (and association is to the right). In linguistic semantics this would be written without the arrows: $\alpha_1 \ldots \alpha_n t$. So $\langle e \rangle$ is the type of unary and $\langle eee \rangle$ the type of ternary relations in intension of type e objects, while $\langle\langle e \rangle\rangle$ is the type of properties of properties of individuals (quantifiers). The type $\langle\rangle$ is a limiting case. It corresponds to t in the functional set-up. Objects of this type are propositions, and their extensions are truth values. Note that, since (2) fails, there may be many non-identical true propositions in any given model, just as there may be many non-identical but co-extensional relations in other types.

Language. Terms of the logic are built up in the usual way from variables and non-logical constants with the help of application, λ-abstraction and a few logical constants, here $\sqsubseteq$ and $\bot$ ($\sqsubseteq$ is meant to denote inclusion of extensions and $\bot$ will be a proposition that is always false). Typing of terms is as expected, given the correlation between relational and functional types that was just described. For example, $(\lambda x.A)$ is of type $\langle \alpha_1 \alpha_2 \ldots \alpha_n \rangle$ if A is of type $\langle \alpha_2 \ldots \alpha_n \rangle$ and x is of type α_1, while (AB) is of type $\langle \alpha_2 \ldots \alpha_n \rangle$ if A is of type $\langle \alpha_1 \alpha_2 \ldots \alpha_n \rangle$ and B is of type α_1. Successive applications can 'eat up' all the argument places of a relation until $\langle\rangle$ is reached.

2.1 Further Information about ITL

At this point a reader mainly interested in the application of the logic ITL to the theory of names may want to skip to section 3. Those who want slightly more information about the logic may first want to read the rest of this section.

Intensional Models. Models for ITL distinguish between the *intension* of a term (given an assignment) and the *extension* associated with that intension. Fitting (2002) uses a similar way of defining models and the set-up is strongly reminiscent of that of Frege (1892).

[3] Fitting's (2002) interpretation of lambda abstraction is non-standard, while in the theory of Benzmüller et al. (2004) the interpretation of application is.

Table 1. Gentzen rules for ITL

$$\frac{\Pi \Rightarrow \Sigma}{\Pi' \Rightarrow \Sigma'}[W], \quad \text{if } \Pi \subseteq \Pi',\ \Sigma \subseteq \Sigma'$$

$$\frac{}{\Pi, \varphi \Rightarrow \Sigma, \varphi}[R] \qquad \frac{}{\Pi, \bot \Rightarrow \Sigma}[\bot\mathsf{L}]$$

$$\frac{\Pi, A\{x := B\}\vec{C} \Rightarrow \Sigma}{\Pi, (\lambda x.A)B\vec{C} \Rightarrow \Sigma}[\lambda\mathsf{L}] \qquad \frac{\Pi \Rightarrow \Sigma, A\{x := B\}\vec{C}}{\Pi \Rightarrow \Sigma, (\lambda x.A)B\vec{C}}[\lambda\mathsf{R}]$$

if B is free for x in A — if B is free for x in A

$$\frac{\Pi, B\vec{C} \Rightarrow \Sigma \qquad \Pi \Rightarrow \Sigma, A\vec{C}}{\Pi, A \subset B \Rightarrow \Sigma}[\subset\mathsf{L}] \qquad \frac{\Pi, A\vec{c} \Rightarrow \Sigma, B\vec{c}}{\Pi \Rightarrow \Sigma, A \subset B}[\subset\mathsf{R}]$$

if the constants $\vec{c}$ are fresh

Collections of domains will be sets $\{D_\alpha \mid \alpha \text{ is a type}\}$ of pairwise disjoint non-empty sets. There is no further restriction on collections of domains and in particular sets $D_{\langle\alpha_1\ldots\alpha_n\rangle}$ need *not* consist of relations over lower type domains, as is the case in the usual Henkin models. *Assignments* and notation for assignments are defined as usual. *Intension functions* are defined to be functions that send terms and assignments to elements of the D_α and respect the following constraints.

- $I(a, A) \in D_\alpha$, if A is of type α
- $I(a, x) = a(x)$, if x is a variable
- $I(a, A) = I(a', A)$, if a and a' agree on all variables free in A
- $I(a, A\{x := B\}) = I(a[I(a, B)/x], A)$, if B is free for x in A

These constraints are still very liberal and do not amount to the constraints imposed by the usual Tarski definition.

The next step associates extensions with intensions. For each $\alpha = \langle\alpha_1 \ldots \alpha_n\rangle$, a function $E_\alpha : D_\alpha \to \mathcal{P}(D_{\alpha_1} \times \cdots \times D_{\alpha_n})$ is called an *extension function.* A triple consisting of a collection of domains, an intension function, and a family of extension functions is called a *generalised frame.* (Note that in a generalised frame $E_{\langle\rangle} : D_{\langle\rangle} \to \{0, 1\}$, if some standard identifications are made.) Generalised frames are *intensional models* if, for all $\alpha = \langle\alpha_1 \ldots \alpha_n\rangle$, and for all terms A of type α, the extensions $E_\alpha(I(a, A))$, for which we write $V(a, A)$, satisfy the following constraints.

- $V(a, \bot) = 0$
- $V(a, AB) = \{\langle\vec{d}\rangle \mid \langle I(a, B), \vec{d}\rangle \in V(a, A)\}$
- $V(a, \lambda x_\beta.A) = \{\langle d, \vec{d}\rangle \mid d \in D_\beta \text{ and } \langle\vec{d}\rangle \in V(a[d/x], A)\}$
- $V(a, A \subset B) = 1 \iff V(a, A) \subseteq V(a, B)$

Table 2. Some classical rules derivable in ITL

$$\frac{}{\Pi \Rightarrow \Sigma, \top}\ [\top\mathsf{R}]$$

$$\frac{\Pi, \psi \Rightarrow \Sigma \qquad \Pi \Rightarrow \Sigma, \varphi}{\Pi, \varphi \to \psi \Rightarrow \Sigma}\ [\to\mathsf{L}] \qquad \frac{\Pi, \varphi \Rightarrow \Sigma, \psi}{\Pi \Rightarrow \Sigma, \varphi \to \psi}\ [\to\mathsf{R}]$$

$$\frac{\Pi, \varphi\{x := A\} \Rightarrow \Sigma}{\Pi, \forall x\varphi \Rightarrow \Sigma}\ [\forall\mathsf{L}] \qquad \frac{\Pi \Rightarrow \Sigma, \varphi\{x := c\}}{\Pi \Rightarrow \Sigma, \forall x\varphi}\ [\forall\mathsf{R}]$$

where c is fresh

$$\frac{\Pi, A \doteq B \Rightarrow \Sigma, \varphi\{x := A\}}{\Pi, A \doteq B \Rightarrow \Sigma, \varphi\{x := B\}}\ [=\mathsf{L}] \qquad \frac{}{\Pi \Rightarrow \Sigma, A = A}\ [=\mathsf{R}]$$

where $A \doteq B$ is $A = B$ or $B = A$

These last clauses constrain extensions to behave as in the usual Tarski value definition. For the treatment of abstraction and application in a relational setting, see also Muskens (1995).

Entailment is defined in the usual way, with the help of intensional models. The rules for λ-*conversion*, (α), (β) and (η), do not automatically hold (they preserve extension, but not necessarily intension), but it is possible to consistently add them to the logic. Extensionality is not universally valid. This is because the functions $E_\alpha\colon D_\alpha \to \mathcal{P}(D_{\alpha_1} \times \cdots \times D_{\alpha_n})$ need not be injective. In fact, intensional models in which all extension functions are injective essentially are Henkin's general models, while a further requirement of surjectivity will give full models.

Proofs. The Gentzen calculus in Table 1 is generalised complete for the semantic notion of entailment just defined (see Muskens (2007) for a proof). Table 2 gives derived rules for some operators defined from the two primitives $\subset$ and $\bot$. The identity here is Leibniz identity again, having the same properties, i.e. $A = B$ is short for $\forall Z(ZA \to ZB)$.

3 Names in a Truly Intensional Setting

Given a truly intensional logic such as the one just defined, a theory of names can take the following form.

- Ordinary proper names are *predicates*.
- They are *singular* in the sense that their extensions are either empty or singletons.
- Meanings are represented by lambda terms and combine with the help of application and *type shifters*.

– Among the type shifters is Partee's type shifter A, i.e. $\lambda P'P.\exists x(P'x \wedge Px)$ (Partee, 1986).[4]

In defending a theory of names as predicates I side with Aristotle, I think. More recent authors who have defended the names-as-predicates view in one form or another are Quine (1948), Quine (1960), Burge (1973), Muskens (1995), Matushansky (2006), and Fara (2011), among others. The idea seems linguistically natural, as names accept modification, combine with determiners, etc., just like common nouns.

Singularity can be enforced by adopting the following constraint, for all names N.[5]

(3) $\forall xy((\mathsf{N}x \wedge \mathsf{N}y) \rightarrow x = y)$

The type shifter A provides the glue that is needed to get predication going. (4) provides a simple example. Let's say *Zeus* translates as the predicate Z (4a); then combining with Partee's type shifter $\lambda P'P.\exists x(P'x \wedge Px)$ leads to the translation in (4b) and a further combination with the translation of *smiles*, S say, to that in (4c) (here it is assumed that the rules of λ-conversion have indeed been added to the logic).

(4) a. Zeus $\rightsquigarrow$ Z
b. Zeus $\rightsquigarrow \lambda P.\exists x(\mathsf{Z}x \wedge Px)$
c. Zeus smiles $\rightsquigarrow \exists x(\mathsf{Z}x \wedge \mathsf{S}x)$

(4c) also illustrates how non-referring names are dealt with. Atheists denying the existence of Zeus can consistently claim the statement $\exists x(\mathsf{Z}x \wedge \mathsf{S}x)$ to be false, a possibility that was also provided for in Russell (1905), but is not available in theories that translate names as individual constants.

Let us look at identity statements, such as the infamous *Hesperus is Phosphorus* case. In (5a) we translate *Phosphorus* as Φ, a translation that, I take it, can be inherited by *is Phosphorus.*[6] The translation in (5c) is then obtained in a way analogous to the one in (4).

---

[4] We generally use Q as a variable over type $\langle\langle e\rangle\rangle$ (quantifiers), P as a variable of type $\langle e\rangle$ (properties of individuals), R as a variable of type $\langle ee\rangle$ (binary relations in intension of individuals), and x, y and z as variables of type e (individuals).

[5] Another way to obtain singularity is to work with Partee's type shifter THE, $\lambda P'P.\exists x(\forall y(P'y \leftrightarrow x = y) \wedge Px)$. The application of THE can then be restricted to names in *argument* positions. Adopting such a theory would bring us closer to the theory of Fara (2011), for example.

[6] A traditional way to obtain the translation of *is Phosphorus* from that of *Phosphorus* is to start with (5a), to then observe that Partee's A shifter allows for an interpretation of *Phosphorus* as $\lambda P.\exists x(\Phi x \wedge Px)$, as in the *Zeus* case. To the latter we could apply the linear combinator $\lambda Q\lambda R\lambda x.Q(\lambda y.Ryx)$, which is generally useful for combining transitive verb meanings with the meanings of their direct objects. This would result in a translation $\lambda R\lambda x.\exists z(\Phi z \wedge Rzx)$, which, combined with the translation of *is*, $\lambda xy.x = y$, would lead to $\lambda x.\exists z(\Phi z \wedge z = x)$. The latter is extensionally, but not intensionally, equivalent to Φ.

(5) a. Phosphorus $\rightsquigarrow \Phi$
b. is Phosphorus $\rightsquigarrow \Phi$
c. Hesperus is Phosphorus $\rightsquigarrow \exists x(\mathsf{H}x \wedge \Phi x)$
d. Hesperus is Hesperus $\rightsquigarrow \exists x(\mathsf{H}x \wedge \mathsf{H}x)$

Note that it is consistent to assume that the semantic value of *Hesperus is Phosphorus*, $\exists x(\mathsf{H}x \wedge \Phi x)$, and that of *Hesperus is Hesperus*, given in (5d) as $\exists x(\mathsf{H}x \wedge \mathsf{H}x)$, are completely distinct propositions, even if the first is true. Also, while $\forall x(\mathsf{H}x \leftrightarrow \Phi x)$ follows from $\exists x(\mathsf{H}x \wedge \Phi x)$ and the singularity requirement for name denotations (from which we get $\forall xy((\mathsf{H}x \wedge \mathsf{H}y) \rightarrow x = y)$ and $\forall xy((\Phi x \wedge \Phi y) \rightarrow x = y)$), it does *not* follow that $\mathsf{H} = \Phi$ and H may well have properties that Φ lacks or vice versa. Co-extensionality crucially does not entail identity, having the same properties, in our theory and the theory allows for the possibility that Phosphorus has, but Hesperus fails to have, the property (λX.we do not know *a priori* that Hesperus is X), as in (1).

3.1 Worlds, Necessity, and Rigidity

Possible worlds are not needed to obtain true intensionality, and in fact cannot provide it, but they are immensely useful for modeling all kinds of *modal* phenomena. Here we construct them as certain properties of propositions (see Muskens (2007) for more details). Propositions have type $\langle\rangle$, so properties of propositions have type $\langle\langle\rangle\rangle$, and the property of being a world, a property of properties of propositions has type $\langle\langle\langle\rangle\rangle\rangle$. We will write Ω for this special property and stipulate the following.

W1 $\forall w(\Omega w \rightarrow \neg w\bot)$
W2 $\forall w(\Omega w \rightarrow (w(A \subset B) \leftrightarrow \forall \vec{x}(w(A\vec{x}) \rightarrow w(B\vec{x}))))$

W1 requires world extensions to be consistent while addition of W2 makes worlds 'distribute over logical operators'. Statements such as the following become derivable.

a. $\forall w(\Omega w \rightarrow (w(\neg\varphi) \leftrightarrow \neg(w\varphi)))$
b. $\forall w(\Omega w \rightarrow (w(\varphi \wedge \psi) \leftrightarrow ((w\varphi) \wedge (w\psi))))$
c. $\forall w(\Omega w \rightarrow (w(\forall x\varphi) \leftrightarrow \forall x(w\varphi)))$
d. $\forall w(\Omega w \rightarrow (w(\exists x\varphi) \leftrightarrow \exists x(w\varphi)))$

The first of these statements says that worlds are complete, while the last two are 'Henkin properties' that enforce, for example, that if an existential proposition is an element of the extension of a given world some proposition witnessing the existential must also be an element. In general, given W1 and W2, worlds single out sets of propositions that could be simultaneously true. The term $\lambda p.p$ (with p of type $\langle\rangle$) will be a world if we assume $\Omega(\lambda p.p)$ and it will then have the function of the *actual* world, as, in any model, $\lambda p.p$ will hold of φ iff φ is indeed true. Let us make $\Omega(\lambda p.p)$ into an official postulate and let's consider two more.

W3 $\Omega(\lambda p.p)$
W4 $\forall ww'((\Omega w \wedge \Omega w') \rightarrow (w(w'\varphi) \leftrightarrow (w'\varphi)))$
W5 $\forall w(\Omega w \rightarrow \forall w'(\Omega w' \leftrightarrow w(\Omega w')))$

W4 says that whether a proposition holds in a world is a global property, and W5 says something similar about the question whether a property of propositions is a world.

Once worlds are introduced, it becomes useful to associate *domains* with them. Some objects may exist in some worlds but not in others. We introduce a constant E of type $\langle e\rangle$ that will function as an existence predicate. Quantification over *existing* objects can then be obtained by relativizing to E. For example, the type shifter A may now be redefined as $\lambda P'P.\exists x(\mathsf{E}x \wedge P'x \wedge Px)$. This will lead to slightly revised translations, e.g. *Hesperus is Phosphorus* will now go to $\exists x(\mathsf{E}x \wedge \mathsf{H}x \wedge \Phi x)$.

Having worlds at our disposal, we can now express that φ is globally necessary by writing $\forall w(\Omega w \rightarrow w\varphi)$ and we may abbreviate this as $\Box\varphi$.[7] The following scheme says that names have singleton extensions in all worlds.

(6) $\Box\exists x\forall y(\mathsf{N}y \leftrightarrow y = x)$

This entails (3) but no longer leaves open the possibility of empty denotation that was useful for non-referring names. Since we now have an existence predicate at our disposal, that possibility is no longer needed.

We now come to rigidity. There are various ways to model variants of the notion. Here is a strong and straightforward one.

(7) $\exists x\Box\forall y(\mathsf{N}y \leftrightarrow y = x)$

The idea is that for all names there is a possible object o such that the name's extension is $\{o\}$ across all possible worlds. Clearly, in the presence of this requirement $\exists x(\mathsf{E}x \wedge \mathsf{H}x \wedge \Phi x)$ will entail $\Box\exists x(\mathsf{H}x \wedge \Phi x)$, so if Hesperus is Phosphorus, it is necessary that Hesperus is Phoshorus wherever it exists and the usual Kripkean intuitions are formalised.

On the other hand codesignating names cannot be replaced for one another in arbitrary contexts. While Hesperus and Phosphorus have the same extension in all possible worlds, they may still have distinct intensions, as intension is not determined by extension, not even by extension in all possible worlds. And since $\exists x(\mathsf{E}x \wedge \mathsf{H}x \wedge \Phi x)$, and $\exists x(\mathsf{E}x \wedge \mathsf{H}x \wedge \mathsf{H}x)$ are completely distinct propositions it is possible, for example, to bear the relation of belief to the second but not to the first.

4 Conclusion

In this paper I have shown that Kripke's intuitions with respect to the rigid designation of proper names can be formalised in a way that does not result in

[7] Muskens (2007) discusses modalities based on accessibility relations, but here we can make do without these.

a theory predicting the intersubstitutivity of codesignating names in arbitrary contexts. This means that this intersubstitutivity does not follow from the intuitions. The theory I have developed accepts rigidity of names, but rejects the Millian idea of direct reference, the idea that the meaning of a name is its bearer or at least is determined by its bearer. In the present theory a person can have many names, all with different intensions.

References

Benzmüller, C., Brown, C.E., Kohlhase, M.: Higher Order Semantics and Extensionality. Journal of Symbolic Logic 69 (2004)

Burge, T.: Reference and Proper Names. Journal of Philosophy 70, 425–439 (1973)

Graff Fara, D.: Names as Predicates (2011) (unpublished manuscript)

Fitting, M.: Types, Tableaus, and Gödels God. Kluwer Academic Publishers, Dordrecht (2002)

Frege, G.: Über Sinn und Bedeutung. In: Patzig, G. (ed.) Funktion, Begriff, Bedeutung. Fünf Logische Studien. Vanden Hoeck, Göttingen (1892)

Henkin, L.: Completeness in the Theory of Types. Journal of Symbolic Logic 15, 81–91 (1950)

Kripke, S.: Identity and Necessity. In: Munitz, M. (ed.) Identity and Individuation, pp. 135–164. New York University Press (1971)

Kripke, S.: Naming and Necessity. In: Davidson, D., Harman, G. (eds.) Semantics of Natural Language, pp. 253–355. Reidel, Dordrecht (1972)

Matushansky, O.: Why Rose is the Rose: On the use of definite articles in proper names. In: Bonami, O., Cabredo Hofherr, P. (eds.) Empirical Issues in Formal Syntax and Semantics, vol. 6, pp. 285–307. CSSP, Paris (2006)

Muskens, R.A.: Intensional Models for the Theory of Types. Journal of Symbolic Logic 72(1), 98–118 (2007)

Muskens, R.A.: Meaning and Partiality. CSLI, Stanford (1995)

Partee, B.: Noun Phrase Interpretation and Type Shifting Principles. In: Groenendijk, J., de Jongh, D., Stokhof, M. (eds.) Studies in Discourse Representation and the Theory of Generalized Quantifiers, pp. 115–143. Foris, Dordrecht (1986)

Quine, W.V.O.: From a Logical Point of View. Harper and Row, New York (1953)

Quine, W.V.O.: On What There Is. Review of Metaphysics 2, 21–28 (1948); Reprinted in Quine (1953)

Quine, W.V.O.: Word and Object. MIT Press (1960)

Russell, B.: On Denoting. Mind 14(56), 479–493 (1905)

Whitehead, A.N., Russell, B.: Principia Mathematica. Cambridge University Press (1913)

Multiple Foci in Japanese Clefts and the Growth of Semantic Representation

Tohru Seraku

St. Catherine's College, Manor Road, Oxford, UK. OX1 3UJ
tohru.seraku@stcatz.ox.ac.uk

Abstract. There are two types of Japanese clefts, depending on the presence of a case particle attached to a focus; multiple foci are possible only in clefts with particles. This paper proposes that this distinct behavior with respect to multiple foci emerges as an outcome of the incremental growth of semantic structure in Dynamic Syntax. The analysis enables a uniform treatment of the two types of clefts, and handles a set of new data concerning partially case-specified foci, which poses a puzzle for previous works in Principles and Parameters Theory.

Keywords: incrementality, structural underspecification, Dynamic Syntax.

1 Introduction

Japanese clefts are divided into two types, depending on whether a focus item has a case particle or not [5], as shown in (1), where the focus is *ringo* and the case particle is *o*. In this paper, clefts with a case particle (e.g. *ringo-o*) are called "clefts$_{+P}$", while clefts without a case particle (e.g. *ringo*) are called "clefts$_{-P}$".

(1)	[*Tom-ga*	*kat-ta*	*no*]-*wa*	*ringo*(*-o*)	*da.*
	[Tom-NOM	buy-PAST	NO]-TOP	apple(-ACC)	COP
	'It is an apple that Tom bought.'				

As observed in [12, p.239], multiple foci are possible only in clefts$_{+P}$.

(2)	[e$_i$ e$_j$	*purezento-o*	*age-ta*	*no*]-*wa*
	[	present-ACC	give-PAST	NOM]-TOP
	Tom$_i$*(*-ga*)	*Mary*$_j$*(*-ni*)	*da.*	
	Tom-NOM	Mary-DAT	COP	
	Lit. 'It is Tom$_i$ to Mary$_j$ that e$_i$ gave a present e$_j$.'			

Within Principles and Parameters Theory, the distinct behavior of clefts$_{-P}$/clefts$_{+P}$ with respect to multiple foci has been a challenge to a uniform analysis [3, 4, 11, 12, 13, 15]; the tendency is to assume that the particle *no* in clefts$_{+P}$ is a complementizer, while *no* in clefts$_{-P}$ is a pronominal. Kizu [10] is exceptional in presenting a unitary account, but her account is not applicable to multiple foci, as admitted in [10, p.54].

M. Aloni et al. (Eds.): Amsterdam Colloquium 2011, LNCS 7218, pp. 450–459, 2012.
© Springer-Verlag Berlin Heidelberg 2012

There is also an empirical puzzle that has been overlooked in the literature. Though case particles have been presumed to be obligatory in multiple foci, a second focus, but not a first focus, may be case-less, as shown in the contrast between (3) and (4).

(3) [e_i e_j *purezento-o* *age-ta* *no*]-*wa*
[present-ACC give-PAST NO]-TOP
Tom$_i$-*ga* *Mary*$_j$ *da.*
Tom-NOM Mary COP
Lit. 'It is Tom$_i$ to Mary$_j$ that e_i gave a present e_j.'

(4) *[e_i e_j *purezento-o* *age-ta* *no*]-*wa*
[present-ACC give-PAST NO]-TOP
Tom$_i$ *Mary*$_j$-*ni* *da.*
Tom Mary-DAT COP

These data have not been noted elsewhere aside from [4] which makes a passing remark but without analysis. Within Principles and Parameters Theory, it is generally assumed that the derivation of clefts$_{+P}$ involves movement, and that of clefts$_{-P}$ does not [4, 5]. One may then argue that in (3), the gap for *Tom-ga* is a trace of movement while the gap for *Mary* is a small pro. Even if (3) can be accounted for in this fashion, however, the ungrammaticality of (4) remains a mystery.

This paper develops a uniform analysis of clefts in connection with multiple foci, from the perspective of the incremental growth of semantic structure, as modeled in Dynamic Syntax. After Section 2 introduces the framework, Section 3 articulates a unified analysis of clefts, and Section 4 examines the issues in multiple foci. Finally, the analysis is extended to long-distance clefts in Section 5.

2 Framework: Dynamic Syntax

Dynamic Syntax is a grammar formalism that models "Knowledge of Language", which is defined as a set of constraints on the building-up of semantic structure [1, 9]. With such constraints, a parser processes a string of words in order, and builds up semantic structure gradually. In this view, a string is directly mapped onto semantic structure; thus, there is no syntactic structure.

The growth of semantic structure is a combination of (a) general action, (b) lexical action, and (c) pragmatic action, all driven by a range of requirements on successful completion. For illustration, consider how the string (5) is progressively mapped onto its semantic structure word-by-word.

(5) *Mary-ga* *nai-ta.*
Mary-NOM cry-PAST
'Mary cried.'

The first item to be parsed is *Mary*. Japanese allows the permutation of arguments, and thus a parser cannot see at this point whether *Mary* will turn out to be a subject or an object, etc. So, the node for *Mary* is unfixed at this stage. Formally, the general action of LOCAL *ADJUNCTION induces structural underspecification, or an unfixed node, which must be fixed in a local proposition. The lexical action of *Mary* then decorates this unfixed node. These two actions give rise to the semantic structure (6). In (6), the structural underspecification is notated with the dashed line.

(6) Parsing *Mary*

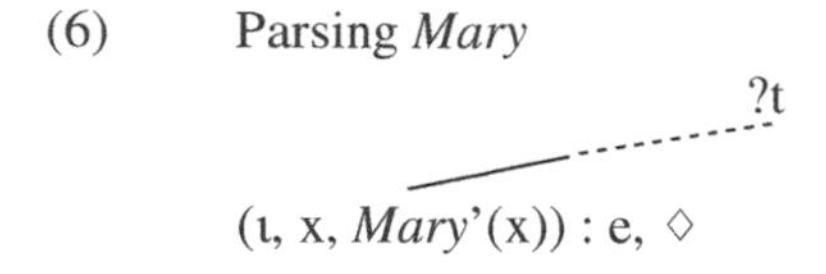

In (6), the top node is decorated with ?t, a requirement that this node be decorated with a type-t content (i.e. proposition). The unfixed daughter node has three pieces of information. First, (ι, x, *Mary*'(x)) is a content of *Mary*. (ι is an iota operator that binds the variable x, which is restricted by *Mary*'.) Second, e states that the content is of semantic type-e. Third, the pointer ◇ indicates a node under current development.

The underspecification in (6) is resolved by the lexical action of the nominative case particle *ga*, which fixes the node for *Mary* as a subject node, as in (7), where the dashed line becomes solid.

(7) Parsing *Mary-ga*

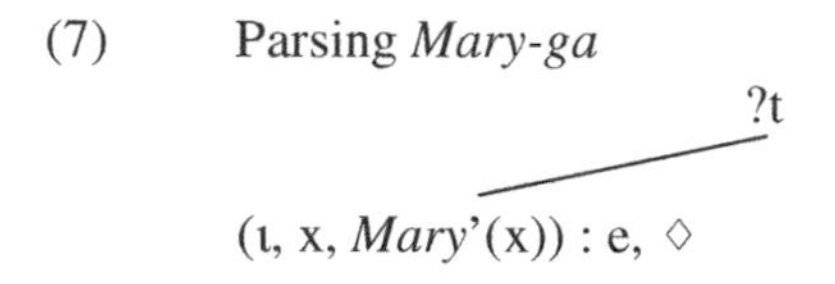

The next item is the predicate *nai*. Japanese being a pro-drop language, a predicate constructs an open proposition with argument slots. Thus, the lexical actions of *nai* update (7) into (8), where a subject slot has been already identified.

(8) Parsing *Mary-ga nai*

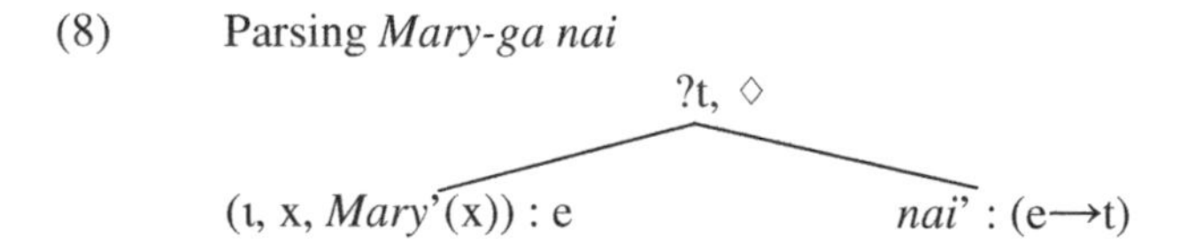

Finally, the general action of functional application composes the content at each daughter node, and the parse of *ta* puts tense information PAST at the root node.[1] The final state is given in (9); this structure is well-formed in the sense that no outstanding requirements ? remain.

(9) Parsing *Mary-ga nai-ta*

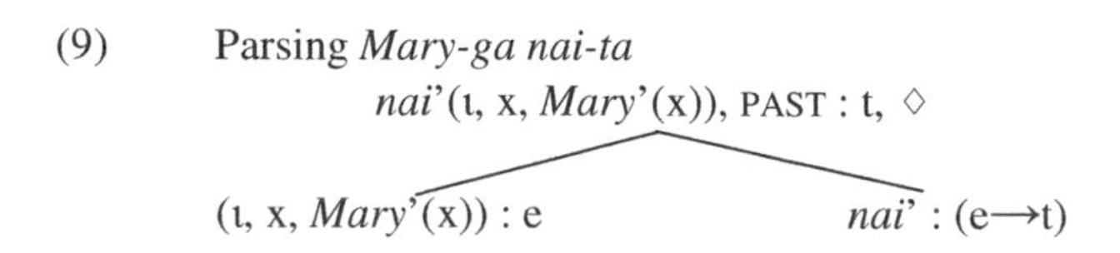

[1] In a fully articulated structure, tense is represented as an event term in Epsilon Calculus [2].

This section has introduced the framework of Dynamic Syntax: a parser constructs semantic structure incrementally on the basis of left-to-right word-by-word parsing.

3 Uniform Analysis of Clefts

3.1 Clefts$_{+P}$

This sub-section provides a tree transition for simple cases of clefts$_{+P}$, such as (10).[2]

(10) [*Nai-ta* *no*]-*wa* *gakusee-ga* *san-nin* *da*.
 [cry-PAST NO]-TOP student-NOM 3-CL COP
 'It is 3 students that cried.'

First, the parse of *Nai-ta* yields a proposition; the gap is notated as (ε, x, P(x)), where ε is an existential operator, and P is an abstract restrictor [8, p.65]. Next, following [1, p.285], Seraku [14] regards *no* as a nominalizer that copies a type-e term within a proposition, and pastes it at a newly created node that is LINKed to the proposition. In (11), the LINK relation is notated by the curved arrow, and what is copied/pasted is the type-e term (ε, x, P(x)).

(11) Parsing *Nai-ta no*

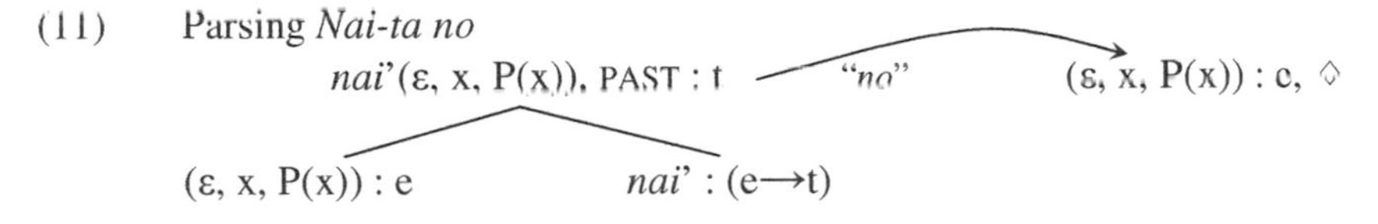

What comes next is the topic marker *wa*. The standard analysis of *wa* [1, p.268] is to put at a new type-t-requiring node the requirement ?<D>(ε, x, P(x)), which requires that a node below this node should be decorated with (ε, x, P(x)). (This requirement will be met in (13) below, since the term (ε, x, *gakusee*'(x)) is stronger than the term (ε, x, P(x)).)

(12) Parsing *Nai-ta no wa*

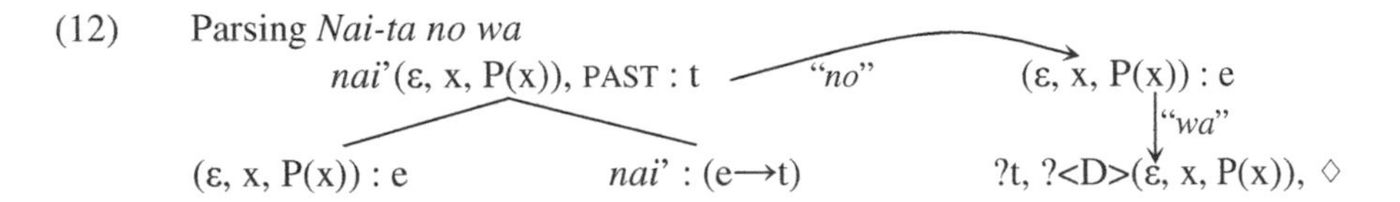

[2] For many speakers, if a focus has a nominative case particle *ga*, the string is not completely fine. As noted in [11], the string becomes acceptable if a numeral quantifier, such as *san-nin* in (10), is used. Also, a *ga*-marked focus is acceptable in multiple foci, as in (2) in Section 1. This tendency is stronger in Korean, where clefts with a nominative case-marked focus are completely unacceptable [6]. This paper sets aside the speaker variation concerning the acceptability judgments of case-marked foci in Japanese clefts.

In order to parse the next item *gakusee*, structural underspecification is required. As mentioned in (6) in Section 2, the general action of LOCAL *ADJUNCTION induces an unfixed node (marked by the dashed line), which allows *gakusee* to be parsed.

(13) Parsing *Nai-ta no wa gakusee*

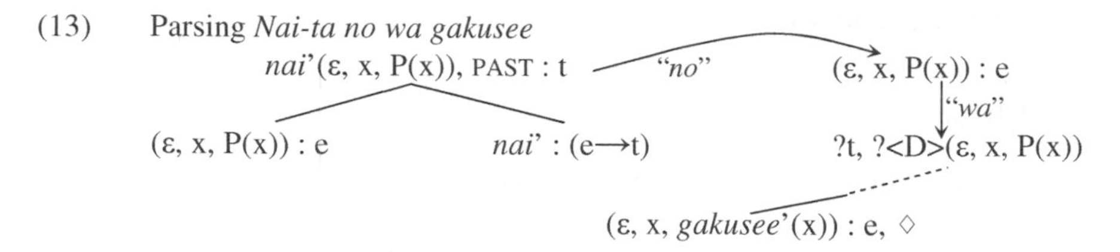

So far, the tree transition is neutral between clefts$_{-P}$ and clefts$_{+P}$.

At this stage, the transition starts to differ between clefts$_{-P}$ and clefts$_{+P}$, since only the latter has a case particle. In the cleft$_{+P}$ (10), the case particle *ga* is parsed, which fixes the unfixed node as a subject node, as indicated by the solid line in (14). (In (14), the quantifier *san-nin* has been parsed; note the strengthened operator ε_3.)

(14) Parsing *Nai-ta no wa gakusee-ga san-nin*

nai'(ε, x, P(x)), PAST : t —"*no*"→ (ε, x, P(x)) : e

(ε, x, P(x)) : e *nai'* : (e→t)

↓"*wa*"

?t, ?<D>(ε, x, P(x))

(ε_3, x, *gakusee'*(x)) : e, ◇

Finally, the copula *da* is parsed. Seraku [14] argues that *da* provides a type-t meta-variable, which licenses the re-use of previous representation or actions. In the case of (15), the representation (i.e. the structure built up by the parse of *Nai-ta*) is re-used, where a subject node has been already present due to the parse of *gakusee-ga*.

(15) Parsing *Nai-ta no wa gakusee-ga san-nin da*

nai'(ε, x, P(x)), PAST : t —"*no*"→ (ε, x, P(x)) : e

(ε, x, P(x)) : e *nai'* : (e→t)

↓"*wa*"

nai'(ε_3, x, *gakusee'*(x)), PAST : t, ◇

(ε_3, x, *gakusee'*(x)) : e *nai'* : (e→t)

This semantic representation is the final state of the tree transition for the cleft$_{+P}$ (10).

3.2 Clefts$_{-P}$

The analysis of clefts$_{+P}$ is straightforwardly applicable to clefts$_{-P}$, like (16).

(16) [*Nai-ta* *no*]-*wa* *gakusee* *san-nin* *da.*
[cry-PAST NO]-TOP student 3-CL COP
'It is 3 students that cried.'

Given the transition above, the parse of *Nai-ta no wa gakusee san-nin* yields (17).

(17)

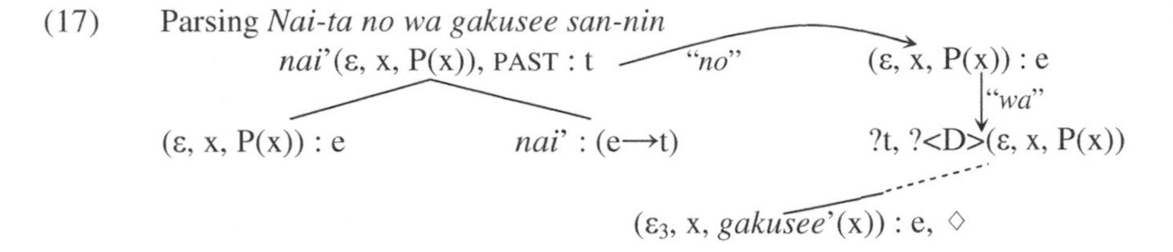

This time, the cleft$_{-P}$ (16) has no particle, and the transition proceeds without fixing the unfixed node. The parse of *da* then licenses the re-use of the previous structure.

(18)

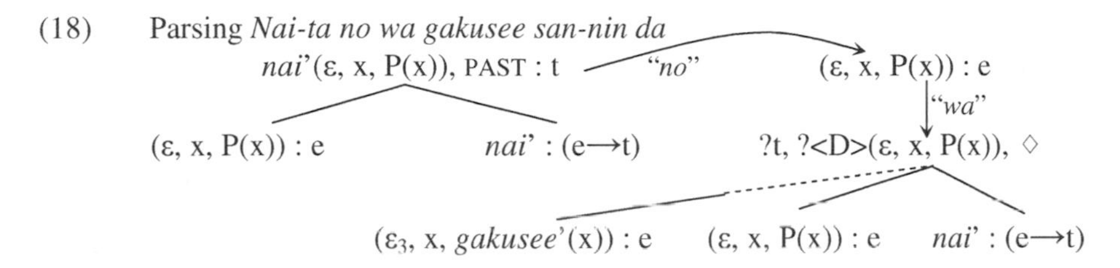

The subject node then unifies with the unfixed node by the general action of MERGE [1], engendering the final state (19).[3]

(19) Parsing *Nai-ta no wa gakusee san-nin da*

nai'(ε, x, P(x)), PAST : t "*no*" (ε, x, P(x)) : e

"*wa*"

(ε, x, P(x)) : e *nai'* : (e→t) *nai'*(ε_3, x, *gakusee'*(x)), PAST : t, ◇

(ε_3, x, *gakusee'*(x)) : e *nai'* : (e→t)

This structure of the cleft$_{-P}$ (16) is identical to that of the cleft$_{+P}$ (10). This ensures that the two types of clefts are truth-conditionally equivalent.

Based on [1, 14], this section has sketched the unitary account of the two types of clefts in Japanese. The next section examines issues posed by multiple foci.

4 Multiple Foci

In Section 3, when a focus item was parsed, an unfixed node was introduced as a result of the general action of LOCAL *ADJUNCTION. This general action may be conducted more than once from the same node only if an unfixed node has been fixed before another unfixed node is introduced [1, p.235]: this is because multiple nodes that are unfixed from the same node are indistinguishable, leading to inconsistency.

---

[3] MERGE outputs (ε_3, x, *gakusee'*(x)&P(x)), but it is simply notated as (ε_3, x, *gakusee'*(x)). The omission of the abstract restrictor P makes no difference if there is a concrete restrictor.

This single-unfixed-node constraint accounts for the difference in the availability of multiple foci between the two types of clefts. First, in clefts$_{+P}$, multiple foci are licensed, since an unfixed node for each focus is fixed by each case particle. Second, in clefts$_{-P}$, the only means of fixing an unfixed node is unification (i.e. MERGE). In order for unification to occur, however, a proposition must have been built up, and, as shown in Section 3, such a proposition is created only after the parse of *da*. Thus, more than one item cannot be clefted in clefts$_{-P}$.

Within Dynamic Syntax, there are two other types of structural underspecification: *ADJUNCTION and GENERALIZED ADJUNCTION [1]. A question, then, is: can these be utilized more than once from the same node? The answer is no, since an unfixed node induced by these actions is not fixed by the parse of a particle [1]. Thus, if these types of underspecification are induced, multiple foci are prohibited in clefts$_{+P}$ and clefts$_{-P}$. One may object that this wrongly predicts that both clefts$_{+P}$ and clefts$_{-P}$ do not allow multiple foci. Nevertheless, the availability of multiple foci in clefts$_{+P}$ is adequately modeled, as there is a successful tree transition based on LOCAL *ADJUNCTION.

This analysis accounts for a further range of data. First, it predicts that the order of foci does not matter as long as each focus has an appropriate case particle. Compare (20) with the example (2) in Section 1, where the order of foci is reversed.

(20) [e_i e_j *purezento-o* *age-ta* *no*]-*wa*
     [present-ACC give-PAST NO]-TOP
     Mary$_j$*(*-ni*) *Tom*$_i$*(*-ga*) *da*.
     Mary-DAT Tom-NOM COP
     Lit. 'It is Tom$_i$ to Mary$_j$ that e_i gave a present e_j.'

Second, the analysis explains the partially case-specified foci in (3) in Section 1, repeated here as (21). In this case, the unfixed node for the second focus may be fixed by unification, and thus does not need to be fixed by the parse of a case particle.

(21) [e_i e_j *purezento-o* *age-ta* *no*]-*wa*
     [present-ACC give-PAST NO]-TOP
     Tom$_i$*-ga* *Mary*$_j$ *da*.
     Tom-NOM Mary COP
     Lit. 'It is Tom$_i$ to Mary$_j$ that e_i gave a present e_j.'

For some speakers, (21) may not be completely fine, but it is much better than (22).

(22) *[e_i e_j *purezento-o* *age-ta* *no*]-*wa*
     [present-ACC give-PAST NO]-TOP
     Tom$_i$ *Mary*$_j$*-ni* *da*.
     Tom Mary-DAT COP

In (22), *Tom* has no case particle, and the underspecification for *Tom* is not resolved. Thus, another structural underspecification cannot be induced for *Mary*.

Finally, it is further predicted that a cleft is acceptable where there are three foci and only the final focus lacks a case particle. This is because unification requires that a proposition be present, and, given that a proposition is constructed by the parse of *da*, unification is applicable only to a final focus.

(23) [e$_i$ e$_j$ e$_k$ *age-ta* *no*]-*wa*
[give-PAST NO]-TOP
Tom$_i$-*ga* *Mary*$_j$-*ni* *purezento*$_k$ *da.*
Tom-NOM Mary-DAT present COP
Lit. 'It is Tom$_i$ to Mary$_j$ a present$_k$ that e$_i$ gave e$_k$ e$_j$.'

Again, (23) may not be completely acceptable, but it is much better than the other six possible examples: examples where 1) only *Tom* is case-less, 2) only *Mary* is case-less, 3) only *Tom* and *Mary* are case-less, 4) only *Tom* and *purezento* are case-less, 5) only *Mary* and *purezento* are case-less, 6) all foci are case-less. These six possible examples are all unacceptable, since the tree transitions for such examples necessarily lead to the multiplication of unfixed nodes from the same node.

5 Long-Distance Clefts

The analysis developed so far can be extended to long-distance clefts, like (24). Based on the treatment of long-distance scrambling in [7], the tree transition for (24) is as in (25); after *wa* is parsed, a type-t-requiring node is introduced by *ADJUNCTION, and the foci are parsed by LOCAL *ADJUNCTION below this node.[4]

(24) [*Mary-ga* [e$_i$ e$_j$ *purezento-o* *age-ta* *to*]
[Mary-NOM [present-ACC give-PAST COMP]
it-ta *no*]-*wa* *Ken*$_i$-*ga* *Tom*$_j$-*ni* *da.*
say-PAST NO]-TOP Ken-NOM Tom-DAT COP
Lit. 'It is Ken$_i$ to Tom$_j$ that Mary said that e$_i$ gave a present e$_j$.'

(25) Parsing [*Mary-ga* [*purezento-o age-ta to*] *it-ta no*]-*wa Ken-ga Tom-ni*

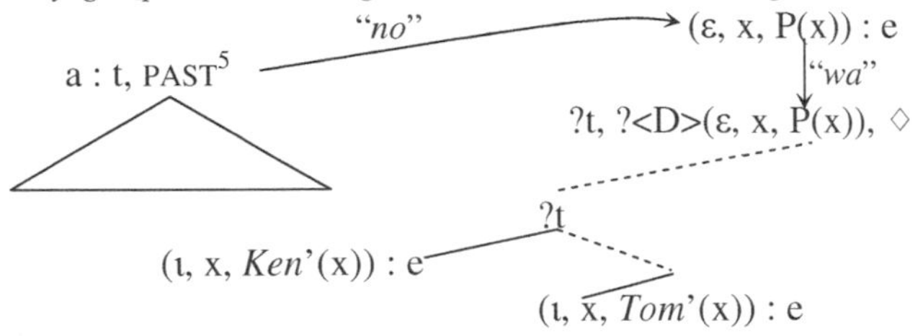

---

[4] The node for *Tom* in (25) is not fixed at this point, since the presence of an unfixed node enables the pointer ◊ to move back to the top node [7, p.180].

[5] A stands for *it'*(*age'*(ε, x, *purezento'*(x))(ε, x, P(x))(ε, x, Q(x)))(ι, x, *Mary'*(x)). This is the proposition expressed by the pre-*no* string: *Mary-ga purezento-o age-ta to it-ta*.

The rest of the tree transition is as usual; *da* licenses the re-use of the representation (i.e. the structure built up by the parse of *Mary-ga purezento-o age-ta to it-ta*), and the nodes for *Ken* and *Tom* are identified as the arguments of the predicate *age*. After functional application, the complete semantic structure emerges.

Interestingly, the string (24) is ambiguous; the following reading is also possible.

(26) [e_i e_j [*Mary-ga* *purezento-o* *age-ta* *to*]
[[Mary-NOM present-ACC give-PAST COMP]
it-ta *no*]-*wa* *Ken*$_i$-*ga* *Tom*$_j$-*ni* *da*.
say-PAST NO]-TOP Ken-NOM Tom-DAT COP
Lit. 'It is Ken$_i$ to Tom$_j$ that e_i said e_j that Mary gave a present to someone.'

In this case, the foci are parsed without the use of *ADJUNCTION; that is, the foci are parsed by LOCAL *ADJUNCTION directly below the node decorated with ?t, ?<D>(ε, x, P(x)). The parse of *da* then licenses the re-use of the propositional structure, into which the content of each focus is reflected.

In the string in question, what is not possible is the reading (27).

(27) *[e_i [*Mary-ga* e_j *purezento-o* *age-ta* *to*]
[[Mary-NOM present-ACC give-PAST COMP]
it-ta *no*]-*wa* *Ken*$_i$-*ga* *Tom*$_j$-*ni* *da*.
say-PAST NO]-TOP Ken-NOM Tom-DAT COP
Lit. 'It is Ken$_i$ to Tom$_j$ that e_i said that Mary gave a present e_j.'

This indicates that foci must belong to the same local clause. This restriction has been already observed [11], and called the "clause-mate" condition. There have been other attempts to account for this restriction, but they are not fully persuasive; Takano [15], one of the most advanced works on this matter, ends up positing the extra mechanism to create "surprising constituents".

In my analysis, however, the clause-mate condition emerges as a by-product of the gradual updating of semantic representation. First, in (24), the foci are parsed below the type-t-requiring node that is created by *ADJUNCTION. Second, in (26), the foci are parsed below the type-t-requiring node that is decorated by the parse of *wa*. Thus, the content of each focus is reflected into the same local proposition, and the reading in (27) is not possible. The upshot is that what is called the clause-mate condition is a result of the progressive development of semantic representation, which renders the condition merely an epiphenomenon.

As a further confirmation, if *kinoo* (= 'yesterday') is inserted between the foci *Ken* and *Tom* in (24)/(26), as in *Ken-ga kinoo Tom-ni*, the expected pattern occurs. In (24), *kinoo* can modify only the most embedded clause; in (26), *kinoo* can modify only the clause whose main predicate is *it* (= 'say'). This is because, in each case, *kinoo* is parsed below the same type-t-requiring node as the foci are parsed.

Thus, my analysis captures all possible and only possible readings of long-distance clefts from the basic architecture of the framework.

6 Conclusion

This paper has argued that Japanese clefts are uniformly characterized, and the issues raised by multiple foci are directly addressed by the incremental growth of semantic structure that involves "structural underspecification and subsequent resolution". The analysis succeeds in four ways. First, the single entry of *no* as a nominalizer that copies a type-e term leads to the uniform treatment of clefts_{-P}/clefts_{+P}. Second, the output of the tree growth captures the parallelism of the two types of clefts: truth-conditional equivalence. Third, the process of tree growth captures the difference: availability of multiple foci. Fourth, restrictions on multiple foci (i.e. partially case-specified foci and the clause-mate condition) are explained as a by-product of the tree growth. These four achievements show clear advantages over previous studies within Principles and Parameters Theory, where, as stated in Section 1, multiple foci have been an obstacle to a unified treatment, and the data such as partially case-specified foci and the clause-mate condition have posed empirical puzzles.

Acknowledgments. I would like to thank Ash Asudeh, David Cram, Mary Dalrymple, Ruth Kempson, Jieun Kiaer, and the audience of my talk at the 18th Amsterdam Colloquium for their valuable suggestions. All remaining inadequacies are my own.

References

1. Cann, R., Kempson, R., Marten, L.: The Dynamics of Language. Elsevier, Oxford (2005)
2. Gregoromichelaki, E.: Conditionals in Dynamic Syntax. In: Kempson, R., et al. (eds.) The Dynamics of Lexical Interfaces, pp. 237–278. CSLI, Stanford (2011)
3. Hasegawa, N.: A Copula-based Analysis of Japanese Clefts. In: Inoue, K. (ed.) Grant-in-Aid for COE Research Report (1), pp. 15–38. KUIS, Chiba (1997)
4. Hiraiwa, K., Ishihara, S.: Syntactic Metamorphosis. Syntax (to appear)
5. Hoji, H.: Theories of Anaphora and Aspects of Japanese Syntax. Ms., USC (1990)
6. Kang, B.: Some Peculiarities of Korean *Kes* Cleft Constructions. Studia Linguistica 60, 251–281 (2006)
7. Kempson, R., Kiaer, J.: Multiple Long-distance Scrambling. Journal of Linguistics 46, 127–192 (2010)
8. Kempson, R., Kurosawa, A.: At the Syntax-Pragmatics Interface. In: Hoshi, H. (ed.) The Dynamics of Language Faculty, Kuroshio, Tokyo (2009)
9. Kempson, R., Meyer-Viol, W., Gabbay, D.: Dynamic Syntax. Blackwell, Oxford (2001)
10. Kizu, M.: Cleft Constructions in Japanese Syntax. Palgrave, New York (2005)
11. Koizumi, M.: String Vacuous Overt Verb Raising. Journal of East Asian Linguistics 9, 227–285 (2000)
12. Kuroda, S.-Y.: Tokoro-Clause (Tokoro-Setsu). In: Kuroda, S.-Y.: Generative Grammar from the Perspective of Japanese. (Nihongo kara Mita Seesee Bunpoo), Iwanami, Tokyo (2005)
13. Kuwabara, K.: Multiple Wh-Phrases in Elliptical Clauses and Some Aspects of Clefts with Multiple Foci. MITWPL 29, 97–116 (1996)
14. Seraku, T.: On the Polyfunctionality of Copula Sentences in Japanese. In: Proceedings of the Sixth Cambridge Postgraduate Conference in Language Research (2011)
15. Takano, Y.: Surprising Constituents. Journal of East Asian Linguistics 11, 243–301 (2002)

Focus, Uniqueness and Soft Presupposition Triggers*

Andreas Walker

University of Konstanz

Abstract. Von Heusinger (2007) shows that the definite article's uniqueness presupposition causes problems in a standard account of focus alternatives. He solves this problem by proposing a new lexical entry for the definite article, a solution adopted by Riester and Kamp (2010). This paper shows that the observed behavior is not limited to the definite article, making such a solution undesirable. Our argument is based on two observations: (1) The definite article is a soft presupposition trigger with respect to uniqueness and (2) soft presupposition triggers have a special behavior in focus alternatives.

1 The Problem

Von Heusinger (2007) observes that in a standard account of focus alternatives, the sentence in (1) receives the following analysis: the uttered sentence (excluding *only*) expresses the proposition in (2), and focus gives rise to the set of alternatives in (3).

(1) John only talked to the GERman$_F$professor.
(2) $\lambda w : \exists! x[\text{prof}(x, w) \wedge \text{german}(x, w)].\exists y[\text{P}(y, w) \wedge \text{G}(y, w) \wedge \text{T}(j, y, w)]$
(3) $\{[\lambda w : \exists! x[\text{P}(x, w) \wedge \text{G}(x, w)].\exists y[\text{P}(y, w) \wedge \text{G}(y, w) \wedge \text{T}(j, y, w)]]\}$

That is, the uttered sentence roughly conveys (blending together presuppositions and assertions for the sake of readability) that there is a unique German professor such that John talked to him. The alternative set only contains the singleton of this proposition, and other alternatives, i.e. the propositions that there are unique professors of other nationalities (French, Dutch, ...) such that John talked to them, are not included as they suffer presupposition failure. While this analysis might yield the correct reading in some contexts, von Heusinger (2007) notes that this analysis becomes problematic in the scenario described in (4).

(4) *Scenario:* At a party there are one German professor, two Dutch professors and two French professors.

* I thank Maribel Romero, two anonymous reviewers, the participants of the Amsterdam Colloquium 2011 and the Research Colloquium in Konstanz for comments.

M. Aloni et al. (Eds.): Amsterdam Colloquium 2011, LNCS 7218, pp. 460–469, 2012.
© Springer-Verlag Berlin Heidelberg 2012

The prediction would be that (1) is always true if John talked to the German professor, even if he talked to one of the other professors. This results from the fact that the set of alternatives only includes the German professor, since none of the other professors is unique. What we observe instead is that

1. the sentence is felicitous and true in (4) if John talked to the German professor but none of the other professors.
2. the sentence is felicitous and false in (4) if John talked to any other professor, whether they are unique (i.e. Dutch) or non-unique (i.e. French).

Von Heusinger (2007) suggests solving this problem by assuming a different lexical entry for the definite article in the alternative semantics, namely $[\![\text{the}]\!]^f = \cup$ (the generalized union). A computation in the alternative semantics of Rooth (1992) is given in (5). As the reader can see, the generalized union yields the set of all professors, which is passed up the tree in the alternative semantics computation without any uniqueness presupposition. The set of resulting propositional alternatives is then contextually restricted by Rooth's (1992) squiggle operator, yielding propositional alternatives concerning the contextually relevant professors, e.g. the ones at the party, regardless of their nationality or uniqueness. This way, the set of alternatives consists of all professors, rather than of all unique professors of a certain nationality. Von Heusinger's (2007) solution is adopted by Riester and Kamp (2010).

(5) $[\![\text{GERman}_F\text{professor}]\!]^f = \{\lambda x.[\text{Q}(x) \wedge \text{PROF}(x)] | Q \in D_{\langle e,t \rangle}\}$
$[\![\text{the}]\!]^f = [\lambda P. \cup P]_{\langle\langle\langle e,t\rangle,t\rangle,\langle e,t\rangle\rangle}$
$[\![\text{the GERman}_F\text{professor}]\!]^f = \cup\{\lambda x.[\text{Q}(x) \wedge \text{PROF}(x)] | Q \in D_{\langle e,t \rangle}\}$
$= \{x | \exists Q[Q(x) \wedge PROF(x)]\}$
$= \{x | PROF(x)\}$
$[\![\text{John talked to the GERman}_F\text{professor}]\!]^f = \{\text{TALK(j,x)} | \text{PROF(x)}\}$

While this approach arrives at the correct reading, it misses an important generalization. The observed behavior is in fact not unique to the definite article, but can be shown to arise for the larger class of Soft Presupposition Triggers. In Sect. 2 we will discuss some of the data that has been presented on these Soft Presupposition Triggers in the literature, in order to provide the background for the observations presented in this paper. Two observations will be presented in Sect. 3: We will show that the definite article does in fact belong into the class of Soft Presupposition Triggers (Sect. 3.1), and that while Hard Presupposition Triggers must project their presuppositions in focus alternative sets, Soft Presupposition Triggers do no need to (Sect. 3.2). Further, we will point out that the non-projective behavior of Soft Presupposition Triggers in the alternative set does not amount to local accommodation, but to the eliminiation of the presupposed proposition as a whole. This is different from the behavior of Soft Presupposition Triggers in the environments investigated in the previous literature, where non-projection amounted to local accommodation (Sect. 3.3). Once the special behavior of Soft Presupposition Triggers is established, von Heusinger's (2007) example will be shown to be just one case of the general pattern (Sect. 3.4). Section 4 concludes the paper.

2 Background on Soft Presupposition Triggers

The literature on presupposition triggers distinguishes between Soft and Hard Presupposition Triggers. Soft Presupposition Triggers (SPT) and Hard Presupposition Triggers (HPT) show different behavior in a number of contexts, a phenomenon that has been described by Simons (2001) and Abusch (2002) amongst others. In these contexts, the presuppositions of SPTs do not need to project, while projection is obligatory for HPTs. We are going to examine two such environments, disjunction and ignorance of the speaker with respect to the presupposition.

The first context, involving disjunction, is given in the following examples by Abusch (2002). Consider first a simple example of presupposition with a Soft Presupposition Trigger such as (6).

(6) John will continue missing meetings.

The presupposition that John has missed meetings before, triggered by *continue*, here projects to the whole sentence. However, in (7) the presuppositions of *continue* do not project to the whole sentence, but rather stay within the disjuncts, yielding a reading paraphrased in (8).

(7) After the first meeting, John will either continue missing meetings
or continue attending meetings.
(8) "After the first meeting, John will either *have missed the first meeting and* continue missing meetings or *have attended the first meeting and* continue attending meetings."

The same projection behavior as in (6) can be shown for the HPT *too* in the simple case, where the presupposition that John missed the first meeting projects to the whole sentence.

(9) John missed the second meeting too.

However, HPTs do not show the non-projection behavior observed for SPTs under disjunction. In (10), the conflicting presuppositions of *too* must project, rendering the sentence pragmatically odd.

(10) # After the first meeting, John will either miss the second meeting too or attend the second meeting too.

The second context, involving the speaker's explicite ignorance with respect to the presupposition, is given by Simons (2001). The presupposition of *stop* projects to the whole sentence as normal in (11), so that the sentence presupposes that John smoked before. The presupposition does not project in (12) where it instead becomes part of the question, as paraphrased in (13)

(11) Did John stop smoking?
(12) I see that you keep chewing on your pencil. Have you recently stopped smoking?

(13) "Have you *been a smoker and* recently stopped smoking?

However, when a HPT is used, the same contrast as under disjunction arises. That is, the presupposition of *again* projects in the simple example (14) as well as in (15) where the speaker's ignorance is made explicit. Since the presupposition that Jane rented Manhattan before is incompatible with this context, (15) is rendered pragmatically odd.

(14) Jane is renting Manhattan again.

(15) # I have no idea whether Jane ever rented "Manhattan", but perhaps she's renting it again.

Based on these contexts, several presupposition triggers have been classified as either hard or soft:

1. Soft Presupposition Triggers: e.g. factive verbs, change of state verbs
2. Hard Presupposition Triggers: e.g. *again*, *even*, *too*, *also*, negative polarity *either*, *it*-cleft

Missing from this list is the definite article. Its status is disputed in the literature, with no clear classification arising. Abusch (2002) describes it as "unclear". Abbott (2006) argues for treating the definite article as a HPT based on examples like (16). Here, the presupposition that the book has an owner, triggered by *the* must project. Since the speaker's ignorance with respect to the presupposition is explicitely stated, the sentence is pragmatically odd.

(16) # Possibly no one owns this book, but if I find the owner I will return it.

However, these discussions are only concerned with the definite article's existential presupposition. The problematic presupposition in von Heusinger's (2007) puzzle is the uniqueness presupposition. The question arises whether the definite article is a soft or a hard presupposition trigger with respect to uniqueness.

3 Observations

The following sections present data on two observations: (a) The definite article is a SPT for uniqueness, and (b) SPTs show the same behavior as the definite article with respect to focus alternatives, but HPTs do not. Taken together, these two observations allow us to locate the original problem within a larger class of problems that need to be addressed together.

3.1 Observation A: The Definite Article Is a Soft Presupposition Trigger for Uniqueness

In the contexts described by Simons (2001) and Abusch (2002), the definite article behaves like a SPT, i.e. its uniqueness presupposition does not need to project in these contexts. Similar to the behavior of *continue* in (7), the definite article's presupposition does not project under disjunction in (18), given the scenario described in (17). Instead it is locally accommodated within the disjuncts.

(17) *Scenario:* In a historical setting with a pope and a counterpope, a council is being held in order to settle the conflict by agreeing on one unique pope, but it is yet unclear whether this will succeed
(18) After the council, either the pope will unite Rome, or the popes will tear it apart.

This gives rise to the reading paraphrased in (19), i.e. the uniqueness is not a presupposition of the sentence but is only locally assumed in the disjunction.

(19) "After the council, either *there will be a unique pope and this* pope will unite Rome or *there will be more than one pope and these* popes will tear it apart."

In a scenario like (20) where we can assume the speaker's ignorance with respect to the presupposition, it is not projected either. Instead, it is locally accommodated, giving rise to the reading paraphrased in (22).

(20) *Scenario:* Pina needs a linear equation A to have exactly one solution. I do not know how many solutions A has. When I meet Pina she seems happy.
(21) Did you find the solution for A?
(22) "*Is there a unique solution for A and* did you find it?"

Since the definite article behaves as a Soft Presupposition Trigger in these contexts, we can assume that it is in fact a SPT with respect to uniqueness.[1]

3.2 Observation B: Soft Presupposition Triggers and Focus Alternatives

We will now explore the behavior of Soft and Hard Presupposition Triggers with respect to focus alternatives. As expected, both SPTs and HPTs have the potential to project in focus alternatives. (24) demonstrates this for the SPT *aware*.

(23) *Scenario:* Three soccer players, A(lexandra), B(irgit) and C(elia), are pregnant. John believes that Birgit is pregnant.
(24) It's a good thing John is only aware that BIRgit$_F$ is pregnant.

Here, the presuppositions in the focus alternatives, i.e. that Alexandra is pregnant and that Celia is pregnant, do project as expected. A rough paraphrase (conflating assertions and presuppositions for the sake of readability) is given in (25).

[1] Note that not all SPTs are prone to non-projection to the same extent in all environments. For example, it is harder to construct examples of non-projection for the definite article's presupposition under negation than under questions or disjunction.

(25) "B is pregnant and John believes that B is pregnant, A is pregnant and John doesn't believe that A is pregnant, and C is pregnant and John doesn't believe that C is pregnant."

The same can be shown for the HPT *also* in (27). Note that the relevant reading is the one in which *only* goes together with *Birgit*, and *also* with *conjunctivitis*.

(26) *Scenario:* John believes that all three players have a cold. On top, he thinks that Birgit has conjunctivitis.
(27) John only thinks that BIRgit also has conjunctivitis.

The presupposition - i.e. that John thinks that the players have a condition other than conjunctivitis - projects, and we obtain the rough paraphrase given in (28).

(28) "John thinks B has a condition other than conjunctivitis, and John thinks that B has conjunctivitis, and John thinks A has a condition other than conjunctivitis, and John doesn't think that A has conjunctivitis, and John thinks C has a condition other than conjunctivitis, and John doesn't think that C has conjunctivitis."

Interestingly, we note that a difference between the SPT and the HPT arises when the context does not support the presupposition. In the case of the SPT, the presupposition does not need to project, and the sentence is still acceptable.

(29) *Scenario:* There are rumours that the three players will not be able to play in the semi-final due to a pregnancy. It is not known whether any of the three actually is pregnant.
(30) The team's doctor is only aware that BIRgit$_F$ is pregnant.

In (30) the sentence as a whole does not presuppose that Alexandra and Celia are pregnant. Instead we obtain the reading roughly paraphrased in (31)[2].

(31) "B is pregnant and the doctor believes that B is pregnant, the doctor does not believe that A is pregnant, and the team's doctor does not believe that C is pregnant."

If we construct a similar example with the HPT *also*, we notice that the presupposition projection is obligatory here. We obtain the rough paraphrase in (34) which is at odds with the scenario in (32), rendering the sentence in (33) pragmatically odd.[3]

[2] Note that the presupposition is not locally accommodated here, which we will discuss in more detail in Sect. 3.3.

[3] It has been pointed out by one reviewer that the two examples do not form a minimal pair, as only one of the scenarios is agnostic with respect to the presupposition. While we agree that the acceptability of the sentences is generally higher in agnostic contexts, we still observe a contrast between the examples. A more in-depth exploration of the effects of the different context is needed.

(32) *Scenario:* It has been established that only Birgit has the flu and the doctor knows that.
(33) # The team's doctor only thinks that BIRgit$_F$ also has conjunctivitis.
(34) "The doctor thinks that B has a condition other than conjunctivitis and the doctor thinks that B has conjunctivitis, the doctor thinks that A has a condition other than conjunctivitis and the doctor does not think A has conjunctivitis, and the doctor thinks that C has a condition other than conjunctivitis and the doctor does not think that C has conjunctivitis."

In order to show that this observed difference is a general pattern for all SPTs and HPTs, let us briefly consider a different example that exhibits the same behavior, i.e. that the SPT does not need to project in whereas the HPT does. (35) provides a scenario in which the SPT *stop* in (36) does not project.

(35) *Scenario:* The Lamb & Lion serves beer and wine, but the Rose & Crown only serves beer. New legislation makes it illegal to serve either beer or wine but pubs will be financially compensated for this, based on what kind of alcohol they served before.
(36) The Lamb & Lion stopped serving wine and beer. They will get 540£. The Rose & Crown only stopped serving BEER$_F$. They will get 250£.

Here, the presupposition that the Rose & Crown served wine before does not appear in the alternative set.

In contrast, in the case of a HPT, the presupposition must project. This is shown in (39).

(37) *Scenario:* Continuing the scenario in (35), the legislation is lifted.
(38) The Lamb & Lion only serves BEER$_F$ again.
(39) # The Rose & Crown only serves BEER$_F$ again.

In both (38) and (39) the presupposition of *again* projects. That is, (38) and (39) presuppose that the Rose & Crown and the Lamb & Lion respectively served both wine and beer before. This is the case for the Lamb & Lion, and thus (38) is acceptable in this scenario. But this presupposition is not satisfied for (39) in the present scenario, as the Rose & Crown served beer but not wine before. This renders sentence (39) unacceptable in this scenario.

3.3 Local Accommodation or Cancellation?

We haved observed that both in the contexts presented by Simons (2001) and Abusch (2002) and within focus alternatives, SPTs do not need to project. However, there is a difference between the exact ways in which this lack of projection comes about. The cases discussed in the literature assume that the presupposition is locally accommodated, i.e. instead of projecting it becomes part of the assertion. This can be seen both in the original examples such as (7) and (12), and in the examples we provided for the definite article, such as (18) and (21). In the case of focus alternatives however, the presuppositions of the Soft Presupposition Triggers simply do not appear in the alternative set. No local accommodation can be observed here. The following examples rule out local accommodation as the source of SPTs behavior with respect to focus alternatives. Instead, the empirical result is that the presuppositions are simply eliminated from the alternative set.

(40) *Scenario:* Only Birgit is pregnant. The pregnancy tests are faulty, leading the doctor to falsely believe that Alexandra and Celia are pregnant as well.
(41) I think there is something wrong with the pregnancy tests. After using them, the doctor believed that three players are pregnant. But he got it right for only one of them: # The doctor is only aware that BIRgit$_F$ is pregnant.

If local accommodation of the presupposition within each focus alternative was an option, (41) would have (42) as one of its possible readings, thus making the sentence felicitous in scenario (40). Since the sentence is infelicitous, this reading is not available for the sentence.

(42) "B is pregnant and the doctor believes that B is pregnant, and it is not the case that (*A is pregnant and* the doctor believes that A is pregnant), and it is not the case that (*C is pregnant and* the doctor believes that C is pregnant).

Instead of (42), the only reading available where presuppositions do not project is (43), where the presuposed proposition has been entirely eliminated from the focus alternatives.

(43) "B is pregnant and the doctor believes that B is pregnant, and the doctor does not believe that A is pregnant, and the doctor does not believe that C is pregnant."

It follows that the paraphrase in (43) should indeed be formalized as the set of alternatives in (44).

(44) $\{\lambda w : \text{PREGNANT}(a, w).\text{believe}(d, \text{PREGNANT}(a), w)),$
$\lambda w : \text{PREGNANT}(b, w).\text{believe}(d, \text{PREGNANT}(b), w)),$
$\lambda w : \text{PREGNANT}(c, w).\text{believe}(d, \text{PREGNANT}(c), w))\}$

This provides a challenge to current accounts for SPTs, such as Romoli (2011) and Abrusán (2011). Since the literature usually assumes the local accommodation behavior this new observed behavior may proof difficult to integrate with these approaches[4].

3.4 Observation A+B and the Original Problem

The observations made above now present us with a new perspective on the original problem presented in (1). The phenomenon there is not due to some peculiar property of the definite article's lexical entry in the alternative semantics, but rather due to its status as a SPT and the behavior of SPTs with respect to focus alternatives. The reading that we obtain for (45) (repeated from (1) above) is roughly paraphrased in (46).

(45) John only talked to the GERman$_F$professor.
(46) "There is a unique German professor and John spoke to the German professor, and John did not speak to a French professor and John did not speak to a Dutch professor."

As in the examples above, we obtain the correct formalization of this reading if we construct the alternative set as normal, but do not include the presuppositions in the alternatives.

(47) $\{[\lambda w : \cancel{\exists!x[\mathrm{P}(x,w) \wedge \mathrm{G}(x,w)]}.\exists y[\mathrm{P}(y,w) \wedge \mathrm{G}(y,w) \wedge \mathrm{T}(j,y,w)]],$
$[\lambda w : \cancel{\exists!x[\mathrm{P}(x,w) \wedge \mathrm{F}(x,w)]}.\exists y[\mathrm{P}(y,w) \wedge \mathrm{F}(y,w) \wedge \mathrm{T}(j,y,w)]],$
$[\lambda w : \cancel{\exists!x[\mathrm{P}(x,w) \wedge \mathrm{D}(x,w)]}.\exists y[\mathrm{P}(y,w) \wedge \mathrm{D}(y,w) \wedge \mathrm{T}(j,y,w)]], ...\}$

We then arrive at the following truth conditions for (45):

(48a) *Presupposition:* $\lambda w.\exists!x[\mathrm{P}(x,w) \wedge \mathrm{G}(x,w)] \wedge \exists x[\mathrm{P}(x,w) \wedge \mathrm{G}(x,w) \wedge \mathrm{T}(j,x,w)]$
(48b) *Assertion:* $[\lambda w.\neg[\exists y[\mathrm{P}(y,w) \wedge \mathrm{F}(y,w) \wedge \mathrm{T}(j,y,w)]] \wedge$
$\neg[\exists y[\mathrm{P}(y,w) \wedge \mathrm{D}(y,w) \wedge \mathrm{T}(j,y,w)]]$

Arriving at this reading by changing the lexical entry for the definite article overlooks the crucial fact that this behavior can be systematically observed for all Soft Presupposition Triggers when they appear together with focus alternatives.

4 Conclusions

Considering these observations, we arrive at the following conclusions:

1. The definite article is a Soft Presupposition Trigger for uniqueness.
2. Soft Presupposition Triggers and Hard Presupposition Triggers show different behavior not only in the contexts provided by Simons (2001) and

[4] Note however that in some cases, e.g. (36), the local accommodation behavior seems to be available with focus alternatives. A possible explanation for this is that *stop P*, besides presupposing that P was true before, asserts that there was a change from P to $\neg P$. If so the correct reading for the pub examples is still obtained by completely eliminating the presuppositions from the alternative set.

Abusch (2002), but also within focus alternative sets: While Hard Triggers must obligatorily project their presuppositions, Soft Triggers do not need to.
3. There is no need for a special lexical entry for the definite article in alternative semantics. Instead, a more general account for the interaction between Soft Presupposition Triggers and focus alternatives is necessary.
4. Any account of Soft Presupposition Triggers must be able to explain
   (a) the behavior of Soft Presupposition Triggers (now including the definite article) within the previously known contexts, where the non-projection corresponds with readings that resemble a form of local accommodation,
   (b) the behavior of Soft Presupposition Triggers within focus alternatives, where non-projection seems to amount to complete disappearance of the presupposition.

Further research will be needed to show how these conclusions interact with current accounts for Soft Presupposition Triggers, and whether these can be modified to account for the new behavior observed as well as the new Soft Trigger. The answers found there should then provide us with an answer to von Heusinger's (2007) puzzle as well.

References

Abbott, B.: Where Have Some of the Presuppositions Gone? In: Birner, B., Ward, G. (eds.) Drawing the Boundaries of Meaning. Neo-Gricean Studies in Pragmatics and Semantics in Honor of Laurence R. Horn. Benjamins, Philadelphia (2006)

Abrusán, M.: Triggering Verbal Presuppositions. In: Li, N., Lutz, D. (eds.) Semantics and Linguistic Theory (SALT) 20. Vancouver (2011)

Abusch, D.: Lexical Alternatives as a Source of Pragmatic Presuppositions. In: Jackson, B. (ed.) SALT XII. CLC, New York (2002)

Riester, A., Kamp, H.: Squiggly Issues: Alternative Sets, Complex DPs, and Intensionality. In: Aloni, M., Bastiaanse, H., de Jager, T., Schulz, K. (eds.) Logic, Language and Meaning. LNCS, vol. 6042, pp. 374–383. Springer, Heidelberg (2010)

Romoli, J.: The Presuppositions of Soft Triggers Are Not Presuppositions. In: Ashton, N., Chereches, A., Lutz, D. (eds.) Proceedings of SALT 21. Rutgers University (2011)

Rooth, M.: A Theory of Focus Interpretation. Natural Language Semantics 1, 75–116 (1992)

Simons, M.: On the Conversational Basis of Some Presuppositions. In: Hastings, R., Jackson, B., Zvolensky, Z. (eds.) Proceedings of Semantics and Linguistic Theory, vol. 11. CLC, New York (2001)

Von Heusinger, K.: Alternative Semantics for Definite NPs. In: Schwabe, K., Winkler, S. (eds.) On Information Structure, Meaning and Form. Generalizations Across Languages. Benjamins, Amsterdam (2007)

You Again: How Is Its Ambiguity Derived?

Ting Xu

University of Connecticut, Storrs, USA
xuting.thu@gmail.com

Abstract. It is well-known that an English sentence with a complex predicate modified by *again* displays a repetitive vs. restitutive ambiguity. Like English *again*, Chinese *you* 'again' modifying a resultative verb compound also exhibits a repetitive vs. restitutive ambiguity. However, Chinese differs from English in that the position of *you* 'again' is relatively fixed: it can only occur preverbally but not postverbally. This study examines how the ambiguity of Chinese *you* is derived. Investigating the scope interaction between *you* 'again' and an indefinite object, I argue that the ambiguity of *you* 'again' is structural but not lexical. I further propose that *you* 'again' moves overtly as a last resort to satisfy a PF requirement specific to Chinese.

Keywords: ambiguity, restitutive *again*, scope, movement.

1 Introduction

It is well-known that a sentence with a complex predicate modified by *again* (1) is ambiguous between a repetitive reading and a restitutive reading. The former presupposes that the subject has previously performed the action denoted by the VP (1a). The latter presupposes that the result state has held before. It either held from the very beginning (1b) or came into being as a result of someone else performing the action (1c).

(1) Sally painted the wall white again.
  a. Sally had painted the wall white before.
  b. The wall had been white before.
  c. Someone else other than Sally painted the wall white before.

There exist different approaches to account for the ambiguity of English *again*: structural and lexical. The structural approach suggests that the ambiguity arises because (1) has different possible syntactic structures (see von Stechow, 1995, 1996; among others). The lexical approach instead argues that the ambiguity has nothing to do with the structure, but is a result of *again* having different meanings (e.g. Dowty, 1979; among others).

Most of the previous studies focus on English *again* and German *wieder*, which can occur in different positions of a sentence. In this paper, I examine *you* 'again' in Mandarin Chinese. Like English, Chinese is also a head-initial language.

A. Aloni et al. (Eds.): Amsterdam Colloquium 2011, LNCS 7218, pp. 470–479, 2012.
© Springer-Verlag Berlin Heidelberg 2012

Nevertheless, Chinese differs from English in that the position of *you* 'again' is relatively fixed: it can only occur preverbally but not postverbally (2).[1]

(2) ta you tu-hong le (*you) na-ge beike (*you).
she again paint-red Asp (again) that-CL shell (again).
She painted two of the shells red again.

Like English *again*, however, Chinese *you* 'again' exhibits an ambiguity between a repetitive and a restitutive reading. Yet the preverbal position of *you* 'again' implies that it is adjoined at the vP level or higher, and thus seems to rule out a syntactic analysis involving multiple attachment points. In this paper, I address the question how the ambiguity of *you* is derived. I argue that despite the preverbal position of *you* 'again', its ambiguity is structural rather than lexical. The crucial evidence comes from the scope interaction between *you* 'again' and an indefinite object.

2 Previous Analyses of English *Again*

There are two main approaches to account for the ambiguity of English *again* in previous literature: structural analysis and lexical analyses.[2]

2.1 The Structural Analysis

Under the structural analysis (see von Stechow, 1995, 1996, among others), *again* always denotes repetition, as shown in (3). It introduces a presupposition that the proposition (a set of events) expressed by P was already true of another event e'.

(3) Let P be a property of eventualities and let e be an eventuality.
[[again]](P)(e) is defined only if $\exists e'[P(e')=1 \& e'<e]$.
Where defined, [[again]](P)(e)=1 iff P(e)=1. (adapted from von Stechow, 1996)

A complex predicate like *painted the wall white* in (4a) can be decomposed into an agentive event (wall painting) and a resultative state (the wall becoming white). The LF of (4a) is presented in (4b), in which the object NP *the wall* raises to semantically bind the empty pronominal (PRO) subject of AP.

[1] The generalization requires *you* to occur before a verb, including a light verb such as *ba* (cf. (9)). But it does not have to immediately precede the highest verb. As we will see later, *you* can occur after *ba* but precede the post-*ba* verb (cf. (19)). It does not necessarily precede the main verb, but can precede a secondary predicate (cf. (18)).

[2] There exists an alternative account for the ambiguity: We can treat the repetitive reading as a special case of the restitutive reading, as the former entails the latter. As far as I know, previous studies did not discuss this pragmatic approach.

(4) a. Sally painted the wall white.
  b. [the wall 1 [$_{vP}$ Sally [v [$_{VP}$ t_1 paint [$_{AP}$ PRO_1 white]]]]]

Sentence-final *again* can potentially adjoin to different positions in (4b), hence the ambiguity of *Sally painted the wall white again*. If *again* adjoins to vP or higher, we get the repetitive reading that Sally has painted the wall white before. If, on the other hand, *again* adjoins to the small clause AP, the restitutive reading is derived.

2.2 The Lexical Analysis

Contrary to the structural analysis, which argues that there is only one *again*, a lexical analysis claims that there is more than one *again*. The idea can be traced back to Dowty (1979), who account for the ambiguity by postulating a repetitive *again*, a restitutive *again*, and some type-shifting in between. According to the type shifting rule, a repetitive *again* automatically yields a corresponding restitutive *again*, which is interpreted as if the repetitive *again* takes narrower scope than CAUSE and BECOME.

Different forms of lexical analysis have been proposed ever since. For example, Fabricius-Hansen (2001) proposes that in addition to expressing repetition, *again* can also express reversal of the direction (counterdirectional *again*), which leads to the restitutive reading.

2.3 Two Kinds of Restitutive Readings

So far we have been focusing on a two-way dichotomy of *again*: repetitive vs. restitutive. Notice that both scenarios in (1b) and (1c) are compatible with the restitutive reading of (1), with (1c) entailing (1b). Some researchers further argue that (1c) is a restitutive reading independent from (1b), and it is derived by adjunction of *again* to VP in (4) (Nissenbaum, 2006; Bale 2007; Dobler, 2008). (1b) and (1c) are called by Nissembaum (2006) the low restitutive reading and the high restitutive reading, respectively.

Here I summarize Nissembaum's (2006) argument in favor of two restitutive readings, which involves the interaction between *again* and an indefinite object. Assuming the LF in (5b), Nissembaum (2006) makes the following predictions about possible scope readings for (5a).

(5) a. I painted one of my trees blue again.
  b.

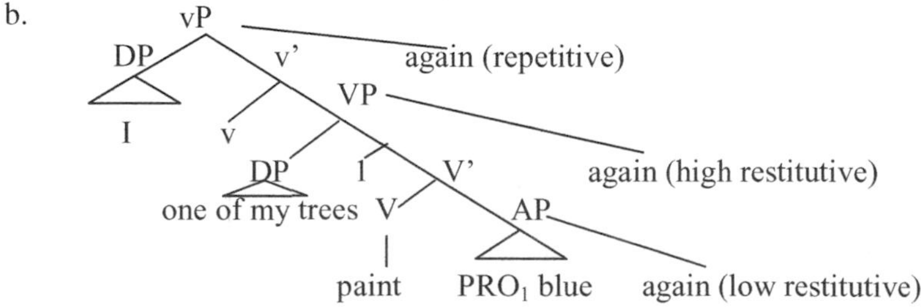

First of all, for a low restitutive reading (with *again* adjoined to AP), the indefinite object can never be interpreted inside the scope of *again*, because it is base generated in Spec, VP. The prediction is borne out: It is felicitous to continue Scenario A with the sentence in (6a). The usage of pronoun in (6a) forces an interpretation in which the same tree was painted blue. As a contrast, (6b) does not serve as felicitous continuation of the scenario. This is expected assuming the structural analysis of *again*: The usage of an indefinite object in (6b) forces an interpretation in which a different tree was painted blue, which cannot be derived via the structure in (5b).

(6) Scenario A: *One of my birch trees came up blue when it was a sapling; it later turned white like the rest. But I liked the idea of a blue birch tree so much that...*
a. I painted it blue again.
b. #I painted one of my trees blue again.

Although for the low restitutive reading, wide scope reading of *again* with respect to the indefinite object is not available, the indefinite object can be interpreted outside the scope of *again*. This is illustrated in (7), in which the tree I painted blue has to be previously blue simply because of the universal quantifier in the scenario.

(7) Scenario B: *All of my birch trees were blue when they were saplings; they later turned white like birch trees are supposed to be. But I liked the idea of a blue birch tree so much that...*
I painted one of my trees blue again.

Contrary to the low restitutive reading, for a high restitutive reading, the indefinite object can be interpreted inside the scope of *again*. (8) contains a scenario in which I painted a different tree blue from someone else. And it is felicitous to continue the scenario with a sentence having *again* and an indefinite object.

(8) Scenario C: *One of my birch trees had been painted blue when I moved in. It later died and had to be cut down. But I liked the idea of a blue birch tree so much that...*
I painted one of my trees blue again.

The scope facts are compatible with the structural analysis of *again* and challenge the lexical analyses. Under a lexicalist account, to the extent that the facts may be described, there is no principled reason why only the low restitutive reading with wide scope of *again* is absent. Under Dowty (1979)'s analysis, to account for the absent reading, we need to explain why the type shifting rule cannot shift a repetitive *again* to a repetitive *again*. Under Fabricius-Hansen's (2001) analysis, it is not clear how the independent restitutive *again* (counterdirectional *again* in her term), disappears when it applies to a predicate with an embedded existential quantifier.

3 The Puzzle: The Ambiguity of *You* in Mandarin Chinese

Like English, a Chinese sentence with a complex predicate modified by *you* 'again' (2) is compatible with all three scenarios in (1). Different from English, *you* 'again' in Chinese has a restricted distribution: it can only occur preverbally (2).

The structural analysis cannot be extended directly to Chinese. On the assumption that the complex predicate ends up in v, the preverbal *you* 'again' would appear to adjoin to *v*P or higher, thus only the repetitive reading is expected.

The Chinese *ba*-construction makes the ambiguity even more puzzling. A *ba*-sentence with *you* preceding *ba* also displays a three-way ambiguity (9). It is generally assumed in the literature that *ba* is an overt realization of a recursive small *v* (see Huang, Li & Li, 2009). When *you* precedes *ba*, it must adjoin to vP or even higher, thus the restitutive reading becomes unexpected.

(9) Zhangsan you ba men da-kai le.
Zhangsan again BA door hit-open Asp.
Zhangsan opened the door again. (repetitive, high restitutive, low restitutive)

Following Nissenbaum (2006), I examine how the ambiguity is derived by investigating the scope interaction between *you* 'again' and an indefinite object.

4 Scope Interaction between *You* and a Quantifier

Since *you* 'again' yields a three-way ambiguity, when it interacts with another scope-bearing element, such as an indefinite object, there are altogether six logical possibilities. Native speakers were presented with scenarios for each of these possibilities and were asked to judge whether bare sentences and *ba*-constructions with *you* were felicitous. The judgment is summarized in Table 1: All readings are available except the low restitutive reading with *you* 'again' taking wide scope.

Table 1. Scope between *you* 'again' and an indefinite object in Chinese

Low restitutive reading	∃> *you* 'again'	# *you* 'again'>∃
High restitutive reading	∃> *you* 'again'	*you* 'again'>∃
Repetitive reading	∃> *you* 'again'	*you* 'again'>∃

For expository convenience, I focus on the following three readings: a) low restitutive reading with *you* taking narrow scope; b) low restitutive reading with *you* taking wide scope; c) high restitutive reading with *you* taking wide scope.

First of all, when the low restitutive reading is intended, the indefinite object can be interpreted outside the presupposition of *you* 'again', i.e. the object in the presupposition and assertion can be the same (10)-(11).

(10) Context: *Lisa had a bunch of red shells. Unfortunately after a while they all got very dusty and the redness faded. In need of two red shells to decorate her Christmas tree, ...*

ta	you	tu-hong	le	qizhong	liang-ge	beike.[3]
she	again	paint-red	Asp	among	two-CL	shell.

She painted two of the shells red again.

(11) Context: *John ordered ten pocket watches. Unfortunately, all of them had always been open due to a manufacturing error. Yesterday he got all his watches fixed and closed them for the first time. Today, ...*

ta	you	da-kai	le	qizhong	yi-kuai	huaibiao.
He	again	hit-open	Asp	among	one-CL	pocket-watch.

He opened one of his pocket watches again.

For the low restitutive reading, although the indefinite object can take wide scope with respect to *you* 'again', but not visc versa. In other words, when a low restitutive reading is intended, the object cannot be interpreted within the presupposition of *you* 'again'. A number of my informants found the use of *you* 'again' odd in scenarios like (12) and (13), especially compared with (10) and (11).

(12) Context: *Zhangsan went to the beach and collected a lot of white shells and two red shells. When his wife cleaned the house, she accidentally broke the two red shells. Worried that Zhangsan would notice the mishap, ...*[4]

#ta	you	tu-hong	le	liang-ge	beike.
she	again	paint-red	Asp	two-CL	shell.

She painted two shells red again.

(13) Context: *John ordered many pocket watches. Unfortunately, two of them had always been open due to a manufacturing error. Yesterday he got them fixed and they closed for the first time. Today...*[5]

#ta	you	da-kai le	yi-kuai	huaibiao.
he	again	hit-open Asp	one-CL	pocket-watch.

He opened a pocket watch again.

[3] The test sentences in (10) and (11) involve a partitive marker *qizhong* 'among'. As a matter of fact, native speakers were asked to judge sentences either with or without a partitive marker for all the scenarios. It turned out when *you* 'again' takes wider scope than the indefinite object, speakers tended to prefer sentences without partitive markers. On the other hand, they preferred sentences with partitive markers when *you* 'again' takes narrow scope. For expository convenience, when concentrating on scenarios in which *you* takes narrow scope, I only present the judgment of sentences with partitive markers; when focusing on scenarios in which *you* takes wide scope, I only present the judgment of sentences without partitive markers.

[4] The scenario is adapted from Dobler (2008).

[5] The scenario is adapted from Bale (2007).

The reason why *you* 'again' sounds infelicitous in (12) and (13) is not because it cannot take wide scope. As shown in (14) and (15), when a high restitutive reading is intended, the indefinite object can be interpreted within the presupposition of *you* 'again'.

(14) Context*: John and Jane had some white shells. Since they needed two red shells to decorate their Christmas tree, John painted two shells red. Unfortunately, Jane accidentally broke the two red shells that John just painted. Therefore, …*
sentence in (12)

(15) Context*: Jane and John bought five pocket watches. Jane picked out one pocket watch, opened it and closed it. Later John needed to check the time. He wanted to open the watch that Jane opened just now, but he couldn't find it. Therefore,…*
sentence in (13)

The scope facts suggest that syntactic structure is playing a role in the ambiguity of *you*. Let us take (16) as an example, whose LF is presented in (17).[6] The scope facts would fall in place if *you* 'again' can adjoin to some lower projections such as VP or XP in (17). If *you* 'again' can indeed adjoin to XP, it will give rise to the low restitutive reading. As the indefinite object is base-generated in Spec, VP, it takes wider scope than *you* 'again', but not the other way around. On the other hand, if *you* 'again' can adjoin to VP, with the high restitutive reading, the indefinite object can be interpreted within the presupposition of *you* 'again'. Under a lexicalist account, to the extent the facts may be described, there is no principled reason why the low restitutive reading with the wide scope of *you* is absent.

(16) Lisi tu-hong le liang-ge beike.
Lisi paint-red Asp two-CL shell.
Lisi painted two shells red.

(17) [$_{IP}$…[$_{vP}$ Lisi v [$_{VP}$ two-shell 1 paint [$_{FP}$ F [$_{XP}$ PRO$_1$ red]]]]

5 Analysis

Following Ernst (2004), who argues that adverbs are licensed in their base positions whenever the relevant semantic rule gives them their proper interpretation, I propose that *you* can adjoin to some lower projections in (17), such as a small clause XP or VP, which gives rise to the low restitutive reading and the high restitutive reading, respectively.

First of all, *you* can modify a small clause. This is attested by Chinese *de*-resultative constructions (18a). Following the analysis of Tang (1997), which is

---

[6] Resultative verb compounds such as *tu-hong* 'paint red' have received much attention in the liteature. For expository convenience, I assume a head-movement-plus-control analysis. The presence of FP in (17) is proposed to capture the connection between sentences with resultative verb compounds and *de*-resultative constructions (18a), in which *de* is the overt realization of F. Other analyses of resultative verb compounds are also compatible with what I am illustrating here.

shown in (18b), particle *de* is base-generated in the head position of the functional projection, and incorporates to V. Then V-*de* as a whole undergoes V-to-v movement. Since *you* 'again' can occur between the object *Zhangsan* and the secondary predicate *ku* 'cry' (18a), it has to adjoin to XP in syntax to derive the correct linear order.[7]

(18) a. Lisi da-de Zhangsan you ku le
Lisi hit-de Zhangsan again cry Asp.
Lisi hit Zhangsan to the extent that again Zhangsan cried.

b. [$_{vP}$ Lisi v+da$_k$-de$_j$ [$_{VP}$ Zhangsan$_i$ t$_k$+t$_j$ [$_{FP}$ t$_j$ **<u>you</u>** [$_{XP}$ PRO$_i$ ku le]]

In addition to result-denoting XP, *you* can adjoin to VP in syntax. This is demonstrated by the Chinese *ba*-construction (19a). Here I follow Kuo (2009), who argues that *ba* is the head of an Applicative phrase, as illustrated in (19b).[8] Not only can *you* 'again' precede *ba* (cf. (9)), it can also occur between the *ba*-NP (the NP immediately following *ba*) and the predicate (19). To derive the correct linear order, *you* 'again' has to adjoin to VP in the structure.[9]

(19) a. wo ba Sara you da shang le
I BA Sara again hit hurt Asp.
I hit Sara so that again she got hurt.

b. [$_{vP}$ wo v+ba$_k$ [$_{ApplP}$ Sara$_i$ t$_k$ **<u>you</u>** [<u>$_{VP}$</u> t$_i$ da [$_{XP}$ PRO$_i$ shang le]]]]]

Furthermore, as Chinese adverbs in general occur preverbally in a bare sentence, I postulate here that there is a PF-requirement in Chinese, which requires adverbs like *you* to occur before a verb.[10]

---

[7] In fact, it is not clear whether *you* adjoins to FP or XP. To answer this question, we need to figure out the denotation of particle *de*. The answer will not affect our analyses. For expository convenience, I assume in this paper that it adjoins to XP, which denotes the result state, and treat the functional head *de* as semantically vacuous. This is in line with the analysis which analyzes the functional head *de* as a complementizer (e.g. Wang, 2010).

[8] Other analyses of *ba*-constructions (for example, treating *ba* as an overt realization of a recursive *v*) are also compatible with my point here.

[9] In both (18) and (19), the base position fully determines the reading available: (18) allows the low restitutive reading and (19) the high restitutive reading. It is hard to tell whether these two sentences have the repetitive reading, because the scenarios for high and low restitutive readings are a superset of those for the repetitive reading. Without evidence of the existence of repetitive reading, we cannot tell whether other readings can become available through LF raising.

[10] It has been widely observed that there exists a restriction in Mandarin that a transitive verb with an object cannot be modified by a post-verbal manner expression (i). The verb has to be doubled (i). The PF requirement proposed here may be part of a broader (but unexplained) generalization about modifiers in Chinese resisting VP-internal surface positions, even when they are interpreted there (Jonathan Bobaljik, pc).

(i) Wo pa shan *(pa) de kuai.
I climb mountain climb DE fast.
I climbed the mountain fast.

I propose a movement analysis of *you* 'again' to account for its ambiguity. In a bare sentence such as (16), when base-generated in a VP- or XP-attached position, *you* 'again' moves overtly as a last resort to satisfy this PF-requirement. When it gets interpreted, it undergoes LF reconstruction or semantic reconstruction (see Cresti, 1995; among others).[11]

6 Further Arguments and Consequences

Under the movement analysis, several additional patterns follow directly. First, only the surface scope reading is available when *you* precedes negation *mei* (20).

(20) Zhe-ge yue ta you mei dasao fangjian.
This-CL month he again Neg clean room.
Again, he didn't clean the room this month. (again>Neg, *Neg>again)

Furthermore, neither high nor low restitutive reading is available when *you* 'again' immediately precedes negation (21a). In fact, to express the restitutive reading, a different lexical item *zai*, which is generally assumed to denote repetition in irrealis context, is used and it has to follow negation (21b).

(21) Scenario: *The door was open before (not opened by John). Then it got closed. John wants to open the door but he was too tired, so...*

a. #ta you mei ba men da-kai.
He again NEG BA door hit-open
b. ta mei zai ba men da-kai.
He NEG again BA door hit-open
He didn't open the door again.

Some researchers treat the negative particle *mei* 'not' as a type of adverb (Li & Thompson 1981; Ernst, 1995), and negative particles may be licensed in an adjoined position like other types of adverb under Tang's (1990, 2001) theory of licensing adverbials (see Ting 2006 for some inconclusive discussion).[12] If this is on the right track, the above patterns follow directly from Relativized Minimality (Rizzi, 1990), a well-established syntactic constraint that bans movement of an adjunct like *you* across another adjunct.

---

[11] As a reviewer correctly pointed out, this PF-driven syntactic movement is somewhat stipulated to get the word order correct, with its motivation being unclear. I leave this issue for future research.

[12] The status of negation in Mandarin is controversial: Some researchers analyze negation as some kind of adverb and deny the existence of NegP; whereas some treat it as the head or specifier of NegP (see Ting, 2006 and references therein). Here I assume the adverb analysis. Readers may refer to Ernst (1995) and Ting (2006) for supporting evidence.

7 Conclusion

In this study, I examine how the repetitive/restitutive ambiguity of *you* 'again' in Mandarin Chinese is derived. On the surface, Chinese poses a problem for the structural analysis because the adverb *you* can only be preverbal. However, the scope interaction between *you* 'again' and an indefinite object suggests that the ambiguity of *you* 'again' must be structural rather than lexical. The puzzle can be resolved if *you* 'again' moves overtly and gets reconstructed either at LF or in semantics.

References

1. Bale, A.: Quantifiers and verb phrases: An exploration of propositional complexity. Natural Language and Linguistic Theory 25(3), 447–483 (2007)
2. Cresti, D.: Extraction and Reconstruction. Natural Language Semantics 3, 79–122 (1995)
3. Dobler, E.: Again and the structure of result states. In: Proceedings of ConSOEL XV 2008, pp. 41–66 (2008)
4. Dowty, D.R.: Word Meaning and Montague Grammar. Reidel, Dordrecht (1979)
5. Ernst, T.: Negation in Mandarin Chinese. Natural Language and Linguistic Theory 13, 665–707 (1995)
6. Ernst, T.: Principles of adverbial distribution in the lower clause. Lingua 114, 755–777 (2004)
7. Fabricius-Hansen, C.: Wi(e)der and Again(st). In: Fery, C., Sternefeld, W. (eds.) Audiatur Vox Sapientiae. A Festschrift for Arnim von Stechow, pp. 101–130 (2001)
8. Huang, C.-T.J., Li, Y.A., Li, Y.: The syntax of Chinese. Cambridge University Press, Cambridge (2009)
9. Kuo, P.: IP internal movement and topicalization. PhD dissertation, University of Connecticut (2009)
10. Li, C., Thompson, S.: Mandarin Chinese: A Functional Reference Grammar. University of California press, Berkeley (1981)
11. Nissenbaum, J.: Decomposing Resultatives: Two kinds of restitutive readings with Again. Poster presented at NELS37. University of Illinois, Urbana-Champain (2006)
12. Rizzi, L.: Relativized Minimality. MIT Press, Cambridge (1990)
13. Tang, C-C.J.: Chinese Phrase Structure and the Extended X′-Theory. Doctoral dissertation, Cornell University (1990)
14. Tang, C.-C.J.: Functional Projections and Adverbial Expressions in Chinese. Language and Linguistics 2(2), 203–241 (2001)
15. Tang, S.: Parametrization of features in syntax. Ph.D. Dissertation, University of California, Irvine (1997)
16. Ting, J.: NegP and the particle *suo* in Mandarin Chinese. Concentric: Studies in Linguistics 32(2), 71–92 (2006)
17. von Stechow, A.: Lexical Decomposition in Syntax. In: Egli, U., et al. (eds.) Lexical Knowledge in the Organization of Language, Benjamins, Amsterdam, pp. 81–118 (1995)
18. von Stechow, A.: The Different Readings of Wieder. Journal of Semantics 13, 87–138 (1996)
19. Wang, C.: The microparametric syntax of resultatives in Chinese languages. PhD dissertation, New York University (2010)

Author Index

CPSIA information can be obtained at www.ICGtesting.com
Printed in the USA
LVOW011432140413

329061LV00003B/58/P